**MARTIN J. PRING**

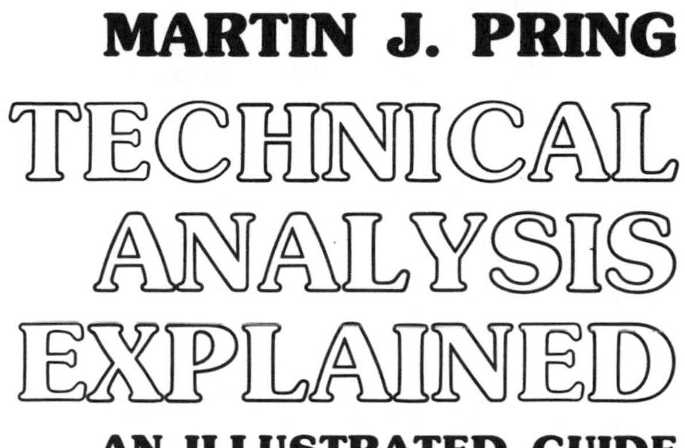

# TECHNICAL ANALYSIS EXPLAINED

## AN ILLUSTRATED GUIDE FOR THE INVESTOR

**McGRAW-HILL BOOK COMPANY**
New York St. Louis San Francisco Auckland Bogotá
Düsseldorf Johannesburg London Madrid Mexico
Montreal New Delhi Panama Paris São Paulo
Singapore Sydney Tokyo Toronto

NOTE: Although masculine pronouns have been used with such terms as "investor" or "analyst," they are intended to cover both sexes. Technical analysis concerns itself with the actions of a large number of market participants; therefore, the importance of women investors and traders cannot be overlooked.

*Library of Congress Cataloging in Publication Data*

Pring, Martin J
Technical analysis explained.

Bibliography: p.
Includes index.
1. Investment analysis. I. Title.
HG4521.P84 332.6'7 79-4394
ISBN 0-07-050871-2



234567890 BPBP 8987654321

The editors for this book were Kiril Sokoloff and Celia Knight, the designer was Elliot Epstein, and the production supervisor was Thomas G. Kowalczyk. It was set in Electra by Progressive Typographers, Inc.

Printed and bound by The Book Press.

# CONTENTS

## PART THREE INTEREST RATES AND THE STOCK MARKET



## PART FOUR OTHER ASPECTS OF MARKET BEHAVIOR

# PREFACE

There is no reason why anyone cannot make a substantial amount of money in the stock market, but there are many reasons why most people will not. As with most endeavors in life, the key to success in the stock market is knowledge and action. This book has been written in an attempt to throw some light on the internal workings of the market and therefore help expand the "knowledge" aspect, leaving the "action" to the patience, discipline, and objectivity of the individual investor.

Over the years many excellent books have been written on the technical approach to investing in the stock market. Since the goal of most investors and speculators is to make money, most of these books have concentrated on short-term trading techniques and stock selection. As a result there is a dearth of material explaining the technical position of "the market" itself. The majority of stocks move in sympathy with the overall trend of the market, so an analysis of the macro picture is a good starting point for any serious investment strategy.

In order to answer the question "What do I need to know to obtain a good working knowledge of the market?" an attempt has been made to combine the more useful facets of technical analysis in one volume, with the historical perspective presented in the appendixes. Since changes in interest rates have a pronounced influence on the stock market, several chapters have been devoted to this subject, although strictly speaking it is not "technical analysis" in the sense understood by this book.

Much of the technical work published today tends to develop indicators geared to giving exact timing signals. The reputation of such indicators is often based on a relatively short 5- to 10-year statistical history. When researched back over a considerably longer period, the consistent success of such indicators invariably proves to be misleading over the many different market conditions such as inflation, deflation, war, and peace. This is because no "perfect" indicator has ever been developed and one probably never will be.

To be successful, technical analysis should be regarded as an art, not a science. While many of the mechanistic techniques described in subsequent pages offer reliable indications of changing market conditions, all suffer from the common characteristic that they do sometimes fail to operate satisfactorily. This characteristic presents no problem to the consciously disciplined investor, since a good working knowledge of the principles underlying major movements in the market, and a balanced view of its overall technical position, offer a much superior framework within which to operate.

There is, after all, no substitute for independent thought. The action of the technical indicators portrays the underlying characteristics of the market, and it is up to the analyst to put the pieces of the jigsaw puzzle together and come up with a working hypothesis.

The task is by no means easy, as initial success can lead to overconfidence and arrogance. Charles H. Dow, the father of technical analysis, once wrote that "pride of opinion caused the downfall of more men on Wall Street than all the other opinions put together. . . ."

This is because the market is essentially a reflection of people in action. Normally such activity develops on a reasonably predictable path. Since people can—and do—change their minds, price trends in the market can deviate unexpectedly from their anticipated course. If serious trouble is to be avoided, investors must adjust their attitudes as changes in the technical position emerge.

In addition to pecuniary rewards, a study of the market can also reveal much about human nature, both from observation of other people in action and from the aspect of self-development. As the investor reacts to the constant struggle through which the market will undoubtedly put him, he will also learn a little about his own makeup. Washington Irving might well have been referring to this challenge of the stock market when he wrote: "Little minds are taxed and subdued by misfortune but great minds rise above it."

*Martin J. Pring*

# ACKNOWLEDGMENTS

The material for this book has been gathered from a substantial number of sources, and I am greatly indebted to the many organizations that have given their permission to reproduce charts and diagrams. Special thanks go to my associates at *The Bank Credit Analyst* (3463 Peel Street, Montreal, Quebec, Canada H3A 1W7), in particular to Tony Boeckh for his encouragement in the development of many of the charts and techniques illustrated in the book and for permission to reproduce them. Thanks also to Linda LaRoche and Cindy Jones for their help, patience, and encouragement in the development of many of the charts and to Ekow Otto for his fine computer work.

Acknowledgment is also due to Bill DiIanni for his useful suggestions on Chapter One, Dow Theory, and to Ian Notley, North America's leading cycles analyst, for his comments on Chapter 9, "Time."

This book would not have been possible without the help and encouragement of my wife, Danny, who not only typed the manuscript but also translated it into readable English.

*To my wife, Danny.*

# INTRODUCTION

To investors willing to buy and hold common stocks for the long term, the stock market has offered excellent rewards over the years in terms of both dividend growth and capital appreciation. The market is even more challenging, fulfilling, and rewarding to resourceful investors willing to learn the art of cyclical timing through a study of technical analysis.

The advantages of cyclical investing over the "buy and hold" approach have been particularly marked since 1966.

The market made no headway at all—as measured by the Dow Jones Industrial Average (DJIA)—in the 12 years between 1966 and 1978. Yet there were some substantial price fluctuations.

The potential rewards of cyclical investing are better appreciated when it is realized that even though the DJIA was unable to record a net advance between 1966 and 1978, the period did encompass three cyclical advances totaling over 1100 Dow points.

Had a cyclically oriented investor been fortunate enough to sell at the three tops in 1966, 1968, and 1973 and reinvest his money at the troughs of 1966, 1970, and 1974, his investment (excluding transactions costs and capital gains tax) could have grown from a theoretical $1000 (i.e., $1 for every Dow point) in 1966 to almost $4000 by September 1976. In contrast, the "buy and hold" approach would have realized a mere $20 gain over the same period.

In practice, of course, it is impossible to consistently buy and sell at exact cyclical market turning points, but the enormous potential of this approach still leaves substantial room for error even when commission costs and taxes are taken into consideration. For example, using the rather conservative assumption that a cyclically oriented investor would have required a 15 percent price movement from the primary peak or trough before the indicators on which he based his judgment signaled a change in investment posture, the results would still show a substantial gain over the "buy and hold" strategy.

Consequently the rewards can be substantial to those who can identify major market junctures and take the appropriate action based on such knowledge.

If it is to be successful, the approach involves taking a contrary position to the majority or consensus view of the outlook for the market. This requires patience, objectivity, and discipline, since it means acquiring stocks at a time of depression and gloom and liquidating them in an environment of euphoria and excessive optimism. The aim of this book is to explain the technical characteristics which may be expected at major market turning points so that they may be assessed as objectively as possible. It has therefore been designed to give a better understanding of market action, thus enabling the investor to minimize future mistakes and capitalize on major swings.

## TECHNICAL ANALYSIS DEFINED

The technical approach to investment is essentially a reflection of the idea that the stock market moves in trends which are determined by the changing attitudes of investors to a variety of economic, monetary, political, and psychological forces. The art of technical analysis, for it is an art, is to identify changes in such trends at an early stage and to maintain an investment posture until a reversal of that trend is indicated.

Human nature remains more or less constant and tends to react to similar situations in consistent ways. By studying the nature of previous market turning points, it is possible to develop some characteristics which can help identify major market tops and bottoms. Technical analysis is therefore based on the assumption that people will continue to make the same mistakes that they have made in the past. Human relationships are extremely complex and are never repeated in identical combinations. The stock market, which is a reflection of people in action, never repeats a performance exactly, but the recurrence of similar characteristics is sufficient to permit the technician to identify major juncture points.

Since no single indicator has signaled or indeed could signal every cyclical market juncture, technical analysts have developed an arsenal of tools to help identify these points.

## THREE BRANCHES OF TECHNICAL ANALYSIS

Technical analysis can essentially be broken down into three areas: sentiment indicators, flow-of-funds indicators, and market structure indicators.

### Sentiment Indicators

*Sentiment* or *expectational indicators* monitor the actions of certain market participants—for example, the so-called "odd lotter," mutual funds, floor specialists, and many others. Just as the pendulum of a clock is continually moving from one extreme to another, so the sentiment indexes (which monitor the emotions of those various investors) move from one extreme at a bear market bottom to another at a bull market top. The logic behind the use of such indicators is that different groups of investors are consistent in their actions at major market turning points. For example, insiders (i.e., key employees or major stockholders of a company) and New York Stock Exchange (NYSE) members as a group have a tendency to be correct at market turning points; in aggregate, their transactions are on the buy side toward market bottoms and on the sell side toward tops.

Conversely, mutual funds and advisory services as a group are wrong at market turning points, since they consistently become bullish at market tops and bearish at market troughs. Indexes derived from such data are able to show that certain readings have historically corresponded with market tops and others with market bottoms. Since the consensus or majority opinion is normally wrong at market turning points, these indicators of market psychology are a useful basis on which to form a contrary opinion.

There are two basic disadvantages to the sentiment approach. First, with very few exceptions (such as odd-lot statistics), the data on which these indexes are based are available only for the relatively brief period of 10 to 15 years. In the history of the stock market this is a very short period indeed, and conclusions drawn from such meager observations can prove misleading.

Second, it is unclear to what extent the advent of the options markets in 1973 has affected those indexes for which data are available over a long period, especially those based on short selling. There are few market observers who would disagree that some distortions have taken place, but no documented proof as to the degree of the distortion has yet been offered. While it is useful to observe the trends in the sentiment indexes, it is probably wiser to treat most of them as an adjunct rather than as an integral part of the analysis. For this reason they will not be discussed in depth but will be briefly described in Chapter 15.

### Flow-of-Funds Indicators

The second area of technical analysis involves what are loosely termed *flow-of-funds indicators*. This approach analyzes the financial position of various investor groups in an attempt to measure their potential capacity for buying or selling stocks. Since there has to be a purchase for each sale, the "ex post" or actual dollar balance between supply and demand for stock must always be equal. The price at which a stock transaction takes place has to be the same for the buyer and the seller, so naturally the amount of money flowing out of the market must equal that which is being put in. The flow-of-funds approach is therefore concerned with the before-the-fact balance between supply and demand, known as the "ex ante" relationship. If at a given price there is a preponderance of buyers over sellers on an ex ante basis, it follows that the actual (ex post) price will have to rise to bring buyers and sellers into balance.

The *short interest ratio* is perhaps the most widely used indicator of this type. It is calculated by taking the monthly NYSE short interest position (i.e., the number of NYSE shares that have been sold short) and dividing it by the average daily volume for the month in question. Since every share sold short must eventually be repurchased, a high short interest of 1.8 to 2.0 or more is considered to be bullish, since it represents 1.8 to 2.0 days of potential buying power. (The short interest ratio is also a measure of sentiment, since high readings represent an extremely bearish feeling among investors.)

Flow-of-funds analysis is also concerned with trends in mutual fund cash positions and other major institutions such as pension funds, insurance companies, foreign investors, bank trust accounts, and customers' free balances, which are normally a source of cash on the buy side; and new equity offerings, secondary offerings, and margin debt on the supply side.

This money flow analysis also suffers from disadvantages. While the data measure the availability of money for the stock market, e.g., mutual fund cash position or pension fund cash flow, they give no indication of the inclination of these market participants to use this money for the purchase of stocks, nor of the elasticity or willingness to sell at a given price on the sell side. The data for the major institutions and foreign investors are not sufficiently detailed to be of much use, and in addition they are reported well after the fact. In spite of these drawbacks, flow-of-funds statistics may be used as background material.

A superior approach to flow-of-funds analysis is derived from an examination of liquidity trends in the banking system, which measures financial pressure not only on the stock market but on the economy as well.

## Market Structure Indicators

The final area of technical analysis is the one with which this book is mainly concerned, embracing *market structure* or the *character of the market indicators*. These indications monitor the trend of various price indexes, market breadth, cycles, stock market volume, etc., in order to evaluate the health of bull and bear markets.

In a general sense, "the market" refers to the 30 stocks that make up the DJIA or to some other index such as the Standard and Poor 500; these account for a substantial amount of the outstanding capitalization on the NYSE. Most of the time the majority of the other market averages and indicators of internal structure will rise and fall with the DJIA, but toward the end of major market movements the paths of many of these indexes diverge from the senior average. Such divergences offer signs of technical deterioration during advances, and technical strength following declines. Through judicious observation of these signs of latent strength and weakness, the technically oriented investor is alerted to the possibility of a reversal in the trend of the market itself.

Since the technical approach is based on the theory that the stock market is a reflection of mass psychology ("the crowd") in action, it attempts to forecast future price movements on the assumption that crowd psychology moves between panic, fear, and pessimism on one hand and confidence, excessive optimism, and greed on the other. As discussed here, the art of technical analysis is concerned with identifying such changes at an early phase, since these swings in emotion take several years to accomplish. The technically oriented investor is able to buy or sell stocks with greater confidence, on the principle that once a trend is set in motion it will perpetuate itself.

Price movements in the market may be classed as minor, intermediate, and major. Minor movements which last less than 3 or 4 weeks tend to be random in nature and are not discussed in this book.

Intermediate movements usually develop over a period of 3 weeks to as many months, sometimes longer. While not of prime importance, they are nevertheless useful to identify. It is clearly important to distinguish between an intermediate reaction in a bull market and the first down leg of a bear market, for example.

Major movements (sometimes called primary or cyclical) typically work themselves out in a period of 1 to 5 years.

## DISCOUNTING MECHANISM OF THE MARKET

All price movements have one thing in common: they are a reflection of the trend in the hopes, fears, knowledge, optimism, and greed of the investing public.

The sum total of these emotions is expressed in the price level, which is, as Garfield Drew[1] noted, ". . . never what they (stocks) are worth but what people think they are worth. . . ."

This process of market evaluation was well expressed by an editorial in *The Wall Street Journal:*[2]

> The stock market consists of everyone who is "in the market" buying or selling shares at a given moment, plus everyone who is not "in the market" but might be if conditions were right. In this sense, the stock market is potentially everyone with any personal savings.
>
> It is this broad base of participation and potential participation that gives the market its strength as an economic indicator and as an allocator of scarce capital. Movements in and out of a stock, or in and out of the market, are made on the margin as each investor digests new information. This allows the market to incorporate all available information in a way that no one person could hope to. Since its judgments are the consensus of nearly everyone, it tends to outperform any single person or group. . . . [The market] measures the after-tax profits of all the companies whose shares are listed in the market, and it measures these cumulative profits so far into the future one might as well say the horizon is infinite. This cumulative mass of after-tax profits is then, as the economists will say, "discounted back to present value" by the market. A man does the same thing when he pays more for one razor blade than another, figuring he'll get more or easier shaves in the future with the higher-priced one, and figuring its present value on that basis.
>
> This future flow of earnings will ultimately be affected by business conditions everywhere on earth. Little bits of information are constantly flowing into the market from around the world as well as throughout the United States, and the market is much more efficient in reflecting these

[1] Garfield Drew, *New Methods for Profit in the Stock Market*, Metcalfe Press, Boston, 1968, p. 18.

[2] *The Wall Street Journal*, Oct. 20, 1977. *Reprinted by permission of The Wall Street Journal. Copyright Dow Jones and Co. Inc. 1977. All rights reserved.*

> bits of news than are government statisticians. The market relates this information to how much American business can earn in the future. Roughly speaking, the general level of the market is the present value of the capital stock of the U.S.

This implies that investors are looking ahead 6 months or more, and buying their stocks now so that they can liquidate at a higher price when the anticipated news or development actually takes place. If expectations concerning the development are better or worse than originally thought, then through the market mechanism investors sell either sooner or later, depending on the particular circumstances. Thus the familiar maxim "sell on good news" applies only when the "good" news is right on or below the market's (i.e., the investor's) expectations. If the news is good but not as favorable as expected, a quick reassessment will take place, and the market (other things being equal) will fall. If the news is better than anticipated, the possibilities are obviously more favorable. (The reverse would, of course, be true of a declining market.) This process explains the paradox of markets peaking out when economic conditions are strong and forming a bottom when the outlook is most gloomy.

The reaction of the market to new events can be most instructive, for if the market, as reflected in the averages, ignores supposedly bullish news about the economy or a large corporation and sells off, it is certain that the event was well discounted, i.e., already built into the price mechanism. Such a reaction should therefore be viewed bearishly. If the market reacts more favorably to bad news than might be expected, this in turn should be interpreted as a positive sign. There is a good deal of wisdom in the expression "a bear argument known is a bear argument understood."

## THE STOCK MARKET AND THE BUSINESS CYCLE

The major movements in stock prices are caused by major trends in the emotions of the investing public. These emotions are themselves a reflection of the anticipated level and growth rate of future corporate profits, and the attitude of investors toward those profits.

There is a definite link between primary movements in the stock market and cyclical movements in the economy, since on most occasions trends in corporate profitability are an integral part of the business cycle. If the stock market were influenced by basic economic forces only, the task of determining the changes in the primary movements of the market would be relatively simple. In practice it is not, and this is due to several factors.

First, changes in the direction of the economy can take some time to develop. During that period other psychological considerations—for example, political developments or purely internal factors such as a speculative buying wave or selling pressure from margin calls—can affect the equity market and result in misleading rallies and reactions of 5 to 10 percent or more.

Second, while changes in the market usually precede changes in the economy by 6 to 9 months, the lead time can sometimes be far shorter or longer. In 1921 and 1929, the economy turned before the market.

Third, even when an economic recovery is in the middle of its cycle, doubts about its durability can quite often arise. When these are accompanied by political or other adverse developments, rather sharp and confusing corrections can be set off.

Fourth, even though profits may increase, investors' attitudes toward those profits may change. For example, in the spring of 1946 the Dow Jones Industrial Average stood at a 22 times price/earnings ratio. By 1948, the comparable ratio was 9.5 when measured against 1947 earnings. In this period profits had almost doubled and price/earnings ratios had fallen, but stock prices were lower.

## TECHNICAL ANALYSIS—TREND DETERMINATION

Because technical analysis involves a study of the action of the market, it is not concerned with the extremely difficult and subjective tasks of forecasting trends in corporate profitability or of assessing the attitudes of investors toward those profits. Technical analysis is concerned only with the identification of major turning points in *the market*'s assessment of these factors. Since "the market" is a reflection of changes in the balance of opinion between buyers and sellers as expressed in the price mechanism, the essence of technical analysis is to identify important changes in the trends of these prices.

The approach taken here differs from that found in standard presentations of technical analysis. The various techniques used to determine trends and identify their reversals will be examined in Part 1, which deals with moving averages, rates of change, trendlines, price patterns, etc. Following this is a more detailed explanation of the various indicators and indexes themselves and of how they can be combined to build a picture from which the quality of the internal structure of the market can be determined. A study of the market character is a cornerstone of technical analysis, since reversals of price trends in the major averages are almost always preceded by latent strength or weakness in the market structure. Just as a careful driver does not judge the performance of his car from the speedometer alone, so technical analysis looks farther than the price trends of the popular averages.

The character or health of the market will therefore be analyzed in Part 2, "Market Structure." Since trends of investor confidence are responsible for price movements, this emotional aspect is examined from four viewpoints, namely:

Price

Time

Volume

Breadth

Changes in stock prices reflect changes in investor attitude, and *price* indicates the level of that change.

*Time*, the second dimension, measures both the recurring cycles in investor psychology and their length. Changes in confidence go through distinct cycles, some long and some short, as investors swing from excesses of optimism toward deep pessimism. The degree of price movement in the market is usually a function of the time element. The longer it takes for investors to move from a bullish to a bearish extreme, the greater the ensuing price change is likely to be.

*Volume* reflects the intensity of changes in investor attitudes. For example, if stock prices advance on low volume, the enthusiasm implied from the price rise is not nearly as strong as that present when a price rise is accompanied by very high volume.

The fourth dimension, *breadth*, measures the extent of the emotion. This is important, for as long as stocks are advancing on a broad front, the trend in favorable emotion is dispersed among most stocks and industries, thereby indicating a broad economic recovery in general and a widely favorable attitude toward stocks in particular. On the other hand, when interest has narrowed to a few blue-chip stocks, the quality of the trend has deteriorated, and a continuation of the bull market is highly suspect.

Technical analysis measures these psychological dimensions in a number of ways. Most indicators monitor two or more aspects simultaneously; for instance, a simple price chart measures both price (on the vertical axis) and time (on the horizontal axis). Similarly, an advance/decline line measures breadth and time.

Parts 3 and 4 deal with the important subjects of interest rates and their relation to movements of equity prices; speculation; and other aspects of market sentiment. A small glossary at the end of the book explains the terms used but not defined in the text.

## CONCLUSION

Stock prices move in trends caused by the changing attitudes and expectations of investors with regard to the business cycle. Since investors continually make the same type of mistake from cycle to cycle, an understanding of the historical behavior and relationships of certain price averages and stock market indicators can be used to identify major market turning points. Since no one indicator can ever be expected to signal all such trend reversals, it is essential to use a number of them at a time so that an overall picture can be built up.

This approach is by no means infallible, but a careful, patient and objective use of the principles of technical analysis will put the odds of success very much in favor of the investor who incorporates it into his overall investment strategy.

# PART ONE

# TREND DETERMINATION TECHNIQUES

# 1

# DOW THEORY

The Dow theory is the oldest and by far the most publicized method of identifying major trends in the stock market. An extensive account will not be necessary here, as there are many excellent books on this subject. Nevertheless, a brief explanation of the theory is in order since the basic principles incorporated in the Dow theory are used in other branches of technical analysis.

The object of the theory is to determine changes in the primary or major movement of the market. Once a trend is established it is assumed to exist until a reversal is proved. The theory is concerned with the *direction* of a trend and has no forecasting value as to its ultimate duration or size.

If an investor had purchased the stocks in the Dow Jones Industrial Average (DJIA) following each Dow theory "buy" signal, and if he had liquidated his position on "sell" signals and reinvested his money on the next "buy" signal, his original investment of $44 in 1897 would have risen to about $14,500 by December 1977.[1] If instead he had held onto his original $44 investment throughout the 80 years, his investment would also have grown, but only to $800. In reality the substantial profit earned by following the Dow theory would have been trimmed by transaction costs and capital gains taxes. Even if a wide margin for error is allowed the investment performance using this approach would still have been far superior to a "buy and hold" strategy.

It should be recognized that the theory does not always keep pace with

[1] This assumes the averages were available in 1897. Actually, Dow theory was first published in 1900.

events; it occasionally leaves the investor in doubt, and it is by no means infallible, since small losses are sometimes incurred. These points emphasize the fact that while mechanical devices can be useful for forecasting the stock market, there is no substitute for additional supportive analysis on which to base sound, balanced judgment.

Dow theory evolved from the work of Charles H. Dow, which was published in a series of *Wall Street Journal* editorials between 1900 and 1902. Dow used the behavior of the stock market as a barometer of business conditions rather than as a basis for forecasting stock prices themselves. His successor William Peter Hamilton developed Dow's principles and organized them into something approaching the theory as we know it today. These principles were outlined rather loosely in his book *The Stock Market Barometer*, published in 1921. It was not until Robert Rhea wrote *Dow Theory* in 1932 that a more complete and formalized account of the principles was finally published.

The theory assumes that the majority of stocks follow the underlying trend of the market most of the time. In order to measure "the market," Dow constructed two indexes: the (Dow Jones) Industrial Average, a combination of 12 blue-chip (now 30) stocks, and the (Dow Jones) Rail Average, comprising 12 railroad stocks. Since the Rail Average was intended as a proxy for transportation stocks, the evolution of aviation and other forms of transportation has necessitated modifying the old "Rail" Average in order to incorporate other forms of this industry. Consequently, the name of this index has been changed to the "Transportation Average."

## INTERPRETING THE THEORY

In order to interpret the theory correctly, it is necessary to have a record of the daily closing[2] prices of the two averages and the total of daily transactions on the New York Stock Exchange (NYSE). The six basic tenets of the theory are as follows:

### 1. The Averages Discount Everything

Changes in the daily closing prices reflect the aggregate judgment and emotions of all stock market participants, both current and potential. It is therefore assumed that this process discounts everything known and pre-

[2] It is important to use closing prices, since intraday fluctuations are more subject to manipulation.

dictable that can affect the demand/supply relationships of stocks. Although acts of God are obviously unpredictable, their occurrence is quickly appraised and their implications discounted.

## 2. The Market Has Three Movements

There are simultaneously three movements in the stock market.

*Primary movement* The most important is the primary or major trend, more generally known as a bull (rising) or bear (falling) market. These movements last from less than one year to several years.

A *primary bear market* is a long decline interrupted by important rallies. It begins as the hopes on which the stocks were first purchased are abandoned. Its second phase evolves as the level of business activity and of profits declines. The bear market reaches a climax when stocks are liquidated regardless of their underlying value (because of the depressed state of the news or because of forced liquidation due, for example, to margin calls). This represents the third stage of the bear market.

A *primary bull market* is a broad upward movement, averaging at least 2 years, which is interrupted by secondary reactions. The bull market begins when the averages have discounted the worst possible news and confidence begins to revive about the future. The second stage of the bull market is a response of equities to known improvements in business conditions, while the third and final phase evolves from overconfidence and speculation when stocks are advanced on projections that usually prove to be unfounded.

*Secondary reactions* A secondary or intermediate reaction is defined as ". . . an important decline in a bull market or advance in a bear market, usually lasting from three weeks to as many months during which interval the movement generally retraces from 33 to 66 percent of the primary price change since the termination of the last preceding secondary reaction."[3] This relationship is shown in Figure 1-1.

Occasionally a secondary reaction can retrace the whole of the previous primary movement, but normally the move falls in the $\frac{1}{2}$ to $\frac{2}{3}$ area, often at the 50 percent mark. As discussed in greater detail below, the correct differentiation between the first leg of a new primary trend and a secondary movement within the existing trend provides Dow theorists with their most difficult problem.

[3] Robert Rhea, *Dow Theory*, Barron's, New York, 1932.

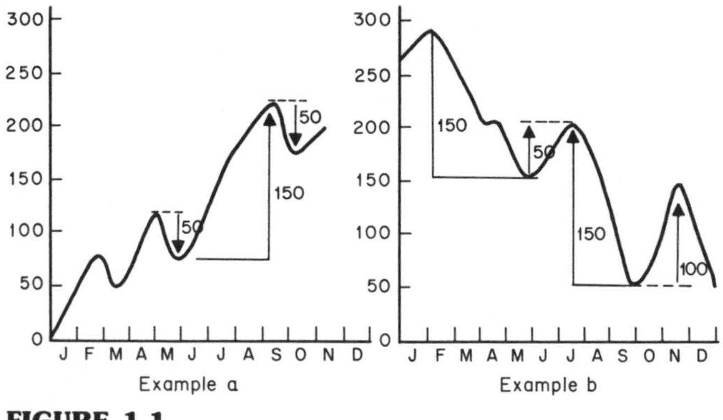


**FIGURE 1-1**

*Minor Movements* The third movement lasts from a matter of hours up to as long as 3 weeks. It is only important in that it forms part of the primary or secondary moves; it has no forecasting value for longer-term investors. This is especially important since short-term movements can be manipulated to some extent, unlike the secondary or primary trends.

### 3. Lines Indicate Movement

Rhea defined a line as ". . . a price movement two to three weeks or longer, during which period the price variation of both averages moves within a range of approximately 5 percent (of their mean average). Such a movement indicates either accumulation [*stock moving into strong and knowledgeable hands and therefore bullish*] or distribution [*stock moving into weak hands and therefore bearish*]."[4] (Words in italics added by the present author.)

An advance above the limits of the "line" indicates accumulation and predicts higher prices, and vice versa. When a line occurs in the middle of a primary advance, it is really forming a horizontal secondary movement and should be treated as such.

### 4. Price/Volume Relationships Provide Background

The normal relationship is for volume to expand on rallies and contract on declines. If volume becomes dull on a price advance and expands on a decline, a warning is given that the prevailing trend may soon be reversed. This principle should be used as background information only,

[4] Ibid.

since the conclusive evidence of trend reversals can be given only by the price of the respective averages.

## 5. Price Action Determines the Trend

Bullish indications are given when successive rallies penetrate peaks while the trough of an intervening decline is above the preceding trough. Conversely, bearish indications arise from a series of declining peaks and troughs.

Figure 1-2 shows a theoretical bull trend interrupted by a secondary reaction. In example *a* the index makes a series of three peaks and troughs, each higher than its respective predecessor. Then, following the third decline, the index rallies but is unable to surpass its third peak. The ensuing decline takes the average below its low point, confirming a bear market as it does so at point X. Similarly in example *b*, following the third peak in the bull market, a bear market is indicated as the average falls below the previous secondary trough. In this instance the preceding secondary was part of a bull market, not the first trough in a bear market, as shown in example *a*. Many Dow theorists do not consider penetration at

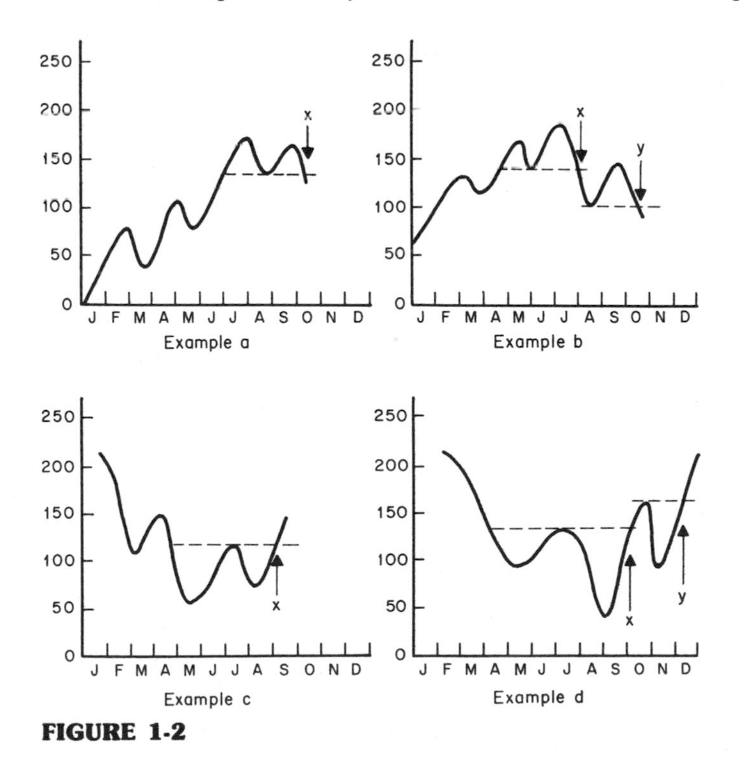


**FIGURE 1-2**

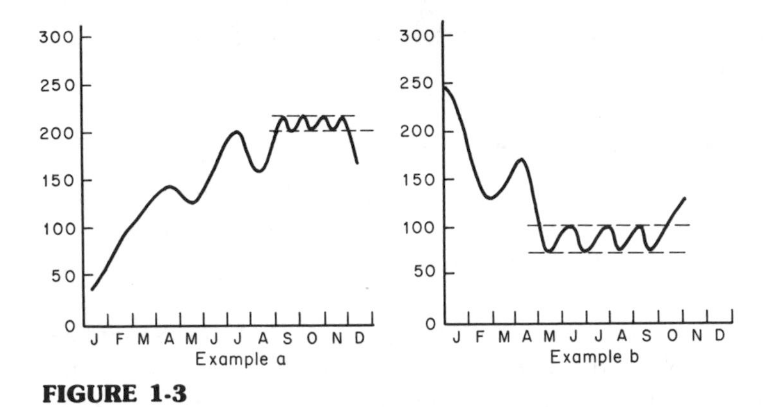


**FIGURE 1-3**

point X in example *b* to be a sufficient indication of a bear market. They prefer to take a more conservative stance by awaiting a rally and subsequent penetration of that previous trough marked as point Y in example *b*.

In such cases, it is wise to approach the interpretation with additional caution. If a bearish indication is given from the volume patterns and if a clearly identifiable speculative stage for the bull market has already materialized, it is probably safe to assume that the bearish indication is valid. In the absence of such characteristics it is wiser to give the bull market the benefit of the doubt and adopt a more conservative position.

Examples *c* and *d* represent similar instances at the bottom of a bear market.

The examples in Figure 1-3 show how the primary reversal would appear if the average had formed a "line" at its peak or trough.

The importance of being able to distinguish between a valid secondary movement and the first leg of a new primary trend is now evident. This is perhaps the most difficult part of the theory to interpret, and unquestionably the most critical.

It is first essential to establish that the secondary reaction has retraced at least one-third of the ground of the preceding primary movement as measured from the termination of the preceding secondary. The secondary should also extend for at least 3 to 4 weeks.

Vital clues can also be obtained from volume characteristics and from an assessment of the maturity of the prevailing primary trend. The odds of a major reversal are much greater if the market has undergone its third phase, characterized by speculation and false hopes during a primary upswing, or a bout of persistent liquidation and widespread pessimism during a major decline. While a change in the primary trend can occur without the existence of a clearly identifiable third phase, generally such reversals prove to be relatively short-lived. On the other hand, the largest primary swings usually develop when the characteristics of a third phase are especially marked during the preceding primary movement. Hence, the excessive bouts of speculation in 1919 and 1929 were followed by particularly sharp setbacks.

## 6. The Averages Must Confirm

One of the most important principles of Dow theory is that the movement of the Industrial and Transportation Averages should always be considered together, i.e., the two averages must confirm each other.

The need for confirming action by both averages would seem fundamentally logical, for if the market is truly a barometer for future business conditions, in an expanding economy investors should be bidding up the prices of both companies that produce goods and those that transport them. It is not possible to have a healthy economy where goods are being manufactured but not sold (i.e., shipped to market). This principle of confirmation is shown in Figure 1-4.

In example *a*, the Industrial Average is the first to signal a bear trend (point A), but the actual bear market is not indicated until the Transportation Average confirms at point *B*. Example *b* shows the beginning of a new bull market. Following a sharp decline the industrials make a new low. A rally then develops, but the ensuing reaction holds above the previous low. When prices push above the preceding rally, a bull signal is given by the industrials at point A. In the meantime the Transportation Average makes a series of two succeeding lows. The question arises as to which average is correctly representing the prevailing trend. Since it is always assumed that a trend is in existence until a reversal is proved, the

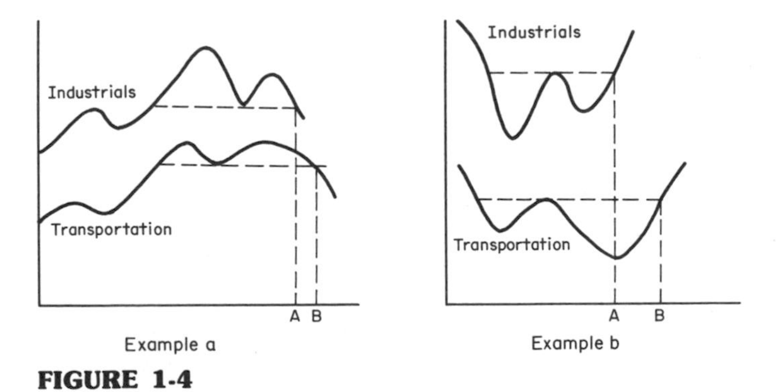


**FIGURE 1-4**

conclusion should be drawn at this point that the Transportation Average is portraying the correct outcome.

It is only when this average exceeds the peak of the preceding secondary at point *B* that a new bull market is confirmed by both averages, thereby resulting in a Dow theory "buy" signal.

The movement of one average unsupported by the other can often lead to a false and misleading conclusion. This fact is well illustrated in Figure 1-5 by the following example, which occurred in 1930.

The 1929–1932 bear market had begun in September 1929 and was confirmed by both averages in late October. In June 1930 both averages made a new low then rallied and reacted in August. Following this correction the industrials surpassed their previous peak. Many believed that this signaled the end of a particularly sharp bear market and that it was only a matter of time before the rails would follow suit. As it turned out, the action of the industrials was totally misleading; the bear market still had another two years to run.

Dow theory does not specify a time period beyond which a confirmation of one average by the other becomes invalid. Generally, the closer the confirmation the stronger the ensuing move is likely to be. For example, confirmation of the 1929–1932 bear market was given by the Rail Averagc just one day after the Industrial Average. The sharp 1962 break was confirmed on the same day.

One of the major criticisms of Dow theory is that many of its signals have proved to be late, often 20 to 25 percent after a peak or trough in the averages has occurred. One rule of thumb that has enabled Dow theorists to anticipate probable reversal at an earlier date is to observe the dividend

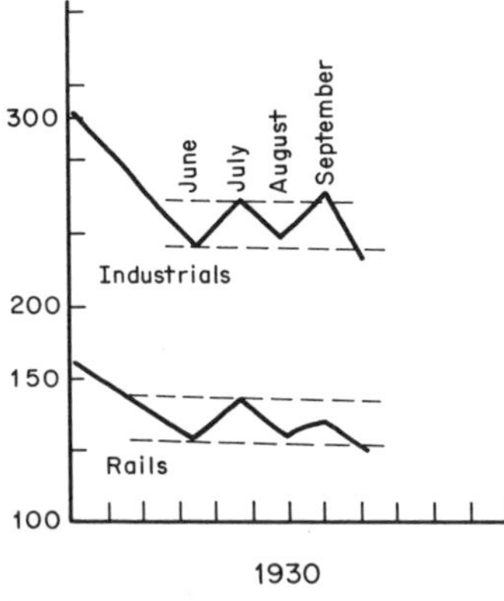


**FIGURE 1-5**

**TABLE 1-1 80-year Record of the Dow Theory**

| Buy Signals | | | Sell Signals | | |
|---|---|---|---|---|---|
| **Date of signal** | **Price of Dow** | **% gain from sell signal when short** | **Date of signal** | **Price of Dow** | **% gain from buy signal** |
| Jul. 1897 | 44 | | Dec. 1899 | 63 | 43 |
| Oct. 1900 | 59 | 6 | Jun. 1903 | 59 | 0 |
| Jul. 1904 | 51 | 14 | Apr. 1906 | 92 | 80 |
| Apr. 1908 | 70 | 24 | May 1910 | 85 | 21 |
| Oct. 1910 | 82 | 4 | Jan. 1913 | 85 | 3 |
| Apr. 1915 | 65 | 24 | Aug. 1917 | 86 | 32 |
| May 1918 | 82 | 5 | Feb. 1920 | 99 | 22 |
| Feb. 1922 | 84 | 16 | Jun. 1923 | 91 | 8 |
| Dec. 1923 | 94 | loss 3 | Oct. 1929 | 306 | 226 |
| May 1933 | 84 | 73 | Sep. 1937 | 164 | 95 |
| Jun. 1938 | 127 | 23 | Mar. 1939 | 136 | 7 |
| Jul. 1939 | 143 | 5 | May 1940 | 138 | loss 7 |
| Feb. 1943 | 126 | 8 | Aug. 1946 | 191 | 52 |
| Apr. 1948 | 184 | 4 | Nov. 1948 | 173 | loss 6 |
| Oct. 1950 | 229 | loss 32 | Apr. 1953 | 280 | 22 |
| Jan. 1954 | 288 | loss 3 | Oct. 1956 | 468 | 63 |
| Apr. 1958 | 450 | 4 | Mar. 1960 | 612 | 36 |
| Nov. 1960 | 602 | 2 | Apr. 1962 | 683 | 13 |
| Nov. 1962 | 625 | 8 | May 1966 | 900 | 43 |
| Jan. 1967 | 823 | 9 | June 1969 | 900 | 9 |
| Dec. 1970 | 823 | 9 | Apr. 1973 | 921 | 12 |
| Jan. 1975 | 680 | 26 | Oct. 1977 | 801 | 18 |
| Average of all Cycles, 11% | | | Average of all Cycles, 36% | | |

yield on the industrials. When the yield on the Industrial Average has fallen to 3 percent or below, this has historically been a good indication of the maturity of the bull market and therefore representative of the so-called third stage. Similarly a yield of 6 percent has been a reliable indicator at market bottoms.

The Dow theorist would not necessarily use these levels as actual buying or selling points, but would probably consider altering the percentage of his equity exposure if a significant nonconfirmation were to develop between the Industrial and Transportation Averages when the yield on the Dow reached these extremes. Such a strategy would help improve the investment return of the Dow theory but would not always result in a superior performance. At the 1976 peak, for example, the yield on the Dow never reached the magic 3 percent level, and prices fell 20 percent before a mechanical signal was confirmed by both averages. Moreover, in the postwar period long-term interest rates have reached higher and higher peaks in each succeeding cycle, so that if inflationary pressures continue and this secular trend of higher rates is extended, it may not be unreasonable to expect future bear markets to end with yields well in excess of 6 percent on the DJIA.

Over the years many criticisms have been leveled at the theory on the basis that from time to time (as in periods of war) the rails have been overregulated, or that the new Transportation Average no longer reflects investor expectations about the future movement of goods. The theory has stood the test of time, however, as the following table indicates. Indeed, the existence of criticism is perfectly healthy, for were the theory to gain widespread acceptance and were its signals purely mechanistic instead of requiring experienced judgment, they would be instantly discounted, rendering Dow theory useless for profitable investment.

## SUMMARY

Dow theory is concerned with determining the direction of the primary trend of the market, not the ultimate duration or size of the trend. Once confirmed by both averages, the new trend is assumed to be in existence until an offsetting confirmation by both averages has taken place.

Major bull and bear markets each have three distinct phases. The identification of these phases and the appearance of any divergence in the normal volume/price relationship both offer useful indications that a reversal in the major trend is about to take place. Such supplementary evidence is particularly useful when the action of the price averages themselves is inconclusive.

# 2

# PRICE PATTERNS

The techniques discussed in this and the next few chapters are concerned with the analysis of either a market average or an indicator of the market's internal structure, but these basic principles can just as validly be applied to individual stocks.

The concept of price patterns is demonstrated in Figures 2-1 and 2-2. Figure 2-1 represents a typical stock market cycle in which there are three trends—two basic and one minor. The two basic trends are the primary bull or rising trend (*A* to *B*) and the bear market or declining phase (*C* to *D*). The minor trend is essentially a horizontal or transitional one which separates the two major market movements. Sometimes a highly emotional market can change without warning, as in Figure 2-2, but this rarely happens. Consider a fast-moving train, which takes a long time to slow down and then go into reverse; the same is normally true of the stock market.

To the market technician, the transitional phase takes on great significance since it marks the turning point between a rising or falling market. If prices have been advancing, the enthusiasm of the buyers has outweighed the pessimism of sellers up to this point, and prices have risen accordingly. During the transition phase, the balance becomes more or less even until finally, for one reason or another, it is tipped in a new direction as the relative weight of selling pushes the trend (of prices) down. At the termination of the bear market the reverse process occurs.

From a technical viewpoint these transition phases are almost invariably signaled by clearly definable price patterns and formations whose successful completion alerts the technician to the fact that a major reversal in trend has taken place.

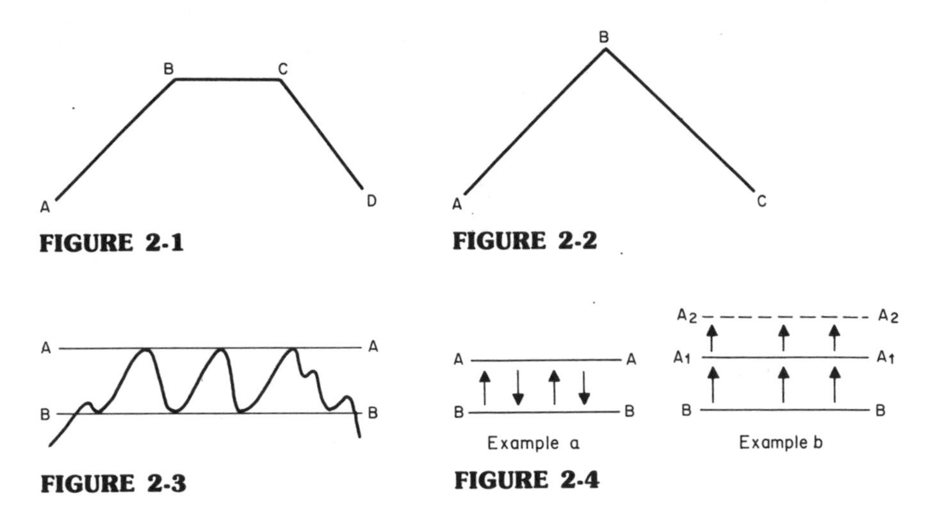


**FIGURE 2-1**

**FIGURE 2-2**

**FIGURE 2-3**

**FIGURE 2-4**

This phenomenon is illustrated in Figure 2-3, which represents the price action at the end of a long rising trend. As soon as the index rises above line *BB* it is in the transitional area, although this is only apparent some time after the picture has unfolded.

Once into the area the index rises to line AA, which is technically termed a *resistance area*. "Resistance" derives from the fact that at this point the index shows a resistance to a further price rise. This arises because as the supply/demand relationship comes into balance at AA, the market quickly turns in favor of the sellers as prices react. This temporary reversal may occur because buyers refuse to pay up for stock, or because the higher price attracts more sellers, or through a combination of both. The important fact is that the relationship between the two groups is temporarily reversed at this point.

Following the unsuccessful assault on AA, prices turn down until line *BB*, known as a *support level*, is reached. Just as the price level at AA reversed the balance in favor of the sellers, so the support level *BB* alters the balance yet again. This time the trend moves in an upward direction, for at *BB* prices again become relatively attractive for buyers who "missed the boat" on the way up, while sellers who feel that the price will again reach AA hold off. For a while there is a standoff between buyers and sellers within the confines of the area bounded by lines AA and *BB*. Finally the price falls below *BB*, and a new major (downward) trend is signaled.

In order to help explain this concept, the "battle" between buyers and sellers can be thought of as one fought by two armies engaged in trench warfare. In Figure 2-4, example *a*, two armies A and B are facing off. Line

AA represents army A's defense, and *BB* is army B's line of defense. The arrows indicate the forays between the two lines as both armies fight their way to the opposing trench but are unable to penetrate the line of defense. In the second example army B finally pushes through A's trench. Army A is then forced to retreat and make a stand at the second line of defense (line $AA_2$). In the stock market, line AA represents selling resistance which, once overcome, signifies a change in the balance between buyers and sellers in favor of the buyers, so that prices will advance quickly until new resistance is met. The second line of defense, line $AA_2$, represents this resistance to a further advance.

On the other hand, army B might quite easily break through $AA_2$, but the farther it advances without time to consolidate its gains the more likely it is to become overextended, and the greater is the probability of its suffering a serious setback. At some point, therefore, it makes more sense for this successful force to wait and consolidate its gains.

If stock market prices extend too far without time to digest their gains, they too are more likely to face a sharp and seemingly unexpected reversal.

The transitional or horizontal phase separating rising and falling price trends is a pattern known as a ***rectangle***. This formation corresponds to the "line" formation developed from Dow theory. The rectangle in Figure 2-3 marking the turning point between the bull and bear phases is termed a ***reversal*** pattern. Reversal patterns at market tops are known as ***distribution*** areas or patterns, and those at market bottoms are called ***accumulation*** patterns (see Figure 2-5*a*). Had the rectangle been completed with a victory for the buyers as the price pushed through line AA (see Figure 2-5*b*), no reversal of the rising trend would have occurred. The "breakout" above AA would therefore have reaffirmed the underlying trend. In this case the corrective phase associated with the formation of the rectangle would have temporarily interrupted the bull market and become a ***consolidation*** pattern. Such formations are also referred to as ***continuation*** patterns.

During the period of formation, there is no way of knowing in advance

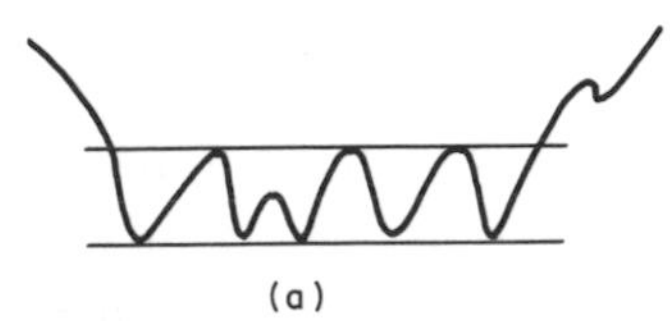


**FIGURE 2-5*a***

A A
B B
(b)

**FIGURE 2-5*b***

which way the price will ultimately break, so it should always be assumed that *the prevailing trend is in existence until it is proved to have been reversed*.

There are certain characteristics that apply to all types of price patterns.

## SIZE AND DEPTH

The significance of a price formation or pattern is a direct function of its size and depth. In other words, the longer a pattern takes to complete and the greater the price fluctuations within the pattern, the more substantial the ensuing move is likely to be. It is as important to build a strong base from which stock prices can rise as it is to build a large, strong, deep foundation for the construction of a skyscraper. In the case of stocks, the foundation is an accumulation pattern which represents an area of indecisive combat between buyers and sellers. The term "accumulation" is used because market bottoms always occur when business news is bad. Such an environment stimulates the sale of stock by uninformed investors who were not expecting developments to deteriorate. During an accumulation phase more sophisticated investors and professionals would be positioning or accumulating stock in anticipation of improved conditions 6 to 9 months ahead. During this period stock is moving from weak, uninformed investors into strong and knowledgeable hands. At market tops the process is reversed as those who accumulated stocks at or near the bottom sell to the less sophisticated market participants, who become more and more attracted to the stock market as prices rise, business conditions improve, and forecasts for the economy are revised upward. Thus the longer the period of accumulation, the greater the amount of stock that moves from weak into strong hands, and the greater is the base from which prices can rise. The reverse is true at market tops, where a substantial amount of distribution inevitably results in a protracted period of price erosion or base building.

The longer the formation of a rectangle pattern takes (see Figure 2-6) and the more often it fails to break through its outer boundaries, the greater is the significance of the ultimate penetration.

The time taken to complete a formation is important because of the amount of stock changing hands, and also because a movement in price beyond the boundaries of a pattern means that the balance between buyers and sellers has altered. When the price action has been in a stale-

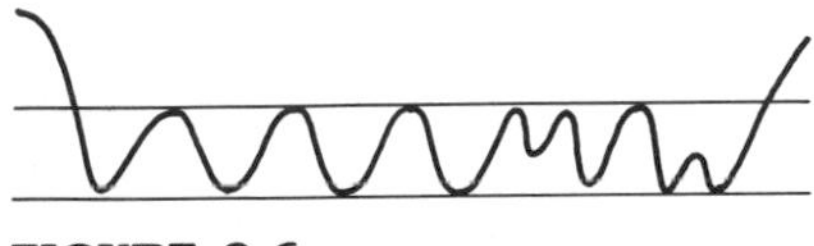

**FIGURE 2-6**

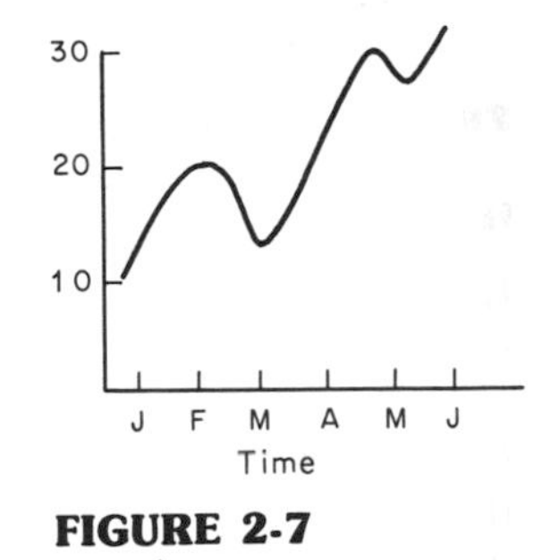


**FIGURE 2-7**

mate for a long time and investors have become used to buying at one price and selling at the other, a move beyond either limit is a fundamental change which has great psychological significance.

## MEASURING IMPLICATIONS

Most of the results obtained with technical analysis procedures do not indicate the eventual duration of a trend. Price patterns are the exception, since their construction offers some limited forecasting possibilities.

There are two alternative methods of charting, and the choice of scale determines the significance of the measuring implications. The two types of graph used in technical analysis are (1) arithmetic and (2) ratio or logarithmic.

### Arithmetic Scale

Arithmetic charts consist of an arithmetic scale on the vertical or *y* axis, with time being shown on the horizontal or *x* axis, as illustrated in Figure 2-7. Each unit of measurement is plotted using the same vertical distance, so that the difference in space between 2 and 4 is the same as that between 20 and 22. This scale is not particularly satisfactory for stock price measurement, since a rise from 2 to 4 represents a doubling of the price, while one from 20 to 22 represents only a 10 percent increase. On an arithmetic scale both moves are represented by the same vertical distance. A relatively large move of 10 points or more was not uncommon from 1966 to 1978, when the DJIA ranged between 600 and 1000. A 10-point move in 1932, though, when the average was as low as 40 to 50, represented a 20 to 25 percent change. For this reason, long-term movements are commonly plotted on a ratio or logarithmic scale.

## Ratio Scale

Indexes or stock prices plotted on this scale show identical distances for identical percentage moves.

In Figure 2-8 the vertical distance between 1 and 2 (a 2:1 ratio) is $\frac{1}{2}$ inch. Similarly, the 2:1 distance between 4 and 2 is also represented on the chart as $\frac{1}{2}$ inch. A specific vertical distance on the chart will represent the same percentage change in the index being measured whatever the price level. For example, if the scale in Figure 2-8 were extended, $\frac{1}{2}$ inch would always represent a doubling, whether from 1 to 2, 16 to 32, 50 to 100, and so on, just as $\frac{1}{4}$ inch would indicate a rise of 50 percent, or 1 inch a quadrupling of prices. Graph paper using ratio or logarithmic scale is easily obtained from any reputable office supplier.

An alternative method of portraying a ratio scale chart is to use arithmetic paper but plot the logarithm of an index instead of the actual index number itself. Hence, the DJIA at 1004 would be plotted at 3.001, which is the base 10 logarithm for 1000, and 758 would be plotted at 2.880, etc.

Figure 2-9 shows a rectangle which has formed and completed a (distribution) top. The measuring implication of this formation is the vertical distance between its outer boundaries, i.e., the distance between lines *AA* and *BB* projected downward from line *BB*. If *AA* represents 100 and *BB* 50, then the downside objective will be 50 percent, using a ratio scale. When projected downward from line *BB*, 50 percent gives a measuring implication of 25. While this measuring formula offers a rough guide, this is usually a *minimum* expectation, and prices often go much farther than their implied objective. In a very high proportion of cases the objective level derived from the measuring formula becomes an area of support or resistance where the price trend is temporarily halted. Should a rectangle appear as a bottom reversal pattern or as a consolidation pattern, the measuring rules remain consistent with the example given for the distribution formation. This is shown in the series in Figure 2-10.

Should the minimal objective prove to be the ultimate extension of the new trend, a substantial amount of accumulation or distribution, whichever is appropriate, will have to occur before prices can move in their original direction. Thus, if a 2-year rectangle is completed and the downward price objective is reached, even though further price erosion does not take place it is still usually necessary for a base (accumulation) to be formed of approximately the same size as the previous distribution (in this case 2 years) before a valid uptrend can take place.

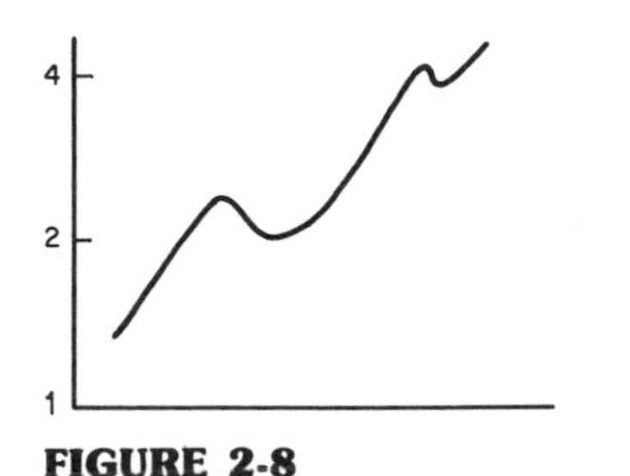


**FIGURE 2-8**

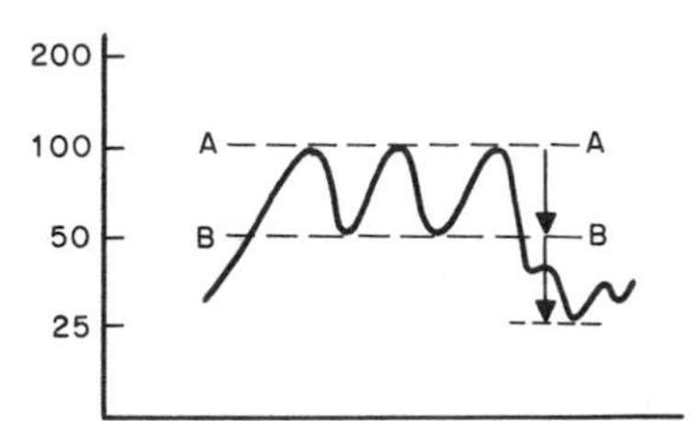


**FIGURE 2-9**

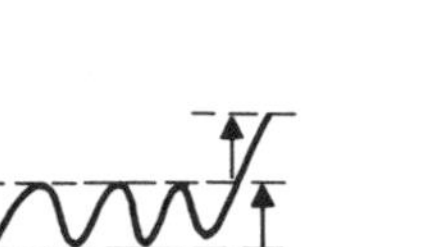

Example a

Example b

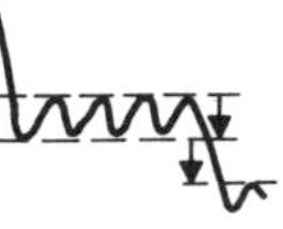

Example c

**FIGURE 2-10**

## CONFIRMATION OF A VALID "BREAKOUT"

### Price

So far it has been assumed that any move out of the price pattern (however small) constitutes a valid signal of a trend reversal (or resumption if the pattern is one of consolidation). Quite often misleading moves known as "whipsaws" occur, so that it is helpful to set up certain criteria to minimize the possibility of such a misinterpretation. In terms of price it has been found useful to await a 3 percent penetration of the boundaries before concluding that the breakout is valid. This filters out a substantial number of misleading moves, even though the resulting signals are less timely.

### Volume

A further confirmation concerning the validity of a breakout can be derived from an examination of the volume characteristics that accompany it. Volume usually goes with the trend, i.e., volume advances with a rising trend of prices and falls with a declining one. This is a normal relationship, and anything which diverges from this characteristic should be considered as a warning sign that the prevailing price trend may be in the

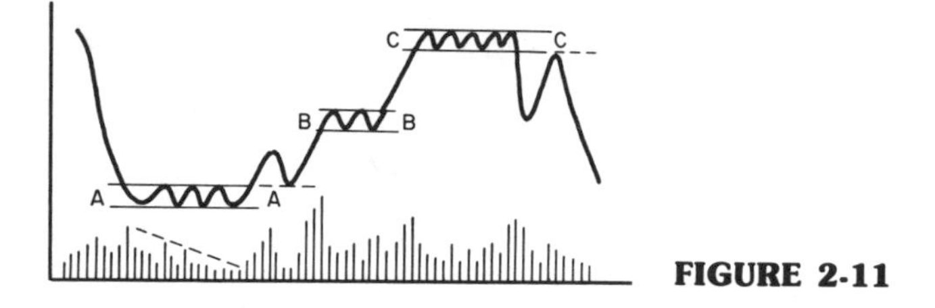


**FIGURE 2-11**

process of reversing. Figure 2-11 shows a typical volume/price relationship.

The volume, that is, the number of shares changing hands, is shown by the vertical lines at the bottom of the chart. Volume expands marginally as the index approaches its low, but as the accumulation pattern is formed, activity recedes noticeably. Volume is always measured on a relative basis, so that heavy volume is "heavy" only in relation to a recent period. As the pattern nears completion, disinterest prevails and volume almost dries up. Then, as if by magic, activity picks up noticeably as the index moves above its level of resistance (bounded by the upper line in the rectangle). It is sometimes possible to draw a trendline joining the lower volume peaks, as shown in Figure 2-11. It is this upward surge in trading activity that confirms the validity of the breakout. A similar move on low volume would have been suspect, as it would have resulted in a failure of volume to move with the trend.

Following the sharp price rise from the rectangle, enthusiasm dies down as prices correct in a sideways movement and volume contracts. This is a perfectly normal relationship, since volume is correcting (declining) with price. Eventually volume and price expand together, and the primary upward trend is once again confirmed. Finally the buyers become exhausted, and the index forms yet another rectangle—characterized as before by falling volume, but this time destined to become a reversal pattern.

It is worth noting that while the volume from the breakout in rectangle *B* is high, it is relatively lower than that which accompanied the move from rectangle *A*. In relation to the overall cycle this is a bearish factor, since volume usually leads price. In this case volume makes its peak just before entering rectangle *B*, while the peak in prices is not reached until rectangle *C*. Chart 2-1 shows how this occurs in the marketplace.

Volume contracts throughout the formation of rectangle C and ex-

**CHART 2-1 Dow Jones Rail Average, 1946**

This chart shows a classic rectangle formation as traced out by the Rail Average at the peak of the1942–1946 bull market. Note the declining trend of vclume, as indicated by the dashed line, during the formation of the rectangle. Worth special mention is the saucer-like formation of the volume during the late July–early August rally. The expansion of activity accompanying the downside breakout in late August completed the bearish implication of the successful completion of this pattern.

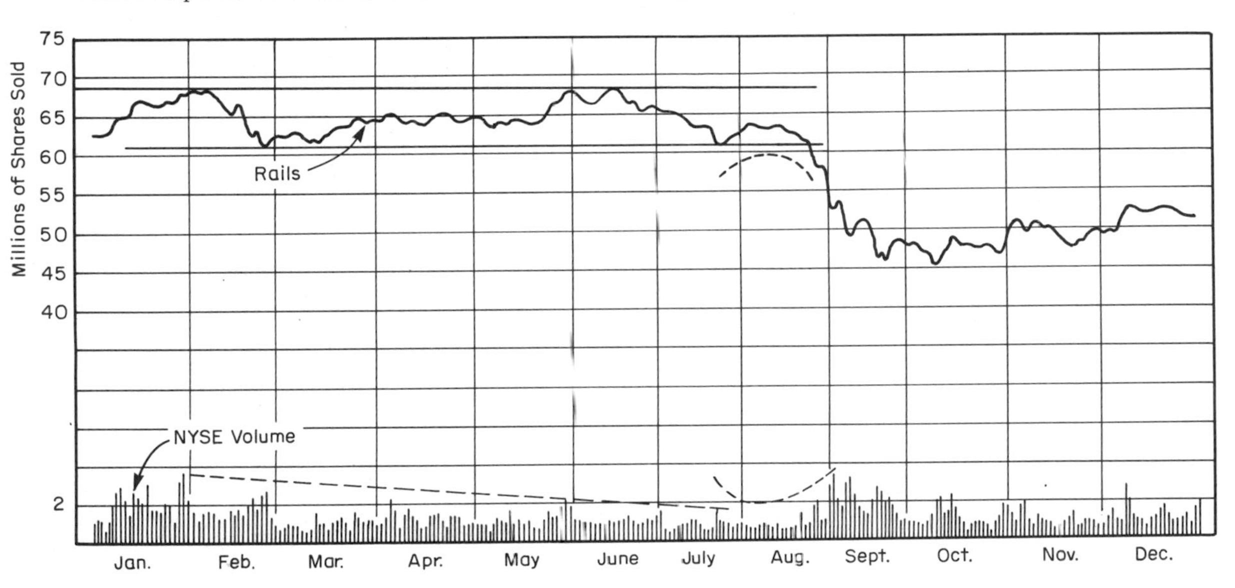

pands as prices break out on the downside. This expanded level of activity associated with the violation of the support at the lower boundary of the rectangle emphasizes the bearish nature of the breakout, although expanding volume is not a prerequisite for a valid signal with downside breakouts as it is for an upside move. More often than not, following the downside breakout prices will reverse and put on a small recovery or retracement rally. This advance is invariably accompanied by declining volume, which itself reinforces the bearish indications. It is halted at the lower end of the rectangle, which now becomes an area of resistance.

Some of the more common reversal formations will now be discussed.

## HEAD AND SHOULDERS

The head and shoulders (H&S) is probably the most reliable of all the chart patterns. These formations occur at both market tops and market bottoms. Figure 2-12 shows a typical head and shoulders distribution pattern.

The pattern consists of a final rally (the head) separating two smaller though not necessarily identical rallies (the shoulders). The first shoulder is the penultimate advance in the bull market, and the second is in effect the first bear market rally.

The volume characteristics are of critical importance in assessing the validity of these patterns. Activity is normally heaviest during the formation of the left shoulder and also tends to be quite heavy as prices approach the peak. The real tip-off that a head and shoulders pattern is being formed comes with the formation of the right shoulder, which is invariably accompanied by distinctly lower volume (see Chart 2-2). The line joining the bottoms of the two shoulders is called the "neckline."

The measuring formula for this price formation is the distance between the head and the neckline projected downward from the neckline, as shown in Figure 2-12. It follows that the deeper the pattern, the greater

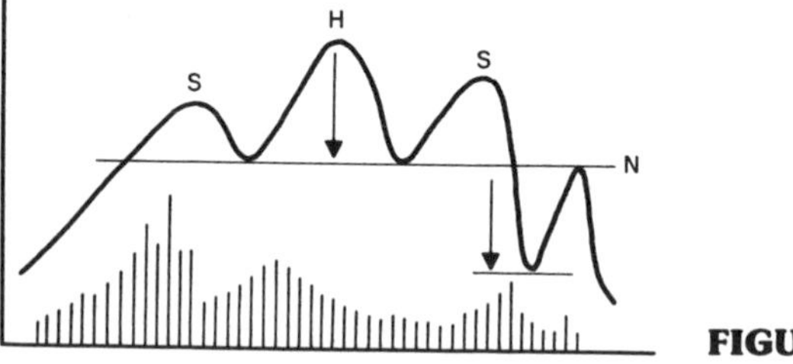


**FIGURE 2-12**

**CHART 2-2** ***New York Times* Average, 1928–1930**

During January and May of 1929 it appeared as if the *New York Times* average, shown below, was forming a broadening top. Three ascending peaks, O, Q, and S, were separated by two declining troughs P and R, but since the third bottom (T) remained above R, no negative signal was given. The break above the line joining O, Q, and S indicated that the "top" would not be completed. A sharp upward move followed as prices ran up to the ultimate peak of the 1921–1929 bull market. Even though this formation was never completed, the very fact that prices moved in such a volatile manner was a strong warning of the underlying weakness. A small right-angled broadening formation seemed to develop in July and August, but this would eventually prove to be the left shoulder of a 2½-month H&S pattern, the completion of which terminated the long bull market.

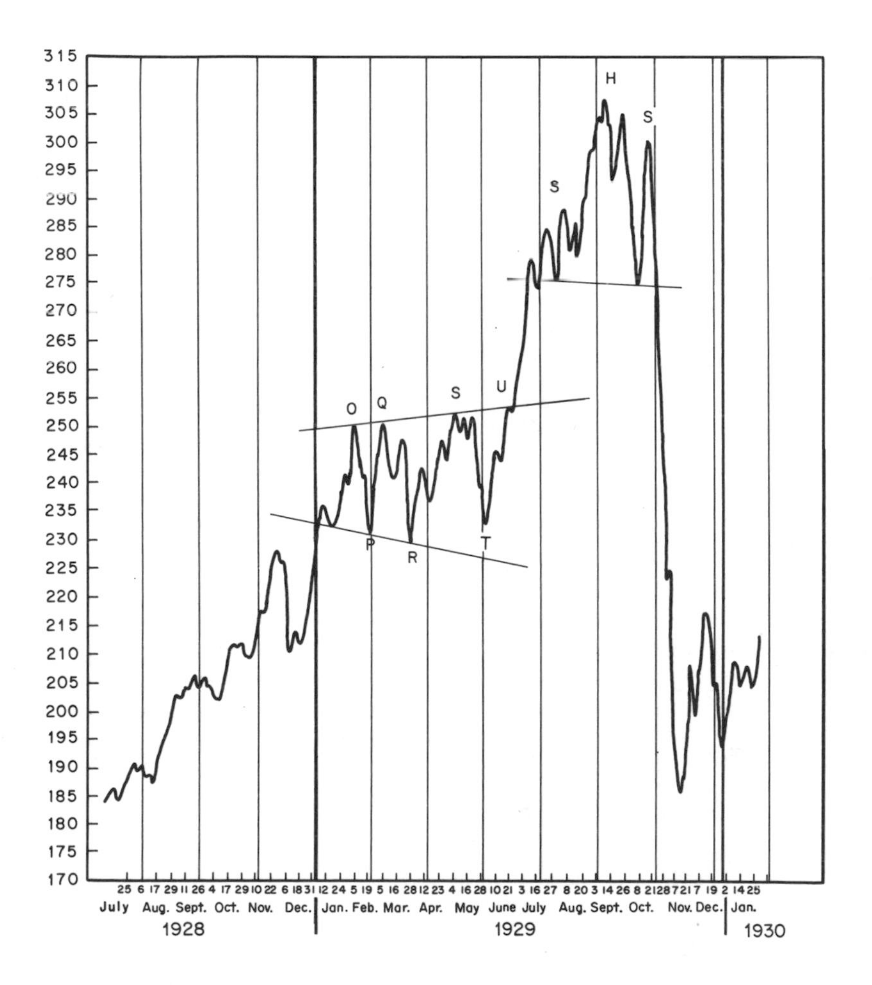

**CHART 2-3 Dow Jones Transportation Average, 1976**

Between mid-April and September 1976, the Dow Jones Transportation Average formed and completed a head and shoulders distribution pattern. The breakout in September proved to be a "bear trap," since the average resumed its advance and reached a major peak at 246 in May 1977. Nevertheless, the 1976 head and shoulders pattern did indicate the presence of distribution, and as a result the progress of the average was halted for a considerable time.

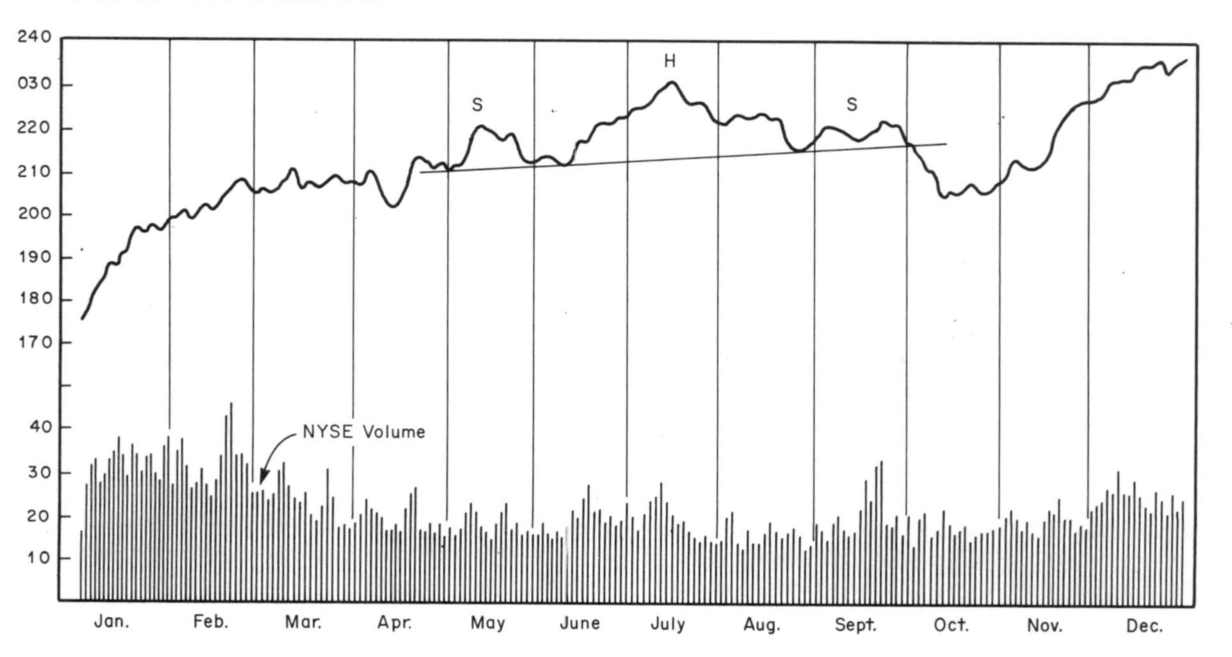

is the (bearish) significance once it has been completed. It is essential to await a 3 percent decline below the neckline, because sometimes the price will immediately reverse itself on reaching the neckline and proceed, usually very sharply, to substantially higher levels. Chart 2-7(*b*) which shows the daily close for the DJIA in 1975, illustrates this phenomenon very clearly as the "failure" of a head and shoulders pattern results in a fairly worthwhile rally. Nevertheless the sign of an H&S that does not "work" is indicative of the fact that while there is still some life left in the situation, the end is not far off. In the example cited above, the rally ended rather abruptly in July (see also Chart 2-3).

Head and shoulders patterns can be formed in a period as short as 3 to 4 weeks or take as long as several years to develop, but generally speaking the longer the period the greater the amount of distribution that has taken place, and therefore the longer the ensuing bear trend is likely to be. The larger head and shoulders formations are often very complex and comprise several smaller ones, as shown in Figure 2-13.

The H&S patterns illustrated in Figures 2-12 and 2-13 have a horizontal neckline, but there are many other varieties (as Figure 2-14 and Chart 2-4 show), all of which possess the same bearish implications as the horizontal variety once they have been completed.

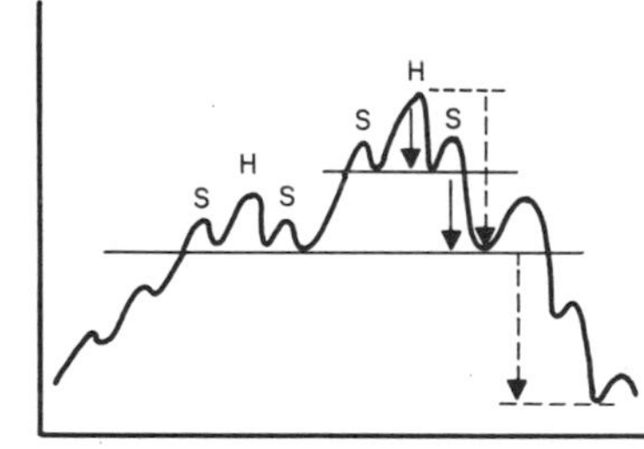


**FIGURE 2-13**

Example a

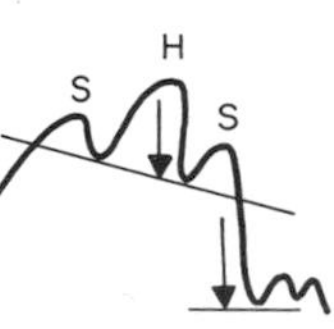


Example b

Example c

**FIGURE 2-14**

**CHART 2-4 *New York Times* Average, 1928**

This chart of the *New York Times* average of 50 railroad and industrial stocks shows the formation of an upward-sloping H&S during March, April, and May 1928. The minimum downside objective of about 182 was achieved fairly quickly, but a 3-month period of base building commensurate with the head and shoulders pattern was still necessary before the effect of the distribution was canceled out and prices were able to resume their primary advance. Note the heavy volume on the left shoulder and head and the relatively low volume on the right shoulder. Also, activity declined substantially during the formation of the triangle but began to expand during the breakout in September.

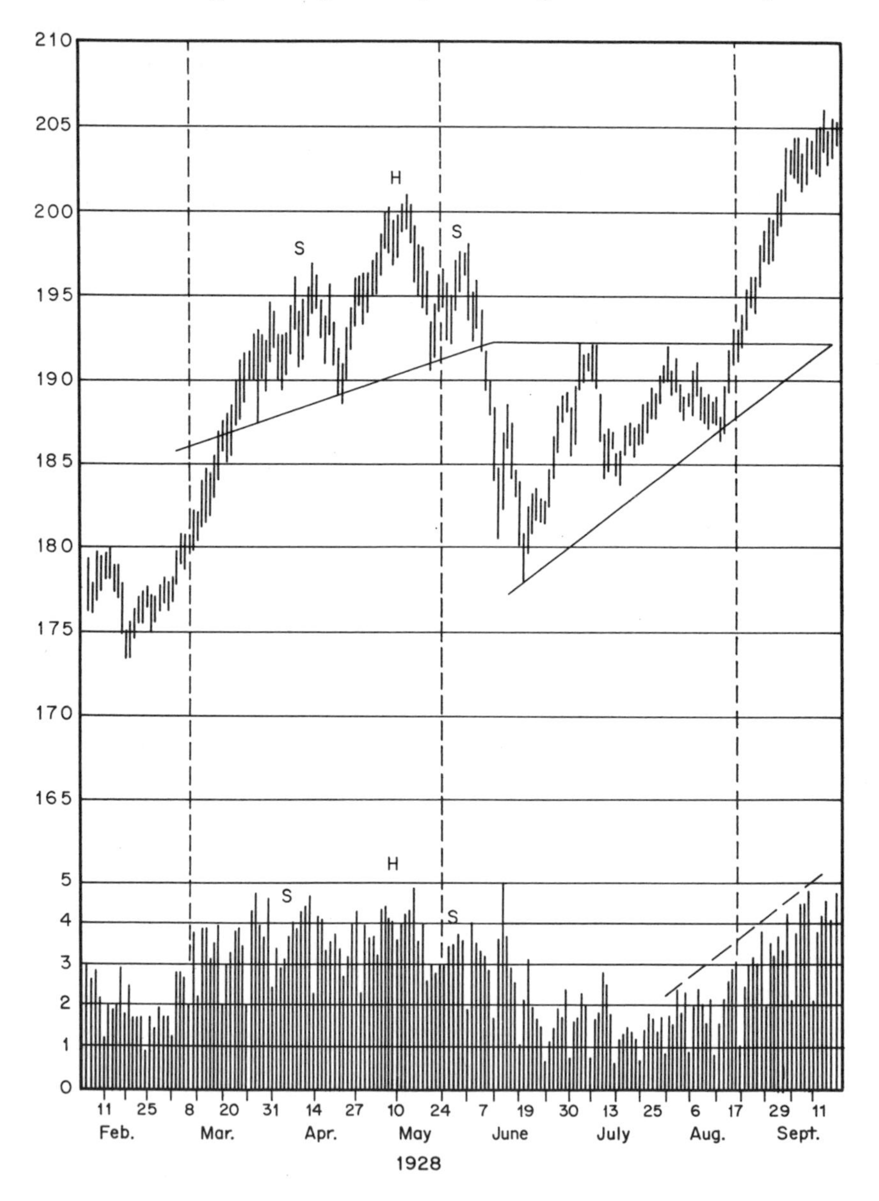

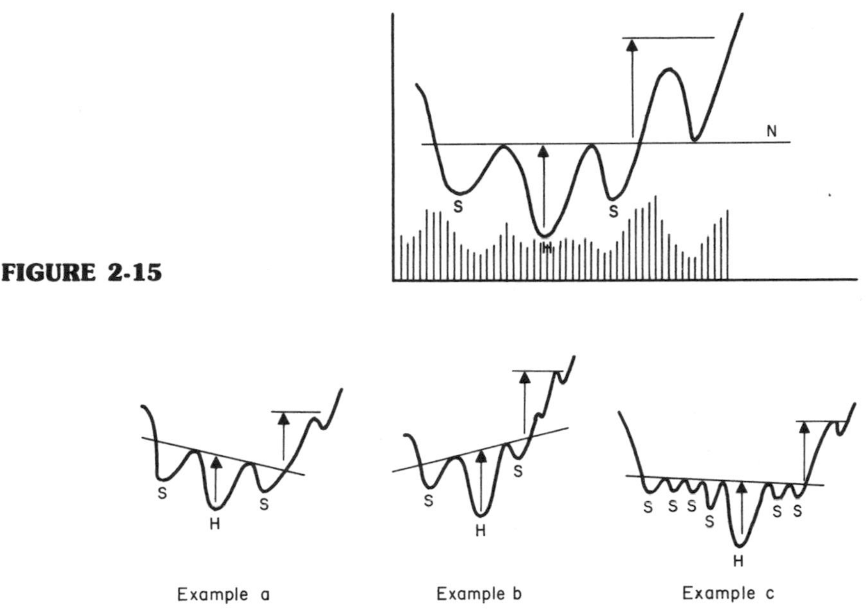


**FIGURE 2-15**

**FIGURE 2-16**

Figure 2-15 shows a head and shoulders pattern at a market bottom; this variety is usually called an "inverse head and shoulders."

Volume is usually relatively high at the bottom of the left shoulder and during the formation of the head. The major factor to watch for is activity on the right shoulder, which should contract during the decline to the trough and expand substantially on the breakout (see Chart 2-5). As with the H&S distribution patterns, the inverse (accumulation) H&S can have a number of variations as to trendline slope, number of shoulders, etc. Some of these are shown in Figure 2-16.

Head and shoulders patterns are extremely reliable formations, and their successful completion usually offers an excellent indication of a trend reversal.

## DOUBLE TOPS AND BOTTOMS

A double top consists of two peaks separated by a reaction or valley in prices. The main characteristic of a double top is that the second top is formed with distinctly less volume than the first (see Chart 2-6).

Minimum downside measuring implications, as shown in Figure 2-17, are similar to the H&S patterns. A double bottom is shown in Figure 2-18.

The pattern is typically accompanied by high volume on the first bot-

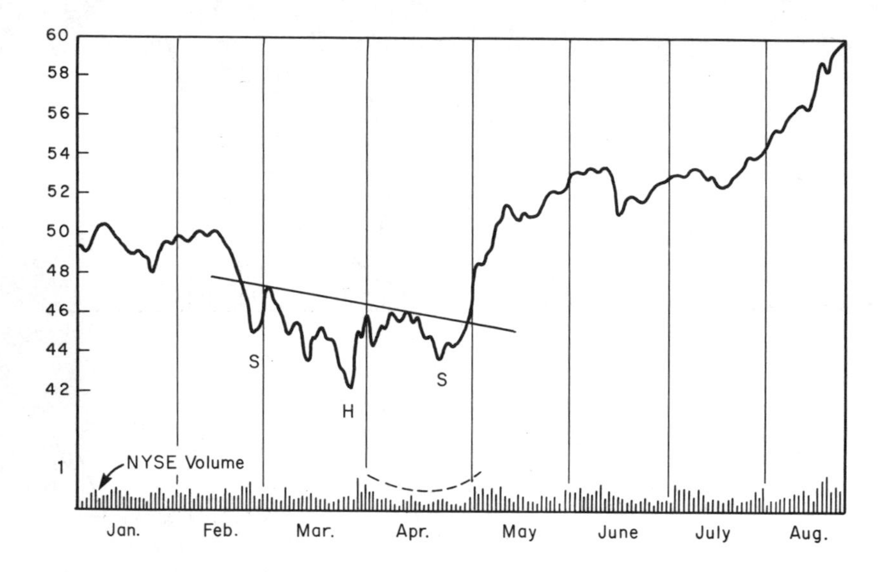
60
58
56
54
52
50
48
46
44
42
S
S
H
1
NYSE Volume
Jan.
Feb.
Mar.
Apr.
May
June
July
Aug.

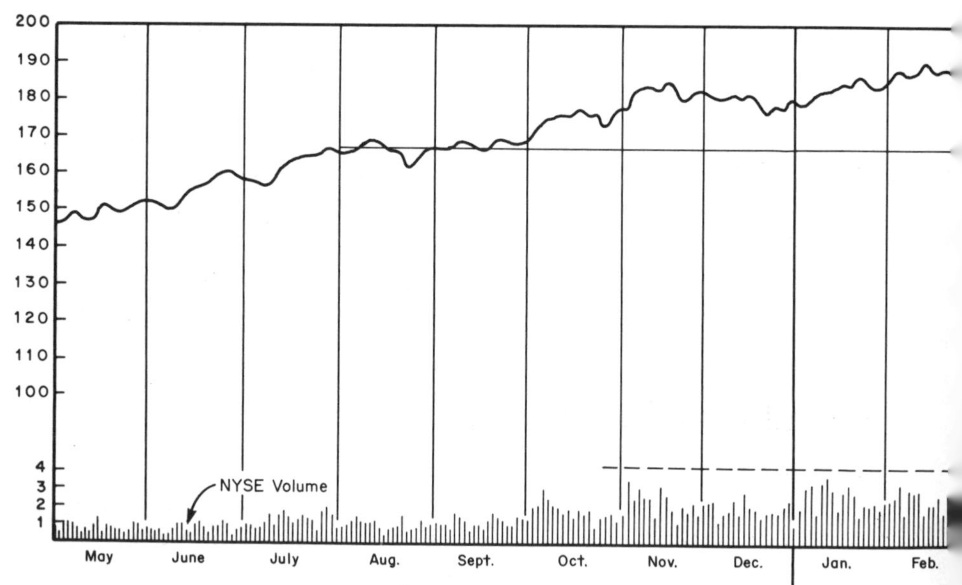
200
190
180
170
160
150
140
130
120
110
100
4
3
2
1
NYSE Volume
May
June
July
Aug.
Sept.
Oct.
Nov.
Dec.
Jan.
Feb.
1936

**CHART 2-5** (*Left*) **Dow Jones Industrial Average, 1898**

This downward-sloping inverse head and shoulders developed in the spring of 1898. Note how the April rally developed on very low volume. The subsequent reaction successfully tested the March low, and the ensuing "breakout" rally was accompanied by a bullish expansion of volume. By August the DJIA had reached 60.97, and by April 1899 it rose to 77.28.

**CHART 2-6** (*Below*) **Dow Jones Industrial Average, 1936–1937**

Following a substantial advance from 1932, the first post-Depression bull market ended in 1937. The chart below shows a classic double top. Note how the volume during the July–August rally was substantially below that of the January–March peak.

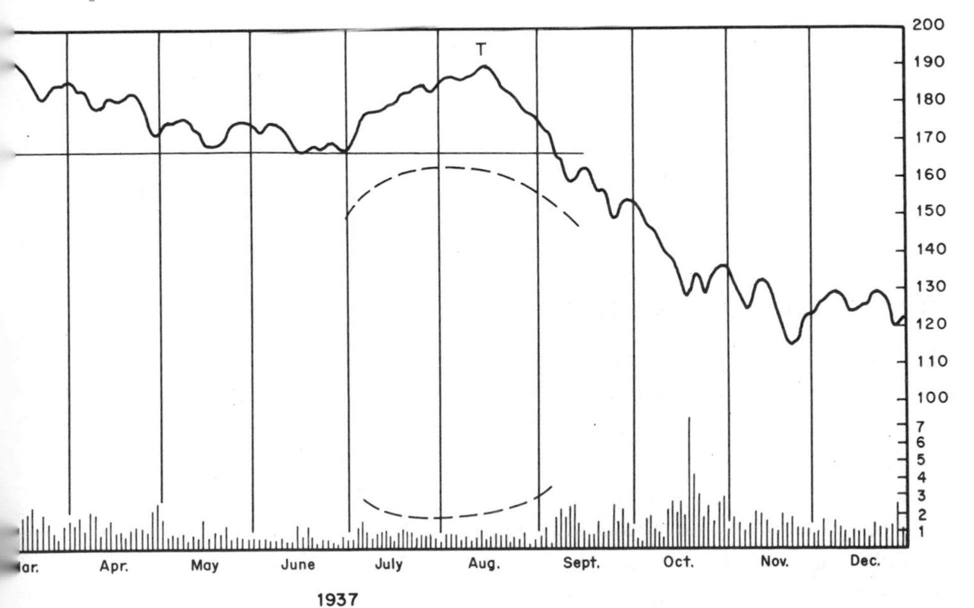

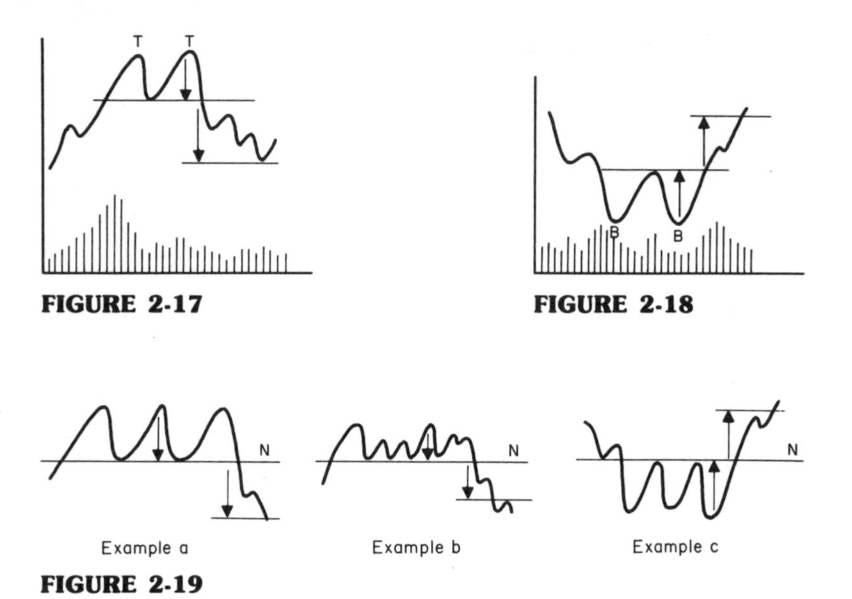


FIGURE 2-17

FIGURE 2-18

FIGURE 2-19

tom, very light volume on the second, and very heavy volume on the breakout. Usually the second bottom is formed above the first, but these formations are equally valid whether or not the second reaction reaches (or even slightly exceeds) the level of its predecessor.

Some "double" patterns extend to form triple tops or bottoms, sometimes even quadruple or other complex formations. Some of the variations are shown in Figure 2-19.

The measuring implications of all such patterns are derived by measuring the distance between the peak (trough) and lower (upper) end of the pattern and projecting this distance from the neckline [see Charts 2-7(*a*) and (*b*)].

## BROADENING FORMATIONS

Broadening formations occur where a series of three or more price fluctuations widen out in size so that peaks and troughs can be connected with two diverging trendlines. The easiest types of broadening formations to detect are those with a "flattened" bottom or top, as shown in example *a*, Figure 2-20.

The pattern in example *a* is sometimes referred to as a right-angled broadening formation. Since the whole concept of widening price swings suggests highly emotional activity, volume patterns are difficult to char-

## CHART 2-7(*a*) Dow Jones Industrial Average, 1962

Charts 2-7(*a*) and (*b*) depict two classic double bottoms in the DJIA formed during 1962 and 1974. Note how the second of each pair was accompanied by lower volume than the first. While volume expanded during the 1962 breakout, the increase in activity was not particularly spectacular. The rise in activity during January 1975 left little doubt that a new bull market had begun.

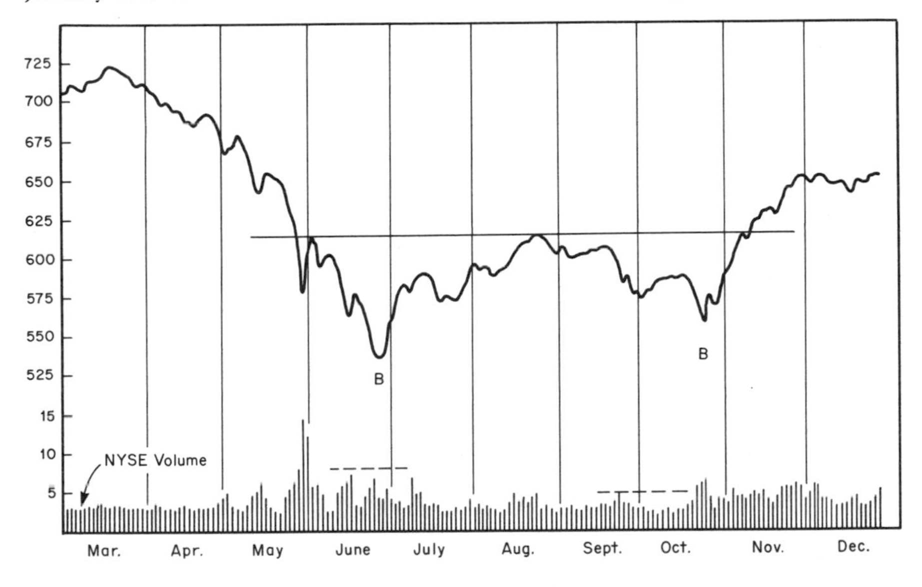

## CHART 2-7(*b*) Dow Jones Industrial Average, 1974–1975

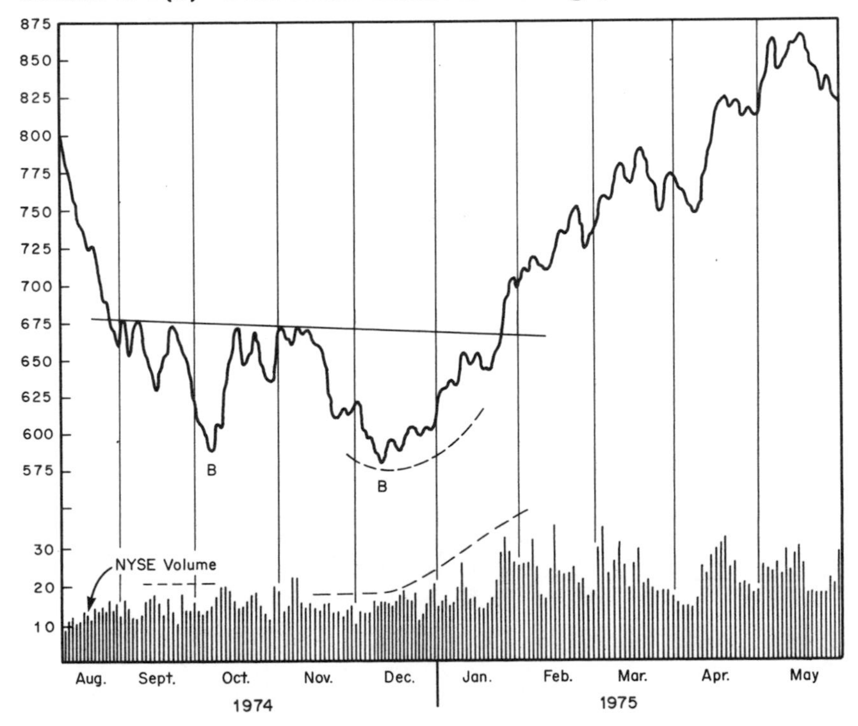

**CHART 2-8 Dow Jones Industrial Average, 1954–1978**

This chart shows the DJIA from 1954 to 1977. Between 1966 and 1977 the progress of this average was halted at the 1000 level. The reactions of 1966, 1970, and 1974 took the DJIA to succeeding new lows, resulting in what would appear to be a broadening formation with a flat top. Should the DJIA rise above the 1000 level (indicated by the dashed line) accompanied by high volume, this would represent an immensely bullish indication that the secular rise of the U.S. stock market had resumed.

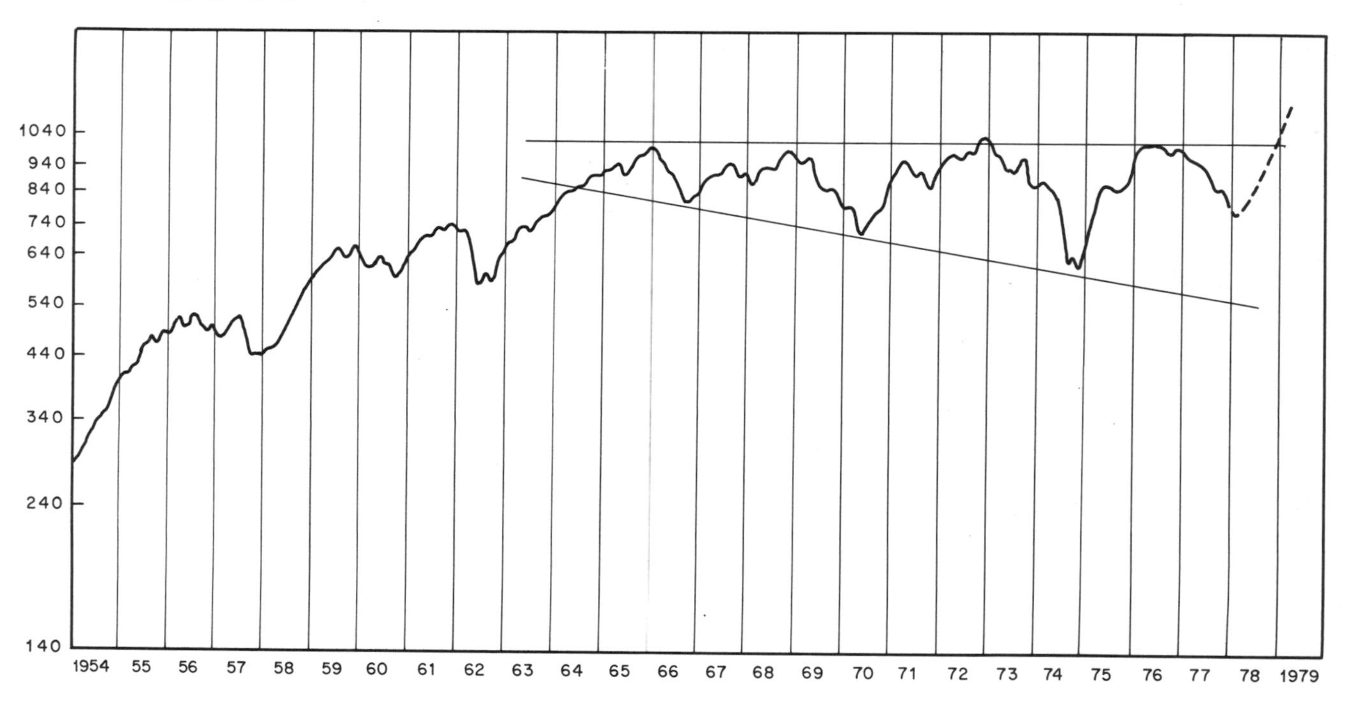

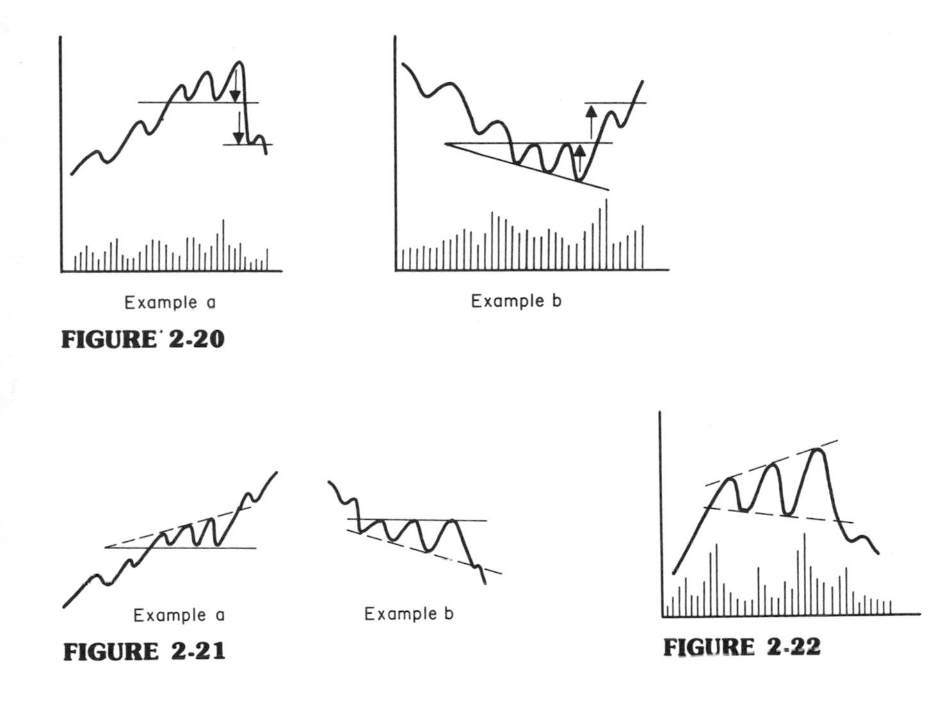


FIGURE 2-20

FIGURE 2-21

FIGURE 2-22

acterize, though at market tops volume is usually heavy during the rally phases. The patterns, both at bottoms and tops, are similar to the H&S variety, except that the "head" in the broadening formation is always the last to be formed. Bearish significance is obtained from a downside breakout confirmed by a 3 percent penetration of the neckline. Volume can be heavy or light, but additional bearish emphasis arises if activity expands at this point (see Charts 2-8 and 2-9).

Since a broadening formation with a flattened top is an accumulation pattern, it is an important requirement that volume expand on the breakout, as shown in example *b*, Figure 2-20.

These two types of broadening formation can also develop as consolidation patterns, as represented in Figure 2-21.

Unfortunately there does not appear to be a reliable point beyond which it is safe to say that the pattern has become one of continuation rather than reversal. The best defense in such cases is to extend the diverging trendlines, i.e., the dashed lines in Figure 2-21, and await a 3 percent penetration by the index as confirmation.

The final type of broadening formation, known as an "orthodox broadening top," is shown in Figure 2-22.

This pattern comprises three rallies, with each succeeding peak higher

**CHART 2-9 Dow Jones Industrial Average, 1919–1920**

This daily chart of the DJIA during 1919 shows a classic right-angled broadening formation. This period marked the peak of a huge run-up in commodity prices and preceded the 1920–1921 depression. The extreme volatility of financial markets at this time is clearly reflected not only in the action of the Dow but also in the volume of stocks traded.

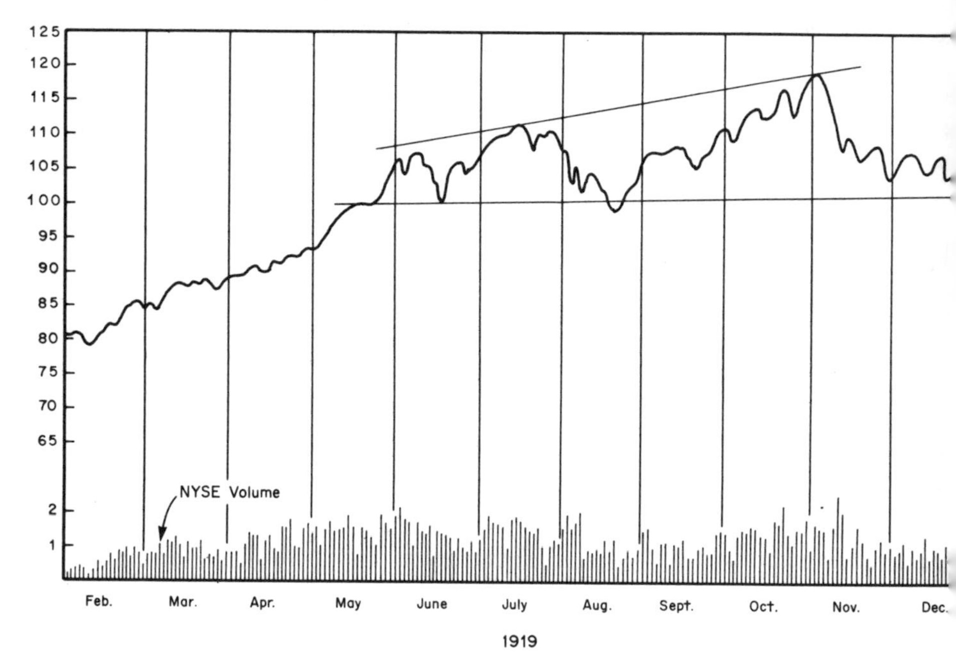


than its predecessor, and each peak separated by two bottoms with the second bottom lower than the first. Orthodox broadening formations are associated with market peaks rather than troughs.

These patterns are extremely difficult to detect until sometime after the final top has been formed, since there is no clearly definable level of support whose violation could serve as a benchmark. The violent and emotional nature of both price and volume swings further compounds the confusion and increases the complexity of defining these situations. Obviously under such conditions a breakout is difficult to pinpoint, but if

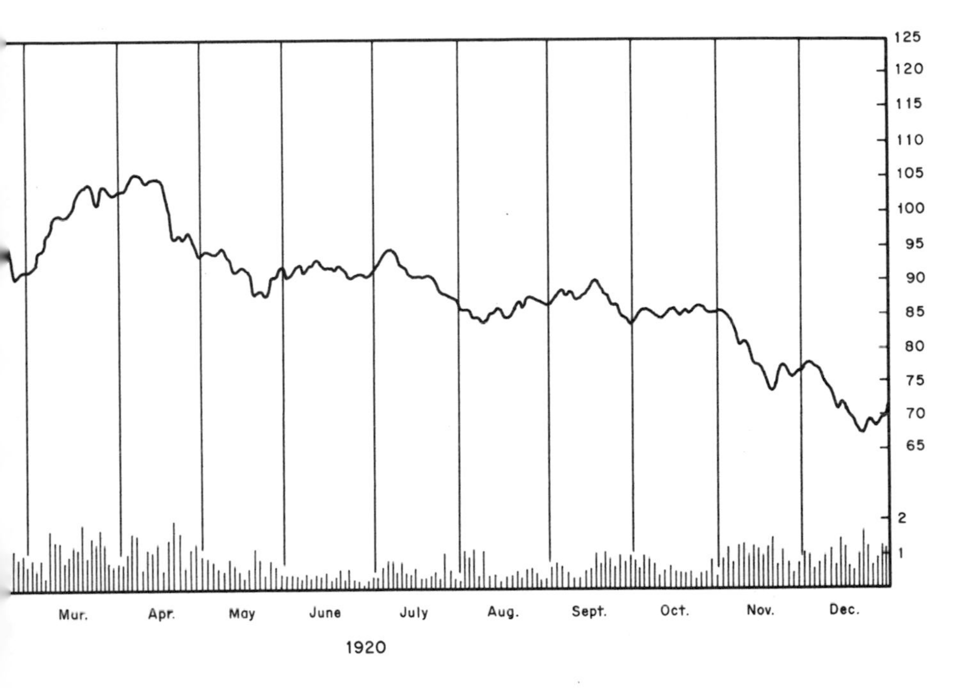


the formation is reasonably symmetrical, a 3 percent move below the descending trendline joining the two bottoms—or even a 3 percent decline below the second bottom—usually serves as a timely warning that an even greater decline is in store.

Measuring implications are similarly difficult to ascertain, but normally the volatile character of a broadening top formation implies the completion of a substantial amount of distribution. Consequently, price declines of considerable proportion usually follow the successful completion of such patterns.

**Chart 2-10 Dow Jones Industrial Average, 1938**

This excellent example of a right-angled triangle occurred at the bottoms of the 1937–1938 bear market. Note the substantial increase in volume that accompanied the upside breakout. Following the breakout, the average traced out a right-angled broadening formation with a flat top. Usually breakouts from these consolidation patterns are followed by a dramatic rise. In this case, however, the 158 level in November was destined to become the high for the 1938–1939 bull market.

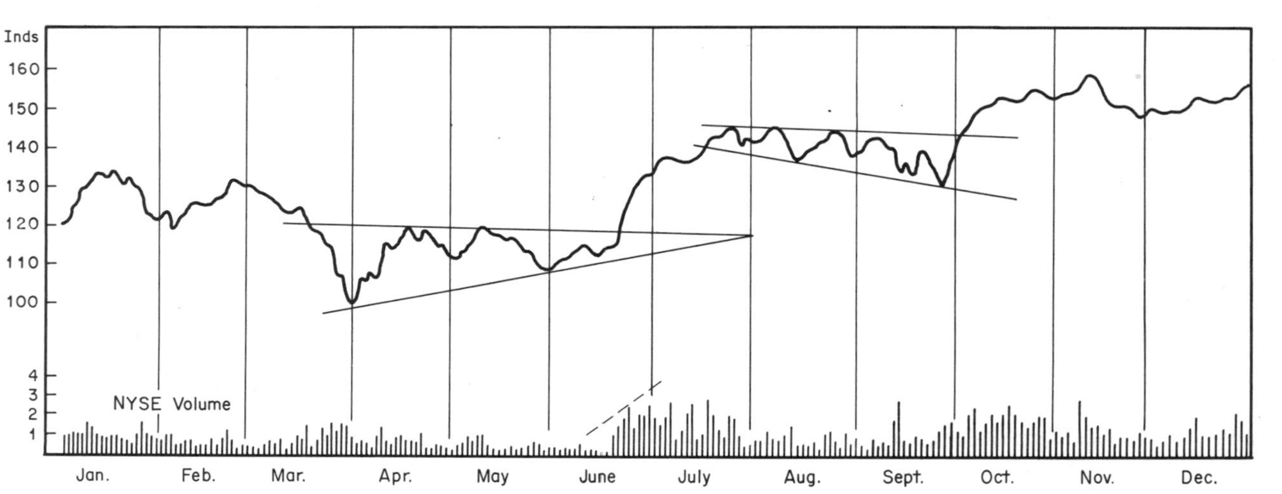

## TRIANGLES

Triangles are the most common of all the price patterns discussed in this chapter but unfortunately the least reliable. Triangles may be consolidation or reversal formations, and they fall into two categories, symmetrical and right-angled.

### Symmetrical Triangles

A symmetrical triangle is composed of a series of two or more rallies and reactions where each succeeding peak is lower than its predecessor and the bottom from each succeeding reaction is higher than its predecessor (see Figure 2-23). A triangle is therefore the opposite of a broadening formation, since the trendlines joining peaks and troughs *converge*, unlike the (orthodox) broadening formation, where they *diverge*.

These patterns are also known as "coils," for the fluctuation in price and volume diminishes as the pattern is completed. Finally, both price and (usually) volume react sharply, as if a coil spring had been wound tighter and tighter and then snapped free as prices broke out of the triangle. Generally speaking, triangles seem to "work" best when the breakout occurs somewhere between half and three-fourths of the distance between the widest peak and rally and the apex (as in Figure 2-24).

The volume and 3 percent confirmation rules used for other patterns are also appropriate for triangles.

### Right-Angled Triangles

Right-angled triangles are really a special form of the symmetrical type in that one of the two boundaries is formed at an angle of 90 degrees, i.e., horizontal to the vertical axis. (These triangle variations are illustrated in Figure 2-25 and Chart 2-10.) Whereas the symmetrical triangle gives no indication of which way it is ultimately likely to break, the right-angled triangle, with its implied level of support or resistance and contracting

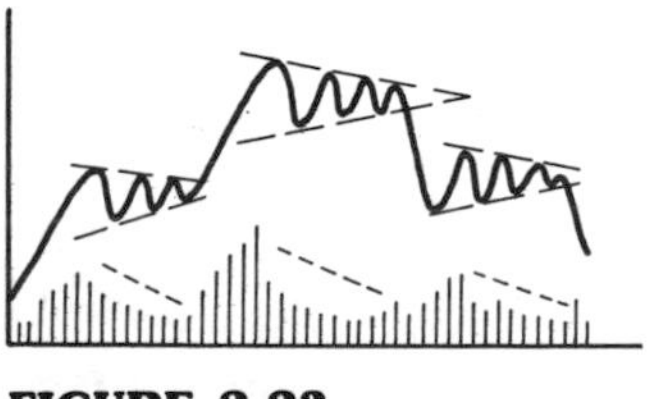

FIGURE 2-23

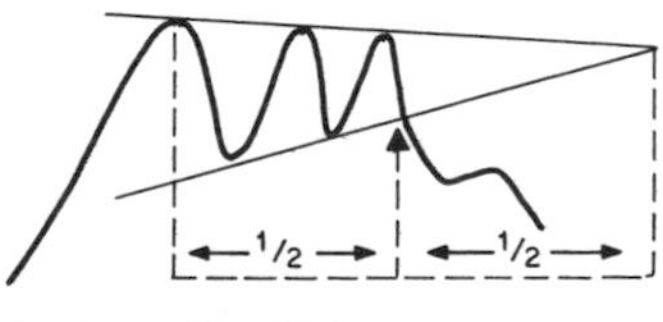

½ ½

FIGURE 2-24

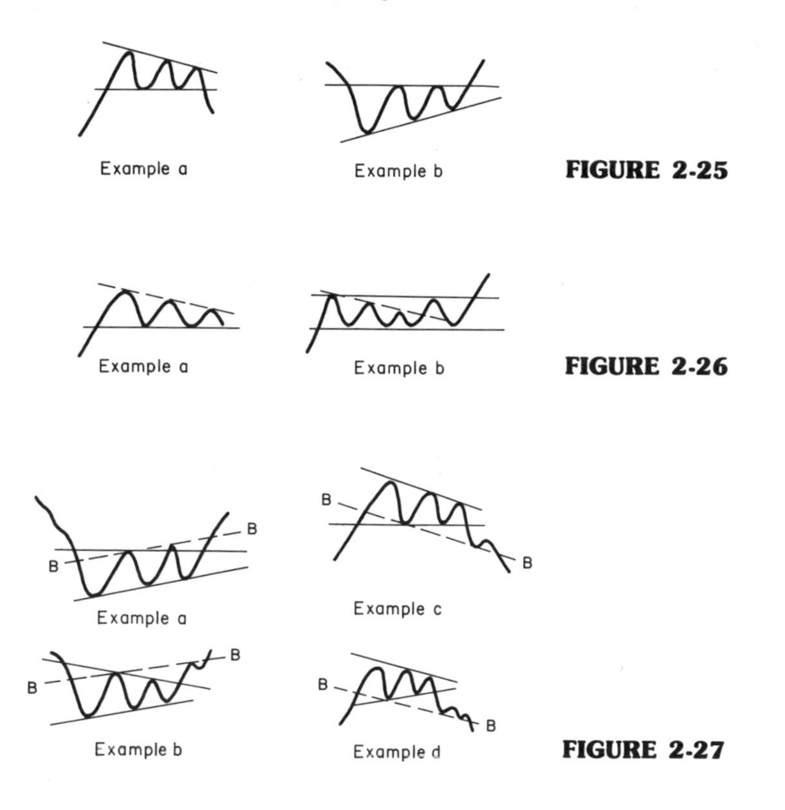


FIGURE 2-25

FIGURE 2-26

FIGURE 2-27

price fluctuations, does. One difficulty of interpreting these formations is that many rectangles begin as right-angled triangles, so considerable caution should be used when interpreting these elusive patterns. Such a situation is shown in Figure 2-26 where a potential downward sloping right-angled triangle in example (*a*) develops into a rectangle in example (*b*).

Measuring objectives for triangles are obtained by drawing a line parallel to the base of the triangle through the peak of the first rally. This line (*BB* in Figure 2-27) represents the price objective which prices may be expected to reach or exceed.

The reverse procedure at market tops is shown in examples *c* and *d*. The same technique is used to project prices when triangles are of the consolidation variety.

## SUMMARY

Stock prices move in trends, and once a trend has been established it is assumed that it will perpetuate itself until a reversal has been proved. Reversals in these price trends take some time to bring about, and they are characterized by a temporary period in which the enthusiasm of buyers and sellers is roughly in balance. Such transitional periods between bull and bear phases can be identified by clearly definable price patterns which, when accompanied by the appropriate trends in volume, offer good and reliable indications that a reversal in trend has taken place. Until the pattern has been formed and completed it should be assumed that the prevailing trend is still operative, i.e., that the pattern is one of consolidation or continuation. This principle takes on added importance when the major trend has been in existence for only a relatively short period, for the more mature the trend the greater is the probability of an imminent reversal.

Price patterns can be formed over periods ranging from 2 to 3 weeks to as many years. Their significance is derived from both their length and depth. The longer the time required to form a pattern and the greater the price fluctuations within a pattern, the more substantial the ensuing price movement is likely to be.

While measuring formulas can be derived for many of the patterns, these are generally minimum objectives, since price trends usually extend much farther than indicated.

# 3

# TRENDLINES

A review of any stock chart will quickly reveal that prices move in trends and that quite often the series of ascending bottoms in a rising market can be joined together by a straight line, just like the tops of a descending series of rally peaks. These lines, known as "trendlines," are a simple but invaluable addition to the technical arsenal (see Figure 3-1).

Since some trends can be sideways, it naturally follows that trendlines can also be drawn horizontally. The neckline of a head and shoulders pattern or the upper or lower boundary of a rectangle is really a trendline, and just as the penetration of those lines warned of a change in trend, so does the penetration of rising or falling trendlines. Also, as certain points in price patterns offer areas of support and resistance, trendlines likewise represent points of support (rising trendline) and resistance (declining trendline).

It may be recalled that the completion of a rectangle pattern can signify either (1) a reversal in the previous trend, in which case it becomes known as a reversal pattern, or (2) a resumption of the previous trend, when it is defined as a consolidation or continuation pattern. Similarly, the penetration of a trendline by an index will result in either a reversal of that trend or its continuation. Figure 3-2 illustrates this point from the aspect of a rising price trend.

In example *a* the trendline joining the series of troughs is eventually penetrated on the downside. Since the fourth peak represented the highest point in the bull market, the downward violation of the trendline signaled that a bear market was under way.

In example *b* the upward price trend and trendline penetration are

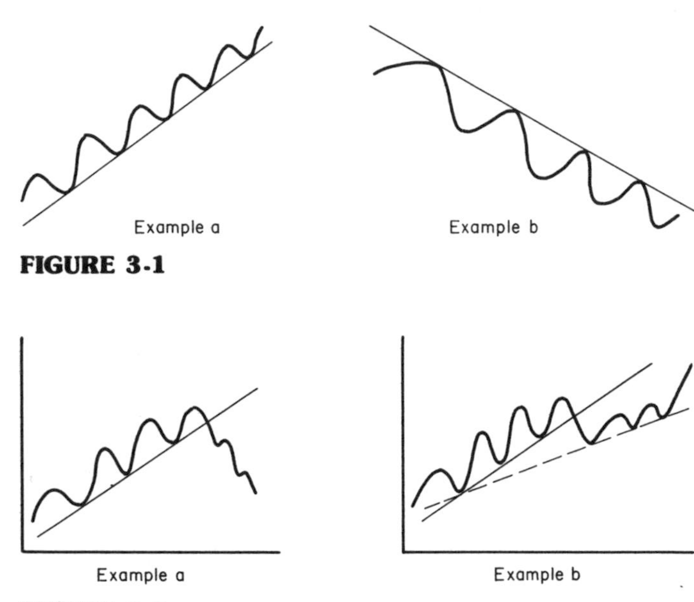


**FIGURE 3-1**

**FIGURE 3-2**

identical to example *a*, but the action following this warning signal is entirely different, since the trendline violation merely signaled that the advance would continue, but at a greatly reduced pace.

Unfortunately there is no way of telling at the time of the violation which possibility will prove to be the eventual outcome. Nevertheless, valuable clues can be obtained from the application of other techniques described in subsequent chapters and from an appraisal of the state of health of the market's overall technical structure, which is examined in Parts 2 through 4. Using the techniques discussed in Chapter 1 can also help. For example, in a rising market, a trendline penetration may occur at the time of or just before the successful completion of a reversal pattern. Some possibilities are shown in Figure 3-3.

In example *a* the rising trendline joins a series of bottoms, but the last two troughs represent reactions from a right shoulder and head which are part of an ascending head and shoulders pattern. Examples *b* and *c* represent a similar situation for a rectangle and broadening top.

Figure 3-4 illustrates the same phenomenon from the point of view of a reversal of a bear market. If the violation occurs simultaneously with or just after the completion of a reversal pattern, the two breaks have the effect of reinforcing each other; but sometimes, as in Figure 3-5, the trendline violation occurs *before* the completion of the pattern. In such cases, since a trend is assumed to continue until a reversal is proved, the violation should be regarded as a sign of interruption of the prevailing movement rather than one of reversal. In the case of an advance, it is necessary to await a decline in the indicator below the previous trough before confirmatory evidence of an actual trend reversal is given.

The opposite will occur in the case of a declining market (see Figure 3-5). In these instances the trendline penetration is treated as an alert. Additional evidence as to the likely significance of a trendline violation can be derived from the level of volume by utilizing the principles outlined in Chapter 1. For example, if a series of ascending peaks and

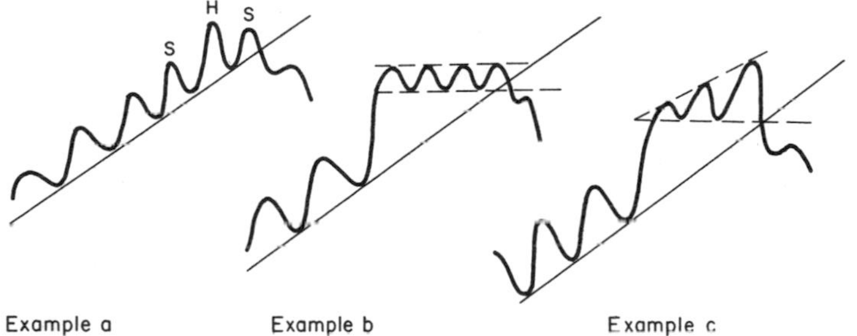


**FIGURE 3-3**

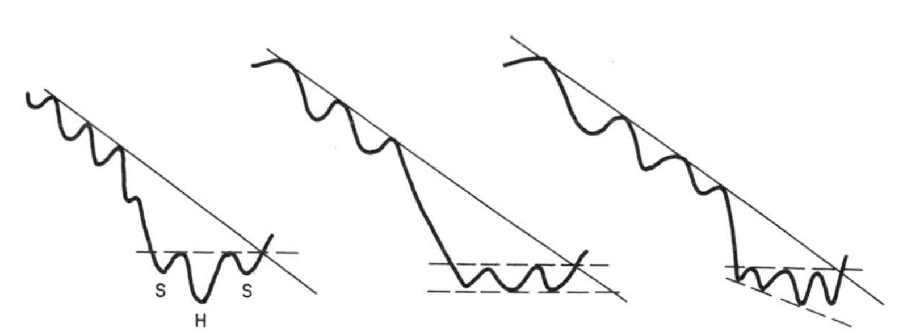


**FIGURE 3-4**

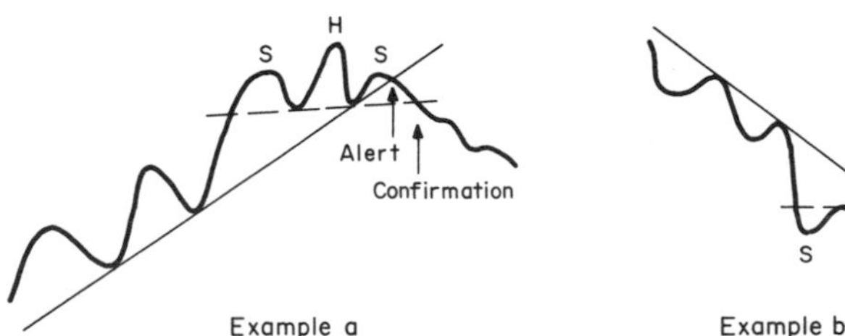


**FIGURE 3-5**

troughs are accompanied by successively less volume, it is a sign that the advance is running out of steam (since volume is no longer going with the trend). Under such circumstances a trendline violation is likely to be of greater significance than if volume continues to expand with each successive rally. Although it is not necessary for a downside penetration to be accompanied by high volume, expansion of activity at such a point is clearly more bearish since it emphasizes the switch in the supply/demand balance in favor of the sellers.

Just as a return move often occurs following a breakout from a price pattern, so a similar move known as a "throwback" sometimes develops following a trendline penetration.

Example *a* in Figure 3-6 shows a trendline reversing its previous role as support while the "throwback" move turns it into an area of resistance. Example *b* shows the same situation for a declining market.

The importance of plotting charts on logarithmic as opposed to arithmetic paper was discussed in Chapter 2. The choice of ratio scale is even more critical for a timely and accurate use of trendline analysis, because at the end of a major movement prices tend to accelerate in the direction of the prevailing trend, i.e., they rise faster at the end of a rising trend and decline more sharply at the termination of a bear market. In a bull market (Chart 3-1) prices rise slowly after an initial burst, then advance at a steeper and steeper angle as they approach the ultimate peak, rather like the left-hand cross section of a mountain.

As Figure 3-7 shows, this exponential movement takes the price well away from the trendline if the index concerned is being plotted on an arithmetic scale. Consequently the price has to fall that much farther before a penetration of the trendline can take place.

## SIGNIFICANCE OF TRENDLINES

It has been established that a break in trend caused by a penetration of a trendline results in either an actual trend reversal or a slowing down in

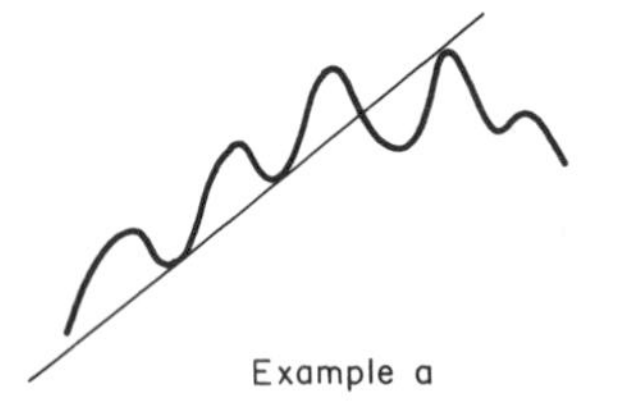


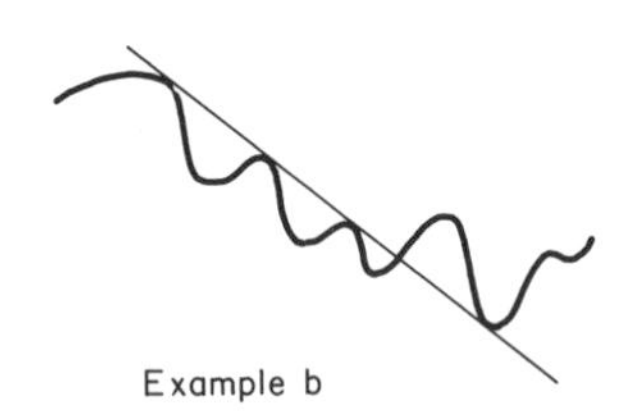


**FIGURE 3-6**

**CHART 3-1 Dow Jones Transportation Average, 1965–1970**

This chart shows the 1966–1968 bull market for the Transportation Average. The downward penetration of the trendline joining the 1966 and early 1968 bottoms signaled the end of the bull market. Note how the index breaks down from what might have been a right-angled triangle consolidation pattern, had it been able to surpass its 1966 and 1967 peaks.

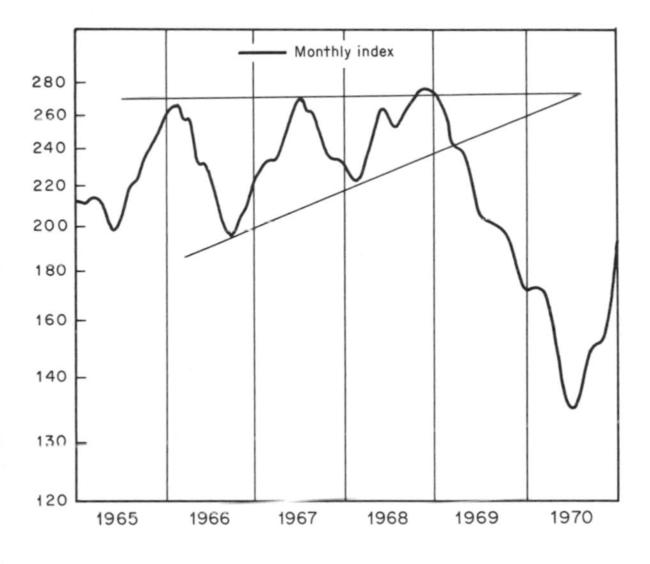


the pace of the trend. While it is not always possible to assess which of these alternatives will develop, it is still important to understand the significance of a trendline penetration; the following guidelines should help in assessing the situation.

## 1. Number of Times the Trendline Has Been Touched

A trendline obtains its authority from the number of times it has been touched, i.e., the larger the number, the greater the significance. This is because a trendline represents an area of support or resistance, so that each successive "test" of the line contributes to the significance of this support or resistance and therefore the authority of the line itself.

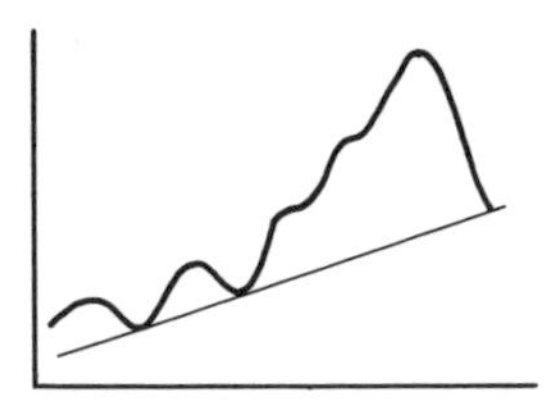

Example a (arithmetic scale) Example b (ratio scale)

**FIGURE 3-7**

## 2. Length of the Line

The size of a trend or its length is an important factor, as with price patterns. If a series of ascending bottoms occurs over a 3- to 4-week span, the resulting trendline is of only minor importance. Should the trend extend over a period of 1 to 3 years, however, its violation marks a significant juncture point. The construction of trendlines is often a matter of judgment established through trial and error. For example, in Figure 3-8 two bottoms are formed at marginally different levels, so that the resulting trendline AA becomes relatively flat; consequently, its penetration is of no use from the point of view of signaling a breakdown or reversal of the major trend. Although trendline *BB* is shorter than AA in this case, it more closely resembles the trend, and its violation is that much more significant, for by the time prices violate AA the bear market is almost over.

A very sharp trend (as in Figure 3-9) is difficult to maintain and is liable to be broken rather easily, even by a short sideways movement.

All trends are eventually violated, but the steeper ones are more likely to be broken more quickly. Consequently the violation of a particularly steep trend is not as significant as that of a slower, more sustainable one. Penetration of a steep line will usually result in a short corrective movement, following which the trend resumes, but at a greatly reduced and more sustainable pace.

## MEASURING IMPLICATION

Trendlines have measuring implications when broken, just like price patterns. For a rising trendline, the vertical distance between the peak in the index and the trendline is measured ($A_1$ in Figure 3-10). This distance is then projected down from the point at which the violation occurs ($A_2$).

The term "price objective" is perhaps a misleading one, for although such objectives are usually reached when a trendline is the reversal type,

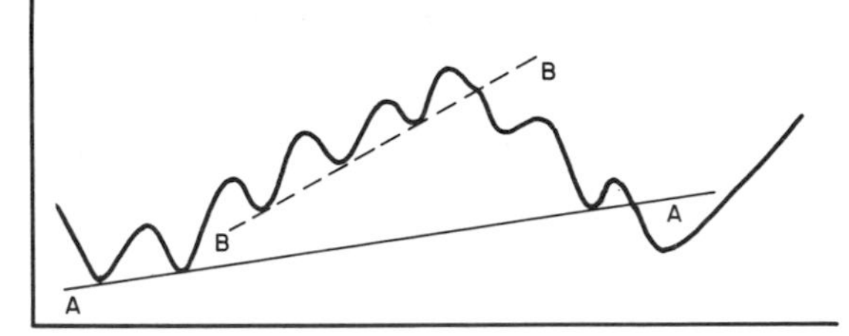

**FIGURE 3-8**

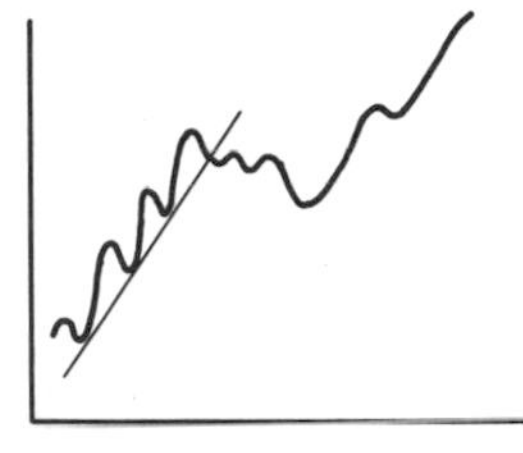

**FIGURE 3-9**

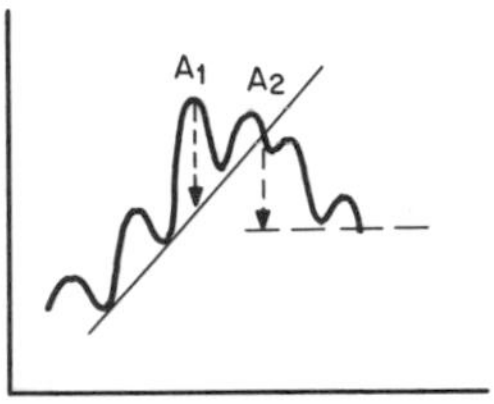

**FIGURE 3-10**

they are more often exceeded, so that (as with price patterns) the objective becomes more of a minimum expectation. When prices fall significantly below the objective, as in Figure 3-11, this area often becomes one of resistance to the next major rally or support for a subsequent reaction (Chart 3-3).

In the case of a falling trendline these principles are reversed (see Figure 3-12).

Time and again these price objective areas prove to be important support or resistance points, but unfortunately there is no way of determining which is likely to prove the actual juncture point for any rally or reaction. This really emphasizes the point made earlier that there is no way of determining the duration of a move in prices. It is only possible to speculate on the probability that a certain area will prove to be an important juncture point. Consequently, when many other technical indicators are

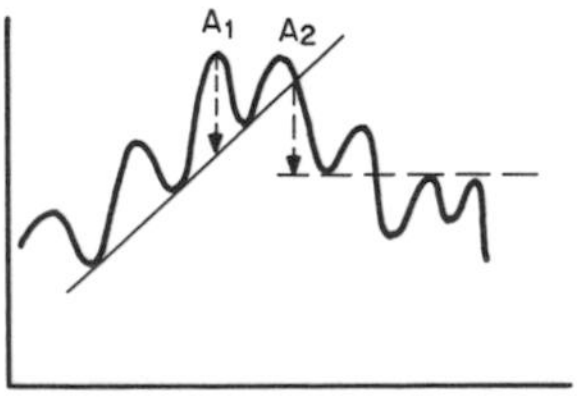

**FIGURE 3-11**

**CHART 3-2 Dow Jones Transportation Average, 1944–1952**

The 1946–1949 bear market trendline was broken on the upside in early 1950. The measuring objective for this monthly index was achieved in late 1950, after which the average paused for some months prior to resuming its cyclical advance.

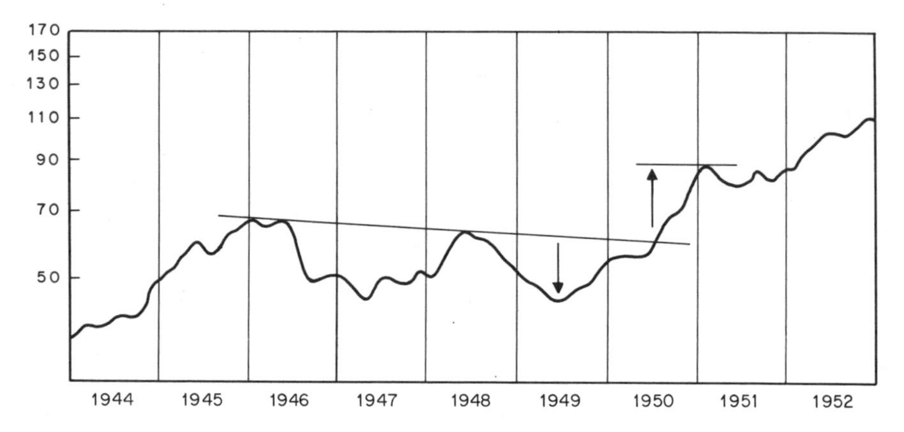


pointing out that a reversal in trend is about to take place, these trendline basing points can be used as a strong confirmation.

## CORRECTIVE FAN PRINCIPLE

At the beginning of a new primary bull market the initial intermediate rally is likely to be relatively steep and the ensuing rate of advance unsustainable. This is because the rally is often a technical reaction to the previous overextended decline, as speculators who were caught short rush to cover their positions. Therefore, the steep trendline constructed from the first minor reaction is quickly violated.

This is represented as line *AA* in Figure 3-13. A new trendline is then constructed, using the bottom of this first intermediate decline (*AB*). This new line rises at a less rapid rate than the initial one. Finally the process is

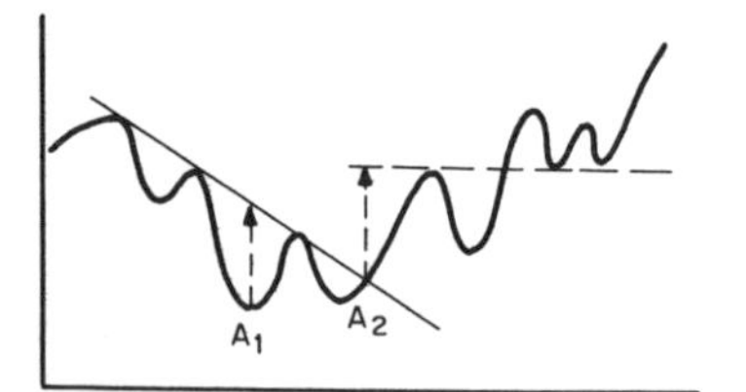


**FIGURE 3-12**

**CHART 3-3 Dow Jones Transportation Average, 1974–1978**

The bull market trendline (AB) joining the 1974 bottoms was violated in 1975. The ensuing rally took this average above its previous high, but the substantial resistance marked by this extended trendline eventually proved too great, so the average was forced to temporarily turn down. A secondary trendline CD was also violated. Even though the average was able to advance to a new cyclical high, an extension of CD also proved to be a formidable resistance barrier.

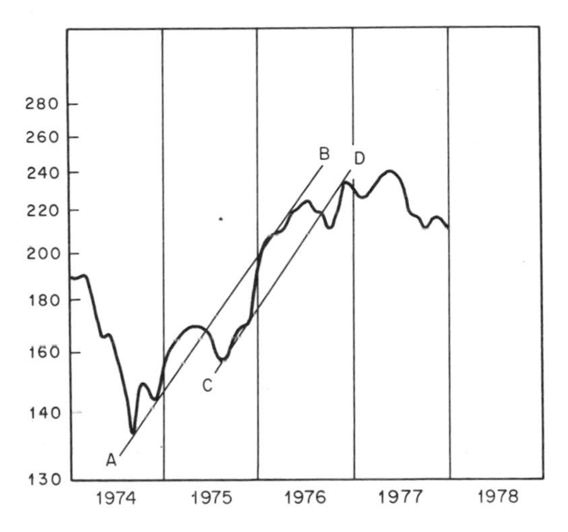


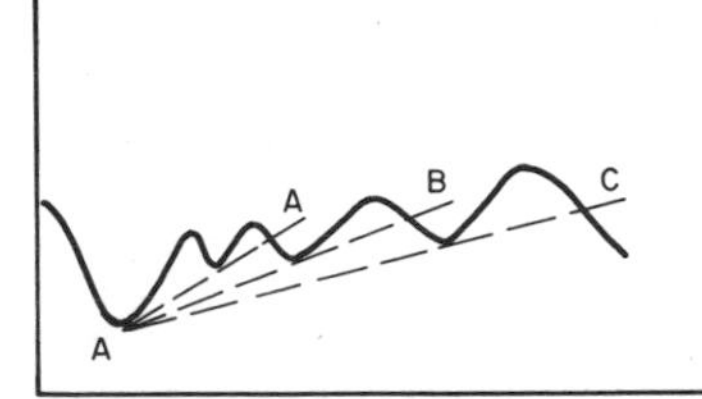


**FIGURE 3-13**

repeated, resulting in the construction of a third line, *AC*. These lines are known as "fan" lines. The principle has been established that once the third trendline has been violated, the end of the bull market is confirmed. In some respects these three rally points and trendlines can be compared to the three stages of a bull or bear market, as outlined in Chapter 1. The fan principle is just as valid for downtrends as it is for uptrends and can also be used for determining intermediate as well as cyclical movements.

## TREND CHANNELS

So far only the possibilities of drawing trendlines joining bottoms in rising markets and tops in declining ones have been examined, but it is also useful to draw lines that are parallel to those "basic" trendlines, as shown in Figure 3-14.

In a rising market the parallel line, known as a "return" trendline, joins the tops of rallies (*AA*, example *a*), and during declines the return line joins the series of bottoms (*BB*, example *b*). The area between these trend extremities is known as a "trend channel."

The return line is useful from two points of view. First, it represents an area of support or resistance, depending on the direction of the trend. Second, and perhaps more important, penetration of the return trendline often signifies reversal of at least a temporary proportion in the basic trend (see Figure 3-15).

In example *a* prices have been rising for some time, then, as is characteristic of the terminal phase of a trend, they accelerate sharply with such force that the return line is penetrated on the upside. This is normally a bearish sign since it indicates exhaustion of a move, especially if such action is accompanied by high volume.

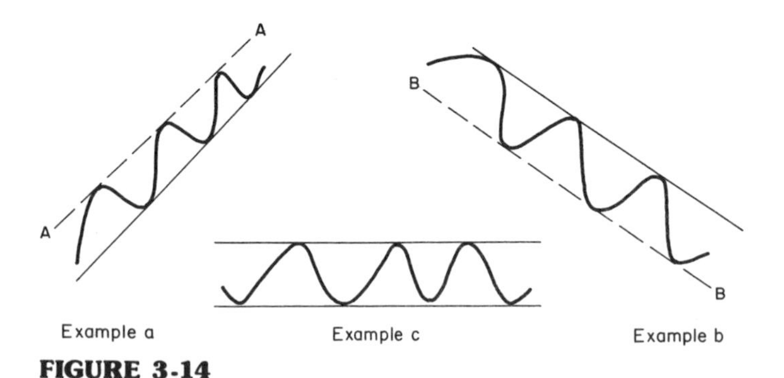


**FIGURE 3-14**

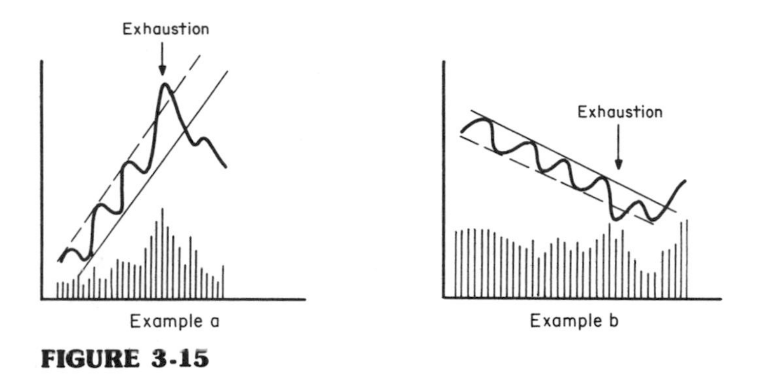


**FIGURE 3-15**

Consider the situation where a man is sawing a thick piece of wood. First of all his sawing strokes are slow but deliberate; gradually he realizes his task is going to take some time, becomes frustrated, and slowly increases the speed of his strokes. Finally he bursts into a frantic effort and is forced to give up his task for at least a temporary period because of complete exhaustion. Example *b* in Figure 3-15 shows an exhaustion move in a declining markct. In this case the expanding volume at the low represents a selling climax.

## SUMMARY

Trendlines are perhaps thc casiest technical tool to understand, but considerable experimentation and practice are required before the art of their interpretation can be successfully mastered. Generally speaking (given the guidelines described in this chapter), the more obvious the construction of a trendline, the greater the significance of its violation is likely to be.

# 4

# MOVING AVERAGES

It is now evident that trends in stock prices can be very volatile, almost taking on a haphazard characteristic at times. One technique for overcoming this phenomenon is the *moving average*.

A moving average attempts to tone down the fluctuations of stock prices into a smoothed trend, so that distortions are reduced to a minimum. There are basically three types of moving average used in the technical analysis of stock trends: simple, weighted, and exponential. Since the construction and use of these averages is different, the discussion of this subject will deal with each type in turn.

## SIMPLE MOVING AVERAGE

A simple moving average is by far the most widely used, but not necessarily the best. It is constructed by totaling a set of data and dividing that total by the number of observations. The resulting number is known as the *average* or *mean average*. In order to get the average to "move," a new item of data is added and the first item on the list subtracted. The new total is then divided by the number of observations and the process repeated ad infinitum.

For example, if the calculation of a 10-week moving average were required, the method would be as shown in Table 4-1.

On March 9 the total of the 10 weeks ending on that date was 966, and 966 divided by 10 results in an average of 96.6. On March 16 the number 90 is added, and the observation of 101 on January 8 is deleted. The new total of 955 is then divided by 10. If a 13-week moving average had been

**TABLE 4-1**

| Date | Index | 10-week total | Moving average |
|---|---|---|---|
| Jan. 8 | 101 | | |
| 15 | 100 | | |
| 22 | 103 | | |
| 29 | 99 | | |
| Feb. 5 | 96 | | |
| 12 | 99 | | |
| 19 | 95 | | |
| 26 | 91 | | |
| Mar. 5 | 93 | | |
| 12 | 89 | 966 | 96.6 |
| 19 | 90 | 955 | 95.5 |
| 26 | 95 | 950 | 95.0 |
| April 2 | 103 | 950 | 95.0 |

required, it would have been necessary to total 13 weeks of data (as of April 2, the thirteenth week of observation) and divide by 13. A 13-week moving average is shown in Figure 4-1, example *a*, by the dashed line. It can be observed that a rising moving average indicates market strength and a declining one denotes weakness.

Reference to the price index and its 13-week moving average shows that the moving average changes direction well after the peak or trough in stock prices and is therefore "late" in changing direction. This is because the moving average is plotted on the thirteenth week, whereas the average price of 13 weeks' observations actually occurs halfway through the observation period, which in this instance is the seventh week. Consequently, if it is to correctly reflect the trend that it is attempting to measure, the plot of the moving average should be centered, i.e., plotted on the seventh week, as shown in Figure 4-1, example *b*.

If the centering technique had been used, it would have been necessary in our example to wait 6 weeks before ascertaining whether the average had changed direction.

A time delay, while an irritant, is not particularly critical when analyzing other time series such as economic data; but given the relatively rapid movement of stock prices and consequent loss of profit potential, a delay here is totally unacceptable. It has thus been found that for the purposes of stock market analysis the best results are achieved by plotting the moving average on the final week.

Changes in the trend of the index or stock being measured are identified not by a change in direction of the moving average, but by a crossover of the moving average by the index itself. A change from a rising to a declining market is signaled when the index moves below its moving average. The reverse set of conditions will confirm the termination of a de-

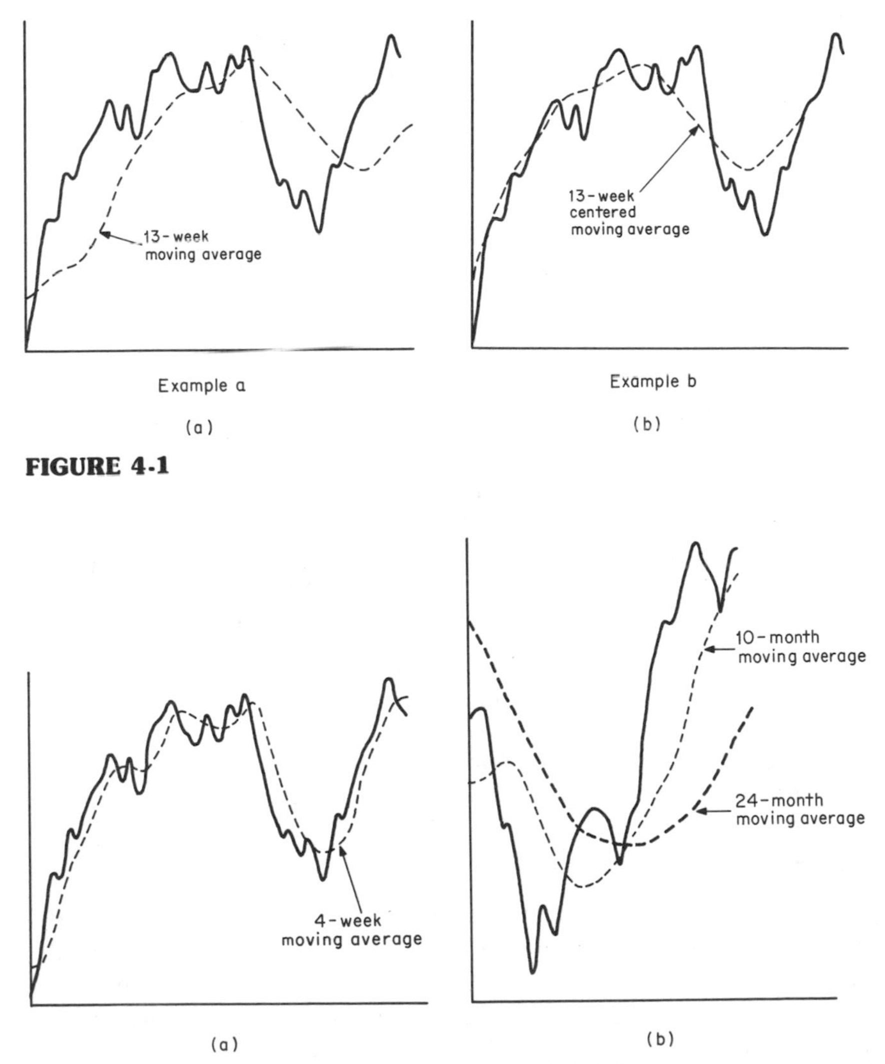


**FIGURE 4-1**

**FIGURE 4-2** (a) Moving average not centered. (b) Moving average centered.

clining trend. Since the use of moving averages gives clear-cut "buy" and "sell" signals, it helps to eliminate some of the subjectivity associated with the construction and interpretation of trendlines.

More often than not it pays to take action based on moving-average crossovers, since this technique is one of the most reliable tools in the technical arsenal. However, the degree of accuracy depends substantially on the choice of moving average. This is discussed later, but some of the characteristics of moving averages should first be outlined in greater detail.

1. Since a moving average is a smoothed version of a trend, the average itself becomes an area of support and resistance. Thus, in a rising market reactions in stock prices are often reversed by finding support at the moving-average level. Similarly, a rally in a declining market often meets resistance at a moving average and turns down.
2. Since a moving average is itself a significant juncture point, its violation warns that a change in trend may already have taken place. If the moving average is flat or has already changed direction, its violation by the index is fairly conclusive proof that the previous trend has been reversed.
3. If the violation occurs while the moving average is still proceeding in the direction of the prevailing trend, this development should be treated as a preliminary warning that a trend reversal has taken place. Confirmation should await a flattening or a change in direction in the moving average itself or should be sought from alternative technical sources.
4. Generally speaking, the longer the time span covered by a moving average, the greater is the significance of a crossover signal. A violation of an 18-month moving average, since it is smoothing a very long-term trend, has substantially more significance than a crossover of a 30-day moving average.

## Choice of Time Span

Moving averages can be constructed for any time period, whether a few days, several weeks, many months, or even years. Choice of the length is very important. For example, if it is assumed that a complete bull and bear cycle lasts for 4 years, a moving average constructed over a period of time greater than 48 months will not reflect the cycle at all, since it smooths out all the fluctuations taking place during this period and will

**CHART 4-1 Comparison of 40-Week Moving Averages, 1966–1978**
[for (*a*) the United States, (*b*) the United Kingdom, and (*c*) the West German Federal Republic]

Charts 4-1(*a*) through (*c*) show the 40-week moving average at work in three different countries. Although crossovers of each index are subject to whipsaws from time to time, the use of this average is still fairly reliable. Note how the 40-week average is continually being used as a support or resistance level.

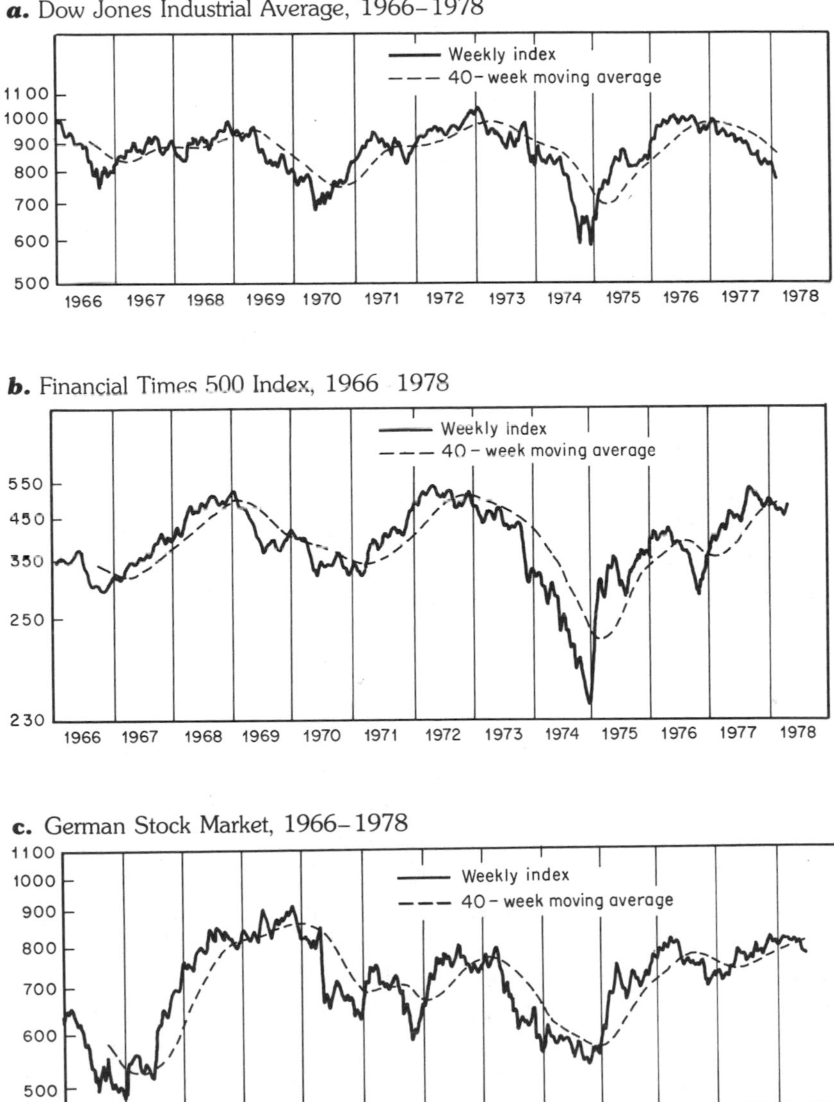

appear more or less as a straight line crossing through the middle of the cycle. On the other hand, a 5-day moving average will catch every minor move in the stock cycle and will also prove to be relatively useless in identifying the actual top and bottom of the overall cycle. Even if the 48-month average were shortened to 24 months and the 5-day average expanded to 4 weeks, for example (see Figure 4-2), use of the crossover signals would still result in the 24-month average giving an agonizingly slow confirmation of a change in trend. The 4-week average would be so sensitive that it would continually give misleading or "whipsaw" signals. Only a moving average that can catch the movement of the actual cycle will provide the optimum tradeoff between lateness and oversensitivity, such as the 10-month moving average in Figure 4-2, example *b*.

The choice of moving average will therefore depend on the type of stock market trend that is to be identified, i.e., a primary or an intermediate trend. For shorter intermediate moves, 10- and 13-week averages seem to offer the best results, whereas for longer primary trends experiments with 20-, 30-, and 40-week moving averages from 1897 to 1967 have shown the 40-week period to be the most helpful.[1]

## MULTIPLE MOVING AVERAGES

Some techniques of trend determination involve more than one moving average at a time. Signals are given by a shorter-term moving average crossing above or below a longer one. This procedure has the advantage of smoothing the data twice and thereby reducing the possibility of a whipsaw, yet warning of trend changes fairly quickly after they have taken place (see Chart 4-2.) Two averages which have been found reliable from the point of view of determining primary market moves are the 10-week and 30-week moving averages when used together. For the purpose of simplifying the calculation, the weekly closing price rather than a 5-day average is used.

Signals are given when the 10-week average moves below the 30-week average and when the 30-week average itself is declining. This development warns that the major trend is down. It is not assumed to be reversed until both averages are rising simultaneously, with the 10-week higher than the 30-week moving average. Should the 10-week average rise above the 30-week average while the longer average is still declining (and vice versa for bull markets), a valid signal is not constituted. By definition,

[1] William Gordon, *Stock Market Indicators*, Investors Press, Palisades Park, N.J., 1968.

**CHART 4-2 Dow Jones Industrial Average versus 30-Week/10-Week Moving Averages**

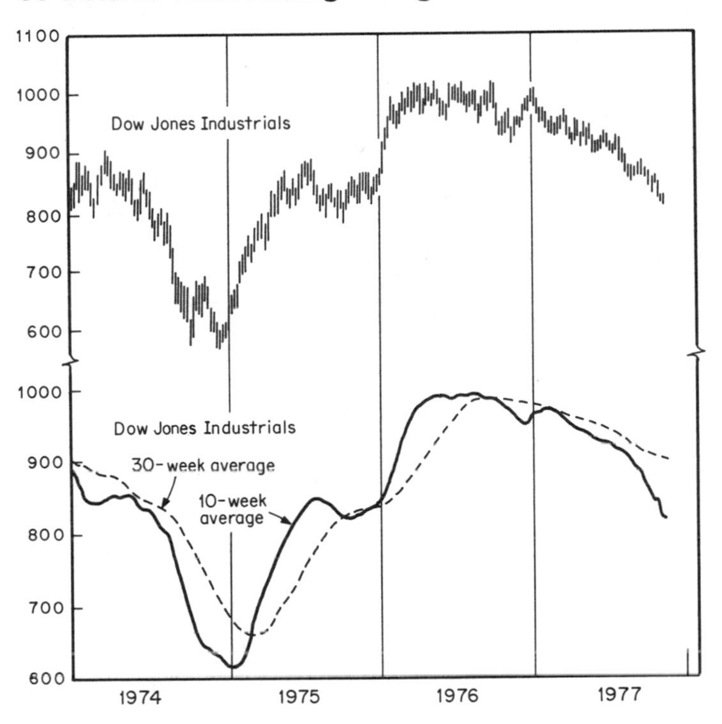


**CHART 4-3 Dow Jones Industrial Average, 1946–1951**

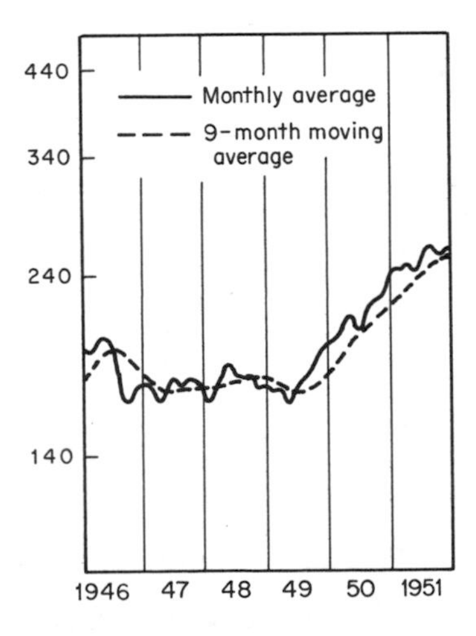

these warning signals always occur after the ultimate peak or trough of stock prices and serve as a *confirmation* of a change in trend rather than being actual juncture points in themselves.

Moving averages should *always* be used in conjunction with other indicators, for sometimes the stock market fluctuates in a broad sideways pattern for an extended period of time, resulting in a series of misleading signals. Chart 4-3, depicting the 1946–1949 bear market, gives an example of such a frustrating period, one in which many misleading crossovers would have been experienced.

## WEIGHTED MOVING AVERAGES

As discussed at the beginning of this chapter, a moving average can correctly represent a trend from a statistical point of view only if it is centered. It was also pointed out that centering a moving average delays the signal of a reversal in the trend of the stock index or indicator being measured. One technique for overcoming this problem is to weight the data in favor of the most recent observations. A moving average constructed in this manner is able to "turn" or reverse direction much more quickly than a simple moving average, which is calculated by treating all the data equally.

There are almost limitless ways in which data can be weighted, but the

**TABLE 4-2**

| Date | | Index (1) | 6 × col. 1 (2) | 5 × col. 1 1 week ago (3) | 4 × col. 1 2 weeks ago (4) | 3 × col. 1 3 weeks ago (5) | 2 × col. 1 4 weeks ago (6) | 1 × col. 1 5 weeks ago (7) | Total cols. 2-7 (8) | Col. 8 ÷ 21 (9) |
|---|---|---|---|---|---|---|---|---|---|---|
| Jan. | 8 | 101 | | | | | | | | |
| | 15 | 100 | | | | | | | | |
| | 22 | 103 | | | | | | | | |
| | 29 | 99 | | | | | | | | |
| Feb. | 5 | 96 | | | | | | | | |
| | 12 | 99 | 594 | 480 | 396 | 309 | 200 | 101 | 2080 | 99.1 |
| | 19 | 95 | 570 | 495 | 384 | 297 | 206 | 100 | 2052 | 97.7 |
| | 26 | 91 | 546 | 475 | 396 | 288 | 198 | 103 | 2006 | 95.5 |
| Mar. | 5 | 93 | 558 | 455 | 380 | 297 | 192 | 99 | 1981 | 94.3 |
| | 12 | 89 | 534 | 465 | 364 | 285 | 198 | 96 | 1924 | 92.5 |

most widely used method incorporates a technique whereby the first time period of data is multiplied by 1, the second by 2, the third by 3, and so on until the latest week, so that a 6-week weighted moving average is multiplied by 6. The resulting amount for each week is then totaled, but instead of dividing by the number of weeks—in this case 6—the divisor becomes the total of the weights, i.e., $1 + 2 + 3 + 4 + 5 + 6 = 21$. For a 10-week weighted moving average, the sum of the weights would be $1 + 2 + 3 + 4 + 5 + 6 + 7 + 8 + 9 + 10 = 55$. Table 4-2 illustrates how the calculations are made.

Another method is to calculate a simple moving average but in doing so to use the most recent observation twice, thereby doubling its weight. The calculation of a 6-week weighted moving average using this approach and the data from the table above would be as follows:

| *Week* | *Index* |
|---|---|
| 1 | 101 |
| 2 | 100 |
| 3 | 103 |
| 4 | 99 |
| 5 | 96 |
| 5 | 96 |
| Total | 595 |

$$\frac{\text{Total}}{\text{Weeks}} = \frac{595}{6} = 99.2$$

The interpretation of a weighted average is different from that of a simple average. In the case of this more sensitive weighted average, a warning of a trend reversal is given by a change in direction of the average rather than a crossover.

## EXPONENTIAL MOVING AVERAGES

Weighted moving averages are very useful for the purpose of identifying trend reversals. However, the time-consuming calculations required to construct and maintain such averages greatly detract from their usefulness (unless, of course, a computer is within easy reach). An exponential moving average is really a short cut to obtaining a form of weighted moving average, since an exponential smoothing also gives greater weight to more recent data. In order to construct a 20-week exponential moving average (EMA), it is first necessary to calculate a simple 20-week moving average, i.e., the total of 20 weeks' observations divided by 20. In Table

4-3 this has been done for the 20 weeks ending January 1, and the result appears as 99.00 in column 6.

This 20-week average becomes the starting point for the exponential moving average and is then transferred to column 2 for the following week. Next, the entry for the twenty-first week (January 8 in the example above) is compared with the moving average, and the difference is added or subtracted and posted in column 3, i.e., 100 − 99 = 1.0. This difference is then multiplied by the exponent, which for a 20-week EMA is 0.1. This exponentially treated difference, 1.0 × 0.1, is then added to the previous week's EMA, and the calculation is repeated each succeeding week. In the example, the exponentially treated difference for January 8 is 0.1, which is added to the previous week's average, 99.0, to obtain an EMA for January 8 of 99.10. This figure in column 6 is then plotted. Figure 4-3 shows an example of a 13-week EMA.

If the difference between the new weekly observation and the previous week's EMA is negative, as in the case of the reading 99.00 versus 99.64 for January 29, the exponentially treated difference is subtracted from the previous week's EMA.

The exponent used varies with the time span of the moving average. The correct exponents for various time spans are as follows:

| | |
|---|---|
| 5-week | 0.4 |
| 10-week | 0.2 |
| 15-week | 0.13 |
| 20-week | 0.1 |
| 40-week | 0.05 |
| 80-week | 0.025 |

**TABLE 4-3**

| Date | | Price (1) | EMA for previous week (2) | Difference (col. 1 − col. 2) (3) | Exponent (4) | Col. 3 × col. 4 +/− (5) | Col. 2 + col. 5 EMA (6) |
|---|---|---|---|---|---|---|---|
| Jan. | 1 | . . . | . . . | . . . | . . . | . . . | 99.00 |
| | 8 | 100.00 | 99.00 | 1.00 | 0.1 | +0.10 | 99.10 |
| | 15 | 103.00 | 99.10 | 3.90 | 0.1 | +0.39 | 00.49 |
| | 22 | 102.00 | 99.49 | 2.51 | 0.1 | +0.25 | 99.74 |
| | 29 | 99.00 | 99.64 | (0.64) | 0.1 | −0.06 | 99.68 |

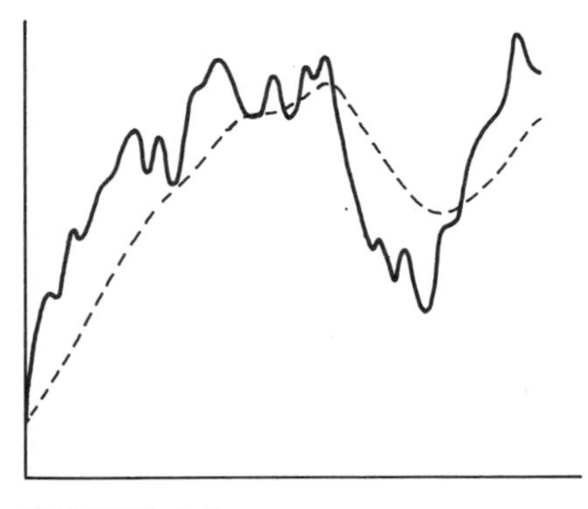

**FIGURE 4-3**

In Table 4-3 the time periods have been described as weekly, but in effect the exponent 0.1 can be used for any measure of 20, whether it is days, weeks, months, or years, etc. Exponents for time periods other than those shown in the table can easily be calculated by dividing 2 by the time span.

For example, a 5-week average will need to be twice as sensitive as a 10-week average, so 2 divided by 5 gives an exponent of 0.4. On the other hand, a 20-week average should be half as sensitive as that for a 10-week period (0.2), so its exponent is halved to 0.1.

If an EMA proves to be too sensitive for the trend being monitored, one solution is to extend its time period. Another is to smooth the EMA

**CHART 4-4 Dow Jones Industrial Average, 1930–1937**

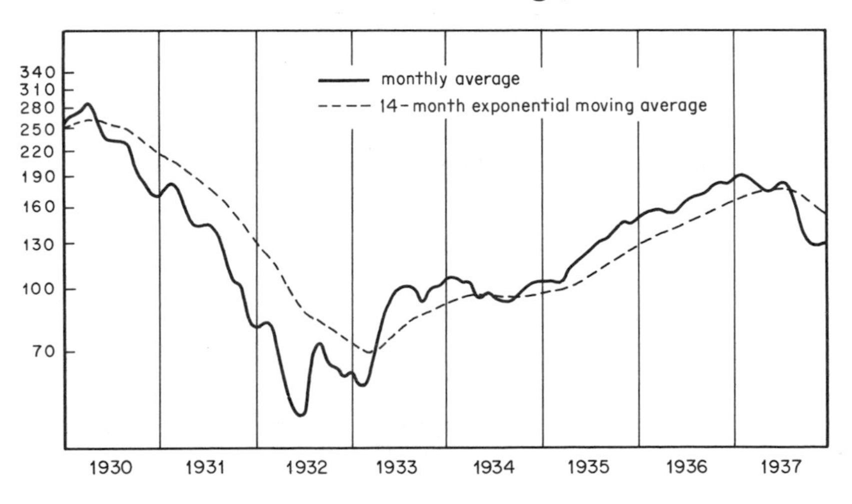

by another EMA. This method uses an EMA, as calculated above, and repeats the process using a further exponent. There is no reason why a third or fourth smoothing could not be done, but the ever-present tradeoff between sensitivity and lateness should be borne in mind.

## ENVELOPES

It has already been mentioned that moving averages often act as important juncture points in their role as support and resistance areas. The longer the time span of the average, the greater will be its significance in this respect. An expansion of this support and resistance principle involves the drawing of symmetrical lines called "envelopes" parallel to a moving average (see Figure 4-4).

This technique is based on the principle that stock prices fluctuate around a given trend in cyclical movements of reasonably similar proportion. In other words, just as the moving average serves as an important juncture point, so do certain lines drawn parallel to that moving average. Looked at in this way, the moving average is really the center of the trend, and the envelope consists of the points of maximum and minimum divergence from that trend.

There is no hard and fast rule as to the exact position at which the envelope should be drawn, since that can only be discovered on a trial-and-error basis with regard to the volatility of the index being monitored and the time span of the moving average. This process can be expanded, as in Figure 4-5, to include four or more envelopes (i.e., two above and two below the moving average), each drawn at an identical distance from its predecessor.

In this example the envelopes have been plotted at 10 percent intervals, so that if the moving average is at 100, the envelopes would be plotted at 90, 110, etc.

The example using the DJIA between 1972 and 1974 (Chart 4-5) shows that the envelope technique can be very useful from two aspects: (1) de-

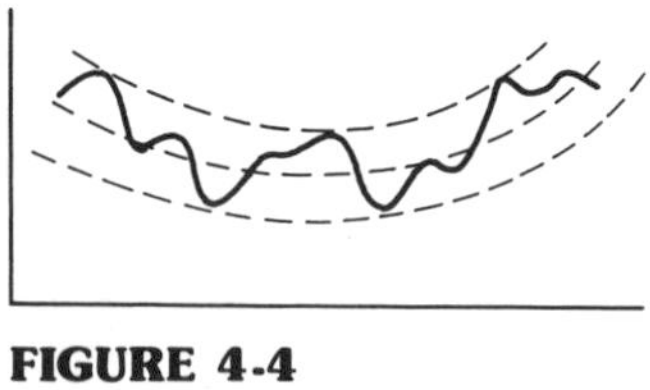

**FIGURE 4-4**

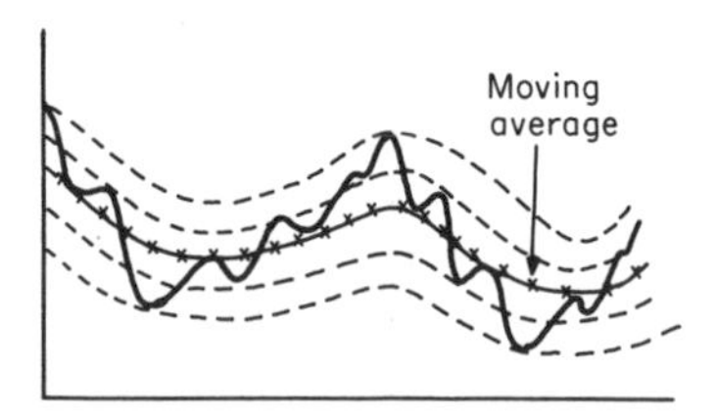


**FIGURE 4-5**

**CHART 4-5 Dow Jones Industrial Average, 1972–1974**

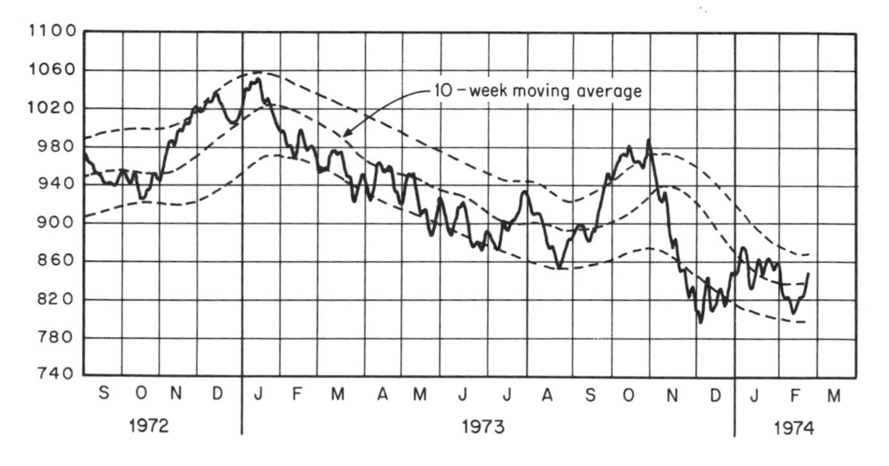


veloping a "feel" for the overall trend and (2) discerning when a rally or reaction has overextended. It suffers from the disadvantage that there is no certainty that the envelope will prove to be the eventual turning point. This method, like all techniques attempting to forecast the duration of a move, should be used on the basis that if the index reaches a particular envelope, there is a good probability that it will reverse course at that juncture. The actual conclusion will be determined by an assessment of a number of characteristics, of which an envelope juncture is a contributing factor.

## SUMMARY

One of the basic assumptions of technical analysis is that stocks move in trends. Since major trends comprise many minor fluctuations in prices, a moving average is constructed to aid in the smoothing of data so that the underlying trend is more clearly visible.

A simple moving average should be plotted at the halfway point in the time period being monitored (a process known as centering), but since this would involve a time lag during which stock prices could change rapidly and lose much of the potential profit of a move, such an average is plotted at the end of the period in question. This drawback has been largely overcome by the use of moving-average crossovers, which provide warnings of a reversal in trend, and by the use of weighted or exponential moving averages, which are more sensitive to changes in the prevailing trend since they weight data in favor of the most recent periods.

# 5

# RATE OF CHANGE: MOMENTUM

The methods of trend determination considered so far have been concerned with an analysis of the movement of an indicator itself through trendlines, price patterns, and moving averages. While such techniques can be extremely useful, they identify a change in trend *after it has taken place* and are helpful only where a trend reversal is detected at a relatively early stage in its development. On the other hand, using the concept of *momentum* can often warn of latent strength or weakness in the indicator being monitored, usually well ahead of its ultimate peak or trough.

## MOMENTUM

The concept of momentum can be understood through a simple example. When a ball is thrown into the air, it begins its trajectory at a very fast pace, i.e., it possesses strong momentum. Gradually the speed with which the ball travels upward becomes distinctly slower, until it finally comes to a temporary standstill before the force of gravity causes it to reverse its course. This slowing-down process, known as a loss of upward momentum, is also a phenomenon of the stock market. If the flight of a ball is equated to a price index of the stock market, the rate of advance in the index begins to slow down noticeably before the ultimate peak in prices is reached.

On the other hand, if a ball is thrown in a room and hits the ceiling while its momentum is still rising, the ball and the momentum will reverse at the same time. Unfortunately momentum indexes in the stock market are not dissimilar, since sometimes momentum and price peak simultaneously as either a ceiling of selling resistance is met, or buying power is temporarily exhausted. In such instances momentum as a lead indicator is not a particularly useful concept, although under such conditions the level of momentum is often as helpful as its direction in assessing the quality of a price trend.

The idea of downward momentum may be better understood by comparing it to a car being pushed over the top of a hill, for example. As the gradient of the hill steepens, the car begins to pick up speed, and finally at the bottom it hits maximum speed. Although its speed decreases, the car continues traveling until it finally comes to a halt. Stock prices in a declining trend act in a similar fashion, since the rate of decline (or loss of momentum) often slows ahead of the final low in stock prices. This is not always the case, since momentum and price sometimes (as at peaks) turn together as prices meet a major level of support. Nevertheless, this concept of momentum leading price occurs sufficiently often to be useful in warning of a potential trend reversal in the indicator or market average which is being monitored.

There are essentially two ways of looking at momentum: as a measure of rate of change, and as a measure of internal market vitality. Rate of change is a more suitable technique for measuring a price average such as the DJIA, while specific measures of internal strength arc better applied to monitoring market indicators such as breadth. For the purpose of clarity internal measures of momentum will be referred to as *oscillators*, even though rates of change also oscillate within a prescribed range, and momentum measures of price averages will be described as *rates of change*.

The following discussion describes the construction and application of momentum indexes from the point of view of determining the trend of the index from which the momentum is derived. The significance for the rest of the market of a change in trend of that particular indicator will be examined in Part 2.

## RATE OF CHANGE

The simplest method of measuring momentum is to calculate the rate at which a market average changes price over a given period of time. If it is desired, for example, to construct an index measuring a 10-week rate of change, the current price is divided by the price 10 weeks ago. If the latest price is 100 and the price 10 weeks ago was 105, the rate of change or momentum index will read 95.2, that is, 100 ÷ 105. The subsequent reading in the index would be calculated by dividing next week's price by the price

9 weeks ago from today (see Table 5-1). The result is an index that oscillates around a central reference point which marks the level at which the price is unchanged from its reading 10 weeks ago (Figure 5-1). Consequently, if a calculation were being made of the momentum of a market average that did not change price, its momentum index would be represented as a straight line.

When a momentum index is above the reference line, the market average which it is measuring is higher than its level 10 weeks ago.

If the momentum index is also rising, it is evident that the difference between the current reading of the market average and its level 10 weeks ago is growing. If a momentum index is above the central line but is declining, the market average is still above its level 10 weeks ago, but the difference between the two readings is shrinking. When the momentum index is below its central line and falling, this indicates that the market

**TABLE 5-1**

| Date | | DJIA (1) | DJIA 10 weeks ago (2) | 10-week rate of change (col. 1 + col. 2) (3) |
|---|---|---|---|---|
| Jan. | 1 | 985 | | |
| | 8 | 980 | | |
| | 15 | 972 | | |
| | 22 | 975 | | |
| | 29 | 965 | | |
| Feb. | 5 | 967 | | |
| | 12 | 972 | | |
| | 19 | 965 | | |
| | 26 | 974 | | |
| Mar. | 5 | 980 | | |
| | 12 | 965 | 985 | 98.0 |
| | 19 | 960 | 980 | 98.0 |
| | 26 | 950 | 972 | 97.7 |
| Apr. | 2 | 960 | 975 | 98.5 |
| | 9 | 965 | 965 | 100.0 |
| | 16 | 970 | 967 | 100.3 |
| | 23 | 974 | 972 | 100.2 |
| | 30 | 980 | 965 | 101.6 |
| May | 7 | 985 | 974 | 101.1 |

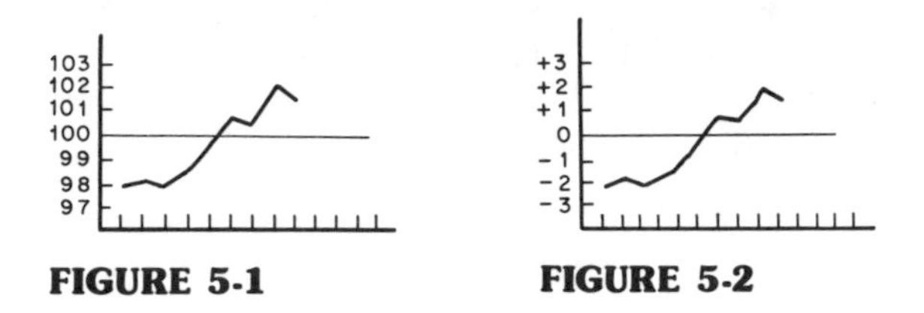


**FIGURE 5-1** **FIGURE 5-2**

average is below its level 10 weeks ago, and the difference between the two is growing. When it is below its central line but rising, the market average is still below its level 10 weeks ago, but its rate of decline is slowing.

In short, a rising rate-of-change index implies a growth in momentum and a falling index a loss of momentum. Rising momentum should be interpreted as a bullish factor and declining momentum a bearish one.

There are two methods of scaling a rate-of-change chart. Since the choice does not affect the trend or level of the index, the method used is not important; but in view of the fact that the two alternatives are often found to be confusing, a brief explanation is in order. The first method is the one described above and shown in Figure 5-1, where 100 becomes the central reference point. In the example, 100 (this week's observation) divided by 99 (the observation 10 weeks ago) is plotted as 101, 100 divided by 98 as 102, 100 divided by 102 as 98, and so on.

The alternative is to take the difference between the index and the 100 level and plot the result as a positive or negative number, using a reference line of 0.

In this case, 101 would be plotted as +1, 102 as +2, 98 as −2, and so on (see Figure 5-2).

## OSCILLATORS (INTERNAL STRENGTH)

When measuring price indexes the rate-of-change method of determining momentum is useful for historical comparative purposes, since it reflects moves of similar proportion in an identical way. On the other hand, this method is not suitable for gauging the vitality of the indicators that monitor internal market structure, such as those that measure volume or breadth, since the construction of such indexes is often started from a purely arbitrary number; under certain circumstances this might require a rate of change to be calculated between a negative and positive number, which would obviously give a completely false impression of the prevailing trend of momentum.

Rather than unduly complicating matters by trying to cover many different types of oscillators for the various stock market indicators, the discussion at this stage can be confined to the construction of one index, since the basic principles are the same for all. The more important oscillators are examined in Part 2.

## Breadth Momentum

Stock market breadth is measured by an *advance/decline line*. This line is constructed by cumulating the difference between the number of stocks advancing over those declining within a given period (usually a day or a week). Generally speaking, the advance/decline line (A/D line) rises and falls with the DJIA, but when their paths diverge it is usually a sign that a reversal in trend in the DJIA may be expected. Since the A/D line is a very useful technical indicator, it is obviously important to derive some measurement of its vitality or momentum so that an advance warning of a reversal in its trend can be obtained.

Momentum of breadth is measured by dividing the total of the number of stocks advancing in a given period against the total of the number declining. For a 10-week oscillator the calculations would be as in Table 5-2.

A momentum oscillator constructed in this way is shown in Figure 5-3.

**TABLE 5-2**

| Date | | No. of stocks advancing (1) | 10-week total of col. 1 (2) | No. of stocks declining (3) | 10-week total of col. 3 (4) | 10-week momentum index (col. 2 ÷ col. 4) (5) |
|---|---|---|---|---|---|---|
| Jan. | 2 | 1346 | . . . | 447 | | |
| | 9 | 1807 | . . . | 165 | | |
| | 16 | 1479 | . . . | 412 | | |
| | 23 | 1451 | . . . | 419 | | |
| | 30 | 1335 | . . . | 547 | | |
| Feb. | 6 | 1086 | . . . | 822 | | |
| | 13 | 1267 | . . . | 631 | | |
| | 20 | 1389 | . . . | 489 | | |
| | 27 | 657 | . . . | 1238 | | |
| Mar. | 5 | 844 | 12661 | 987 | 6157 | 206 |
| | 12 | 1048 | 12363 | 772 | 6482 | 191 |
| | 19 | 721 | 11277 | 1115 | 7432 | 152 |
| | 26 | 1126 | 10924 | 675 | 7695 | 142 |

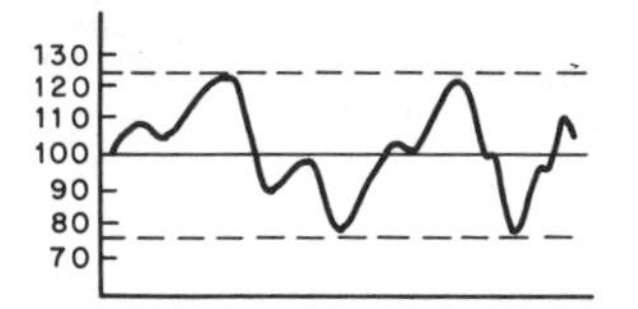


**FIGURE 5-3**

There are, of course, a whole variety of indicators that can be derived from these or similar methods (see Chart 5-1). Some of the variations are examined in more detail in Part 2.

## PRINCIPLES AND APPLICATION OF MOMENTUM INDEXES

The following description of the principles and application of momentum indexes applies equally well to the oscillator type (indexes which measure

**CHART 5-1 Dow Jones Industrial Average versus 10-Day Breadth Oscillator**

The 10-day breadth oscillator shown below is calculated in the same way as a 10-week oscillator (see Table 5-2). Note how the 1978 April–May rally was preceded by declining downward momentum as represented by the fact that the February low in the oscillator was above that of January. This positive divergence between the oscillator and the DJIA itself gave a useful clue to the market's "improving" technical position.

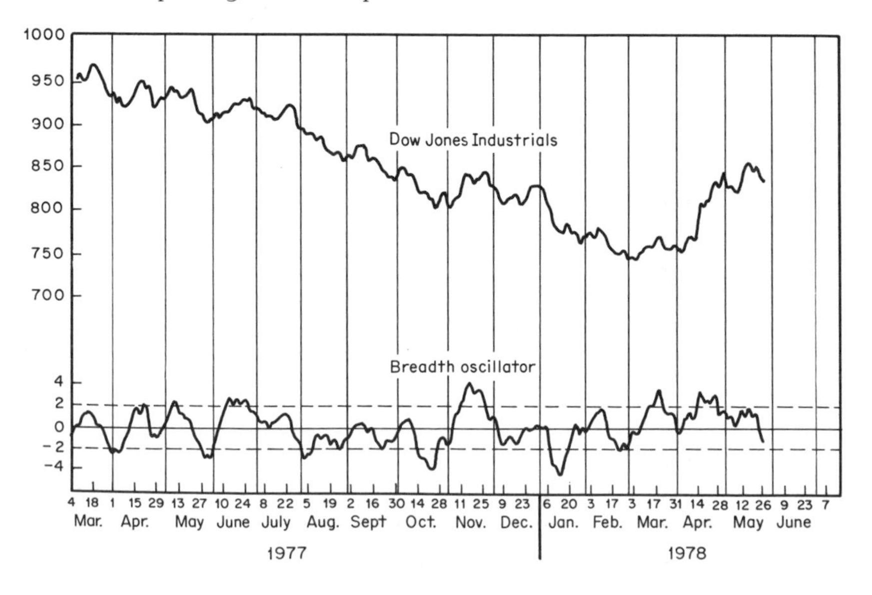

internal market momentum) as it does to the rate-of-change indexes discussed earlier.

The momentum index is usually plotted below the indicator that it is measuring, as in Figure 5-4. The example of the ball used at the beginning of the chapter showed that maximum momentum was obtained fairly close to the point when the ball leaves the hand. In a similar fashion, stock prices usually reach their maximum level of momentum reasonably close to the bear market bottom.

In Figure 5-4 this is shown as point A. If the stock index makes a new high which is confirmed by the momentum index, no indication of technical weakness arises. On the other hand, should the momentum index fail to confirm (point *B*), a negative divergence is set up between the two indexes, and a warning of a weakening technical structure is given. Usually such discrepancies indicate that prices are likely to undergo a corrective process, which can either be sideways or (more likely) downward. However, an index will sometimes continue upward to a third top accompanied by even greater weakness in the momentum index. Alternatively, the third peak in the momentum peak may be higher than the second but still lower than the first. Under either circumstance extreme caution is called for, since this characteristic is a distinct warning that a sharp reversal in price or a long corrective period may soon get under way.

Whenever any divergence between momentum and price occurs, it is essential to wait for a confirmation from the price index that its trend has also been reversed. This confirmation can be achieved by (1) the violation of a simple trendline, as shown in Figure 5-4, (2) the crossover of a moving average, or (3) the completion of a price pattern. This form of insurance is well worth taking, since during a long cyclical advance (such as the 1962–1966 bull market) it is not unknown for an index to continually lose and regain momentum without suffering a break in trend.

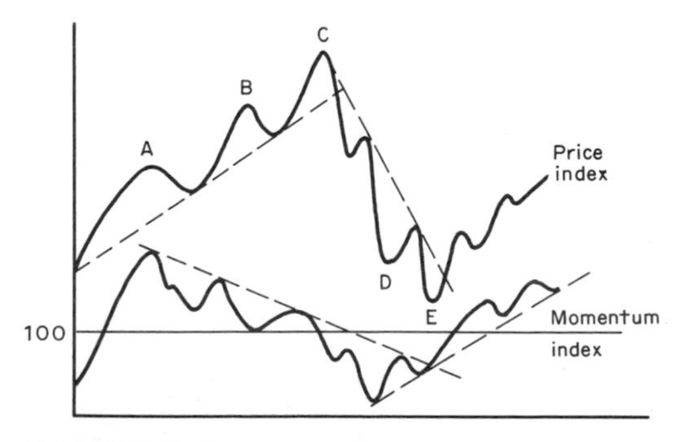


**FIGURE 5-4**

The same principles of divergence are also applicable following price declines. In Figure 5-4 the price index makes a new low at point *E*, but the momentum index does not. This evidence of technical strength was later confirmed when the price index rose above its bear market trendline.

When at times the momentum index peaks simultaneously with price (as shown in Figure 5-5), no advance warning is given that a price decline is imminent. Nevertheless, a clue indicating technical weakness is given when a trendline joining the troughs of the momentum index is penetrated on the downside.

As with any trendline construction, judgment is still required to decide the significance of the break based on the principles outlined in Chapter 3. Moreover, the break in momentum should be regarded as an alert, and action should be taken only when it is confirmed by a break in the price trend itself (indicated by line AA in Figure 5-5).

Momentum indexes are also capable of tracing out price patterns. Usually these are of the accumulation type, although distribution formations are not uncommon. Due to the shorter lead times normally associated with reversals of falling momentum, a breakout from an accumulation pattern when accompanied by a reversal in the downward trend of the index itself is usually a highly reliable indication that a worthwhile move has just begun. Such an example is shown in Figure 5-6.

There is another way in which momentum indexes may be useful, and that is with regard to their level. Since this type of index is an oscillator fluctuating backward and forward across its 0 or 100 reference line, there are clearly definable limits beyond which it rarely goes (see Figure 5-7).

The actual boundaries will depend on the volatility of the index being monitored and the time period on which the rate-of-change period is

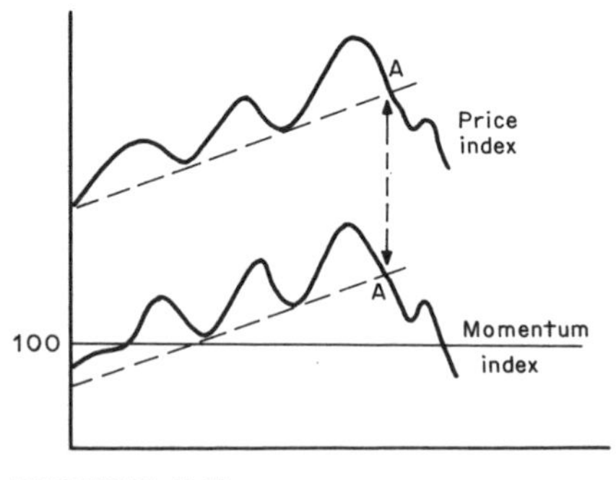


**FIGURE 5-5**

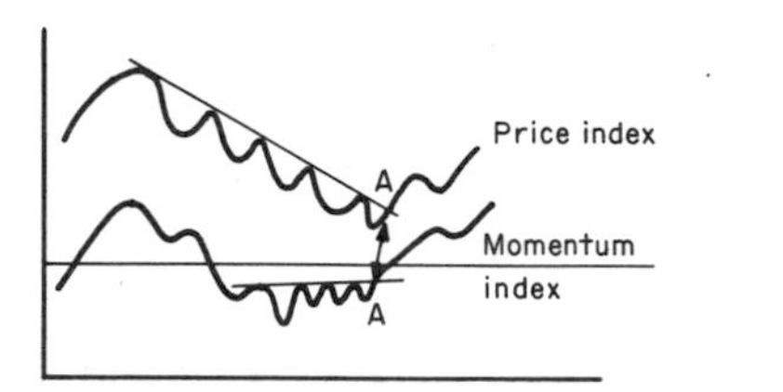

**FIGURE 5-6**

**FIGURE 5-7**

based, since the rate of change of an index has a tendency to vary more over a longer period than a shorter one.

In view of these two variables, there is no hard and fast rule as to what constitutes an unduly high (known as overbought) or low (known as oversold) level. This can be achieved only with reference to the history of the index being monitored and the maturity of the cycle. For example, when a bull market has just begun there is a far greater tendency for an index to move quickly into "overbought" territory and remain at very high readings for a considerable period of time. At such points the "overbought" readings tend to give premature warnings of declines. Consequently, during the early phases of the bull cycle when the market possesses strong momentum, reactions to the "oversold" level are much more responsive to price reversals, and such readings therefore offer more reliable signals. It is only when the bull market is maturing or during bear phases that "overbought" levels indicate that a rally is shortly to be aborted. The fact that an index is unable to remain at such high readings for long periods is itself a signal that the bull market is losing momentum.

A further indication of the maturity of a trend is given when the momentum index moves strongly in one direction but the accompanying move in the price index is a much smaller one. Such a development suggests that the price index is tired of moving in the direction of the prevailing trend, for despite a strong push of energy from the momentum index, prices are unable to respond. This phenomenon is illustrated in Figure 5-8.

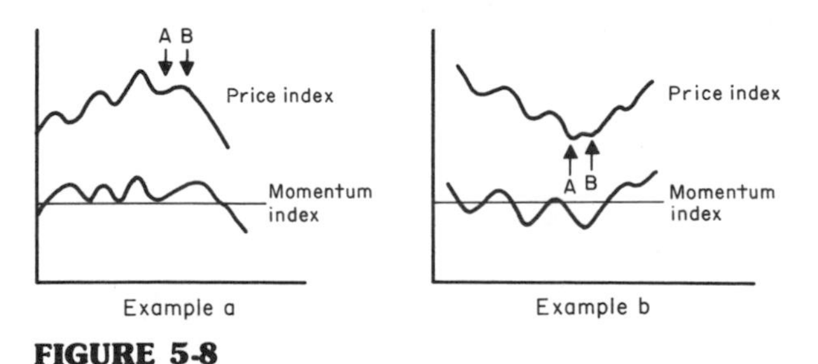


**FIGURE 5-8**

## SMOOTHED MOMENTUM INDEXES

The interpretation of momentum indexes as described above depends to a considerable extent on judgment. One method of reducing this subjectivity is to smooth the rate-of-change index by using a moving average.

Warnings of a probable trend reversal in the price index being monitored would be offered by a reversal in the smoothed momentum index itself (example *a*, Figure 5-9) or by a penetration of the moving average through a designated level (example *b*, Figure 5-9).

The level (see the dashed line) would be determined on a trial-and-error basis by reference to a long historical study of the relationship between the index and the momentum curve. For this purpose any type of moving average can be used, but more timely results have been achieved through the use of a weighted or an exponential average in view of their greater sensitivity (see Chart 5-2).

A further method of filtering out volatility is to construct a momentum index by taking the averages of two or three rates of change and weighting them according to their time span. The calculation using the combination of a 6- and 10-week momentum would be done as in Table 5-3.

A similar momentum index for breadth (or any other indicator measuring internal strength) can also be achieved by combining the ratios of two or more different time periods.

Since these momentum indexes using several periods are still relatively volatile, a moving-average smoothing is normally still necessary. Chart 5-3 indicates how effective such an index can be. In this case the smoothing is a 10-month weighted average of 11- and 14-month rates of change of DJIA monthly closing prices. For timing purposes, this index has been found useful for market bottoms rather than tops, so the momentum curve takes on significance only when it falls below the 0 reference line and rises.

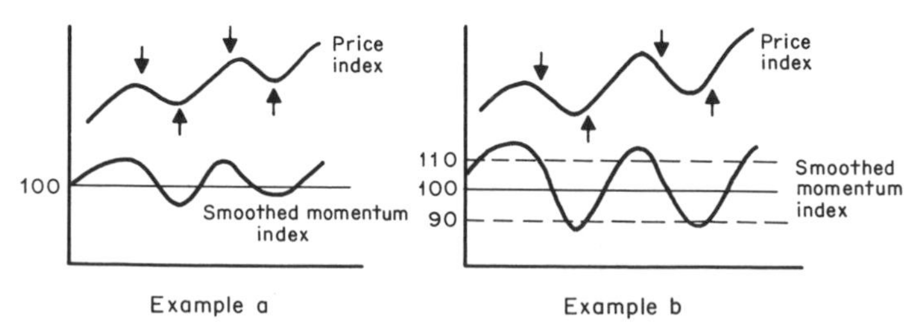


**FIGURE 5-9**

## CHART 5-2 West German Stock Market Averages, 1959–1978

This chart shows the German stock market and a 12-month rate of change. The dashed line is a 14-month EMA of the rate of change. Whenever this EMA rises above the 110 level and falls below it, a useful "sell" signal has resulted.

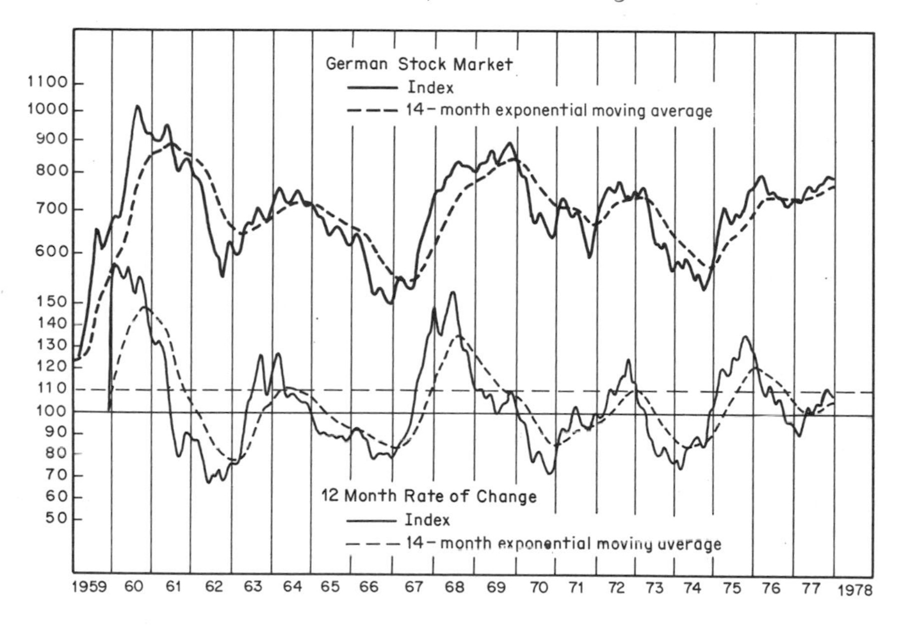


## TABLE 5-3

| Date | | DJIA (1) | 6-week rate of change (2) | 10-week rate of change (3) | Col. 2 × 6 (4) | Col. 3 × 10 (5) | Col. 4 + col. 5) (6) | Col. 6 ÷ 16 (7) |
|---|---|---|---|---|---|---|---|---|
| Jan. | 1 | 900 | | | | | | |
| | 8 | 910 | | | | | | |
| | 15 | 920 | | | | | | |
| | 22 | 918 | | | | | | |
| | 29 | 925 | | | | | | |
| Feb. | 5 | 950 | | | | | | |
| | 12 | 945 | 105 | . . . | 630 | | | |
| | 19 | 936 | 103 | . . . | 618 | | | |
| | 26 | 925 | 101 | . . . | 606 | | | |
| Mar. | 4 | 934 | 102 | . . . | 612 | | | |
| | 11 | 940 | 102 | 104 | 612 | 1040 | 1652 | 103 |
| | 18 | 945 | 100 | 104 | 600 | 1040 | 1640 | 103 |
| | 25 | 951 | 101 | 103 | 606 | 1030 | 1636 | 102 |

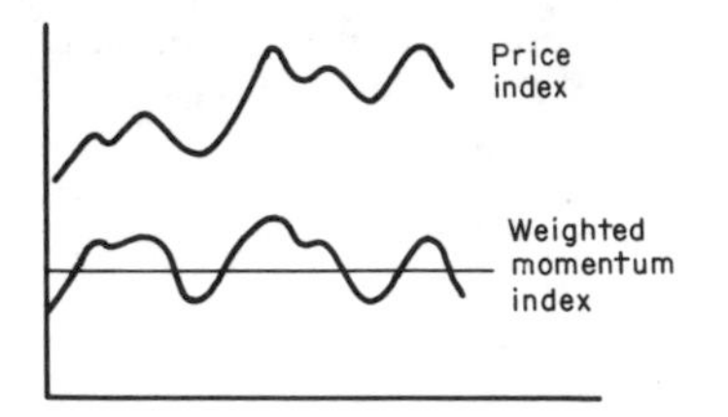


**FIGURE 5-10**

A further variation of constructing a smoothed momentum index is to take the rate of change of a moving average of a price index itself. This method reverses the process described above, for instead of constructing a rate of change and then smoothing the resulting momentum index, the price index itself is first smoothed with a moving average, and a rate of change is taken of that smoothing.

## INDEX TO TREND

A final form of momentum index is obtained by dividing a price index by a moving average of that index. Since the moving average represents the

**CHART 5-3 Long-Term Price Momentum Indicator, 1960–1975**

This momentum indicator has a long history of successful "buy" signals and is plotted up to December 1975 on this chart. It turned up in the following month and successfully confirmed the 1974–1976 bull market.

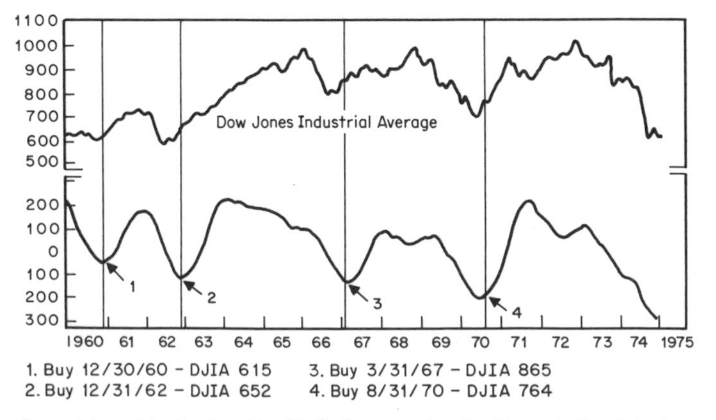


(Derived from data developed by E.S.C. Coppock, Trendex Research, San Antonio, Tex.)

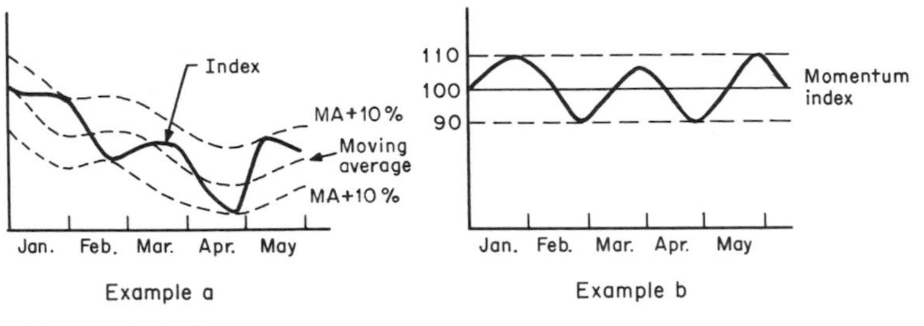


**FIGURE 5-11**

trend of the index being monitored, the resultant momentum indicator shows how fast the index is advancing or declining in relation to that trend.

This is really a horizontal representation of the envelope analysis discussed in the previous chapter. Examples *a* and *b* in Figure 5-11 show these two approaches for the same index. The upper and lower envelopes are both drawn at a level that is 10 percent from the actual moving average, so that when the price index touches the 100 line it is really at the same level as the moving average. When the momentum index is at 110, the price index is 10 percent above its average, and so on.

While this index portrays essentially the same information as the envelope analysis, in many ways it can better illustrate subtle changes in the latent strength or weakness of the price index being monitored.

## SUMMARY

The concept of momentum, which measures the rate at which prices rise or fall, gives useful indications of latent strength or weakness in a price trend. This is because prices usually rise at their fastest pace well ahead of their peak and normally decline at their greatest speed before their ultimate low.

Markets usually spend more time in a rising than a falling phase, so that this lead characteristic is usually greater in bull markets than in bear markets. Since momentum in the vast majority of cases either leads or coincides with price reversals in the trend of the index from which it has been constructed, this concept is useful as supplementary confirmation. Under certain circumstances the level of momentum can also be used to predict a reversal in the price trend itself.

# 6

# MISCELLANEOUS TECHNIQUES FOR DETERMINING TRENDS

Two of the techniques discussed in this chapter, namely, the concepts of support and resistance and of proportion, are concerned with estimating the possible extent or duration of a trend, in contrast to most of the indicators previously examined, which attempt to confirm a change in trend after it has already occurred. Such techniques should be used only as an indication of the *probable* extent of the move, however, not as the basis of an actual forecast, as there is no known method of predicting the duration of a market trend.

The third concept considered in this chapter is that of "relative strength." This is a method which helps to determine the quality of a trend of an index by comparing it with the trend of another index—usually "the market," as measured by the Dow Jones Industrial Average.

Declining relative strength after a long upmove is regarded as a sign of weakness, while improving relative strength after a substantial decline is interpreted as a sign of strength. Since some indexes have a habit of leading the DJIA, changes in the trend of their relative strength offer a good preliminary indication of a reversal in the trend of the index itself.

## SUPPORT AND RESISTANCE

The concept of support and resistance has already been mentioned, but it warrants further consideration. Support has been defined as "buying, actual or potential, sufficient in volume to halt a downtrend in prices for an appreciable period" and resistance as "selling, actual or potential, sufficient in volume to satisfy all bids and hence stop prices from going higher for a time."[1]

A support zone represents a *concentration* of demand, and a resistance zone a *concentration* of supply. The word "concentration" is emphasized since supply and demand are always in balance, but it is their relative strength or concentration which determines trends.

The reason why areas of support and resistance occur, and methods of detecting them in advance, are best illustrated by the use of a simple example.

At the beginning of Figure 6-1 the market is in a downtrend. The bear market is interrupted by a consolidation pattern (rectangle), from which prices ultimately break down. The index reaches its low and then after a brief period of consolidation mounts a rally. The rally is then halted at the dashed line, which is at the same level as the lower end of the rectangle. People who bought during the period when the rectangle was being formed have all lost money up to this point, and many of them are keen to liquidate their positions and "break even." Consequently there is a concentration of supply at this level, transforming what was formerly a level of support to one of resistance. Such levels are often formed at a round number such as 10, 100, or 50, since these represent easy psychological targets for investors to make investment decisions.

Returning to the example, prices then fall off slightly and mount another assault on the resistance level, but still more stock is put up for sale. Finally prices and volume pick up, and the resistance is overcome as the available supply at this level is absorbed.

The first rule in assessing the importance of a support or resistance zone therefore relates to the amount of stock that changed hands in that area. The greater the activity, the more significant the zone.

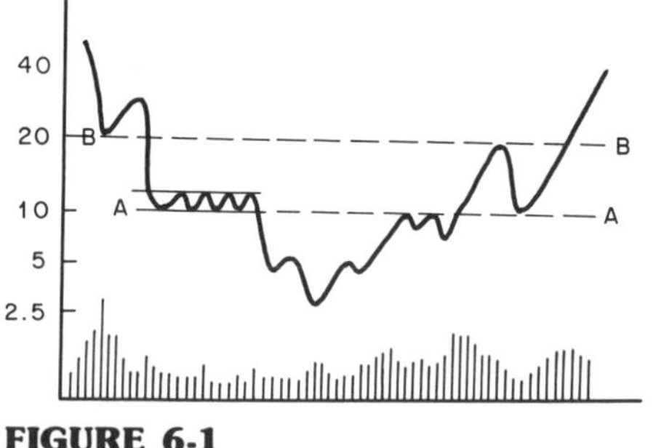


**FIGURE 6-1**

[1] Robert D. Edwards and John Magee, *Technical Analysis of Stock Trends*. John Magee, Springfield, Mass., 1957.

The next level of resistance in Figure 6-1 is around the level of the "panic" bottom (line *BB*) of the previous bear market. The price is forced back at this point since the combination of an overextended rally and substantial level of resistance is too overpowering.

In this instance the attempt of the index to climb through the resistance level can be compared to the efforts of a man who, having just finished a 10-mile run, tries unsuccessfully to push open a heavy concrete trapdoor. The trapdoor would have proved difficult at the best of times, but after such violent physical exercise the man is forced to rest before making another attempt. The same principle can be applied to the market in that a long, steep climb in price is similar to the 10-mile run, and the resistance level resembles the trapdoor. Consequently, a second rule for assessing the significance of a support or resistance zone is the speed and extent of the previous move.

The more overextended the previous price swing is, the less resistance/support that is required to halt it.

Reference to Figure 6-1 shows that the index, having unsuccessfully mounted an assault on the resistance level at *BB*, falls back to the first level of *AA*, which now returns to its previous role of support. This reversal in role arises from the experiences of the investors concerned. In the first place, some who bought at the bear market low may have decided to take profits at *AA*, i.e., $10, and have been so frustrated by subsequently watching the stock rise to $20 that they have decided to repurchase if it ever returns to $10 again. Similarly, an investor who bought at $10 previously and sold at $20 may wish to repeat this shrewd maneuver.

The third rule for establishing the potency of a support or resistance zone is to examine the amount of time that has elapsed between the formation of the original congestion and the nature of general market developments in the meantime. A supply which is 6 months old has greater potency than one established 10 or 20 years previously. In spite of this fact it is almost uncanny how support and resistance levels repeat their effectiveness time and time again even when separated by many years.

The process described above is concerned with decisions on the buying and selling of stocks, but the principle of support and resistance applies equally well to the market averages, since they are themselves constructed from a number of stocks.

## PROPORTION

The law of motion states that for every action there is a reaction. Since the stock market as reflected in the price trends of stocks is really the measurement of crowd psychology in motion, it is also subject to the law. In earlier chapters it was pointed out that at certain times the stock market moves in simple proportion with the prevailing trend or directly against it. The measuring implications of price patterns, trendlines, moving averages, and envelopes are embodiments of this concept of proportion in practice.

Just as support and resistance levels can help give an idea of just where a trend in stock prices may be temporarily halted or reversed, so may the principles of proportion, but these principles go much farther.

When an index is exploring new, all-time high ground, there is no indication of where a resistance level may occur, since there is no record of one. Consequently, the use of the concept of proportion offers a guide as to where a move may terminate.

Perhaps the best-known principle of proportion is the 50 percent rule. Many bear markets, as measured by the DJIA for instance, have often cut prices by half. For instance the 1901–1903, 1907, 1919–1921, and 1937–1938 bear markets recorded declines of 46, 49, 47, and 50 percent respectively. The first leg of the 1929–1932 bear market ended in October 1929 at 195, just over half of the September high. Sometimes the halfway mark in an advance represents the point of balance, often giving a clue as to the ultimate extent of the move in question, or, alternatively, indicating an important juncture point for the return move. Thus, between 1970 and 1973 the market advanced from 628 to 1067. The halfway point in that rise was 848, or approximately the same level at which the first stage of the 1973–1974 bear market ended.

By the same token, rising markets often find resistance after doubling from a low; the first rally from 40 to 81 in the 1932–1937 bull market was a double.

In effect, the 50 percent mark falls halfway between the one-third to two-thirds retracement described in the chapter on Dow theory. These one-third and two-thirds proportions can be widely observed in the stock market and also serve as support or resistance points.

Ratio scale charts are most helpful in determining such points, since moves of identical proportion can easily be projected up and down.

## SPEED RESISTANCE LINES

In recent years the use of speed resistance lines developed by the innovative technician Edson Gould has become popular. This concept uses the one-third and two-thirds proportions, but instead of incorporating them as a base for a probable price objective, they are used in conjunction with the speed of an advance or decline.

During a downward reaction from a price rise, an index may be expected to find support when it reaches a line which is advancing at either two-thirds or one-third of the rate of advance from the previous trough to the previous peak. This is illustrated in Figure 6-2. In examples *a* and *b*, A marks the trough and *B* the peak. The advance from A to *B* is 100 points and takes 100 days, so that the speed of the advance is 1 point per day. A $\frac{1}{3}$ speed resistance line will advance at one-third of that rate (i.e., $\frac{1}{3}$ point per day), and a $\frac{2}{3}$ line will move at $\frac{2}{3}$ point per day.

A rally or decline is measured from the extreme intraday high or low and not the closing price. In order to construct a $\frac{1}{3}$ speed resistance line from example *a* it is necessary to add 33 points (i.e., one-third of the 100-point advance) to the price at A and plot this point directly under *B*. In this case A is 100 points, so a plot is made at 133 under *B*. This point is

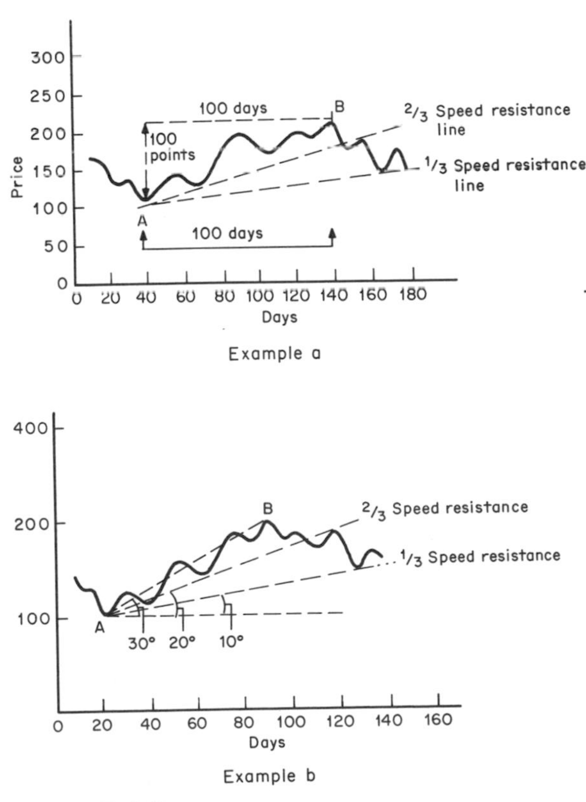

**FIGURE 6-2**

then joined to A and this line extended to the right-hand portion of the graph. Similarly, the $\frac{2}{3}$ line joins A and the 166 level on the same date as *B*. If the chart were plotted on ratio scale, the task would be much easier. All that would be required (as example *b* shows) would be a line joining A and *B*. The angle of ascent is then recorded, in this case 30 degrees. Two lines at one-third (10 degrees) and two-thirds (20 degrees) of this angle are then drawn. Figure 6-3 illustrates the same process for a declining market.

Once constructed, the speed resistance lines act as important support and resistance areas.

More specifically, the application of these lines is based on the following rules:

1. A reaction following a rally will find support at the $\frac{2}{3}$ speed resistance line. If this line is violated, the support should be found at the $\frac{1}{3}$ speed resistance line. Should the index fall

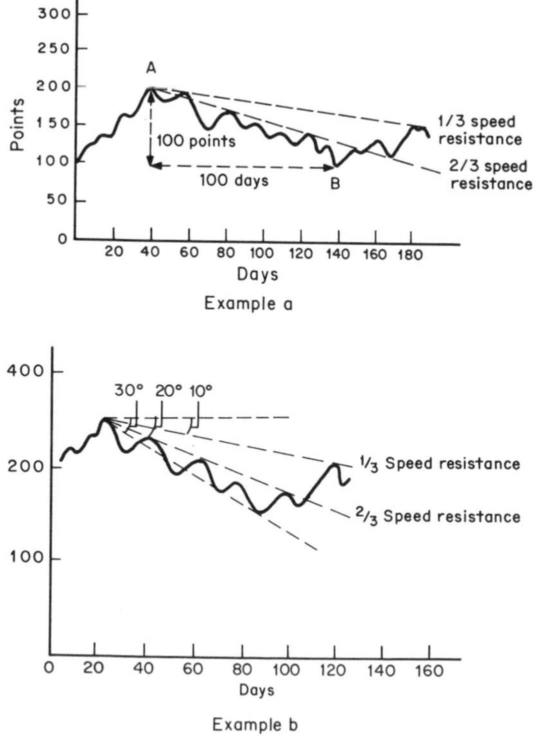

**FIGURE 6-3**

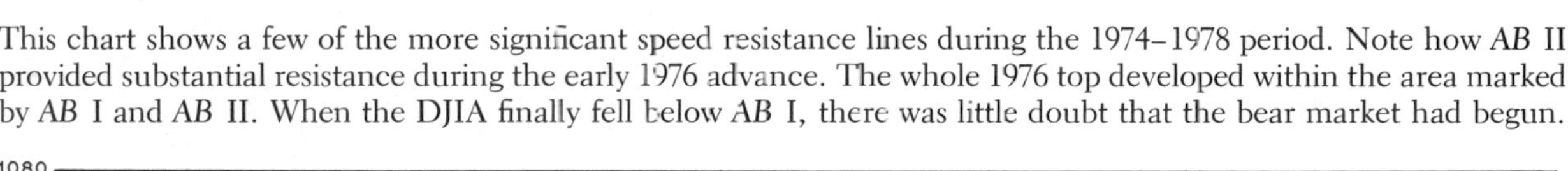

**CHART 6-1 Dow Jones Industrial Average and Speed Resistance Lines**

This chart shows a few of the more significant speed resistance lines during the 1974–1978 period. Note how *AB* II provided substantial resistance during the early 1976 advance. The whole 1976 top developed within the area marked by *AB* I and *AB* II. When the DJIA finally fell below *AB* I, there was little doubt that the bear market had begun.

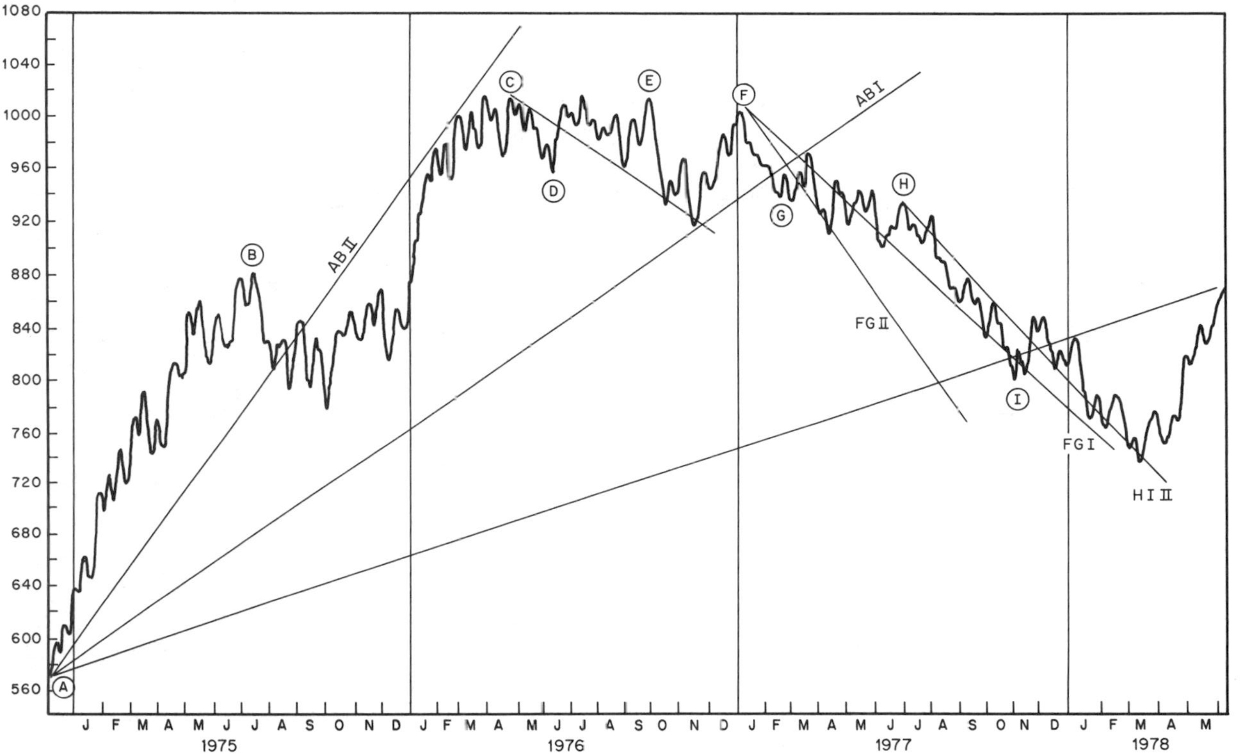

below its $\frac{1}{3}$ line, the probabilities indicate that the rising move has been completed and that the index will decline to a new low, possibly below that upon which the speed resistance lines were based.

2. If the index holds at the $\frac{1}{3}$ line, a resistance to further price advance may be expected at the $\frac{2}{3}$ line. If the index moves above the $\frac{2}{3}$ line, a new high is likely to be recorded.
3. If the index violates its $\frac{1}{3}$ line and then rallies again, it will find resistance to that rally at the $\frac{1}{3}$ line.
4. Rules 1 through 3 apply in reverse for a declining market.

Chart 6-1 shows the application of these rules in the marketplace.

## RELATIVE STRENGTH

Most market averages rise and fall in concert with the general market, and under such conditions no warnings are being given of a probable reversal in their trend. It is only when the market, as measured by the DJIA, makes a new high or low unaccompanied by an index that a signal is given. This discrepancy, known as a divergence, is a useful technique for discovering when various market sectors are moving "out of gear" with the DJIA.

A more subtle approach to the principle of divergence is the concept of *relative strength*. Relative strength is obtained by dividing the price of one index by another. Usually the divisor is a measurement of "the market," for example, the DJIA or the Standard and Poor's 500.[2] When calculated in this way, a rising relative strength line denotes that the index in question is performing better than the market, and a declining line indicates that it is being outperformed by the market.

Relative strength (RS) moves in trends just as stock prices do, and the reversal of these relative strength movements has significance for the particular index concerned.

Relative strength is more normally used for the purpose of deciding *what* stock to buy, for it is obviously much better to choose one that is likely to outperform the market. For the purpose of assessing the overall technical position of the market, RS is still a useful concept when applied to certain market averages, since there are several indexes (such as the

[2] Unless otherwise indicated, the relative strengths of all U.S. stock indexes in this book are calculated against the DJIA.

Utilities) that traditionally lead the DJIA. An assessment of the quality of their RS trends is therefore helpful in determining the overall technical strength of such averages.

Interpretation of trends in relative strength is subject to exactly the same principles as the identification of the trends of the price indexes themselves. An analysis of an RS line, therefore, involves the use of trendlines, moving averages, price patterns, support and resistance, momentum, etc.

When an index and its relative strength are both rising and it appears that the relative strength trend is about to reverse, a warning is given of a possible reversal in the trend of the index itself, since the trend of relative strength typically reverses *ahead* of the trend in the index. During a declining trend, a firming of the relative strength line indicates a position of growing technical strength. Since the lead time given by such signals varies, it is very important to await a confirmation by a similar trend reversal in the index itself. On the other hand, if the RS were being used to monitor a position in a particular stock, a reversal in RS might justify the sale of that particular stock in favor of another one with a rising relative strength trend. However, it would still be important to identify the major trend in the market itself, since a stock with a rising RS could still be falling in price if the rate of decline in the market were even sharper.

The use of the level of the RS index in conjunction with that of the price index itself can also be helpful. In Figure 6-4 it can be seen that the price index rises in a series of three ascending peaks.

The first two of these tops are also confirmed by a new high in the RS line, but the third is not. This represents a negative divergence and is a powerful indication that the positive trend in the index itself is likely to be reversed soon (also see Chart 6-2). Example *b* in Figure 6-4 shows a similar situation, but this time at a market bottom, where a new low in the

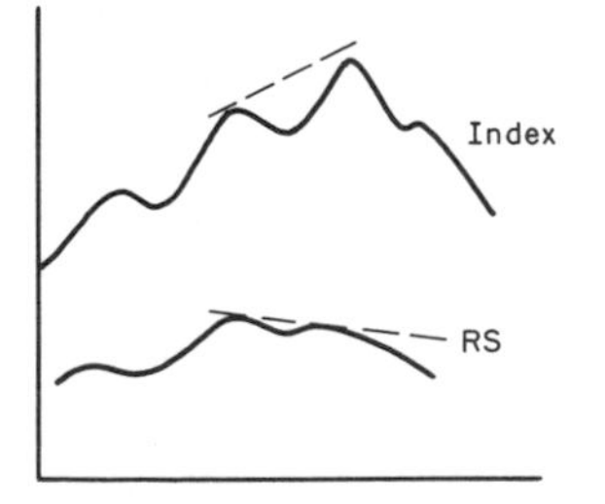


Example a

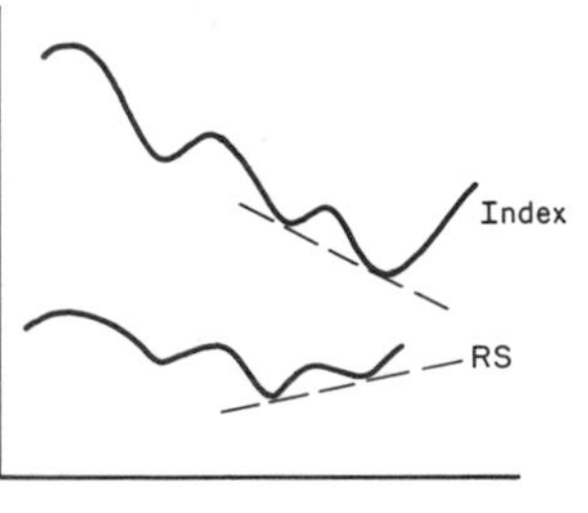


Example b

**FIGURE 6-4**

**CHART 6-2 Dow Jones Transportation Average versus Relative Strength, 1966–1977**

This chart of the Dow Jones Transportation Average and its relative strength shows how in early 1969 both the average and its relative strength broke down from their bull market trendlines, thereby signaling a decline. The same characteristics occurred at the end of 1971.

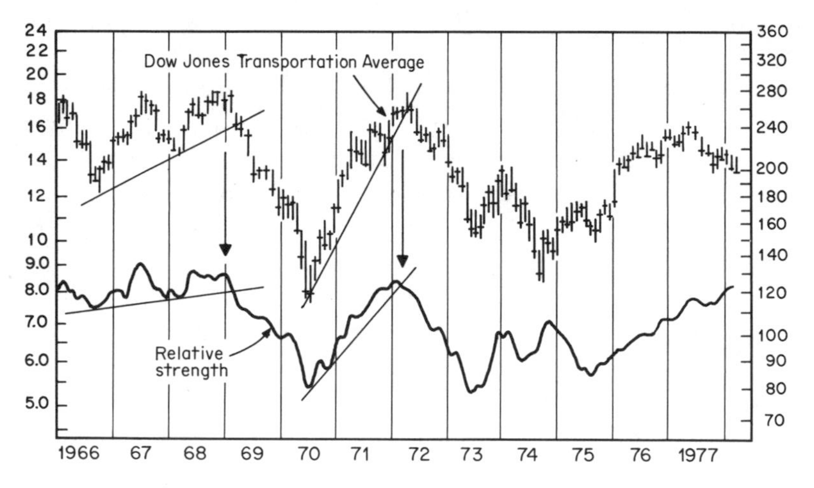


price index is not confirmed by the RS line. This is known as a positive divergence and offers a bullish undertone.

Volume can also be included in this analysis, since an index or stock which is moving up in price on higher volume should be viewed with suspicion if the trend in RS is moving in the opposite direction. A divergence between RS and price volume indicates that although there may be tremendous enthusiasm for a stock (or index), as measured by expanding volume, this high level of activity is still insufficient to permit the index to rise as fast as the market itself.

The concept of RS can be applied to almost any market situation. For example, if a switch between one stock and another were under consideration, a quick calculation of the RS between the two would indicate the trend between them. The trend in gold bullion is often compared with that of gold shares, or the U.S. stock market with that of the United Kingdom or Japan. The variations are almost limitless, but the principles associated with the determination of the trend remain identical.

# 7

# PUTTING THE TECHNIQUES TOGETHER

## An Analysis of the Dow Jones Utility Average, 1962–1969

The Dow Jones Utility Average was first constructed in 1929. It did not record its secular bottom in 1932, as was the case with most of the other indexes, but its low was delayed until April 1942. From that level (10.42) the index rose irregularly for 23 years until April 1965, at which point it reached an all-time high of 164.39; it then proceeded to lose nearly 100 points in the next 9 years (see Chart 7-1). An examination of the period between 1962 and 1969 is valuable since it covers the culmination of the secular advance and its associated distribution as well as several intervening cycles.

In May 1962 the average was at 99.8, which represented the closing low of a 7-month bear market. At this particular juncture there was nothing to suggest that a reversal in trend had taken place. The first significant signs appeared several months after the bottom. At this point the downtrend in the index (trendline AA in Chart 7-2) had been reversed, and the index had crossed above its 40-week moving average. These two signals confirmed the reversal in the downtrend of the weighted momentum index that had taken place earlier (line *AM/AM*). This rate-of-change index was constructed by combining a 6-, 13-, and 26-week rate-of-change index and weighting them according to their respective time periods. These positive signs were also accompanied by the completion of a double bottom by the Utility Average. Since the second decline occurred on lower volume than the first, the breakout to a new recovery high in December indicated that higher prices could be expected.

The average continued to rally until it reached the upside price objective indicated by the double bottom. This level, represented by line *BB* on

**CHART 7.1 Dow Jones Utility Average, 1929–1978**

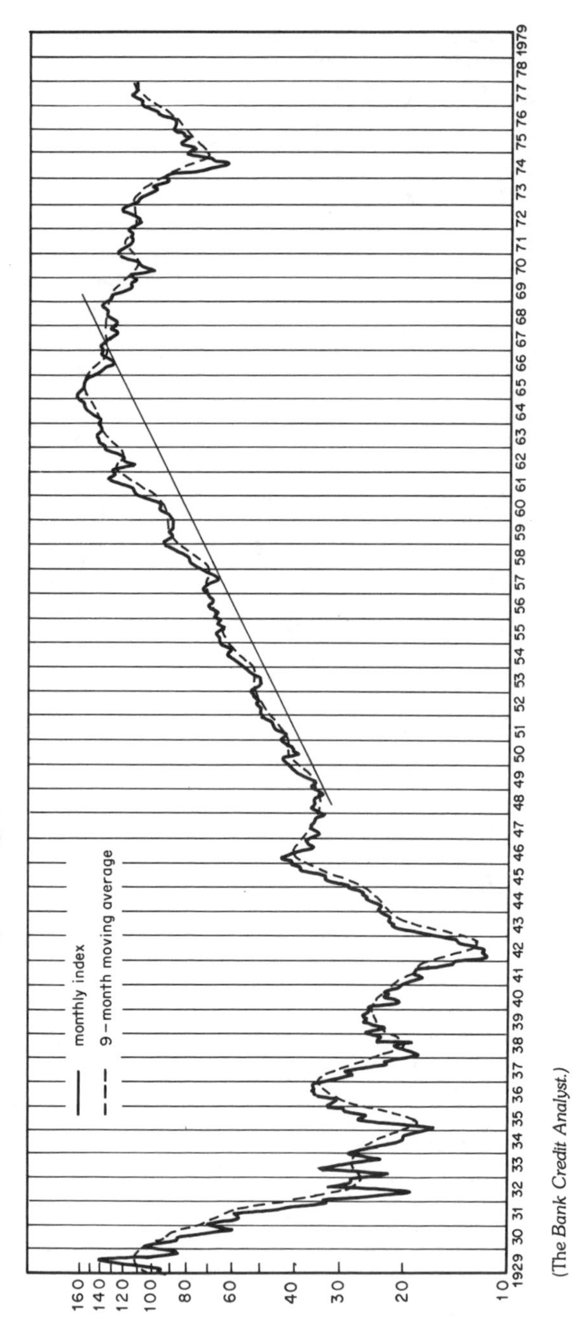


*(The Bank Credit Analyst.)*

**CHART 7-2 Dow Jones Utility Weekly Average, 1959–1978, and 40-Week Moving Average (with Weighted Momentum Index)**

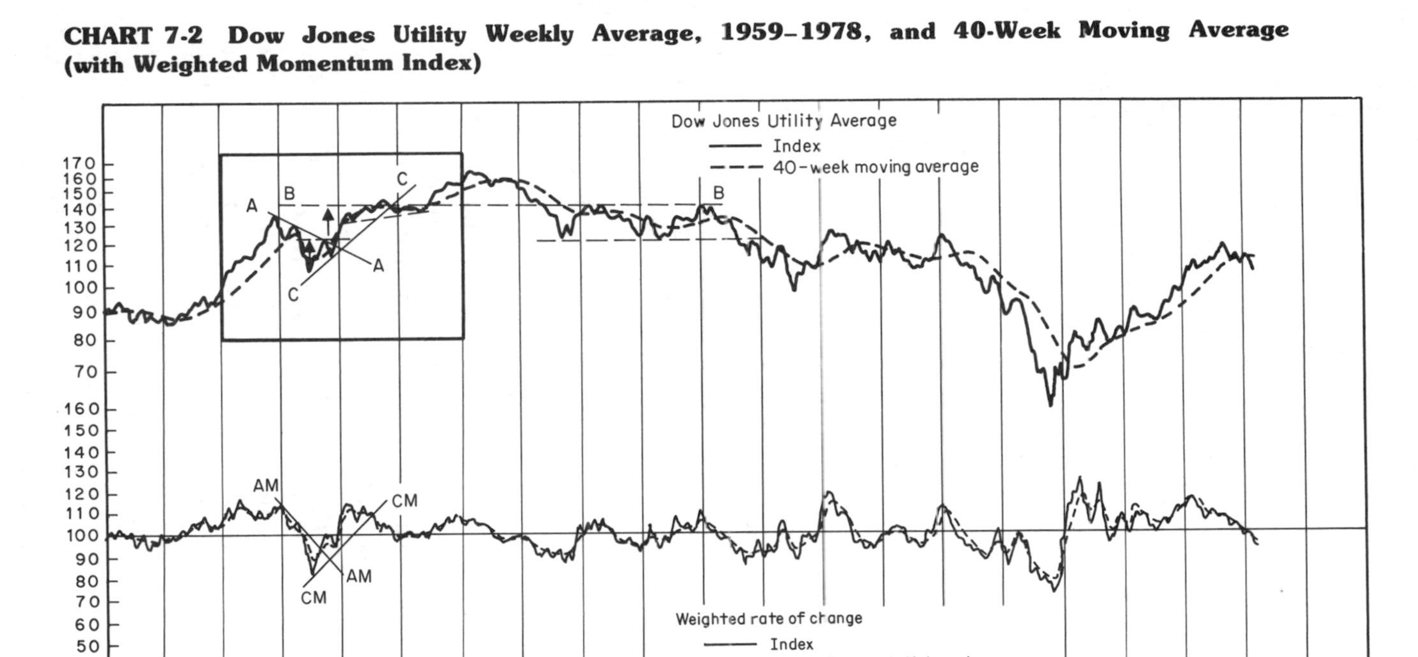


(The *Bank Credit Analyst.*)

the chart, proved to be a major resistance level in the 1963–1964 period and later in 1967 and 1969 as well. During the 12 months following the breakout, the initial bull market trendline CC remained intact, and the average also remained above its 40-week moving average. In late 1963, both the trendline CC and the moving average were violated. This development, combined with the two declining peaks and the break in trend of the momentum index (line *CM/CM*), suggested that further upside potential in the average would be difficult to achieve. Indeed, the index did not decisively break above the September 1963 high for almost a year.

This break in trend was followed by a long sideways movement. The most significant development during this period was the formation of a head and shoulders top between 1963 and 1964. This is shown in more detail in the boxed area of Chart 7-2. The pattern was almost completed, but instead of breaking below the neckline, the index broke above its 40-week moving average and began a sharp advance. This type of move is typical of a potential distribution pattern that is never completed. While the ensuing move is usually long and explosively sharp, the formation of the distribution pattern indicates some form of underlying weakness even though it fails. As it happened, the late 1964 advance was to prove to be the last before the significant 1965 peak.

It should be noted that while the late 1963 break in trend indicated that some form of consolidation was likely, at no time did the 14-month exponential average shown on Chart 7-3 threaten to reverse direction in a way that would have thrown doubt on the entire primary movement. Further reference to Chart 7-3 shows that the early 1964 decline offered a useful benchmark for the construction of a significant bull market trendline (line AA, Chart 7-3) by joining this point with the 1962 bottom. It was an important one, since it lasted a long time (2 years) and rose at a fairly shallow angle. Although this particular bull market had another 2 years to run, the average time span for a primary advance is about 2 years, so a trendline of this type that remains intact for 2 years represents a relatively long period. Its violation would therefore give a strong signal that the overall trend from 1962 had been reversed. In actual fact, the downward penetration of this line occurred in early 1965 just after the peak. Further signs of weakness at this point can be seen from the rather dramatic deterioration in relative strength (see Chart 7-4) that had begun in 1963.

The 12-month momentum index (see Chart 7-3) had also recorded a negative divergence, since it did not confirm the 1965 top in the Utility Average itself.

By late 1965 the technical position had deteriorated even more dramat-

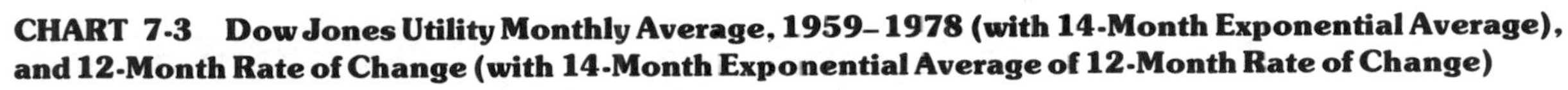

**CHART 7-3 Dow Jones Utility Monthly Average, 1959–1978 (with 14-Month Exponential Average), and 12-Month Rate of Change (with 14-Month Exponential Average of 12-Month Rate of Change)**

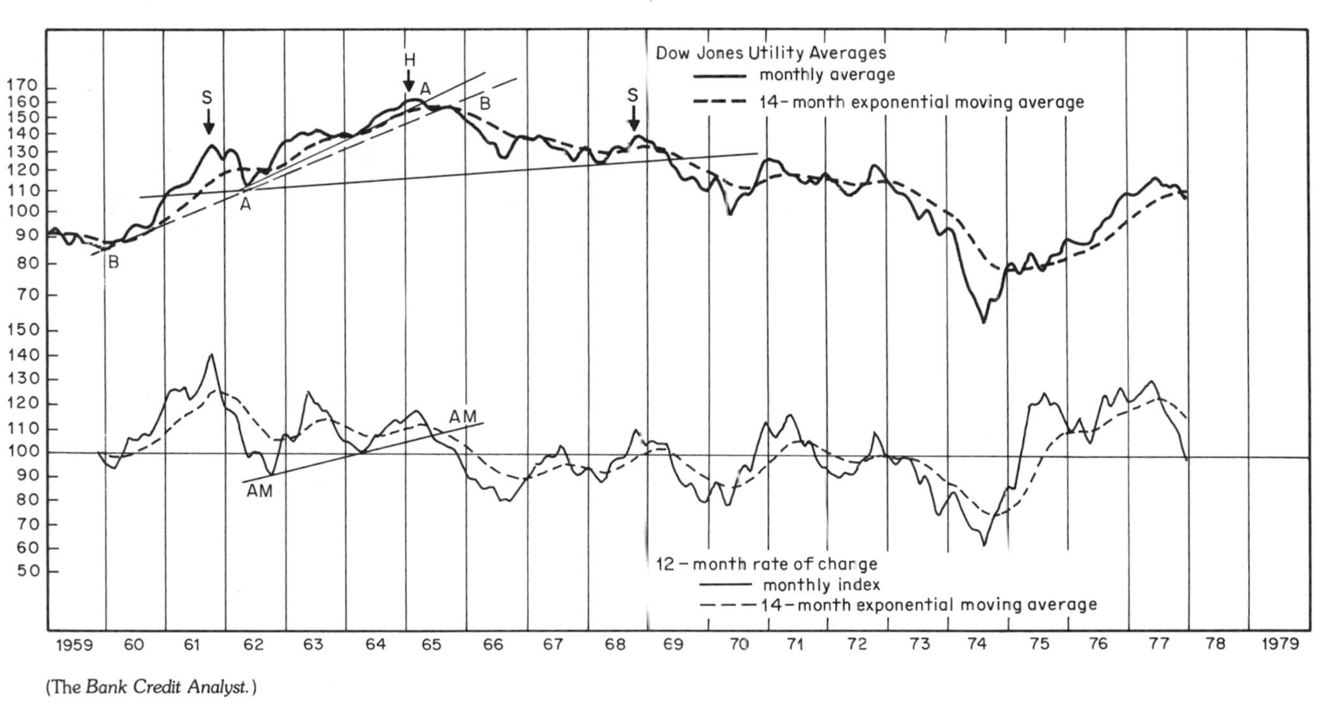


(The *Bank Credit Analyst*.)

**CHART 7.4 Dow Jones Utility Average Relative Strength, 1959–1978**

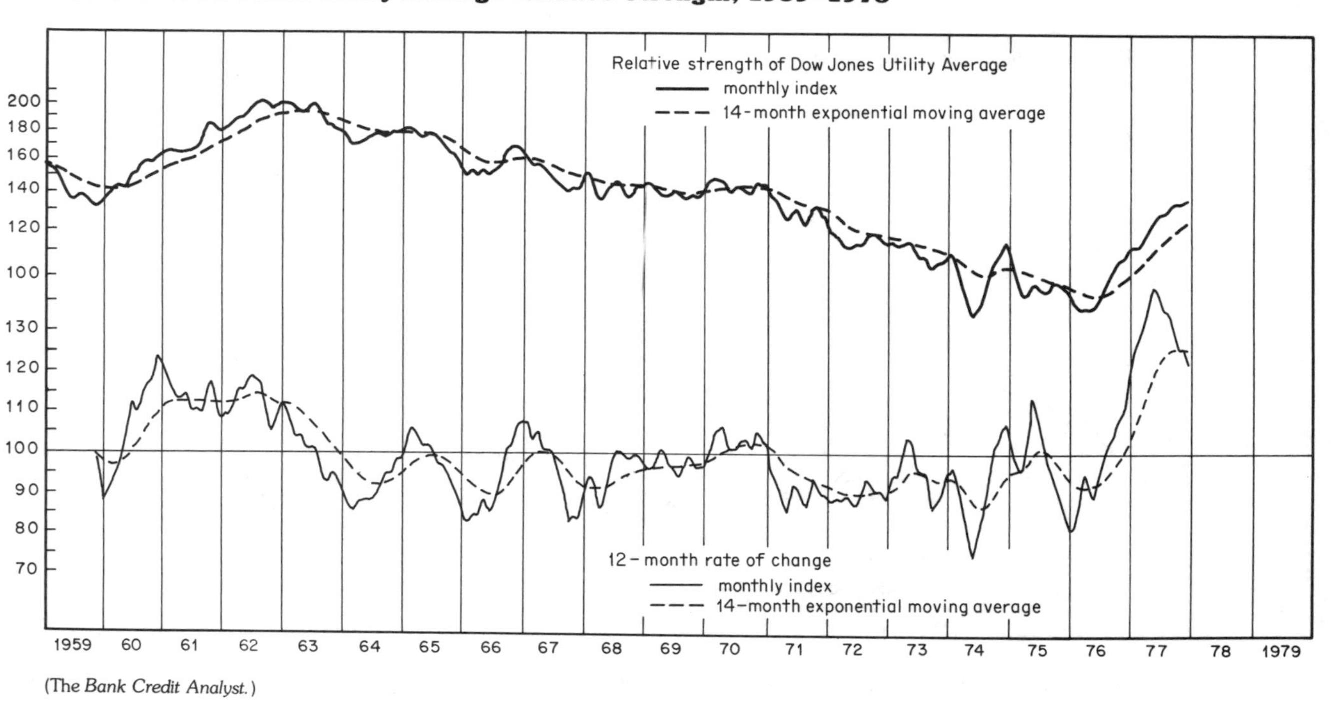


(The *Bank Credit Analyst.*)

ically. First, from a cyclical point of view the 14-month exponential moving average shown on Chart 7-3 had turned down, and the Utilities index had also crossed below this important moving average. The weakening position of the momentum index was also apparent, as the index itself had broken its uptrend established by the 1962 and early 1964 bottoms (line *AM*/*AM*). Moreover, the 14-month exponential moving average of this rate-of-change index had also turned down.

From a longer-term point of view, the most significant development was the fact that the 12-month momentum index had made a series of three declining peaks in 1961, 1963, and 1965. This showed that throughout this period the Utility Average was gradually losing its vitality. As long as the trend in the average itself continued to be positive, there was no danger of falling prices, but by late 1965 the important trendline (*BB*, Chart 7-3) joining the 1960 and 1962 bottoms was decisively penetrated on the downside. While no distribution patterns were evident at this point, the break in this important trendline accompanied by a long-term deterioration in the momentum index and relative strength indicated that the progress of the Utility Average was going to be constrained for a considerable period.

The cyclical decline that began in mid-1965 bottomed out in late 1966. The ensuing bull market was a rather weak one and essentially proved to be a 3-year period of consolidation. In effect the Utility Average, as illustrated by Chart 7-2, was forming a rectangle. By the time the 1969–1970 bear market was under way, it became apparent that the Utility Average had been forming an 8½-year head and shoulders distribution pattern between 1961 and 1969 (Chart 7-3). The completion of this pattern was confirmed when the average broke below its 1968 low in the fall of 1969.

It is worth noting that throughout the 1966–1969 period when the rectangle was being formed, the long-term trend of the Utility Average's relative strength shown in Chart 7-4 was in a sharp downtrend. The break in the secular uptrend of the average (shown on Chart 7-1), which occurred in 1967, gave an additional bearish warning that the ensuing period would be one of substantial constraint for the Utility Average. This indeed proved to be the case, as the average proceeded to lose over 50 percent of its value in the ensuing 5 years.

# PART TWO

# MARKET STRUCTURE

# 8
# PRICE

Price is the most logical starting point for any attempt to analyze the strength of the overall market structure.

There is no ideal index that represents the movement of "the market," since although the majority of stocks move together in the same direction most of the time, there is rarely a period when specific stocks or several groups of stocks are not moving contrary to the general direction of the trend. There are basically two methods of measuring the general level of stock prices. The first, known as an *unweighted index*, takes a mean average of the prices of a wide base of stocks; the second also takes an average of the prices of a number of stocks, but in this case the prices are weighted by the outstanding capitalization (i.e., number and market value of shares) of each company. The first method monitors the movement of the vast majority of listed stocks, but since the second method gives a greater weight to the larger companies, movements in a market average constructed in this way more fairly represent changes in the value of the nation's portfolios. For this reason, and because such averages can be compiled from stocks representing public participation, market leadership, and industry importance, weighted averages are usually used as the best proxy for "the market."

In addition, several price indexes have been developed which measure various segments of the market and whose interrelationship can offer useful clues about the market's overall technical condition. An earlier chapter dealt in detail with the relationship between the Dow Jones Industrial and Transportation averages, but there are many other useful indexes—such as the Utilities and unweighted indexes as well as a few bellwether stocks. These indexes are examined below in the context of their contribution to the market's overall technical structure.

## COMPOSITE MARKET INDEXES

The Dow Jones Industrial Average is the most widely followed stock market index in the world. It is constructed by totaling the prices of the 30 stocks listed in Appendix II and dividing the total by a divisor. The divisor, which is published regularly in the *Wall Street Journal* and *Barron's*, is changed from time to time due to stock splits, stock dividends, and changes in the composition of the average. Strictly speaking it is not a "composite" index, since it does not include such industries as banks, transportation, office equipment, etc. Yet the capitalization of the Dow Jones Industrials is equivalent to approximately 25 to 30 percent of the outstanding capitalization on the New York Stock Exchange (NYSE), and it has proved to be a thoroughly reliable indicator of the general movement of the market. The original reason for including a relatively small number of stocks in an average was convenience, since years ago the averages had to be laboriously calculated by hand. With the advent of the computer, the calculation of a much more comprehensive sample was possible.

One of the drawbacks of this method of construction is that if a stock increases in price and is not split, its influence on the average will become substantially greater, especially if many of the other Dow stocks are growing and splitting at the same time. In spite of this and other drawbacks, the Dow has over the years acted fairly consistently with many of the more widely capitalized market averages.

The Standard and Poor Composite, which comprises 500 stocks representing well over 90 percent of the NYSE market value, is another widely followed bellwether average. It is calculated by multiplying the price of each share by the number of shares outstanding, totaling the value of each company, and reducing the answer to an index number.

Most of the time the DJIA and the S&P 500 both move in the same direction, but there are times when a new high or low is achieved in one index but not the other. Such occasions often warn of a reversal in trend for the overall market. Generally speaking, the greater the divergence, the greater the ensuing move in the opposite direction is likely to be. Reference to Chart 8-1 shows that in late 1968 the S&P 500 reached a new all-time high, unlike the DJIA, which was not able to surpass its 1966 peak. This development helped to signal a bear market that was to wipe nearly 40 percent off the value of both averages. On the other hand, the

**CHART 8-1 Securities Research Chart Showing the Key Market Averages, 1965–1978**

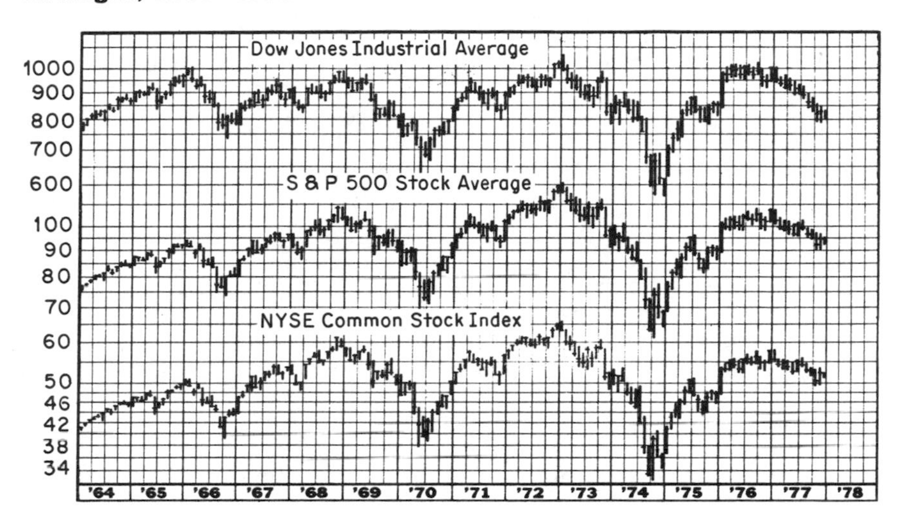


1973–1974 bear market was completed with a double bottom. In the case of the DJIA the second bottom in December 1974 was lower than the October one, yet the S&P 500 failed to confirm the new low in the DJIA. In the space of the next 2 years the DJIA rose by some 80 percent. This is also shown on Chart 8-1.

The New York Stock Exchange compiles an all-encompassing index called the NYSE Composite. In a sense it represents the ideal average, since its value is based on the capitalization of all shares on the exchange. Its movements are very similar to those of the DJIA and the S&P 500. Nevertheless, divergences between the trends of these three averages offer additional confirmation of changes in the overall technical structure.

Chart 9-5 shows a complete history of the DJIA from 1790 to 1976. In reality this is not a 200-year record of the Dow Jones Industrials, since from time to time (through mergers or takeovers, for example) new stocks have been substituted. For example, in the late nineteenth century the index was made up of 12 stocks; this was changed to 20 in 1916 and 30 in 1928. Since records of the original 12 stocks are unavailable for most of the last century, the index portrayed in Chart 9-5 comprises at least two different indexes which have been spliced together to maintain some form of continuity. Consequently, while the interpretation of the actual level of the index in these earlier years should be treated with some caution, the timing and size of the various cyclical swings can be regarded as more reliable.

The average has been plotted on ratio scale, so price movements of identical vertical distance are also of the same proportion. While the wild gyrations of the market in recent years appear substantial to those who have been investing during this period, it is apparent by looking at the chart that such movements have so far been relatively mild by historical standards.

The DJIA is the most widely followed proxy for "the market," so an analysis of the other indexes and indicators described in this and the ensuing chapters is really geared to projecting the performance of the DJIA. Many of the techniques already described in Part 1 are useful for identifying trends in the DJIA itself. The charts in Chapter 2 offered several illustrations of the use of price pattern analysis with the DJIA. Chart 8-2, which portrays the monthly swings in this average from 1929 to 1977, indicates how the reversal or cyclical bull markets can be indicated by the drawing of a simple trendline joining the bear market low with the trough of the first major reaction. This simple technique signaled the termination of nine bull markets at relatively early stages. The 1942–1946 primary uptrend would initially have been broken in 1943, since it was too steep to be maintained. The violation of the 1962–1966 line was late, but it foreshadowed an extended period of consolidation which was later confirmed by a downward penetration of the line joining the 1932 and 1942 bottoms in 1969. Finally, note how these lines, once violated, have become resistance levels to further price advance. This is especially marked for the 1942–1943 and 1962–1966 advances.

Trendlines can also be drawn for bear markets, and these are shown on Chart 8-3. Although this technique provides a reasonable indication of a reversal in trend, some degree of subjectivity and judgment is necessary in the interpretive process. For example, the relatively shallow declining trend in the 1957, 1962, and 1973–1974 bear markets resulted in some unreasonably late signals. In cases where it is obvious that a trendline is going to be "late," it is better to disregard it and fall back on another technique. In the three cases outlined above, for instance, the completion of price patterns could have been used to discern a reversal in trend.

## THE DJIA AND MOVING AVERAGES

When experimenting with a moving average from the point of view of trend determination, it is necessary first to assess the type of cycle to be

**CHART 8-2 Long-Term Dow Showing Important Bull Market Trendlines**

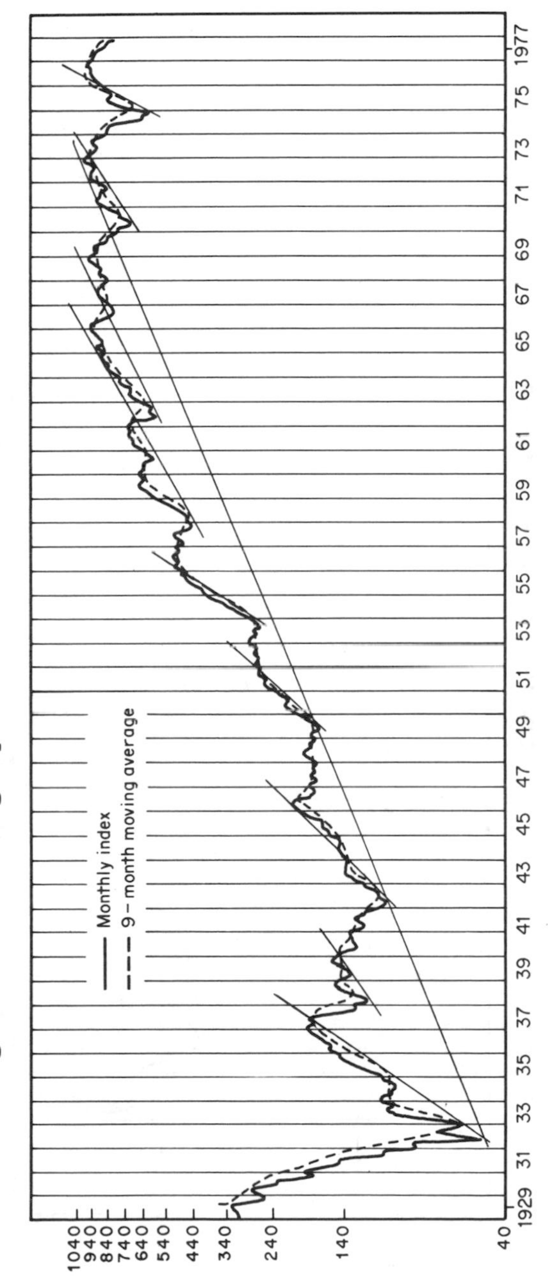

**CHART 8-3 Long-Term Dow Showing Important Bear Market Trendlines**

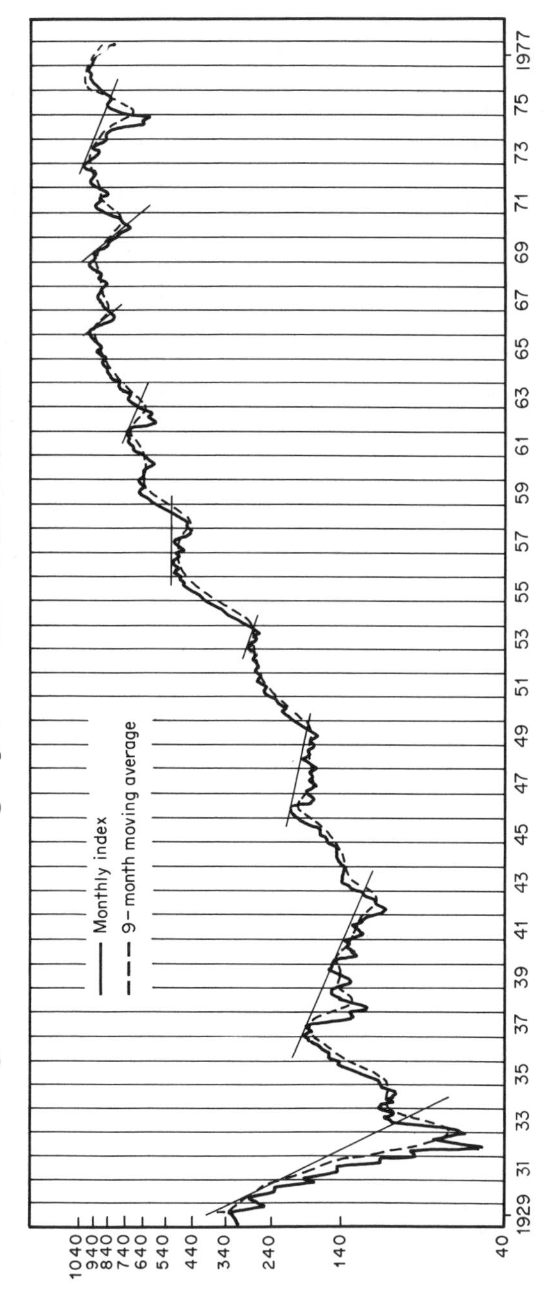

considered. As explained earlier, for example, the 4-year stock market cycle has corresponded with the United States business cycle for many decades. Since the stock market is greatly influenced by business cycle developments, this 4-year (or to place it more exactly, 41-month) cycle is of great significance in trend determination. Consequently, the choice of a moving average to detect such swings is limited to anything less than the full period, i.e., 41 months, since a moving average covering this whole time span would smooth out the complete cycle and theoretically become a straight line. In practice the moving average does fluctuate, since the cycle is rarely limited exactly to its average 41 months and varies in magnitude of price change. Through trial and error it has been found that a 9-month (or 40-week) moving average (approximately a quarter of the time span of the ideal cycle) is the most useful.

In his book *The Stock Market Indicators*, William Gordon calculated that between 1897 and 1967, 29 buy and sell signals were given for the DJIA, using a 40-week crossover as a signal.[1] The average gain for all bull signals (i.e., between the buy and sell signals) was 27 percent, and the average change from sell signals was 4 percent. For investors who had used the buy signals to purchase stocks, nine of the signals would have resulted in some losses, although none greater than 7 percent, while gains were significantly higher. By comparison similar research using a 30-week moving average, also considered by technicians to be a useful average, revealed an average gain for buy signals during this period of 19 percent.

For intermediate swings in the DJIA, crossovers of 13-week and 10-week averages have proved to be useful benchmarks, but naturally a moving average covering such a brief time span can result in many misleading whipsaws and is therefore less reliable than the 40-week average. For even shorter swings a 30-day (6-week) moving average works well, although some technicians prefer a 25-day average. Finally, the use of multiple moving averages is described in Chapter 4—in this case the 30-week/10-week relationship for the DJIA, which has been a reliable indicator.

A simpler technique for identifying intermediate trends is to use a 13-week rate of change of the DJIA in conjunction with trends in the level of the average itself. The technique used in Chart 8-5 involves the drawing of trendlines for both the weekly closing price of the Dow and its 13-week

[1] The actual rule used for buy signals was as follows: "If the 200-day (40-week) average line flattens out following a previous decline, or is advancing and the price of the stock penetrates that average line on the upside, this comprises a major buying signal." The opposite conditions were used for "sell" signals.

**CHART 8-4 Dow Jones Industrial Average versus 4-Week, 13-Week, and 52-Week Moving Averages**

This is a system of three moving averages devised by Richard Russell. Buying (and selling) "alerts" are provided when the 4-week moving average crosses above (or below) the 13-week moving average. Major signals are given when the 13-week average crosses above (or below) the 50-week average. To quote Mr. Russell: "By construction these signals have to come fairly early in a major and extended market move. Since true primary movements tend to be protracted both in duration and extent they are usually signaled in ample time and extent by the 13- and 50-week moving averages."

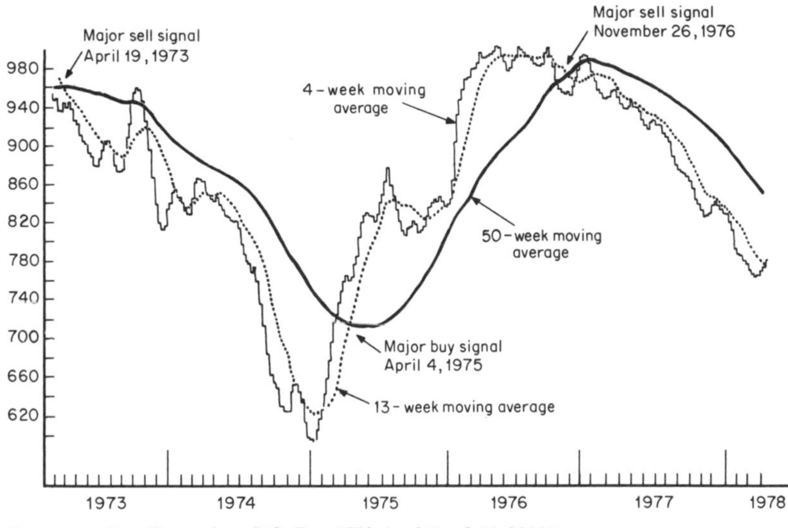


Courtesy of Dow Theory Inc., P.O. Box 1759, La Jolla, Calif. 92038

momentum. When a break in one index is confirmed by the other, a reversal in the prevailing trend usually takes place. Such signals are illustrated in the chart by the use of arrows. This type of analysis should be supported where appropriate with price pattern analysis for the DJIA and with other techniques utilizing the momentum principles described in Chapter 5. Although the method does not always work too well, whenever there are clearly definable trendlines that have been touched three or more times, the conclusions drawn from this technique are usually extremely reliable. A less subjective alternative is to use a moving-average crossover for the DJIA and either a change in direction of a smoothed rate of change for the momentum index or its penetration of a predetermined level.

**CHART 8-5 Dow Jones Industrial Average versus 13-Week Momentum**

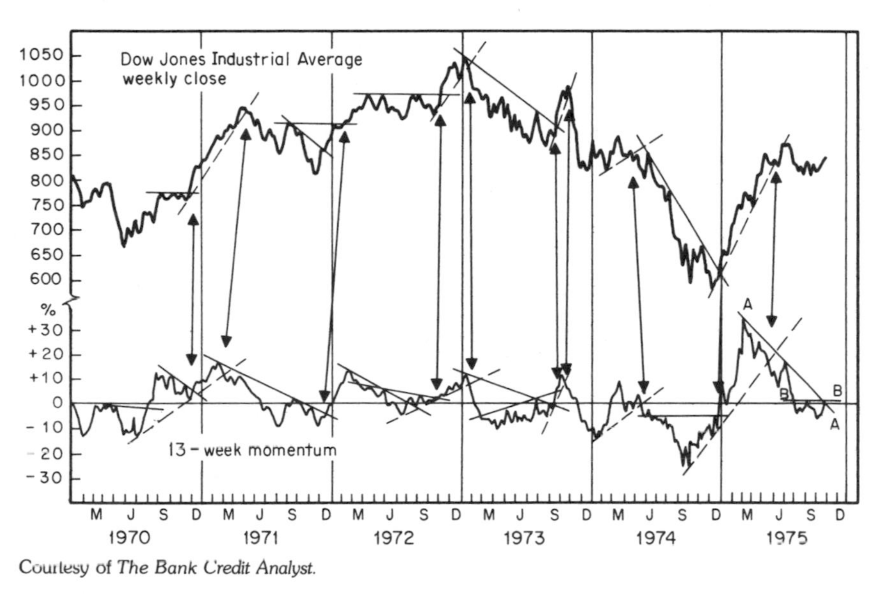


Courtesy of *The Bank Credit Analyst.*

Such a combination is shown on Chart 8-6, which incorporates a 14-month exponential moving average for the DJIA and a 14-month exponential moving average for a 12-month rate-of-change index.

## THE DOW JONES TRANSPORTATION AVERAGE

The Transportation Average was formerly known as the Rail Average before it incorporated other forms of transportation, such as airline and trucking stocks. Originally the Rail Average was devised to reflect the transportation of goods. In the latter part of the nineteenth century and early part of the twentieth century rail was the dominant form of transportation, so an average comprised solely of rails represented a good proxy for transportation stocks. In 1970 the Rail Average was expanded to embrace other transportation segments, and the index was renamed the Transportation Average.

The Transportation Average is basically affected by two factors, volume of business and changes in interest rates. First, as a business recovery gets under way, inventories are low and raw materials are needed to initiate production. Transportation volume picks up, and investors anticipating such a trend bid up the price of transportation shares. Similarly, at

**CHART 8-6 Dow Jones Industrial Average versus 12-Month Rate of Change**

Most of the time a reversal in the 14-month EMA of the DJIA, when confirmed by a similar exponential smoothing for the 12-month rate of change, results in a reliable signal that the new trend will be perpetuated for some time. The experience in 1971, when both moving averages turned down for a short period, is a fine reminder, however, that even the best indicators can give misleading signals at times.

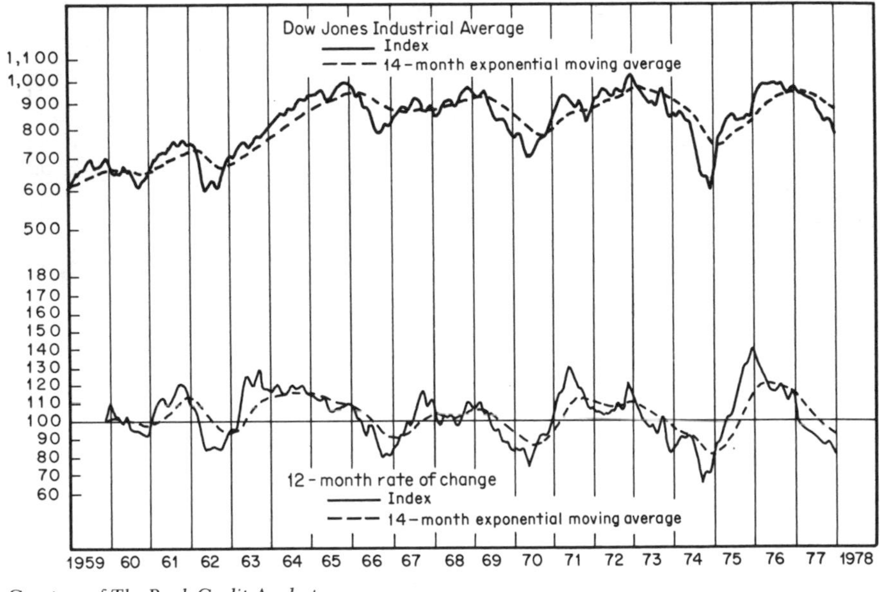


Courtesy of *The Bank Credit Analyst*.

business peaks companies typically overbuild their stocks, and when sales start to fall they reduce their requirements for raw materials. Under such circumstances transportation volume falls sharply, and the stock market reacts accordingly. Secondly, transport companies tend to be relatively more heavily financed with debt than industrials. Because of the leverage of this heavy debt structure, their earnings are also more sensitive to changes in interest rates and business conditions than those of most industrial companies. As a result, the Transportation Index quite often leads the Industrials at important juncture points.

The importance of the Dow theory rule concerning confirmation now takes on added weight, since a move by the producer stocks (the industrials) really has to be associated with an increased volume of transportation to be valid, that is, with a comparable move by the Transports. Similarly, increased business for the transport stocks is likely to be of temporary significance if the industrial companies fail to follow through with a rise in sales and production levels. The longer-term cycles of the Transportation and Industrial Averages are more or less the same as a result of their close association with business conditions. The techniques and choice of time spans for moving averages, rates of change, etc. are therefore similar to those described above for the Industrials.

One principle which is not normally used for the Industrials but can be applied to the Transports is that of relative strength. This technique is particularly useful during periods of nonconfirmation between the two averages, when RS can offer a useful clue as to how the discrepancy will be resolved. One such example occurred in early 1974 and is shown in Chart 8-7. The DJIA made an impressive advance in January 1974, reacted in early February, then rose strongly to a new recovery high in March. The relative strength for the Transports, which had been extremely strong up to this point, topped out in early March, and by the time the Industrials made their second peak the Transports' RS had crossed below its 13-week

**CHART 8-7 Transports and Relative Strength**

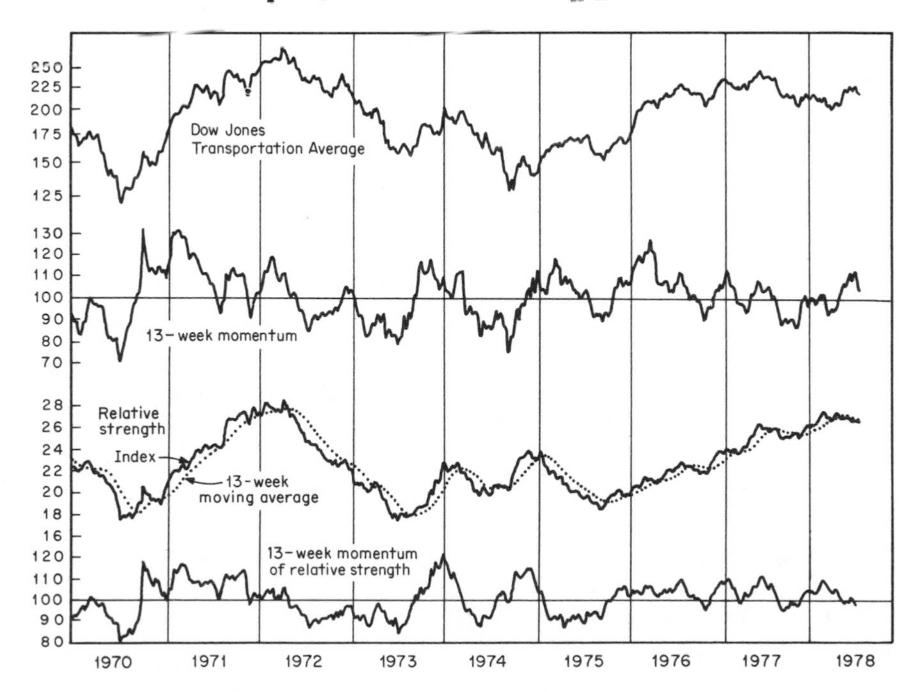

moving average. The 13-week momentum index for the RS, having peaked in January, was also falling sharply during this period. The conclusion that would have been drawn from an analysis of the RS trend at that time would have placed a low probability on the Transports' confirming the March recovery high in the Industrials.

## THE UTILITY AVERAGE

The Dow Jones Utility Average comprises 15 utility stocks drawn from electric utilities, gas pipelines, telephone companies, etc. This average has historically proved to be one of the most reliable barometers of the industrials, because utility stocks are extremely sensitive to changes in interest rates. Interest rate changes are important to utility stocks from two aspects. First, because utility companies require substantial amounts of capital, they are usually highly financed with debt relative to equity. As interest rates rise, the cost of renewing existing debt and raising additional money increases, so that profits are reduced. When interest rates fall, these conditions are reversed and profits rise. Second, utility companies generally pay out most of their earnings in the form of dividends, so that utility stocks are normally bought just as much for their yield as for their potential capital gain. When interest rates rise, bonds—which are also bought for their yield—fall in price, rendering them relatively more attractive than utilities. As a result, investors are tempted to sell utility stocks and buy bonds. When interest rates fall, the money returns once again to utility stocks, which then rise in price.

Since changes in the trend of interest rates usually begin before those of the stock market, the Utility Average more often than not leads the DJIA at both market tops and market bottoms.

Generally speaking, when the Utility Average flattens out after an advance or moves down while the Industrials continue to advance, or when the Utilities, after a decline, refuse to confirm a new low in the Industrials, this is usually a sign of an imminent change in trend for the Industrials. Thus at the 1937, 1946, 1953, 1966, 1968, and 1973 bull market peaks the Utilities led the Industrials. Conversely, at the 1942, 1949, 1953, 1962, 1966, and 1974 bottoms the Utilities made their bear market lows ahead of the Industrials. At most major juncture points the Utilities coincided with the Industrials, and occasionally, such as at the 1970 bottom and the 1976 top, the Utilities lagged.

## NYSE FINANCIAL INDEX

The New York Stock Exchange publishes a series of indexes in addition to the NYSE Composite. They are the NYSE Industrials, the NYSE

**CHART 8-8 Various Measures of the NYSE Financial Index**

This chart shows how the drawing of simple trendlines when used in conjunction with relative strength analysis can offer useful clues as to reversals in trend of a price index—in this case the daily average of the NYSE Financial Index.

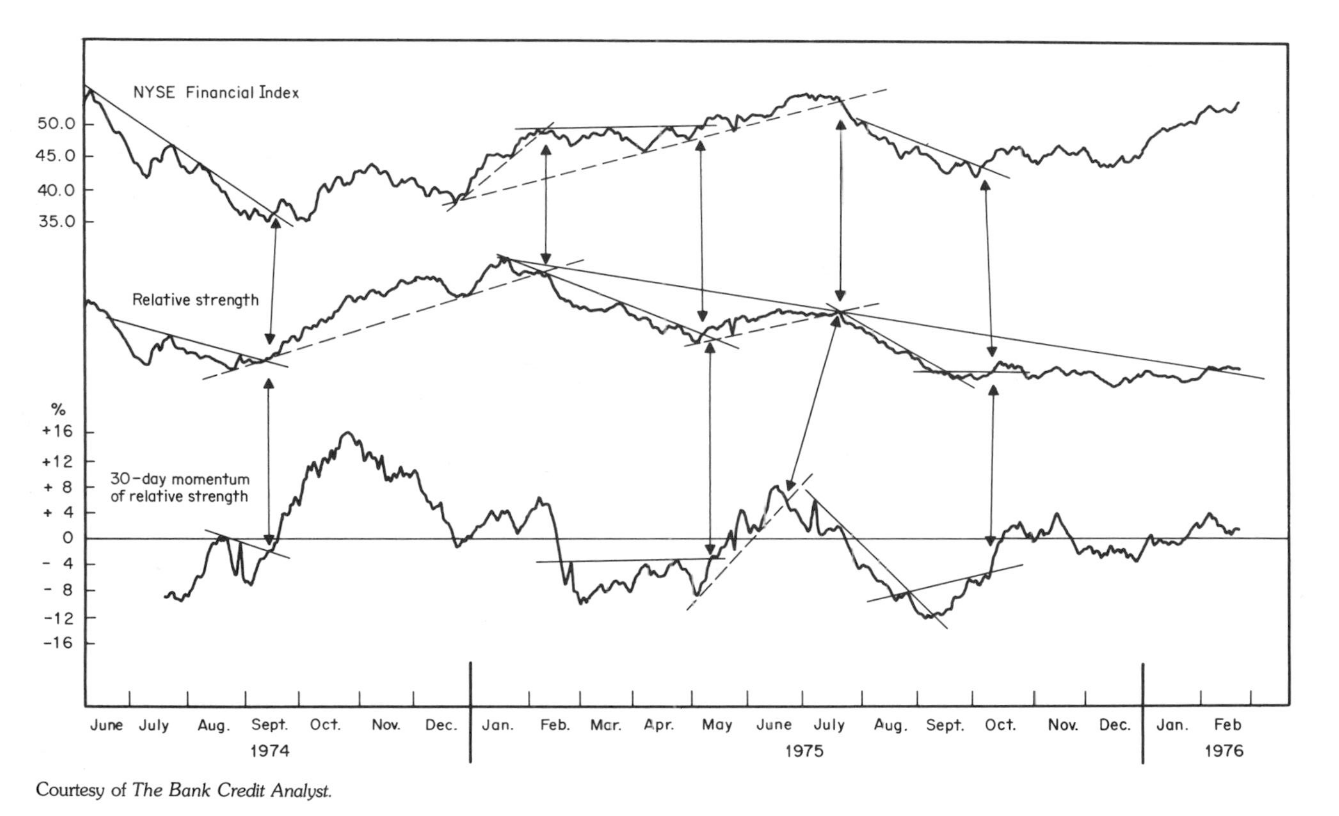


Courtesy of *The Bank Credit Analyst.*

Transportation, and the NYSE Financial, which are all constructed by weighting a series of component stocks and reducing the total to an index. The Financial Index has been singled out since it, too, is a very interest-sensitive indicator; this is because it is constructed using organizations such as banks, savings and loan institutions, and insurance companies whose profits are greatly influenced by changes in interest rates. Unfortunately records of this NYSE index only extend back to the 1960s, so its characteristics cannot really be put to the test with some of the other averages, where the records extend over much longer periods. The Financial Index is shown in Chart 8-8.

## THE UNWEIGHTED INDEXES

An unweighted index is calculated by simply adding up the prices of a number of stocks and dividing the total by that number. The resulting average is therefore one weighted by price rather than by capitalization.

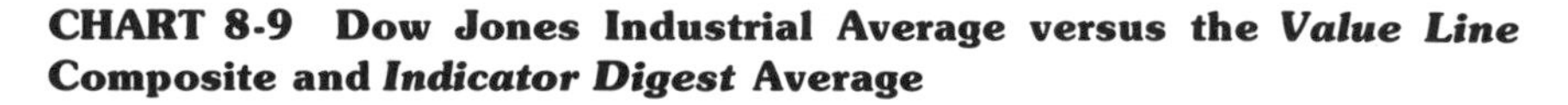

**CHART 8-9 Dow Jones Industrial Average versus the *Value Line* Composite and *Indicator Digest* Average**

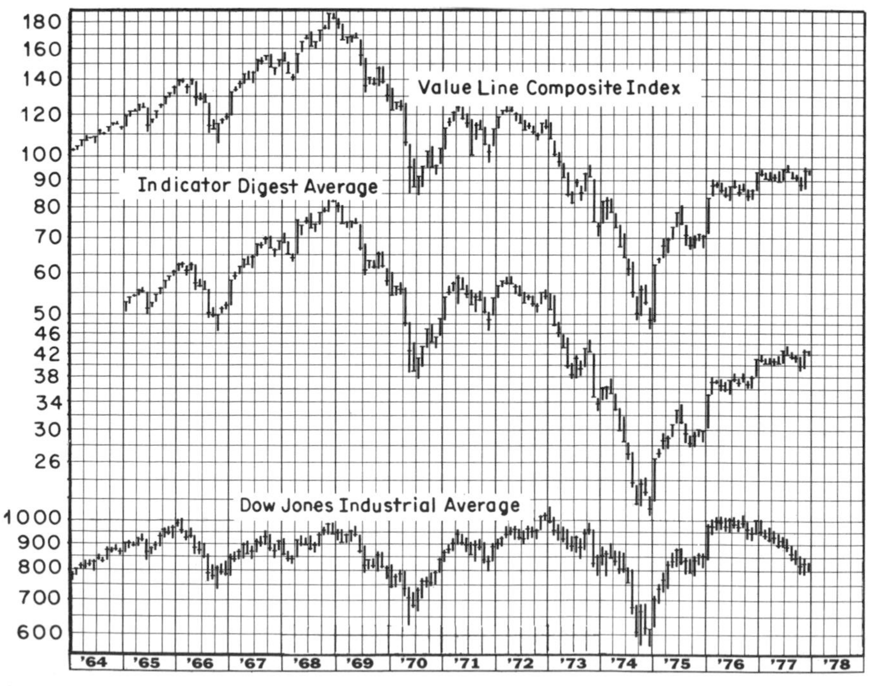


Courtesy of Securities Research.

The 15-year history of these indexes is relatively short because of the laborious task of compilation prior to the use of the computer. The two most widely followed are the *Value Line 1500* and the *Indicator Digest* unweighted index. These unweighted indexes are shown in Chart 8-9. Since their construction is based on more or less the same stocks, their performance over the years has been similar.

Unweighted indexes are useful since they closely represent the price of the "average" stock often found in individual portfolios—as opposed to the blue chips, to which institutional investment is more oriented. Unweighted indexes are also helpful in gaining an understanding of the market's technical structure, since they have a tendency to lead the market (i.e., the DJIA) at market tops (but not bottoms). Only once, in 1977, have they lagged. The performance of these unweighted indicators is almost identical to that of the weekly advance/decline line (see Chart 11-1). The relationship between the broad market and the DJIA is discussed in Chapter 11.

**CHART 8-10*a* Market Profile of General Motors Corporation, 1924–1935, 1948–1975**

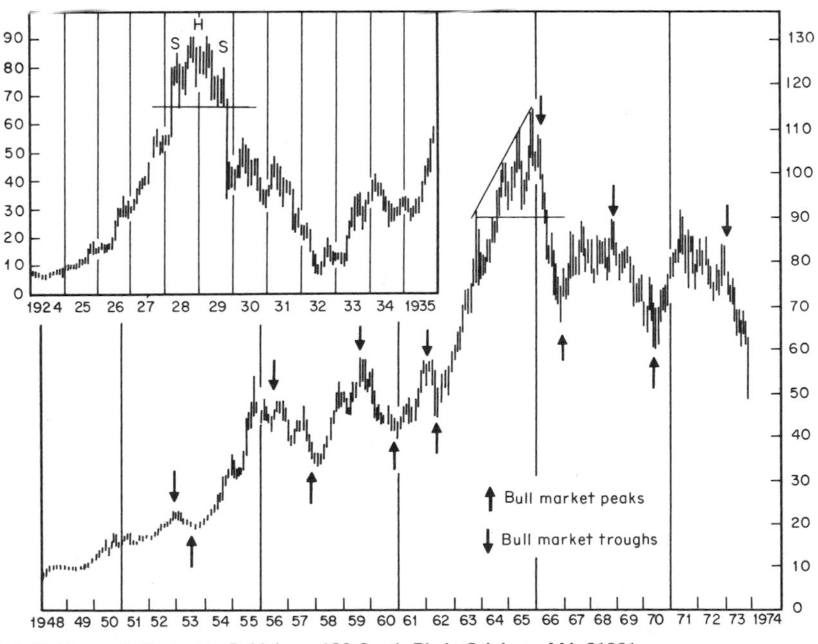


M. C. Horsey & Company, Publishers, 120 South Blvd., Salisbury, Md. 21801.

**CHART 8-10*b* Dow Jones Industrial Average versus General Motors Corporation Stocks, 1966–1978**

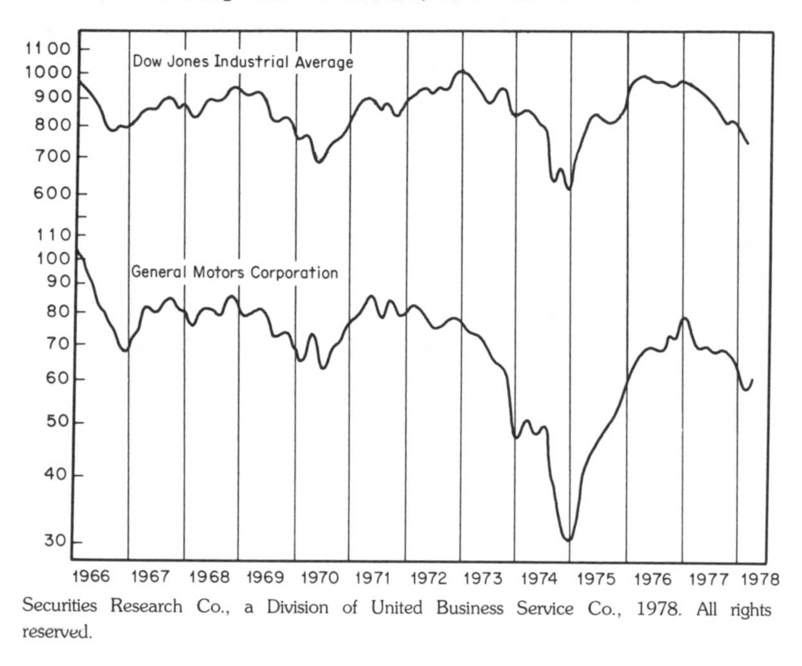




## GENERAL MOTORS

It has been claimed that "what is good for General Motors is good for America," and the record for the last 50 years or so certainly bears out this statement as far as the stock market is concerned, for GM is a bellwether stock par excellence.

General Motors has hundreds and thousands of employees and over one million shareholders, and its business is very sensitive to credit conditions. It is the largest auto manufacturer in the country, and one in six of American jobs is either directly or indirectly dependent on the auto industry.

As a result, GM tends to rise and fall in concert with the trend of the DJIA or S&P 500 most of the time. At market tops GM usually leads the market, so that a new high in the DJIA which is unconfirmed by GM is a warning signal of a reversal in trend. On the other hand, GM is not so helpful at market bottoms since it usually lags the market. The long-term chart of GM (Chart 8-10*a*) shows this quite clearly. It is also worth noting the huge head and shoulders formed between 1928 and 1929 as well as the right-angled broadening top between 1964 and 1966. The completion of both these long-term distribution patterns led to substantial declines in the stock.

Proponents of GM have developed a useful principle. Known as the 4-month rule (although some prefer a 19- to 21-week rule), it states that if in a bull market General Motors fails to make a new high within 4 calendar months of its previous peak, the bullish trend of the market has reversed or is just about to do so. Similarly, during a market decline if GM fails to make a new low within 4 months of its previous trough, a reversal in the downward trend of the markct has already taken place or is just about to occur. The term "month" refers to a calendar month, e.g., February 27 to June 27 or March 31 to July 31, so the period would be regarded as 4 months so far as the GM rule is concerned. The GM rule is not infallible, but its record is extremely good.

# 9

# TIME

Time is represented on the horizontal axis of most technical charts, so it is normally used in conjunction with the other three dimensions of psychology involved in determining trends in the stock market which are measured on the vertical axis. On the other hand, time is often assessed independently through the analysis of cycles.

The previous discussions about the importance of time were related to the fact that the significance of a reversal in a stock market trend depended on the length of time needed for the completion of distribution or accumulation. In other words, the longer the period, the greater the ensuing move in the opposite direction is likely to be. Removal of the speculative excesses of a trend will require a commensurately large corrective movement, just as the discipline of a long period of accumulation provides a sound base from which a substantial and lengthy advance can take place. Although the very long 8-year bull market between 1921 and 1929 was interrupted from time to time with corrective reactions, the substantial increase in stock prices during this period resulted in the development of a considerable amount of excess confidence which could only be erased by a sharp and lengthy decline. Similarly, the 1966 stock market peak, which was preceded by 24 years of basically rising prices, was followed by a long period of consolidation involving widely swinging stock prices. When adjusted for inflation, stock prices peaked in 1965 and have since undergone an extremely severe bear market comparable to the 1929–1932 debacle. This is shown in Chart A3-2 in Appendix 3.

Time, therefore, is concerned with *adjustment*, for the longer a trend takes to complete, the greater is the psychological acceptance of that trend, and the greater will be the necessity for stock prices to move in the opposite direction and adjust accordingly. In a rising market investors become accustomed to rising prices, so that each reaction is reviewed as temporary. When the trend finally has reversed and the first bear market rally takes place, the majority are convinced that this too is a temporary reaction and that the bull trend is being renewed. As prices work their way lower in a bear market, so the adjustment becomes more neutral as the majority of investors forsake their expectations of a rising market and look for prices to move sideways for a time. Finally, the adjustment pendulum switches completely to the other (bearish) side as investors watch prices slip even farther and then become overly pessimistic. At this point sufficient time and downside price action have elapsed to complete this adjustment process, and stocks are then in a position to embark on a new bull market.

Although time has been viewed here in an emotional context in the sense that it is required for investors to adjust to unrealized expectations, time is also deeply bound up with the business cycle. This is because a strong and lengthy recovery such as that between 1921 and 1929 breeds confidence not only among investors but also among businessmen, who tend to become inefficient and overextended as a result of a long period of prosperity. The ensuing contraction in business conditions needed to wipe out such distortions is thus more severe, so that stock prices suffer the double influence of (1) losing their intrinsic value due to the decline in business conditions and (2) being revalued downward from the unrealistically high levels during the period of prosperity. The reverse set of circumstances will apply following a long market decline. This idea of a reaction commensurate with the previous action is known as the *principle of proportionality*.

Measuring time as an independent variable is a complicated process, since prices move in periodic fluctuations known as cycles. These cycles can operate for periods ranging from a few days to many decades. At any given moment a number of these cycles are operating simultaneously, and since they are exerting different forces at different times, the interaction of their changing relationships has the effect of distorting the timing of a particular cycle. The most dominant of the larger cycles is the so-called 4-year cycle, which has a nominal or average length between troughs of 41 months. It corresponds to the business cycle which is reflected in primary bull and bear movements in the stock market. Since several other cycles are all acting at the same time but with different influences, the length of the 4-year cycle has varied from 36 to 54 months.

Cycles are represented on a chart in the form of a sine wave, as in Figure 9-1. These curves are usually based on a rate-of-change index which is then smoothed to iron out misleading fluctuations. Since no two cycles are ever identical in length, an average or *nominal* period is calculated, and this theoretical time span is used as a basis for future forecasting.

In Figure 9-2 this idealized cycle is represented by the dashed line and the actual cycle by the solid line. The arrows indicate the peaks and troughs of the idealized cycle. In actual fact the price trend will rarely reverse exactly at the theoretical point, especially at peaks, where there is often a long lead time. Nevertheless these theoretical points provide a useful guide.

There are three other important principles concerned with cycle analysis in addition to those of proportionality and nominality discussed above. The first is the *principle of commonality*, which states that cyclicality of similar duration exists in the price action of all stocks, indexes, and markets. This means that a 4-year cycle exists not only for the stock market itself but also for each individual stock and for foreign stock markets as well. The second principle, that of *variation*, states that while stocks go through similar cycles, the price magnitudes and durations of these nominal cycles will be different because of fundamental and psychological considerations. In other words, while all stocks, indexes, and markets go through a similar cycle, the timing of both their peaks and their troughs differs, and so does the size of their price fluctuations. For

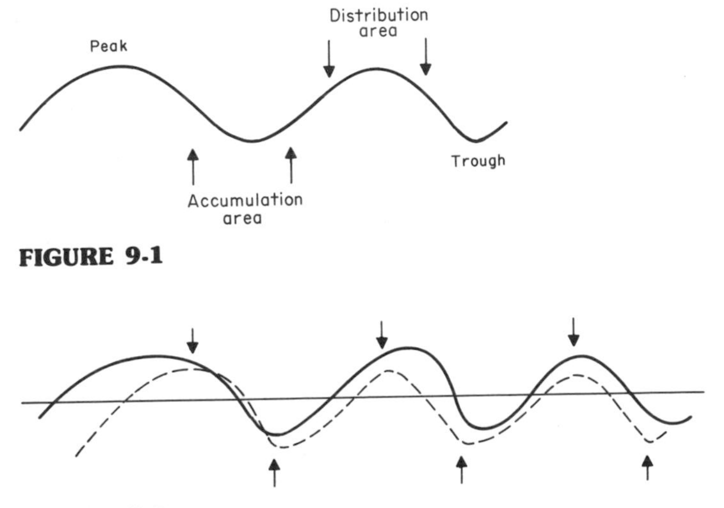


**FIGURE 9-1**

**FIGURE 9-2**

example, the interest-sensitive and cyclical (basic industry) stocks go through a similar cycle, but because interest-sensitive stocks (such as utilities) lead the market, cyclicals (such as steel groups) generally lag behind them. This is shown in Figure 9-3.

This principle is also illustrated in Chart 9-1, which shows the interaction of financial series during a typical business cycle.

Similarly, the interest-sensitive issues may rise by 80 percent from the trough to the peak of their cycle, where the cyclicals advance by only 20 percent.

Each rising part of a cycle usually consists of three stages, which correspond to the three phases described in the Dow theory. It is normal for prices to reach a new cyclical high as each stage unfolds, but sometimes prices do not. Such a development, known as a magnitude failure, is a distinct sign of weakness. It normally results from poor underlying fundamentals for the stock group or index concerned, so that the cycle in effect actually misses a beat. By the same token, strong underlying fundamentals can give rise to a fourth stage where prices have an additional upward leg. Often this final upward surge is associated with an extended period of declining interest rates. Such strong underlying conditions will normally develop when the 4-year cycle is occurring in conjunction with the peak of longer-term cycles such as the Kondratieff (50- to 54-year) and Juglar (9.2-year), which are discussed below.

The third principle is that of *summation* (or summed cyclicality, as it is sometimes known). Summation is really the combination of a number of cycles into one. The result is an idealized cycle such as that shown in Figure 9-4.

For any time series trend such as that of the stock market, there are four influences at any one time—secular, cyclical, seasonal, and random. For the purpose of analyzing primary bull and bear markets, the cyclical trend is the starting point. Specifically, this is the 4-year or Kitchin cycle. The secular influence is a very long-term one that embraces many 4-year cycles. From a stock market point of view, the longest "secular

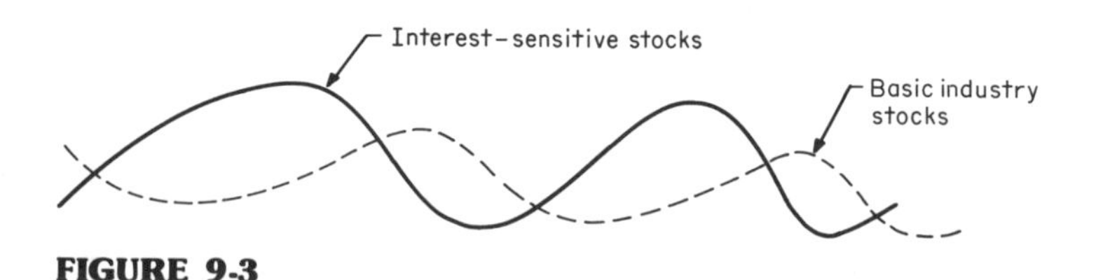


**FIGURE 9-3**

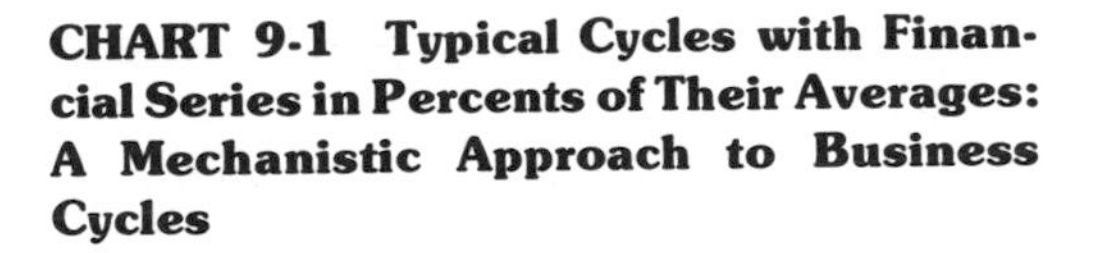

**CHART 9-1 Typical Cycles with Financial Series in Percents of Their Averages: A Mechanistic Approach to Business Cycles**

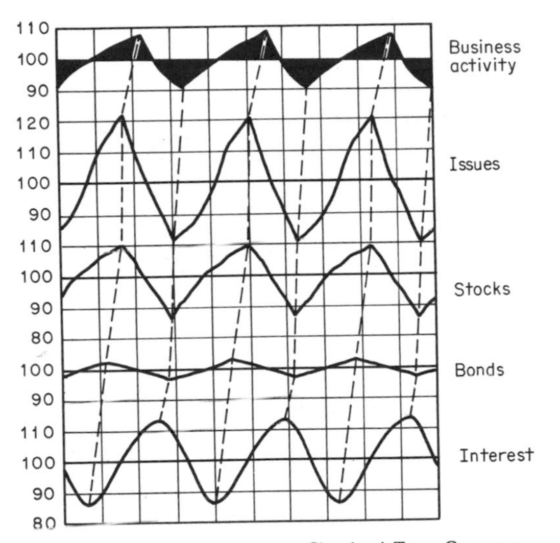


Developed by Leonard P. Ayres, Cleveland Trust Company, 1939.

cycle" is the 50- to 54-year cycle known as the Kondratieff wave (after the Russian economist Nicolai Kondratieff). Two other important cycles in excess of 4 years have also been noted, namely the 9.2-year and 18$\frac{1}{3}$-year cycles. These time periods are not long enough to be defined as secular, but nevertheless they are observable and have an influence on the duration and timing of the 4-year cyclical trend.

Chart 9-2*a*, reproduced from Joseph Schumpeter's book,[1] combines the effect of three observable business cycles into one curve. In effect, it shows the summation principle using three longer-term cycles: the 50- to

**FIGURE 9-4**

[1] Joseph Schumpeter, *Business Cycles*, McGraw-Hill, New York, 1939.

**CHART 9-2a Schumpeter's Model of the Nineteenth-Century Business Cycle**

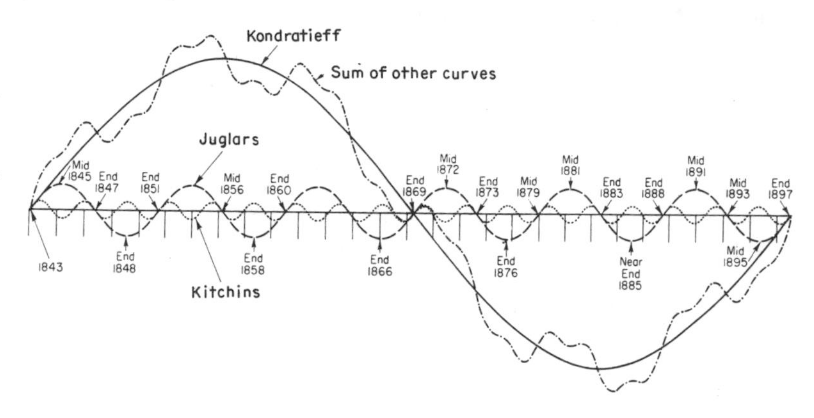


**CHART 9-2b The Twentieth-Century Business Cycle and Crisis Points**
(Calculated path)

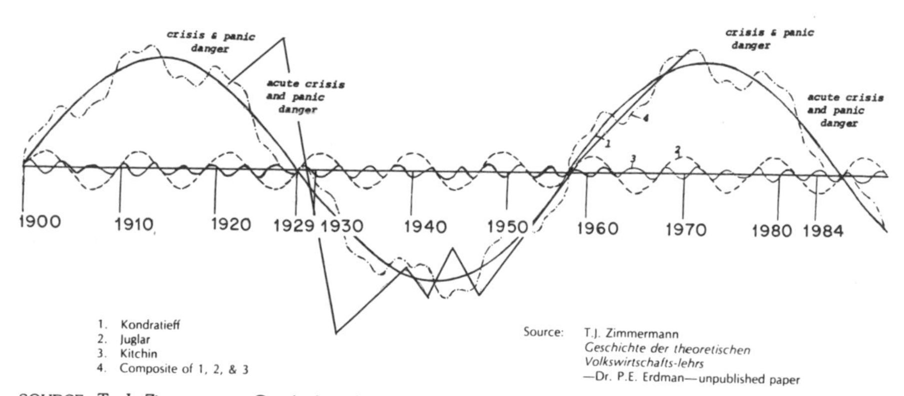


SOURCE: T. J. Zimmerman, *Geschichte der theoretischen Volkswirtschafts-lehrs,* in an unpublished paper by P. E. Erdman.

54-year (Kondratieff), the 9.2-year, and the 41-month (Kitchin). The model is not intended to be an exact prediction of business conditions and stock prices, but rather to indicate the interaction of the shorter cycles with the longer ones. By comparing this model with that in Chart 9-5 it can be seen that the long upmove dating from about 1942 to about 1974 was associated with rising stock prices interrupted by relatively mild cyclical corrections. Between 1974 and 1984 the composite cycle begins its downward path, so that the model predicts lower stock markets and poor business conditions until the mid-eighties.

Since the underlying force of this model is concerned with the 54-year Kondratieff wave, that will be a good starting point.

## KONDRATIEFF WAVE

The 54-year wave is named after a little-known Russian economist who observed in 1926 that the United States had undergone three long economic waves, each lasting between 50 and 54 years.[2] It is worth noting that while only three such cycles have been recorded for the United States, E. H. Phelps Brown and Sheila Hopkins of the London School of Economics have noted a regular recurrence of 50- to 52-year cycles of prices in the United Kingdom, between 1271 and 1954, with the crest for the current cycle projected for the 1974–1978 period. The same cycle has been observed in interest rates, as shown in Chart 9-4*b*.

Kondratieff used wholesale prices as the focal point of his observations, as shown in Chart 9-3. He noted that each wave had three phases: an upwave, lasting about 20 years, a transition or plateau period of 7 to 10 years, and a downwave again of about 20 years. Kondratieff observed that each upwave was associated with rising prices, the plateau with stable prices, and the downwave with declining prices. He also noted that a war was associated with both the beginning and the end of each upwave.

At the start of a cycle, business conditions are very depressed, and since there is considerable excess capacity of plant and machinery, people prefer to save money out of fear rather than invest it. The war at the bottom of the downwave, known as the trough war (see Chart 9-4), acts as a catalyst to get the economy moving again. In view of the tremendous economic slack in the system, this war is not associated with inflation. As time progresses, each cyclical upwave becomes stronger and stronger as confidence returns and business once again reaches full production capacity. Since price inflation is almost absent, rates of interest are very low, so that credit is both abundant and cheap. During this phase businesses not only replace old plant and equipment but also invest in new capacity, thereby improving productivity and creating wealth. This up phase is usually associated with some new technological development, such as canals in the 1820s and 1830s, railroads in the mid-nineteenth century, automo-

[2] The cycle was also noted by Professor Jevons in the latter half of the nineteenth century.

**CHART 9-3 The Kondratieff Wave, Based on Annual Averages with a Ratio Scale of 1967 = 100**

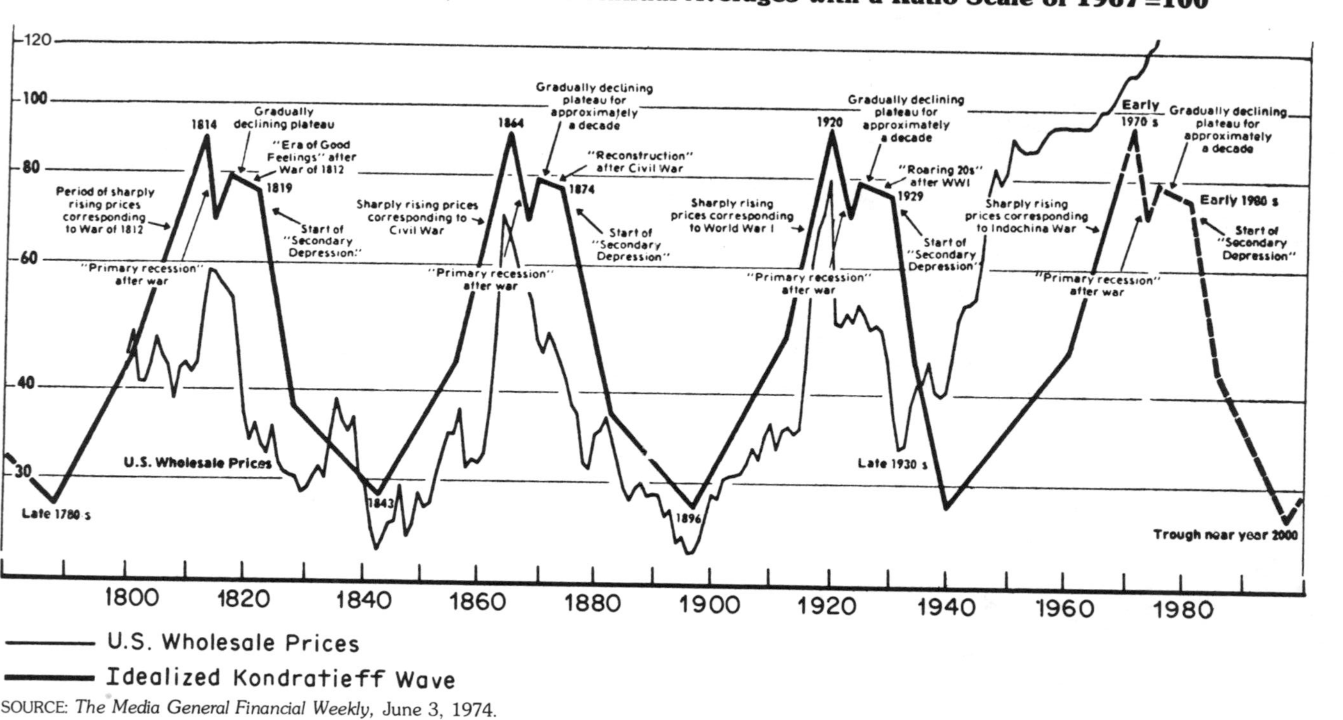

U.S. Wholesale Prices

Idealized Kondratieff Wave

SOURCE: *The Media General Financial Weekly*, June 3, 1974.

**CHART 9-4*a* Index Numbers of 1780–1925 Commodity Prices (1901–1910 = 100)**

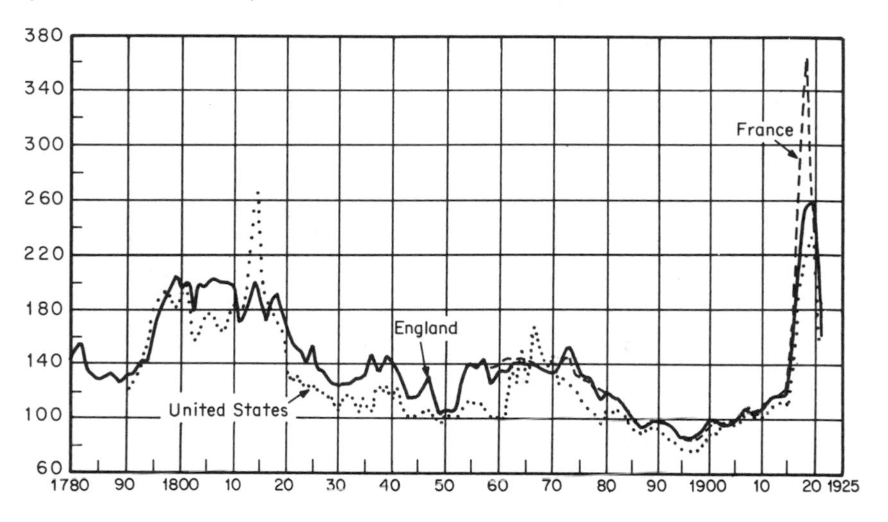


**CHART 9-4*b* Interest Rates for the United Kingdom and France, 1815–1925**

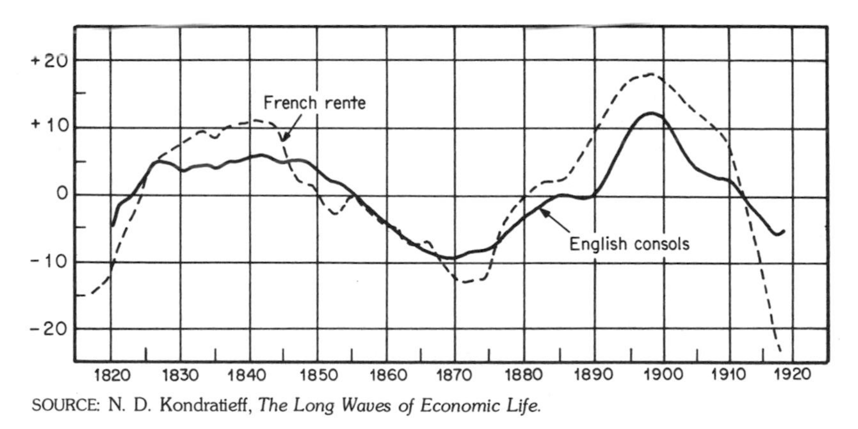


SOURCE: N. D. Kondratieff, *The Long Waves of Economic Life.*

biles in the 1920s, and electronics in the 1960s. As the upwave progresses, distortions arising from overinvestment (which was caused by the long cycle interrupted by mild recessions) begin to develop in the economy, leading to social tensions and economic instability. A common characteristic around this period, known as the peak, is a war accentuating these distortions. This is true of the peaks of 1814, 1864, and 1914.

The above description is important since it is possible to understand the cyclical movements of the market only if they are put into perspective by their position in relation to the longer-term waves (see Chart 9-5). For example, during the up phase and its associated mild recessions, it would be reasonable to expect mild and brief bear markets. During the relatively stable plateau period the possibility of a very strong bull market exists (e.g., 1820, 1860, and 1920), and in the down phase there may be a series of very sharp and devastating bear markets. The interpretation of the technical indicators should be adjusted accordingly. For example, where a normal time span for a bear market might be 12 months during the up phase, with the annual rate of monthly price change being limited to a −20 percent reading, the standards should be different during the Kondratieff down wave, when each business cycle becomes weaker and weaker. Under such conditions, bear markets would be expected to last for considerably more time and be far more devastating. A study and appreciation of the Kondratieff cycle is therefore a fine example of why stock market predictions based on the experience of only the two or three previous cycles are likely to prove unfounded. Since the Kondratieff wave has repeated only three times and each time the conditions have been different, this cycle should be used as a framework from which a better understanding of the forces at work can be derived, rather than as a basis for unqualified prediction.

## THE 18-YEAR CYCLE

Normally the amplitude of a cycle is a function of its duration, i.e., the longer the cycle the bigger the swing.

The 18-year (or, more accurately, the 18$\frac{1}{3}$-year) cycle has occurred fairly reliably in stock market prices since the beginning of the nineteenth century. Greater credibility is given to this cycle through the observation that it has occurred in several other situations, such as real estate activity, loans and discounts, and financial panics.

Chart 9-6 shows a 3-year centered moving average of common stock prices from 1840 to 1974. The use of this average helps to smooth the trend and isolate more clearly the long-term picture. The occurrence of the beginning of the 18-year cycle at major market bottoms is self-evident.

The 3-year moving average bottomed out in 1970 and is due to complete its cycle around 1988. Ideally the peak should occur halfway through 1979. However, a 1- or 2-year difference either way would still be consistent with historical behavior.

The Foundation for the Study of Cycles has spliced together the following series: 1790–1831 Bank & Insurance Companies; 1831–1854 Cleveland Trust Rail Stocks; 1854–1871 Clement-Burgess Composite Index; 1871–1897 Cowles Index of Industrial Stocks; 1897–1974 Dow Jones Industrial Averages. The shaded areas represent the plateau period in the Kondratieff cycle.

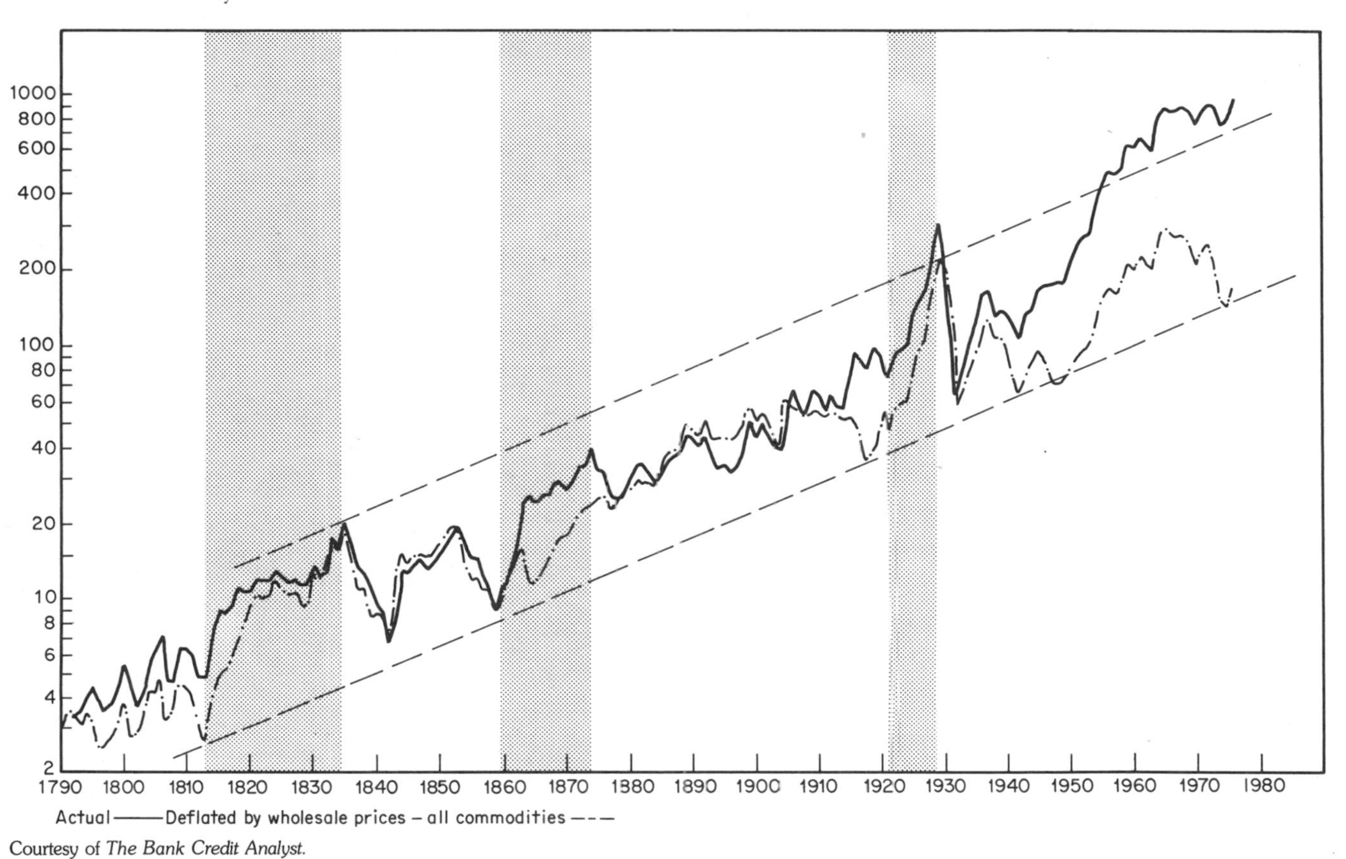


Courtesy of *The Bank Credit Analyst.*

**CHART 9-6 The 18⅓-Year Cycle of Stock Prices, 1840–1974** (3-year centered moving average)

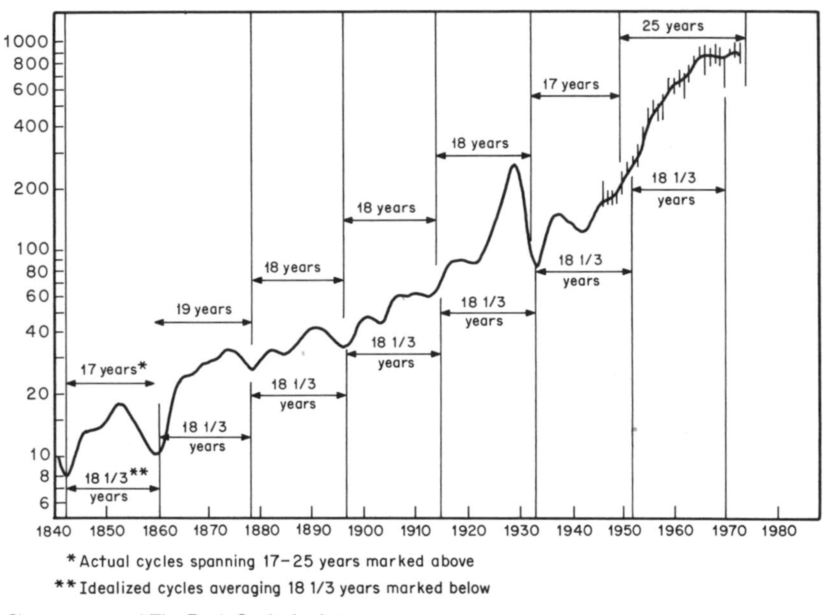


Chart courtesy of *The Bank Credit Analyst.*

While a period of 18⅓ years represents the average cycle, actual cyclical lows can vary 2 to 3 years either way. These troughs are marked on the chart above the 3-year moving average. The increasing interference in the economy resulting from the Keynesian revolution and the postwar commitment to full employment appears to have had two effects on the most recent cycle (i.e., 1952–1970). In the first place, it has stretched it out from 18 to 25 years (1949–1974), and secondly, it has prolonged the upward phase. This is especially noticeable for the 1949 low, which on a 3-year moving average hardly shows as a trough.

The 18-year cycle fits in well with the Kondratieff picture, since three such cycles form one Kondratieff wave. In the last two Kondratieffs, when the upwave part of the 18-year cycle coincided with the plateau period, an explosive bull market and only a mild correction took place. In terms of timing, the projected peak of 1979 for the 18-year cycle is consistent with the suggested peak for the Kondratieff bull market outlined above.

Since 1840, the 18-year cycle has operated fairly consistently. Except for the prolonged nature of the last cycle, there are really no grounds for suspecting that this 18-year periodicity no longer exists.

## THE 9.2-YEAR CYCLE AND THE DECENNIAL PATTERN

Chart 9-7 represents the 9.2-year cycle in stock prices from 1830–1946.[3] The dashed lines represent the ideal cycle if stock prices had reversed exactly on schedule, and the solid line portrays the actual annual average as a percentage of its 9-year moving-average trend.

The cycle has reproduced fourteen times since 1834, and according to the Bartels test of probability, it could not occur by chance more than once in 5000 times. Further evidence of the significance of this cycle is given by the fact that the 9.2-year periodicity has also been observed in other phenomena as far removed as pig iron prices and the thickness of tree rings.

One problem of using the technique portrayed in the chart is that the annual average is expressed as a percentage of a centered 9-year moving average. This means that the trend is not known until 4 years after the fact, so that there is always a 4-year lag in learning whether the 9.2-year cycle is still operating. Neverthcless, if the theoretical crest in 1965.4 is used as a base and the 9.2 years are subtracted all the way back to 1919, the peaks of the 9.2-year cycle correspond fairly closely with major stock market tops. Unfortunately the latest theoretical peak occurred in mid-1974, almost at the bottom of the 1973–1974 bear market, and represents the worst "signal" since 1947.

One interesting characteristic of the 9-year cycle that is probably of greater forecasting significance is the so-called "decennial pattern."

**CHART 9-7 The 9.2-Year Cycle in Stock Prices, 1830–1946**

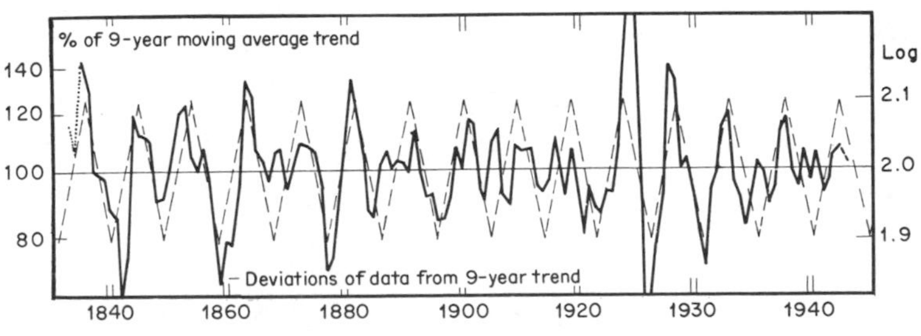


From *Cycles,* by Edward R. Dewey; courtesy of the Foundation for the Study of Cycles.

[3] Reproduced by permission of the Foundation for the Study of Cycles.

**CHART 9-8 The Decennial Patterns of Industrial Stock Prices**

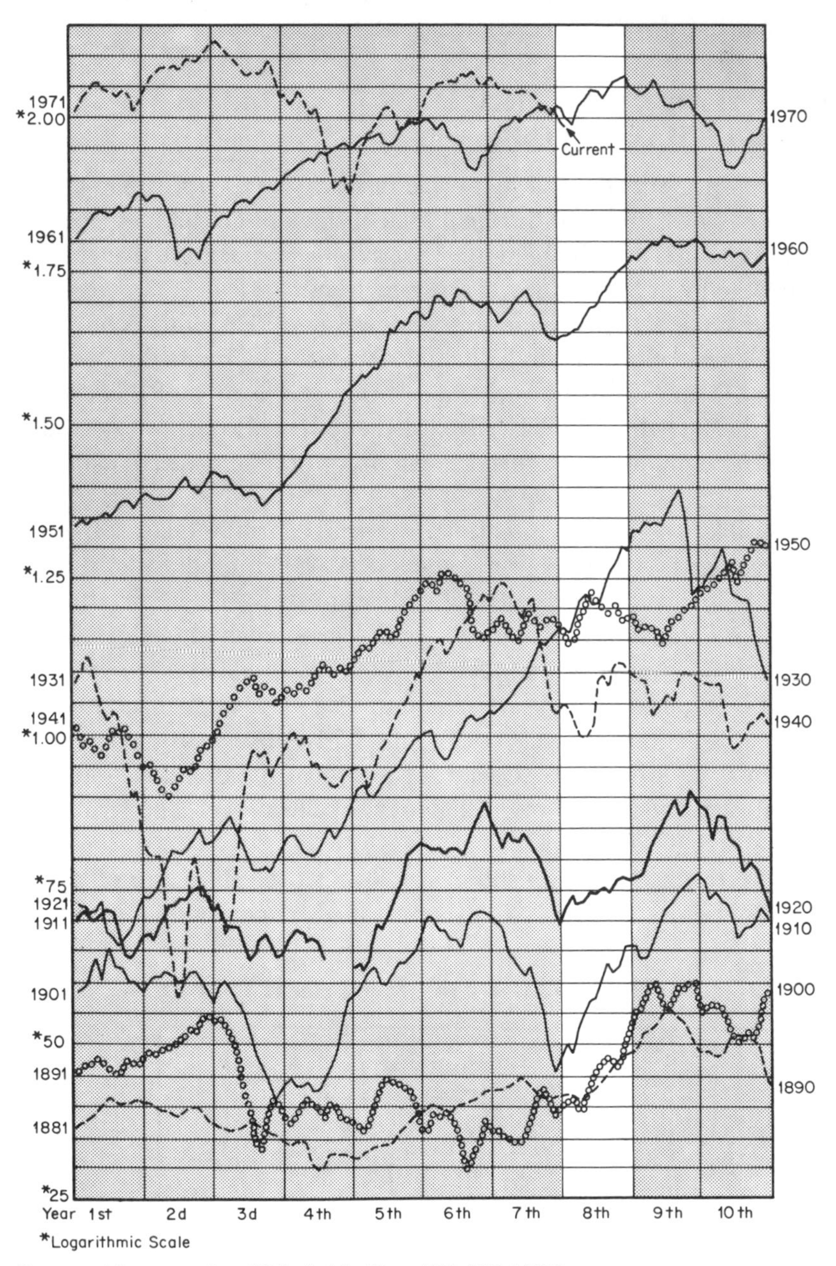


Courtesy of Anametrics Inc., 30 Rockefeller Plaza, N.Y., N.Y. 10020.

**CHART 9.9 Average of Nine Decades, 1881–1970**

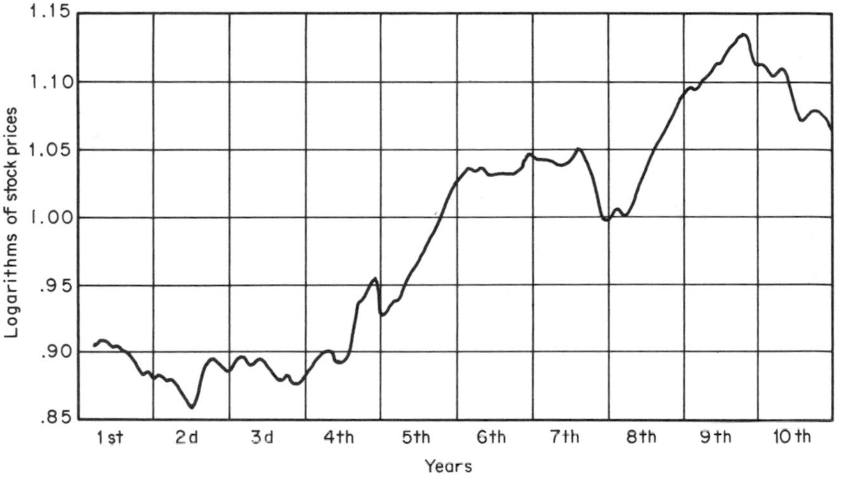


Courtesy of Anametrics Inc., 30 Rockefeller Plaza, N.Y., N.Y. 10020.

This pattern was first noted by Edgar Lawrence Smith, who observed that when the stock market was divided into 10-year segments and these segments were charted above each other, a comparison showed that years ending in the same digit appeared to have similar characteristics.[4] This is represented in Chart 9-8. Thus, years ending in a 5 (1965, 1975, etc.), an 8, and a 9 tend to be consistently good years for investing, while the seventh and tenth years often show weakness. These characteristics are better illustrated in Chart 9-9, which shows the average of the nine decades from 1881–1970.

Smith could find no explanation for this repeating cycle, although in a later book he tried to connect it with human response to weather and sunspot changes.[5]

## THE 41-MONTH (4-YEAR) CYCLE

One interesting observation made by Smith was that the decennial pattern appeared to "contain three separate cycles in a decade, each one lasting for approximately forty months." This is most interesting, as it ties in with the so-called 4-year cycle of stock prices. More precisely, the 4-year

[4] Edgar Lawrence Smith, *Tides in the Affairs of Men: An Approach to the Appraisal of Economic Change*, Macmillan, Toronto, 1939.

[5] *Common Stocks and Business Cycles*, William Frederick Press, New York, 1959.

## CHART 9-10*a* The 41-Month Rhythm in Stock Prices, 1868–1945

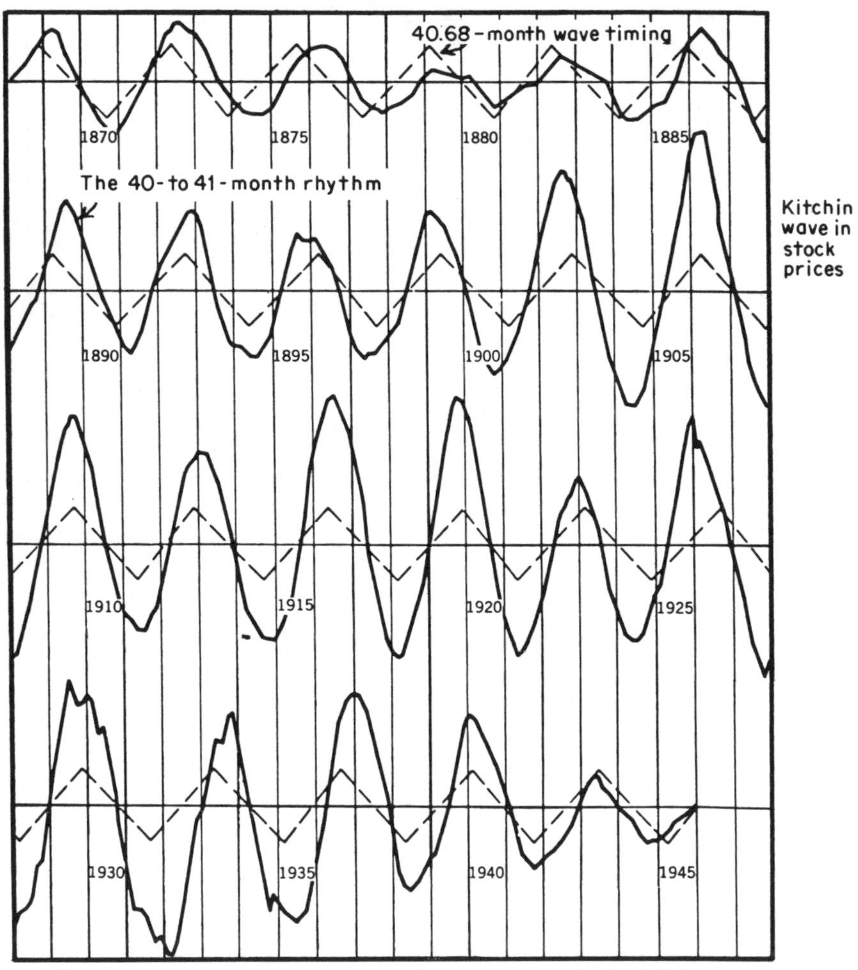

From *Cycles,* by Edward R. Dewey; courtesy of the Foundation for the Study of Cycles.

## CHART 9-10*b* The "Reversed" Cycle from 1946

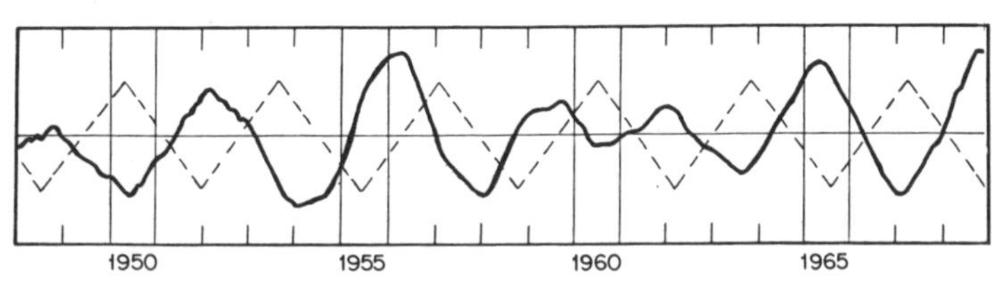


Courtesy of the Foundation for the Study of Cycles.

cycle is a 40.68-month (41-month) cycle. It has been present in stock prices since 1871. Around 1923 Professor Joseph Kitchin was also able to show a cycle of 41 months in bank clearings, wholesale prices, and interest rates in the United States and United Kingdom. This cycle has since carried his name.

The Kitchin cycle applied to stock prices is illustrated in Chart 9-10. Between 1871 and 1946 it has repeated twenty-two times with almost uncanny consistency. Then in 1946, as Edward Dewey describes it, "Almost as if some giant hand had reached down and pushed it, the cycle stumbled, and by the time it had regained its equilibrium it was marching completely out of step from the ideal cadence it had maintained for so many years."[6]

Not only did the cycle reverse, it also extended to an average period of about 50 months, as shown in Chart 9-11*a*. The average of the last six bull phases since 1949 has been 28.7 months, while the corresponding bear phase has lasted 22.1 months. A similar situation applies to the United Kingdom (see Chart 9-11*b*).

This reversal and extension in the Kitchin wave is a fine example of how a cycle that has appeared for a long time to be working consistently can suddenly, for no apparent reason, become totally distorted. Once again, the fact that a particular indicator or cycle has operated successfully in the past is not in itself a guarantee that it will continue to do so, although probability may favor it.

**CHART 9-11*a* Four-Year Cycle in the Postwar Period for the United States**

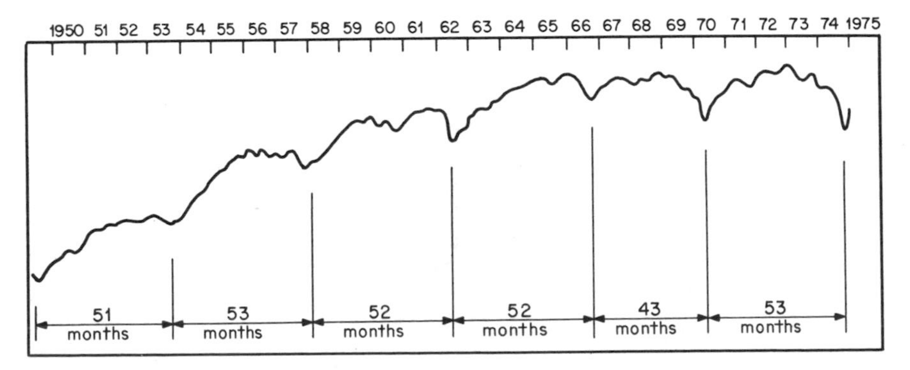


[6] Edward R. Dewey, *Cycles: The Mysterious Forces That Trigger Events*, Hawthorne Books, New York, 1971.

**CHART 9.11*b*  Four-Year Cycle in the Postwar Period for the United Kingdom**

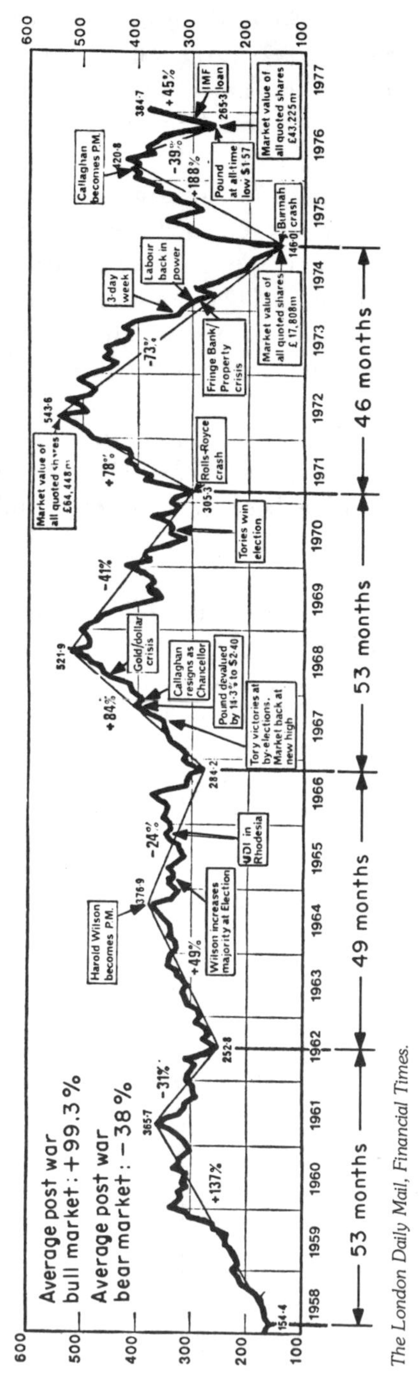


*The London Daily Mail, Financial Times.*

**CHART 9-12 Market Probability Chart: The Chances of the Market Rising on Any Trading Day of the Year** (Based on the number of times the market rose on a particular trading day* during the period May 1952–April 1971)

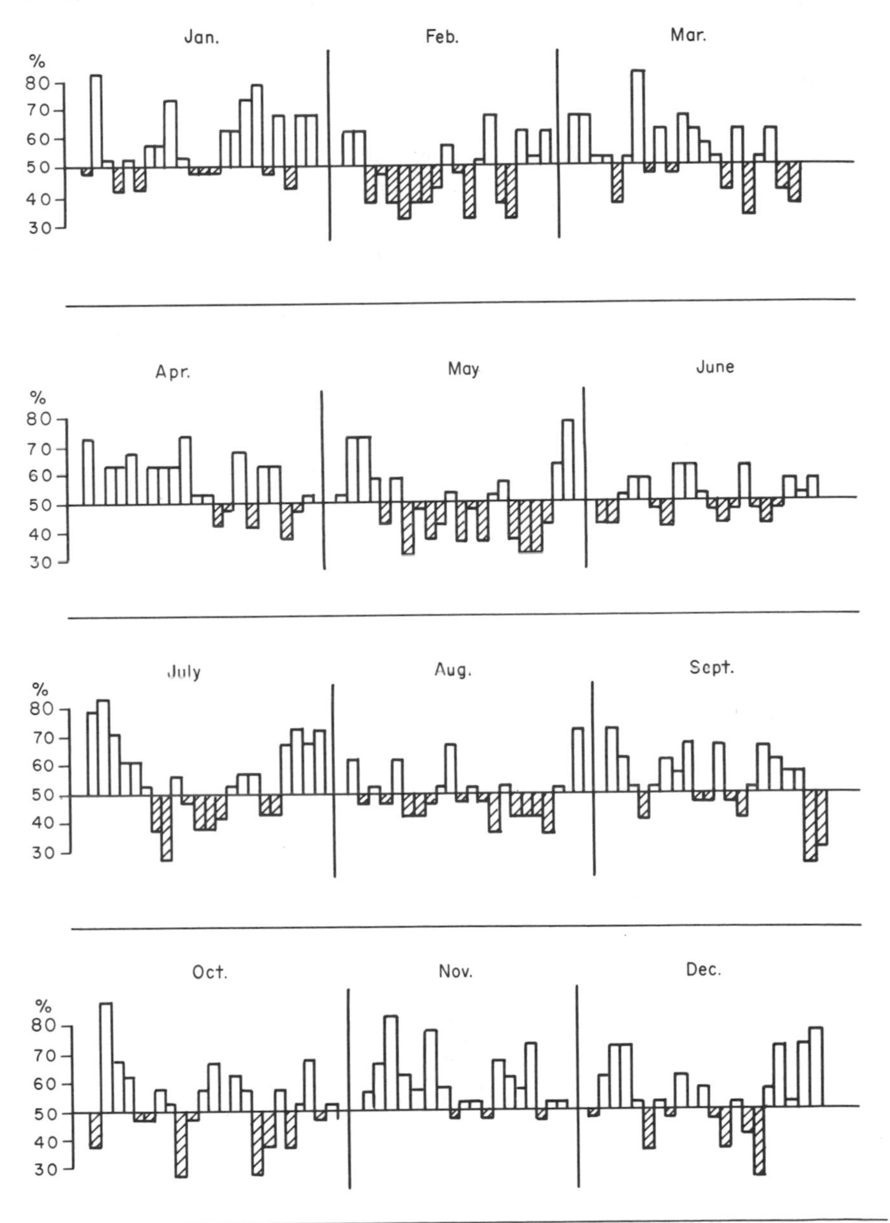


* Shows the usual number of trading days in each month, Saturdays, Sundays, and holidays excluded. *Stock Traders Almanac,* Hirsch Organization, 6 Deer Trail, Old Tappan, N.J.

## SEASONAL PATTERN

There is a distinct seasonal pattern of stock prices that tends to repeat year after year. Stocks seem to have a winter decline, spring rise, late-second-quarter decline, summary rally, and fall decline. The year end witnesses a rally which usually extends into January.

Apart from seasonal changes in the weather which affect economic activity and investor psychology, there are some seasonal patterns in financial activities as well. For example, July and January are heavy months for dividend disbursement. Retail trade around the year-end Christmas period is the strongest of the year, and so on.

Chart 9-12 represents the seasonal tendency of the stock market to rise in any given month. The probabilities were calculated over the 73-year period from 1900–1973. As can be seen, January is normally the strongest month of the year and May the weakest. All movements are relative, since a month with a strong tendency will be accentuated in a bull market and vice versa.

# 10
# VOLUME

Knowledge of the principle that volume goes with the price trend is a useful confirmation of price action, but the study of volume is even more helpful since volume often leads the trend of prices, thereby offering an advance warning of a potential price trend reversal. Before examining the various methods of volume measurement, the general observations made earlier will be briefly reviewed.

1. A price rise accompanied by expanding volume is a normal market characteristic and has no implications so far as a potential trend reversal is concerned.
2. A rally which reaches a new (price) high on expanding volume but whose overall level of activity is lower than the previous rally is suspect and warns of a potential trend reversal (see Figure 10-1, example *a*).
3. A rally which develops on contracting volume is suspect and warns of a potential trend reversal in price (see Figure 10-1, example *b*).
4. Sometimes both price and volume expand slowly, gradually working into an exponential rise with a final blow-off stage. Following this development both volume and price fall off equally sharply. This represents an exhaustion move and is a characteristic of a trend reversal. The significance of the reversal will depend upon the extent of the previous advance and the degree of volume expansion (Figure 10-1, example *c*).

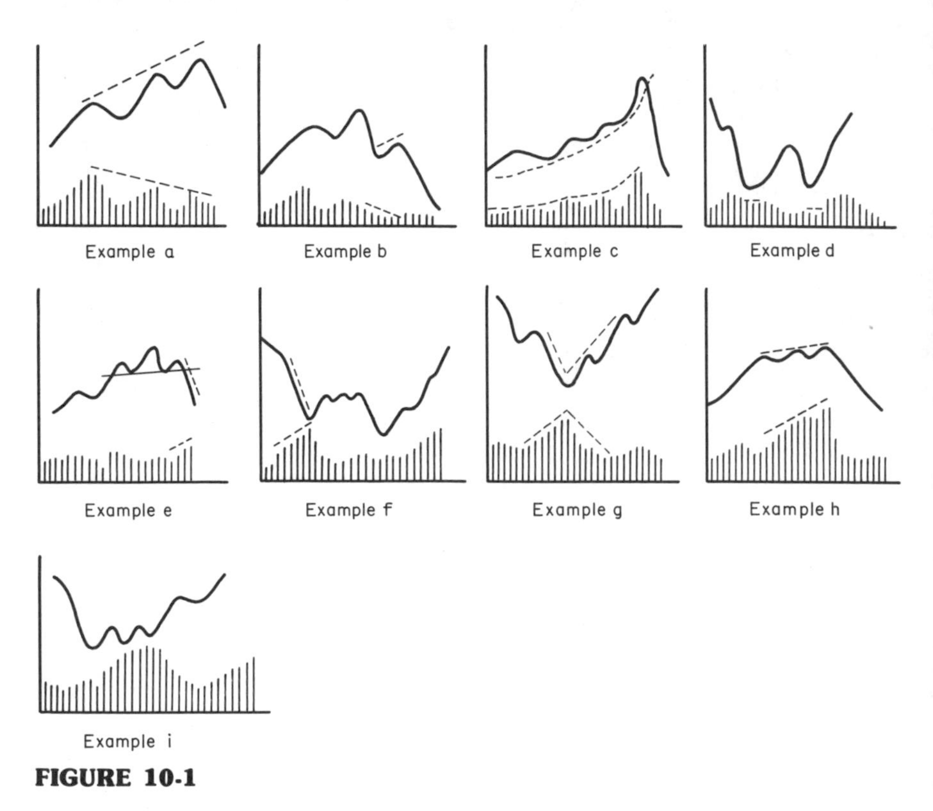


**FIGURE 10-1**

5. A market which rises on expanding volume is a perfectly normal phenomenon and by itself has no warning implications of a trend reversal.
6. When prices advance following a long decline and then react to a level at or above the previous trough, it is a bullish sign when the volume on the secondary trough is lower than the first (Figure 10-1, example *d*).
7. When a downside breakout from a price pattern, trendline, or moving average occurs on heavy volume, this is a bearish sign which emphasizes the reversal in trend (Figure 10-1, example *e*).
8. A selling climax occurs when prices fall for a considerable time at an accelerating pace accompanied by expanding volume. Following a selling climax, prices may be expected to rise, and the low established at the time of the climax is unlikely to be violated for a considerable time. Termination of a bear market

is often, but not always, accompanied by a selling climax (Figure 10-1, examples *f* and *g*).

9. When the market has been rising for many months, an anemic price rise (Figure 10-1, example *h*) accompanied by high volume indicates churning action and is a bearish factor. Following a decline, heavy volume with little price change is indicative of accumulation and is normally a bullish factor (Figure 10-1, example *i*).

## VOLUME LEADS PRICE

In his book *The Stock Market Indicators* William Gordon noted that, between 1877 and 1966, "In 84 percent of the bull markets the volume high did not occur at the price peak but some months before."[1] Of the 18 bull markets, two uptrends ended with volume and price reaching a peak simultaneously (January 1906 and March–November 1916), and in one uptrend (October 1919) volume lagged by a month. The lead time of volume in the remaining 15 cycles was from 2 months (April 1901 and December 1965) to 24 months (January 1951). The average for all cycles was 9 months.

Since volume is a relative concept, i.e., the level of volume is compared to that of a recent period in order to determine its level, it is usually measured on a rate-of-change basis, although some analysts prefer to use moving averages of the actual volume figures.

Chart 10-1 shows the annual rate of price change for the DJIA and an annual rate of change of a 3-month total of volume. In most instances the rate of change of volume led that of price, especially at market bottoms. This method of portraying volume suffers from the fact that the movements are jagged, and it is therefore difficult to interpret when a reversal in trend has taken place. Chart 10-2 attempts to surmount this problem by smoothing the data. The technique used is an annual rate of change of a 6-month total of volume. The resulting index is then smoothed by a 14-month exponential moving average. The price index is a 12-month rate of change smoothed by a 12-month weighted average. When used together the two series are very useful.

First, the volume curve has an almost consistent tendency to peak out ahead of price during both bull and bear phases. Second, when the price index is above its zero reference line and is falling, but volume is rising (e.g., 1966, 1969, and to a lesser extent late 1976), the expanding activity

[1] William Gordon, *The Stock Market Indicators*, Investors Press, Palisades Park, N.J., 1968.

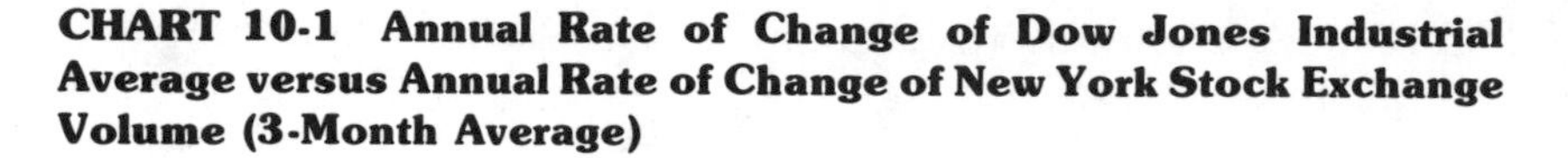

**CHART 10-1 Annual Rate of Change of Dow Jones Industrial Average versus Annual Rate of Change of New York Stock Exchange Volume (3-Month Average)**

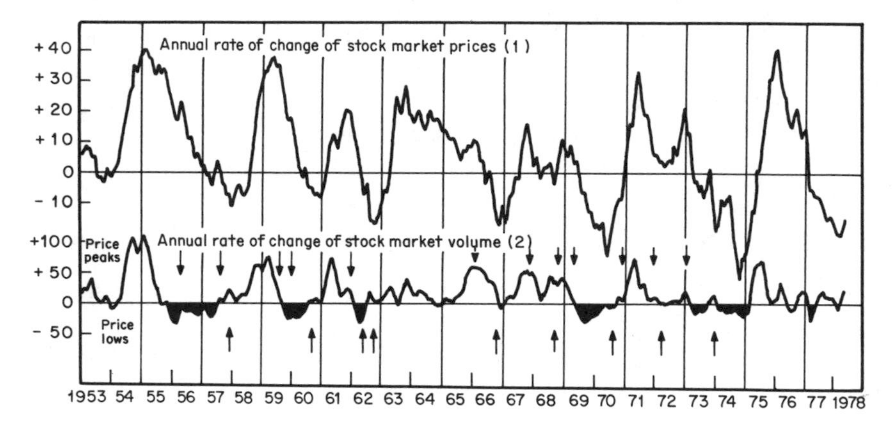


SOURCES: (1) Dow Jones Industrial Averages, (2) New York Stock Exchange 3-Month Averages. Courtesy of *The Bank Credit Analyst.*

appears to be indicative of distribution and should be interpreted as a very bearish factor once the rally has terminated. Third, in view of the long lead times at market tops given by both price and volume, the use of the two indexes for precise timing purposes is best employed at major bottoms. Reference to Chart 10-2 shows that a reversal in one curve unaccompanied by a reversal in the other at market bottoms would have given a premature signal at the end of 1973 and 1977, where volume turned up ahead of price. Consequently it is wiser to await a signal from both, even though this may occur at slightly higher levels.

The monitoring of intermediate market trends involves the use of a rate of change covering a much shorter time span. Chart 10-3 shows a 25-day rate of change of a 25-day moving average of NYSE volume. There is a useful relationship between the 25-day cycles of volume and the corresponding cycle in the DJIA itself, but it should be remembered that a specific time span will be a useful market indicator only as long as the implied cycle in the market itself is still in force. In other words, most of the time the market rises and falls on a fairly consistent basis—for example, 4 years for a cyclical movement, 13 to 26 weeks on an intermediate basis. The moment price movements diverge significantly from these set movements, indicators based on these cycles will fail to operate success-

**CHART 10-2 Long-Term Volume/Price Relationship**

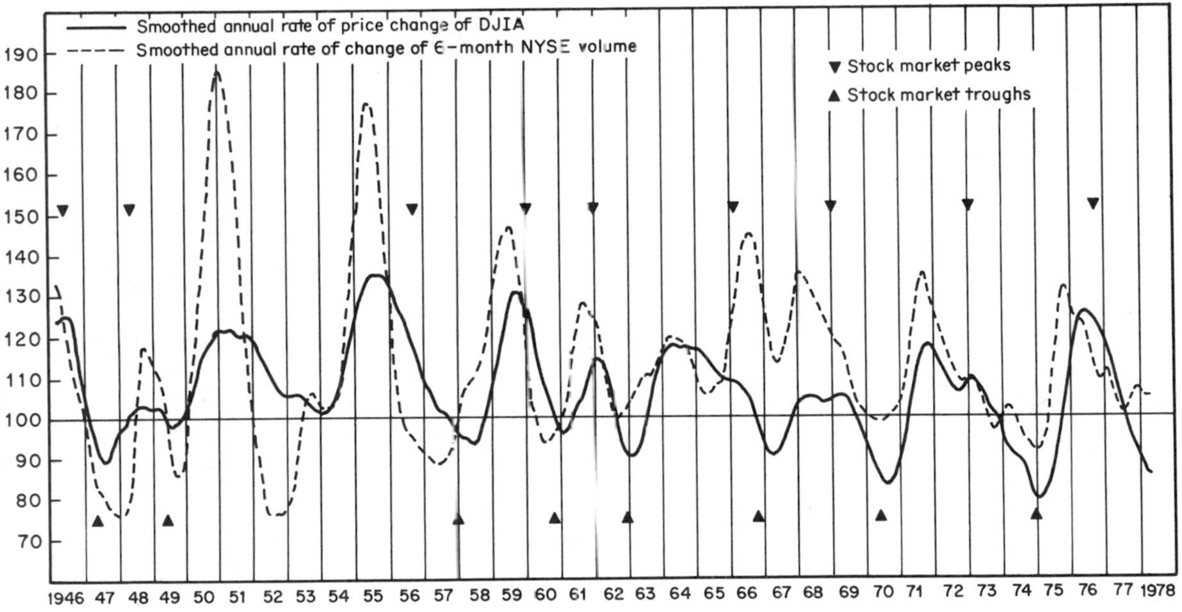

**CHART 10-3 New York Stock Exchange Weekly Advance/Decline Line versus Volume Momentum Indicator**

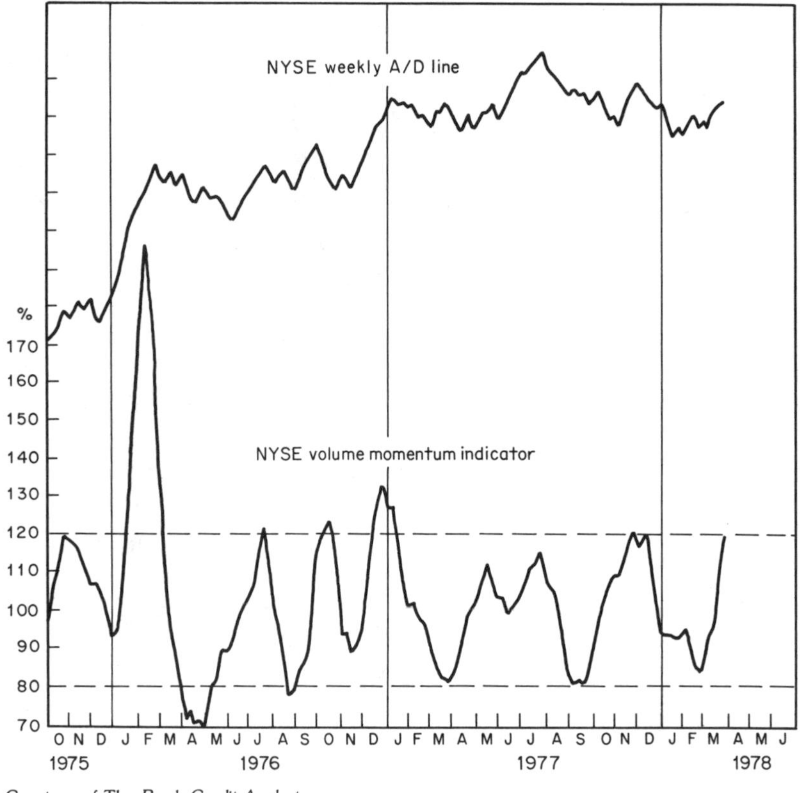


Courtesy of *The Bank Credit Analyst.*

fully. For example, the 1970–1974 cycle was a textbook case of the 4-year cycle in operation. However, the rate-of-change indicators used to successfully signal the peak of this cycle would have been extremely early in forecasting the termination of 1921–1929, 1942–1946, and 1962–1966 bull market peaks. Because intermediate movements are less consistent in either recurrence or magnitude than primary movements, they are far more difficult to predict.

## UPSIDE/DOWNSIDE VOLUME

Measures of upside/downside volume try to separate the volume in advancing and declining stocks. Use of this technique makes it possible to determine in a subtle way whether distribution or accumulation is taking place.

The widespread publishing of these statistics has been available only since 1965, so the history of indicators based on such data is relatively short. The daily volume of advancing and declining stocks on the NYSE is printed in *The Wall Street Journal*; slightly more comprehensive coverage is given in *Barron's*.

Upside/downside volume is measured basically in two ways.

The first is an index known as an *upside/downside volume line*, constructed by cumulating the difference between the daily plurality of the volume of advancing and declining stocks. Since an index of this type is always started from an arbitrary number, it is a good idea to begin with a fairly large one; otherwise there is the possibility that if the market declines sharply for a period, the upside/downside line will become negative, thereby unduly complicating the calculations. If a starting total of 500 million shares were assumed, the line would be constructed as shown in Table 10-1.

These statistics are not published on a weekly or monthly basis, so longer-term analysis should be undertaken by recording the value of the line at the end of each Friday, or taking an average of Friday readings for a monthly plot. From these weekly or monthly observations the appropriate moving average can then be constructed.

It is normal for the upside/downside line to rise during market advances and to fall during declines. When the line fails to confirm a new high (or low) in the price indexes, it warns of a potential trend reversal. The basic principles of trend determination discussed in Part 1 may be applied to the upside/downside line in order to better assess the technical position of this index.

When a market is advancing in an irregular fashion, with successively

**TABLE 10-1**

| Date | Volume of advancing stocks, in millions | Volume of declining stocks, in millions | Difference | Upside/downside line |
|---|---|---|---|---|
| Jan. 1 | 10.1 | 5.1 | +5.0 | 505.0 |
| 2 | 12.0 | 6.0 | +6.0 | 511.0 |
| 3 | 15.5 | 15.5 | 0.0 | 511.0 |
| 4 | 15.0 | 10.0 | +5.0 | 516.0 |
| 5 | 11.1 | 12.0 | −0.9 | 515.1 |

higher rallies interrupted by a series of rising troughs, the upside/downside line should be doing the same. Such action is indicating that the volume of advancing issues is expanding on rallies and contracting during declines. When this trend of the normal price/volume relationship is broken, a warning is given that one of two things is happening. Either upside volume is failing to expand sufficiently, or volume during the decline has begun to expand excessively on the downside. Both are bearish factors. The upside/downside line is particularly useful where prices may be rising to new highs and overall volume is expanding. In such a case, if the volume of declining stocks is rising in relation to that of advancing stocks, this will show up in either a slower rate of advance in the upside/downside line or an actual decline. A hypothetical example of this condition is shown in Figure 10-2.

The upside/downside line from 1972 to 1976 is shown in Chart 10-4. Major peaks and troughs in the DJIA are illustrated by the triangles plotted above and below the line. The initial period covered by the chart occurred during the 1972–1974 bear market. It can be seen that the upside/downside line had already turned down by the time the DJIA made its bull market peak in January 1973. The final run-up by the Dow beginning in November 1972 therefore marked the first bear market rally of the upside/downside line itself. The overall downtrend in this index remained intact until early 1975, at which time a reversal was indicated by both a decisive upside penetration of its bear market trendline and the successful completion of and breakout from a double bottom formation.

Chart 10-5 shows the actual daily plottings of the line with a 30-day moving average for this critical period. During November the sharp downtrend which began in the previous March was broken on the upside and confirmed the crossover of the 30-day moving average which had already taken place in the middle of October. A comparison of Charts 10-4 and 10-5 illustrates well the principle that the significance of a trendline is determined by its length and steepness. This penetration of

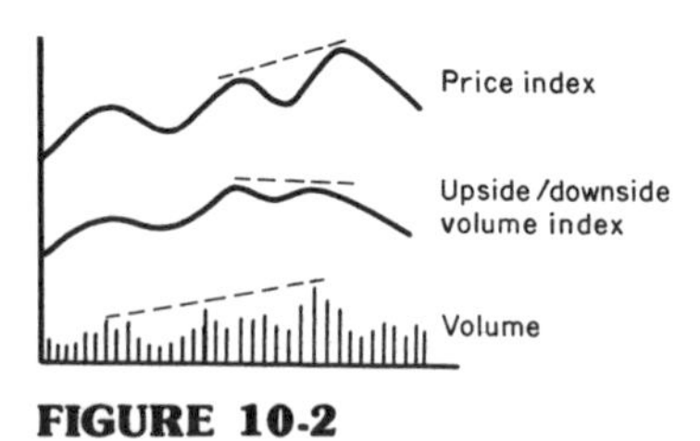


**FIGURE 10-2**

**CHART 10-4 Upside/Downside Volume, 1973–1976**

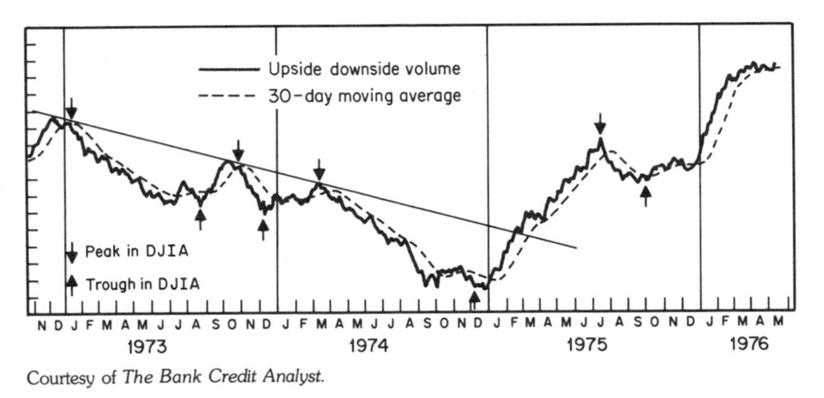


Courtesy of *The Bank Credit Analyst.*

the bear market trendline in Chart 10-4 signaled the end of the bear market, yet the trendline (AA) in the daily chart which covered the March–October 1974 period, because it was so much steeper and shorter, only signaled the end of the intermediate decline. It can also be seen how the extension of this intermediate trendline became a level of support for the

**CHART 10-5 Upside/Downside Volume, 1974–1975** (Closer view)

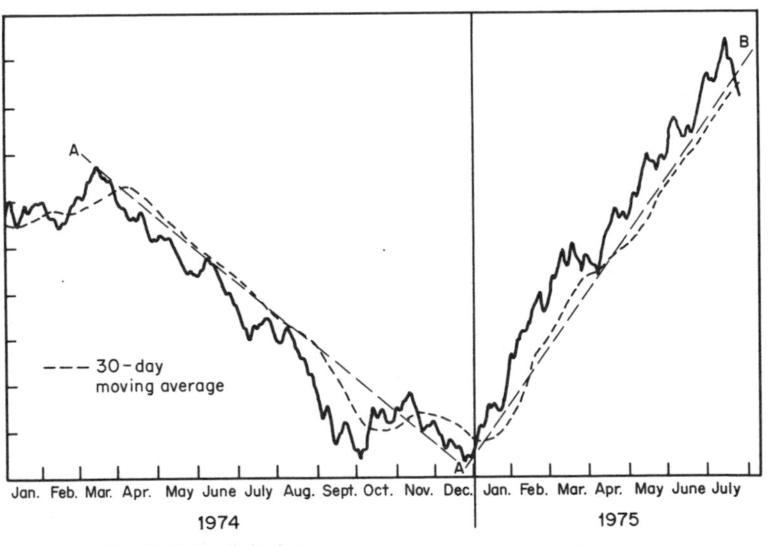


Courtesy of *The Bank Credit Analyst.*

upside/downside line in late November and December. The violation of trendline *AB* and the decisive downside penetration of the 30-day moving average were both useful in signaling that a peak in the DJIA had been reached in mid-July. In view of the steepness and brevity of this sharp uptrend, the breakdown merely signaled that an intermediate top had been reached, so that following a 3-month correction the line was able to resume its primary advance.

The second method of measuring upside/downside data is to compile it from the rate-of-change index, but great care should be taken when comparing levels of the momentum index over different periods. This is because an index constructed from internal data and started from an arbitrary number often bears little relationship to the size of two comparative moves. For example, assuming that the upside/downside line is started at the 100-million-share level, a net decline of 50 million shares over a 10-week period would put the 10-week rate-of-change index at 50 percent, or −50. On the other hand, had the line been started at the 60-million-share level, a net decline of 50 million shares over a 10-week period would result in a rate-of-change reading of 16 percent (or −74). As a result, two entirely different rate-of-change readings would be observed even though the actual market action (i.e., a net decline of 50 million shares) was identical. Consequently, comparisons of levels of rates of change using the upside/downside line, or any other index similarly constructed, should be treated with extreme caution. Such comparisons can be valid only where movement of the index is confined to more or less identical levels.

## BIG BLOCK ACTIVITY

Large volume in a stock or the market has attracted the attention of many technicians and stimulated the development of indicators that can measure big-block activity. Technical research has been heavily concentrated in this area not only because of the emotional appeal of such indicators, but also because it is felt that the large-block traders, i.e., institutions, can have a tremendous effect on market prices through their buying (or selling) power. The ability of an investor (or, for that matter, most technicians) to monitor such activity is limited, since such statistics are not recorded in the popular press. Even the advisory services that do publish such indexes do not have a long historical record from which reliable conclusions can be drawn.

Consequently, most interpretation of block activity is confined to institutional brokerage reports and advisory services. One of the most useful of the block-activity indicators published by the advisory services is that developed by Don Worden.[2] This indicator is constructed by taking the ratio of blocks (i.e., transactions in excess of 100,000 shares) where the price was higher than the previous sale (known as an uptick) and those where it was lower (a downtick). Various moving averages of these ratios are then calculated and plotted as oscillators. Worden's big-block ratios and the DJIA for 1976 and 1977 are shown in Chart 10-6. The relationships are self-evident, and it can be seen that the various ratios rise and fall with the market as they move from "overbought" to "oversold" levels.

## MOST ACTIVE STOCKS

To investors who wish to monitor big-block activity, the most active stocks represent a viable alternative. Statistics on the most active stocks are published in the popular press on both a daily and a weekly basis. Usually the 20 most active NYSE issues are recorded, but results do not seem to differ significantly whether 10, 15, or 20 issues are used. Active stocks are worth monitoring since they not only reflect the actions of institutions but also account for some 20 to 25 percent of total NYSE volume. Since the net price changes of such issues are recorded, it is possi-

**CHART 10-6 Worden $100,000 Indexes versus Dow Jones Industrial Average, 1976–1977**

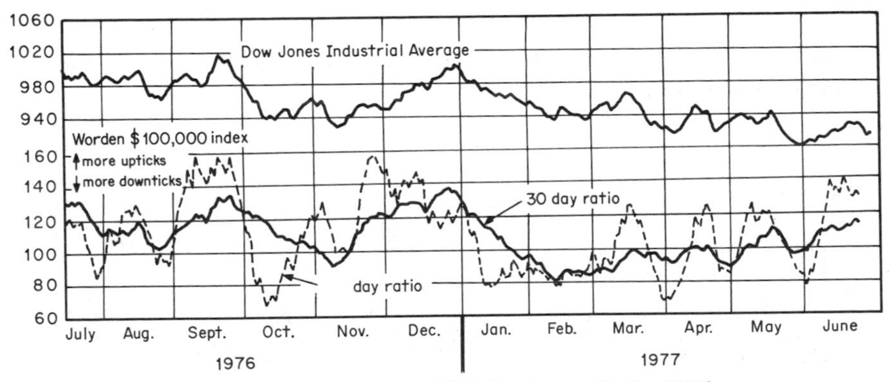


Courtesy Worden Tape Reading Studies, P.O. Box 11878, Fort Lauderdale, Fla. 33339.

[2] Worden Tape Reading Studies, P.O. Box 11878, Fort Lauderdale, Fla. 33339.

**CHART 10-7 New York Stock Exchange Composite versus the 30-Day Most Active Indicator**

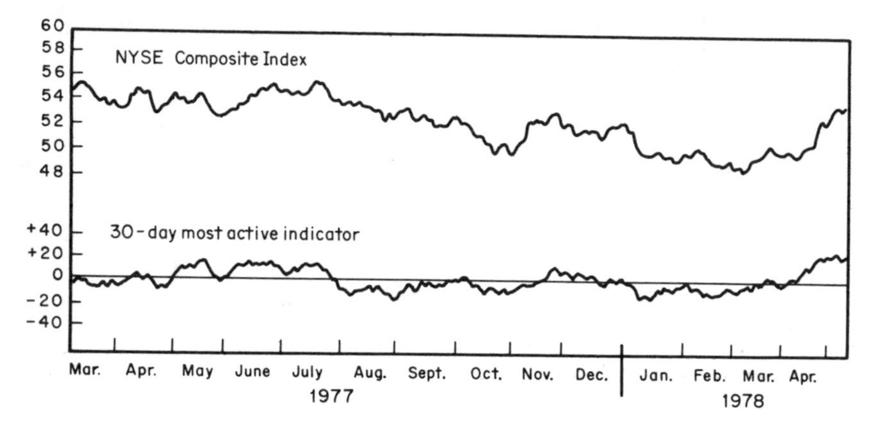


ble to derive an indicator that can confirm movements in the upside/downside line.

The indicators derived from most-active statistics are similar to those derived from upside/downside data. Chart 10-7 shows an oscillator constructed by taking a 30-day moving total of the net difference between the 10 most active advancing and declining stocks. The calculation is made by totaling the number of the 10 most active stocks each day that advance and those that decline. Sometimes the 15 or 20 most-actives are used as a basis for the indicator, but results do not differ appreciably. In order to obtain the 30-day oscillator, the number of advancing stocks over the 30-day period is then totaled. The same calculation is made for the declining

**CHART 10-8 New York Stock Exchange Composite versus the 8-Day Most Active Indicator**

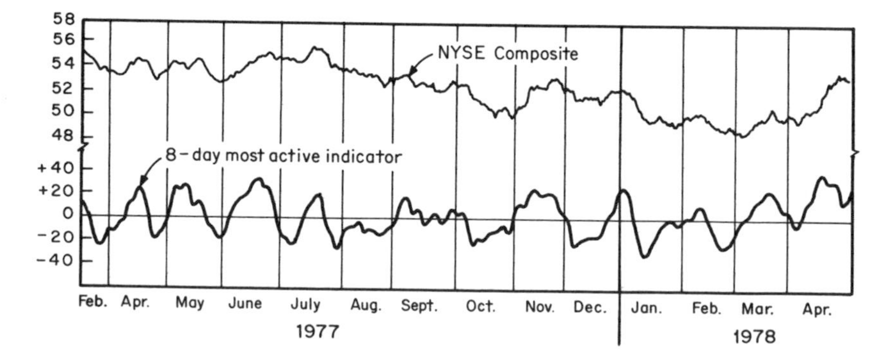

**CHART 10-9 Dow Jones Industrial Average versus the 20 Most Active Stocks**

This chart shows a cumulative A/D line of the 20 most active stocks combined with a 6-week oscillator. Note how neither the A/D line nor its oscillator confirmed the December 1974 low in the Dow.

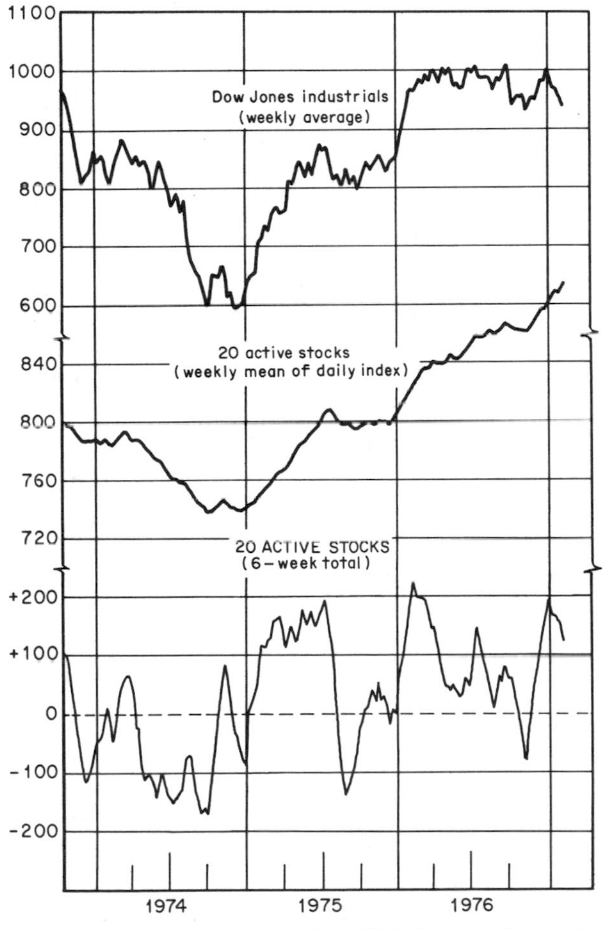


Courtesy of Juncture Recognition, Box 1209, Pompano Beach, Fla. 33061.

stocks, and one total is taken from the other. If the number of advancing stocks over the 30-day period is greater than the number of those declining, the oscillator will have a reading above zero, and vice versa. As new data are received each day they are added to the two totals, while the data for the 31st previous day are deleted. The resulting index appears to be useful for confirming intermediate advances and declines as it crosses above and below its zero reference line. Although the history of this indicator is relatively short, it has already achieved an impressive track record. In order to reduce the possibility of whipsaws it is a good idea to wait for the index to cross its zero reference line more decisively. An alternative method is to construct an oscillator in a similar manner but using weekly data, i.e., the 10, 15, or 20 most active stocks based on weekly volume.

Shorter-term market movements can be indicated by using a shorter time span; in Chart 10-8 an 8-day total has been used.

The second method of using the most active stocks is to take a cumulative difference between the number of stocks advancing each day and those declining. The result is an index which is not dissimilar to the upside/downside line. While most of the time these two indicators move in concert, it is well worth keeping up both series since the junctures when their courses differ are often most informative about a potential reversal in the overall market trend.

# 11

# BREADTH

The degree to which the vast majority of issues are participating in any stock market move is measured by indicators of breadth. Breadth therefore monitors the extent of a market trend. Generally speaking, the fewer the number of issues that are moving with the trend, the greater the probability of an imminent reversal in that trend.

The concept of breadth can probably be best explained using a military analogy. In Figure 11-1, lines *AA* and *BB* mark military lines of defense drawn up during a battle. It might be possible for a few units to cross over from *AA* to *BB*, but the chances are that the *BB* line will hold unless an all-out effort is made. In example *a*, the two units represented by the arrows are quickly repulsed. In example *b*, on the other hand, the assault is successful since many units are taking part, and army B is forced to retreat to a new line of defense at $B_1$.

A narrowly advancing stock market can be compared to example *a*, where it *looks* initially as though the move through the line of defense (in stock market terms, a resistance level) is going to be successful, but because the move is accompanied by such little support, the overall price trend is soon reversed. In the military analogy, even if the two units had successfully assaulted the *BB* defense, it would not be long before army B would have overpowered them, for the further they advanced without broad support, the more vulnerable they would have become to a counter-offensive by army B.

The same is true of the stock market, for the longer a price trend is maintained without a follow-up by the broad market, the more vulnerable is the advance.

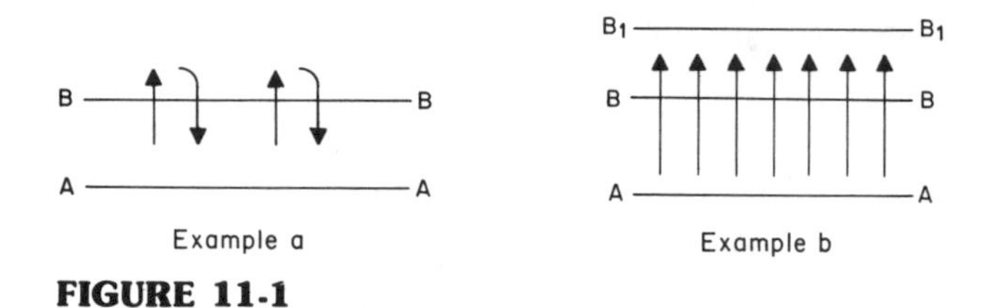


**FIGURE 11-1**

At market bottoms breadth is not such a useful concept for determining major reversals, since the majority of stocks usually coincide with or lag behind the major indexes. On those few occasions when breadth reverses its downtrend before the averages, it can be a useful indicator of market bottoms. For the moment, attention will be concentrated on the reason why the broad market *normally* leads the averages at market tops. The word "normally" is used since, in the vast majority of cases, the broad list of stocks does peak out ahead of the DJIA; this rule is not invariable, however, so it should not be assumed that the technical structure is necessarily sound just because market breadth is strong.

## ADVANCE/DECLINE LINE

The most widely used indicator of market breadth is an *advance/decline line*. This indicator is constructed by taking a cumulative total of the difference (plurality) of the number of NYSE issues advancing over those declining in a particular period (usually a day or a week). Similar indexes may be constructed for the American Exchange, or from the over-the-counter issues. Because the number of issues listed on the NYSE has expanded since breadth records were first kept, an A/D line constructed from a simple plurality of advancing over declining issues gives a greater weighting to more recent years. For the purpose of long-term comparisons it is better to take a ratio of advances versus declines, or a ratio of (advances/declines) ÷ the number of unchanged issues, rather than limiting the calculation to a simple plurality.

One of the most useful measurements of breadth is a cumulative running total of the formula $\sqrt{A/u - D/u}$, where A = the number of stocks advancing, $D$ = the number declining, and $u$ = the number unchanged. Since it is not mathematically possible to calculate a square root of a negative answer, i.e., when the number of declining stocks is greater than the number of those advancing, the $D$ and A are reversed in such cases, so that the formula becomes $\sqrt{D/u - A/u}$. The resulting answer is then subtracted from the cumulative total, as opposed to the answer in the earlier formula, which is added. Table 11-1 illustrates this calculation.

Inclusion of the number of unchanged issues is useful, because at certain points a more reliable advance warning of an imminent trend reversal in the A/D line can be given. This is because the more dynamic the move in either direction, the greater the tendency for the number of unchanged stocks to diminish. Consequently, by giving some weight to the number of unchanged in the formula, it is possible to assess a slowdown in momentum of the A/D line at an earlier date, since an expanding number of unchanged issues will have the tendency to restrain extreme movements.

The A/D line normally rises and falls in sympathy with the DJIA, but at market peaks it usually peaks ahead of the senior average. There appear to be three basic reasons why this is so.

1. The market as a whole discounts the business cycle and normally reaches its bull market peak 6 to 9 months before the economy tops out. Since the peak in business activity is itself preceded by a deterioration of certain leading sectors, such as consumer spending and construction, it is logical to expect that the stocks representing these sectors will also peak out ahead of the general market.
2. Many of the stocks listed on the NYSE, such as preferreds and utilities, are sensitive to changes in interest rates. Since sharply rising interest rates are a phenomenon of a rapidly ma-

**TABLE 11-1**

| Date | | Issues traded (1) | Advances (2) | Declines (3) | Unchanged (4) | Advances ÷ unchanged (5) | Declines ÷ unchanged (6) | Col. 5 − col. 6 (7) | $\sqrt{\text{Col. 7}}$ (8) | Cumulative A/D line (9) |
|---|---|---|---|---|---|---|---|---|---|---|
| Jan. | 7 | 2129 | 989 | 919 | 221 | 448 | 416 | 32 | 5.7 | 2475.6 |
| | 14 | 2103 | 782 | 1073 | 248 | 315 | 433 | −118 | −10.9 | 2464.7 |
| | 21 | 2120 | 966 | 901 | 253 | 382 | 356 | 26 | 5.1 | 2469.8 |
| | 28 | 2103 | 835 | 1036 | 232 | 360 | 447 | −87 | −9.3 | 2460.5 |
| Feb. | 4 | 2089 | 910 | 905 | 274 | 332 | 330 | 2 | 1.4 | 2461.9 |
| | 11 | 2090 | 702 | 1145 | 243 | 289 | 471 | −18.2 | −13.5 | 2448.4 |
| | 18 | 2093 | 938 | 886 | 269 | 349 | 329 | 20 | 4.5 | 2452.9 |
| | 25 | 2080 | 593 | 1227 | 260 | 228 | 472 | 244 | −15.6 | 2437.3 |

turing business cycle, they usually begin to rise before the stock market has reached its peak.

3. Poorer-quality stocks offer the largest gains, but they are also representative of smaller, less well-managed companies that are more vulnerable to reduced earnings (and even bankruptcy) during a recession. On the other hand, since blue-chip stocks normally have good credit ratings, reasonable yields, and sound underlying assets, they are typically the last stocks to be sold by investors during a bull market.

Since the DJIA and other market averages are almost wholly composed of larger companies which are normally in better financial shape, these popular averages continue to advance well after the broad market has peaked out.

Divergences between the A/D line and the DJIA at primary peaks are the most significant ones, but divergences also occur at intermediate tops; because they are generally much smaller than those occurring at primary peaks, the portent of intermediate divergences is usually less ominous. One of the basic principles of the breadth/price relationship is that the longer and greater the divergence, the deeper and more substantial the implied decline is likely to be. For example, Chart 11-1 shows that the weekly A/D line peaked in March 1971, almost 2 years ahead of the DJIA, a very long period by historical standards. The ensuing bear market was the most severe since the Depression. On the other hand, the absence of a divergence does not necessarily mean that a steep bear market cannot take place, as the experience of the December 1968 top indicates. This is also shown in Chart 11-1.

Positive divergences develop at market bottoms where the A/D line refuses to confirm a new low in the Dow. The most significant divergence of this nature occurred in the 1939–1942 period. The DJIA (as shown in Chart 11-2) made a series of lower peaks and troughs between 1939 and 1941, while the A/D line refused to confirm. Finally, in the middle of 1941 the A/D line made a post–1932 recovery high unaccompanied by the DJIA. The immediate result of this discrepancy was a sharp sell-off into the spring of 1942 by both indicators, but even then the A/D line held well above its 1938 bottom, unlike the DJIA. The final low in April 1942 was followed by the best bull market on record (in terms of breadth). This positive action by the broad market is unusual, for typically at market bottoms the A/D line either coincides with or lags behind the low in the DJIA and has no forecasting significance until its downtrend reversal is signaled

**CHART 11-1 Dow Jones Industrials and the Weekly Advance/Decline Line**

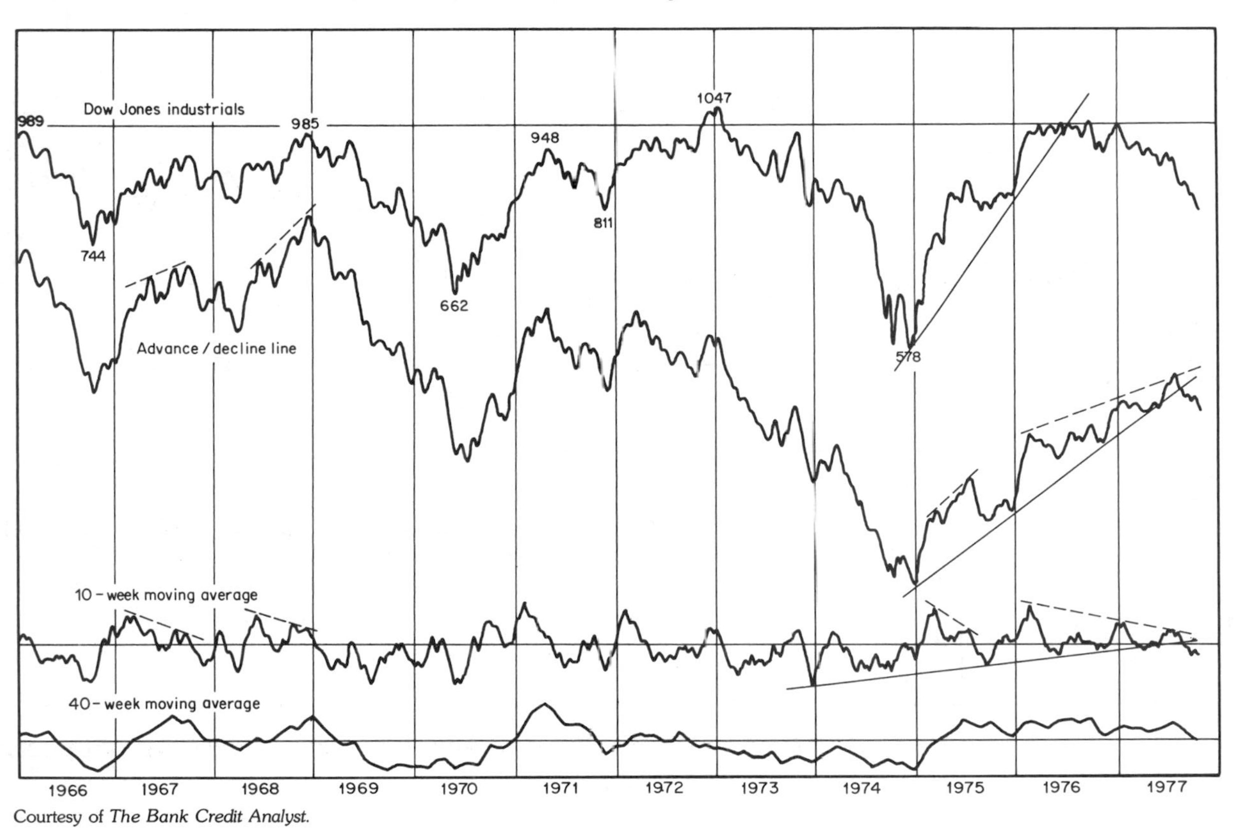


Courtesy of *The Bank Credit Analyst.*

**CHART 11-2 Dow Jones Industrials and the Long-Term Advance/Decline Line**

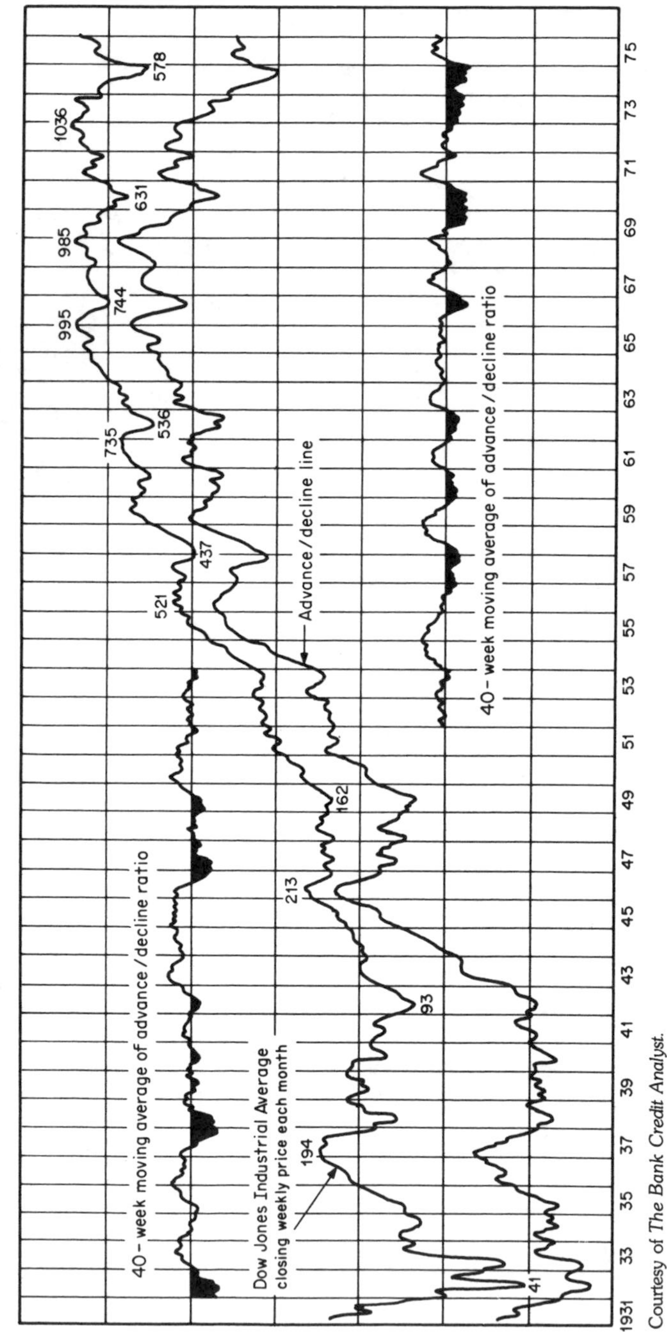


Courtesy of *The Bank Credit Analyst.*

by a breakout from a price pattern, a trendline, or a moving-average crossover, for example.

Chart 11-1 shows the A/D line and its 10-week oscillator. The oscillator is constructed by taking a 10-week moving average of the $\sqrt{A/u - D/u}$ formula. A comparison of the two indicators illustrates the principle of divergence, as evidenced by declining peaks of momentum and rising peaks in the A/D line itself. These discrepancies are shown by the dashed lines just above the two indexes. It is not possible to know at the time how high the A/D line will extend, only that the technical position (indicated by the declining peaks in the 10-week momentum) is deteriorating. The best method of determining when the A/D line has made its final advance is to wait for a downside trendline penetration or a moving-average crossover to confirm the action of the momentum index. Normally, as shown on Chart 11-1, the A/D line will sell off quite sharply following a combination of such trend breaks, but sometimes an extended sideways fluctuation results as the A/D line struggles to regain some momentum. The same principle can also be applied during bear markets, when signals are triggered as a series of higher troughs in the oscillator and lower lows in the A/D line are confirmed by a break in the negative trend of the A/D line itself.

Advance/decline lines have been constructed for most of the world's stock markets, and the same principles of divergence between breadth and the market average seem to apply. The only exception appears to be the Japanese market, where the line seems to have a permanent downward bias and is of relatively little forecasting value. The *International Bank Credit Analyst*[1] publishes a World Stock Index and world A/D line based on the action of 12 industrial countries. Although their relationship has been traced only as far back as 1966, the same principles appear to be in force (see Chart 17-4).

## INDUSTRY BREADTH

An alternate method of measuring breadth is to calculate an A/D line for a number of industry groups. Several organizations publish such data, but those published in *Barron's* will be discussed here since they are the most accessible to the general public. Each week *Barron's* publishes data on 35 industry groups. The industry A/D line is calculated by cumulating the difference between those groups that advanced during the week and those

[1] B.C.A. Publications, 1010 Sherbrooke St. West, Montreal, P. Q. Canada.

that declined. Indexes that remain unchanged are ignored. Some technicians prefer to omit the gold group, since this is considered to be a countercyclical industry. It is doubtful whether the results are meaningfully changed as a result of this exclusion.

The principle behind the interpretation of the industry A/D line is similar to that of the A/D line itself in that a reversal of the prevailing market trend is likely whenever a divergence occurs between it and the DJIA. The industry A/D line is therefore a very useful indicator from the point of view of confirming a trend in the broader A/D line.

The industry group index has a distinct advantage over the A/D line in that it does not include preferred shares and is therefore less interest-sensitive than the A/D line. Normally, as discussed above, interest rates start to rise before the eventual peak in the stock market, so that the inclusion of a substantial amount of interest-sensitive stocks in the A/D line aids the leading characteristics of this index. However, there have been some occasions where interest rates have been stable or falling during the initial phases of a bear market. Between late 1929 and 1930 for instance, interest rates fell by half while stocks also plunged. Similarly, in 1939–1941 interest rates hardly changed at all, yet the DJIA fell. For this reason the industry A/D line usually gives a more accurate reading of what is actually going on in the marketplace than the A/D line itself. Consequently, when the A/D line and the DJIA disagree, the A/D line normally wins. On the few occasions when the industry A/D line and the A/D line diverge, it is the industry A/D line that usually portrays the eventual outcome.

The industry A/D line is useful (1) from the point of view of the divergences which may arise between it and other averages and (2) through an analysis of the quality of its own trend, performed using the trend determination techniques discussed in Part 1.

Chart 11-3 shows the industry A/D line. The 1974 bottom is of particular interest since the industry line did not confirm the December low in the DJIA. Moreover, in 1976 the industry line peaked out in February, while the DJIA made its bull market high in September. By this time, however, the industry A/D line had begun to warn of impending trouble, as it had already traced out a series of declining peaks. Following this divergence the DJIA lost 20 percent in the ensuing 12 months. It is also worthwhile to compare the industry A/D line in this 1976–1977 period with the weekly A/D line shown in Chart 11-1. As noted earlier, the industry line topped out in February 1976, but the A/D line itself reached its bull market high a long time afterwards.

Industry breadth can also be used to construct a rate-of-change indicator.

**CHART 11-3 Industry Advance/Decline Line versus the 13-Week Momentum Index**

Note the extremely bullish positive divergence between the 13-week momentum, which made a decisive bear market low in mid-1974, and the index itself, which almost made a new low in December. The position was reversed in mid-1975, as the July high in the industry A/D line was not confirmed by the 13-week momentum index.

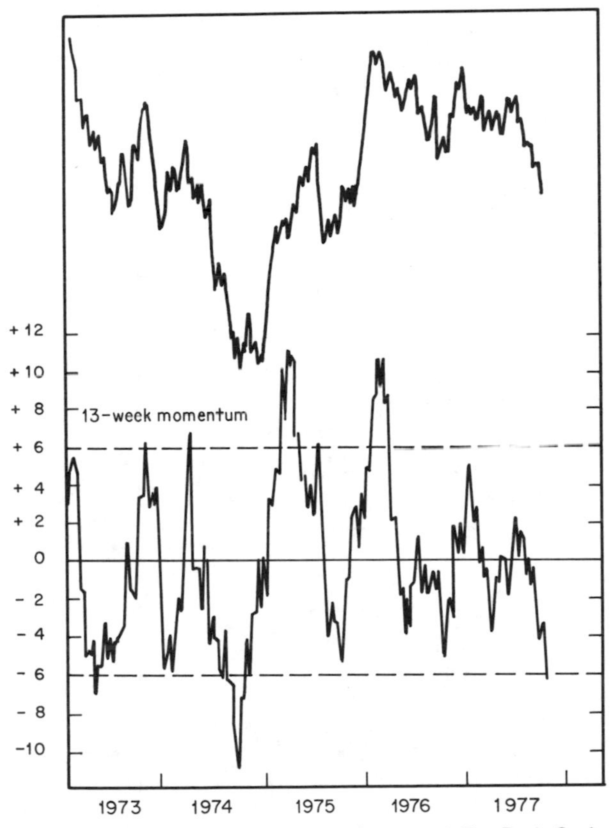


SOURCE: Barron's 35-Industry Groups. Courtesy of *The Bank Credit Analyst.*

**CHART 11-4 Dow Jones Industrial Average versus Industry Breadth, 1974–1977**

In mid-1976, almost at the same time, the bull market trendlines for both the DJIA and the industry advance/decline line were violated on the downside. This joint penetration was a very bearish signal and was one of the first technical warnings of the development of a new bear market.

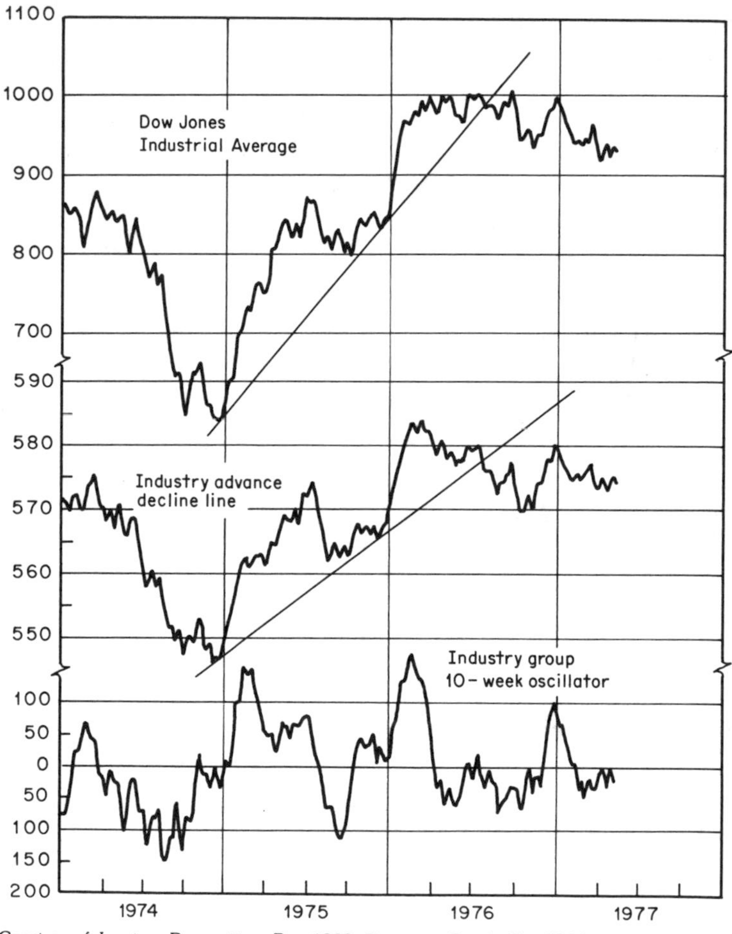


Courtesy of Juncture Recognition, Box 1209, Pompano Beach, Fla. 33061.

As discussed earlier (see Chapter 5), a rate-of-change index based on these data may be used, but only to compare two periods where the level of the indicator has not altered appreciably. Such an index is also shown in Chart 11-3.

The alternate method of constructing a momentum index is to take a total of the net advances over declines for a specific period, such as 10 or 13 weeks, and plot the result. This is illustrated in Table 11-2 for a 10-week oscillator and is also shown in Chart 11-4. The movement of the industry A/D line in Chart 11-4 differs at some points from that shown in 11-3 due to a slightly different construction.

## PERCENTAGE OF STOCKS IN POSITIVE TRENDS

Another useful technique for analyzing stock market breadth is to construct an index based on the number of stocks that may be classified as being in a positive trend. There are basically three ways in which this can be achieved.

### 1. Stocks in Positive Trends

This method involves the classification of stocks according to whether they are in a positive or a negative trend. A stock is considered to be in an

**TABLE 11-2**

| Date | No. of groups advancing (1) | No. of groups declining (2) | Col. 1 ÷ col. 2 (3) | 10-week total of col. 3 (4) |
|---|---|---|---|---|
| Jan. 1 | 1 | 32 | −31 | |
| 8 | 10 | 25 | −15 | |
| 15 | 25 | 10 | +15 | |
| 22 | 17 | 18 | −1 | |
| 29 | 12 | 22 | −10 | |
| Feb. 1 | 35 | 0 | +35 | |
| 8 | 10 | 25 | −15 | |
| 15 | 20 | 15 | +5 | |
| 22 | 5 | 30 | −25 | |
| Mar. 1 | 10 | 25 | −15 | −57 |
| 8 | 18 | 17 | +1 | −25 |
| 15 | 12 | 21 | −9 | −19 |
| 22 | 27 | 8 | +19 | −15 |

uptrend when it rallies following a decline, reacts, and then rallies to a new high above the previous rally peak. This is shown in Figure 11-2, example *a*, at point X. An uptrend is assumed to exist until a previous trough is penetrated on the downside, as in Figure 11-2, example *b*, shown as point Y. Once the number of stocks in an uptrend has been determined, the percentage as a total in the sample is taken and the result plotted. These data are calculated and published by Chartcraft.[2] Chart 11-5 shows this method and similar data for the DJIA stocks and various industry groups. Since this indicator is an oscillator type, its interpretation should be based on the rules and principles established in Chapter 5. It can be seen that the index measuring the percentage of stocks in an uptrend was making a series of declining peaks throughout 1971 and 1972, a period when the DJIA was in a rising trend.

## 2. Percentage of Stocks Over a Moving Average

This method involves the calculation of a specific moving average for a wide number of stocks and then of the percentage of that number that is above the average. The longer the time span chosen for the moving average, the less sensitive the oscillator will be.

Chart 11-6 shows the percentage of the 600 or so stocks monitored by Trendline above their 200-day (30-week) moving average. The data have been smoothed by a 5-week simple moving average to iron out misleading fluctuations.

This indicator usually moves in the same way as the positive trend index discussed above. During the period under consideration, a juncture point worthy of note occurred during the December 1974 low, when the trough in the 200-day indicator was well above that of October. This compares

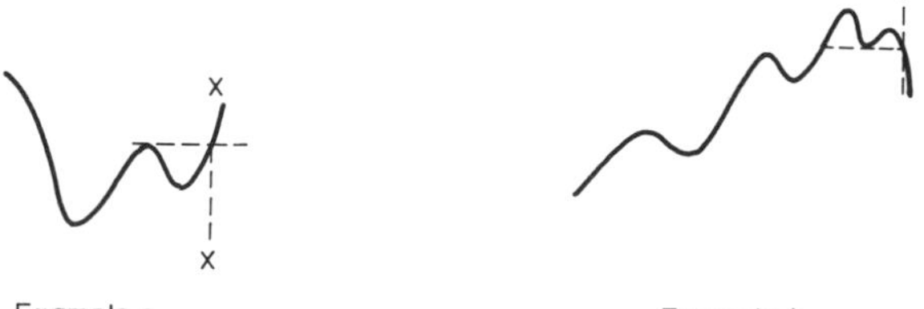

X
X
Example a

Y
Example b

**FIGURE 11-2**

[2] Chartcraft Ltd., Larchmont, N.Y.

**CHART 11-5 Percentage of All Stocks, Dow Stocks, and Industry Groups in Positive Trends**

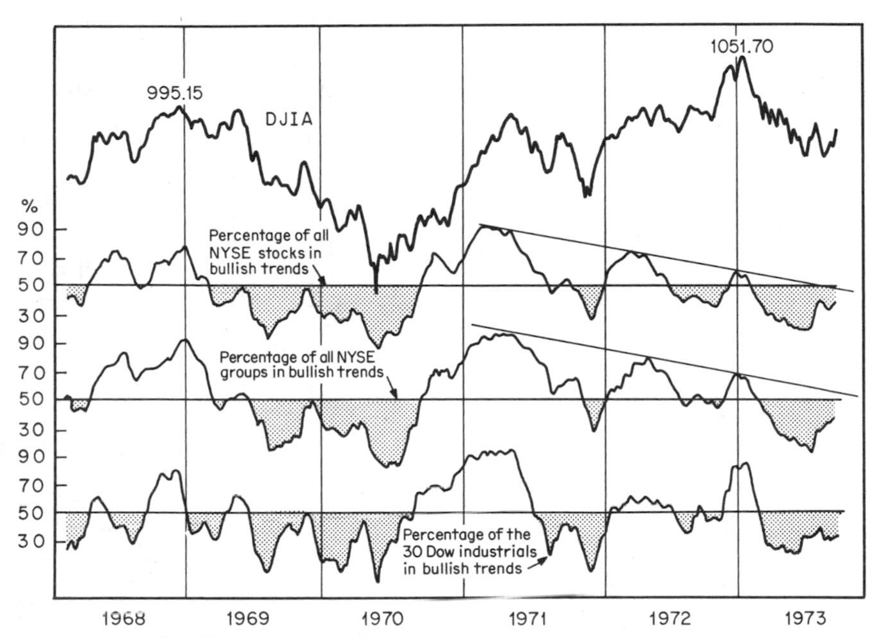


Chart courtesy of Dow Theory Letters Inc., P.O. Box 1759, La Jolla, Calif. 92038, based on data provided by Chartcraft Ltd., Larchmont, N.Y.

**CHART 11-6 Dow Jones Industrial Average versus Over-200-Day Moving-Average Ratio, 1970–1975**

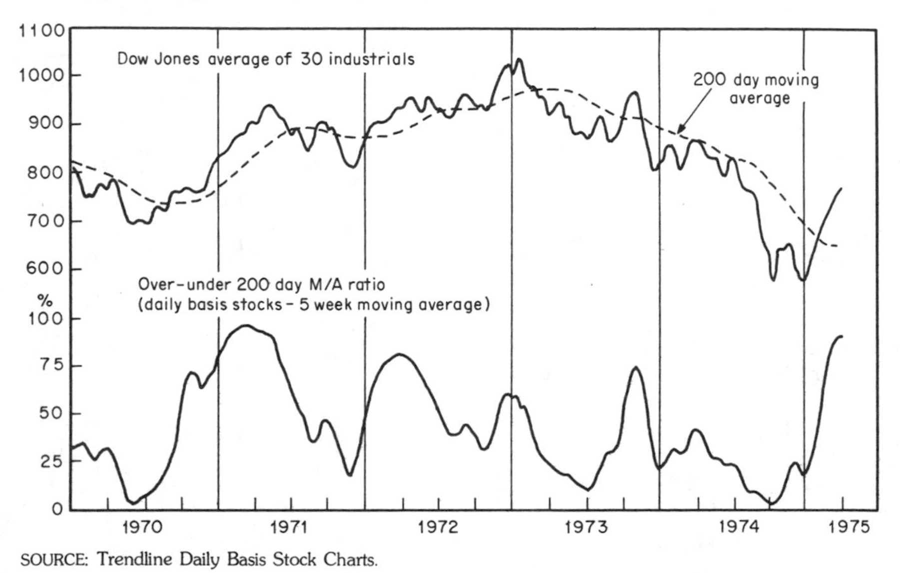


SOURCE: Trendline Daily Basis Stock Charts.

**CHART 11-7 New York Stock Exchange Composite and Two Breadth Ratios**

Chart 11-7 shows two alternative methods of portraying the percentage of stocks above their moving averages. The data were calculated by Chartcraft Ltd. and plotted by the innovative Canadian technician Ian McAvity. The 10-week ratio is useful for identifying intermediate peaks and troughs, while the 30-week average is more useful for determining cyclical bottoms and tops.

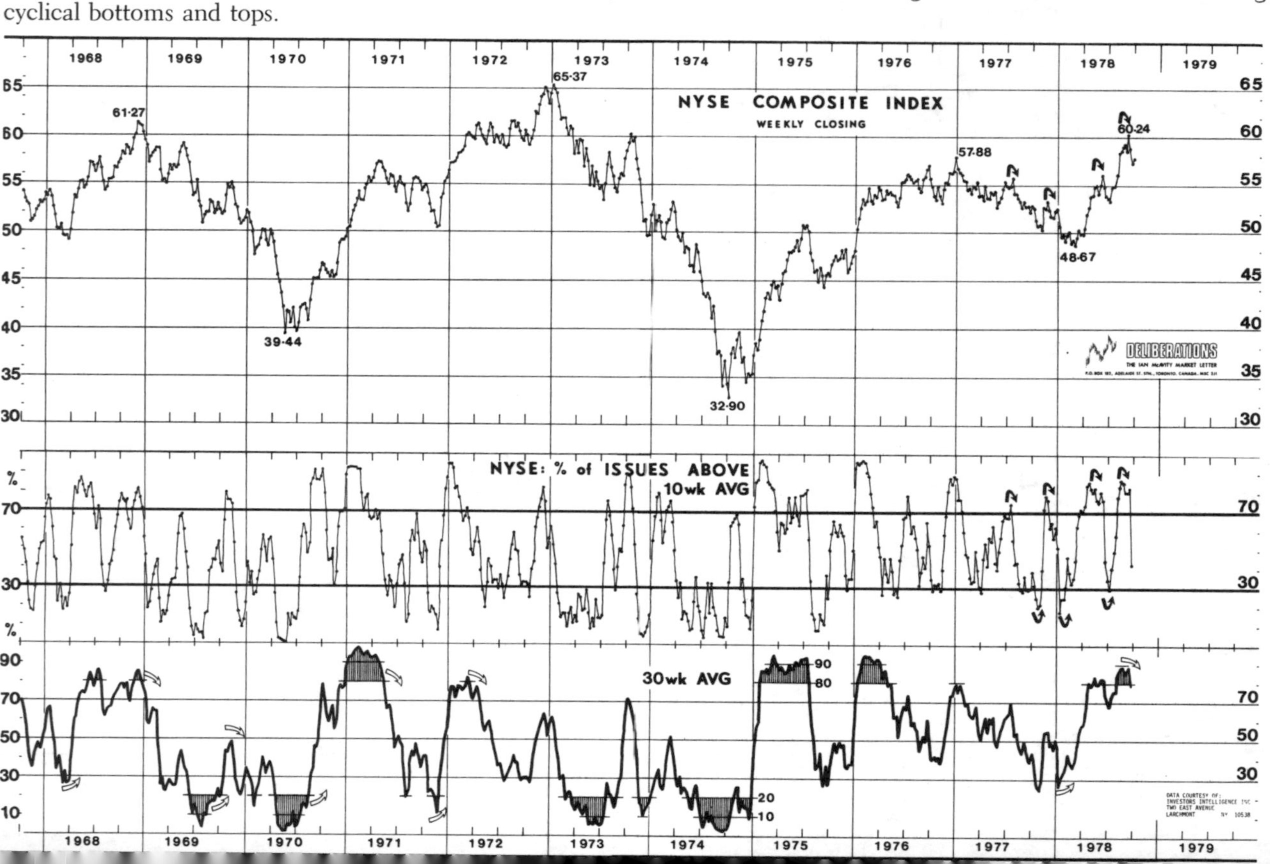

with a marginally higher one for the NYSE Composite Index and one that was lower in the case of the DJIA (shown above the 200-day indicator).

It should also be noted that when this index reaches an extreme of 90 to 100 percent or 10 to 15 percent, it is not necessarily indicating that a sharp reversal is imminent, but it *is* pointing out that a substantial proportion of the move has already taken place. Once the indicator reaches such extremes and starts to reverse its direction, the probabilities favor an immediate reversal in the trend of the market itself (see Chart 11-7). At the beginning of a bull market it is not unknown for this index to remain at a high level for a considerable period of time, as in early 1975, but as the cycle matures, high readings become difficult to reach and maintain. A series of declining peaks in this index is a clear sign that a primary bull market is rapidly running out of steam.

## 3. Percentage of Stocks with Positive Momentum (Diffusion Indexes)

This type of index is calculated by using rate-of-change techniques on either a wide number of stocks or a number of industry indexes, such as

**CHART 11-8*a* Standard & Poor's 500 Stock Index versus *Money Manager* 3-Month Diffusion Index**

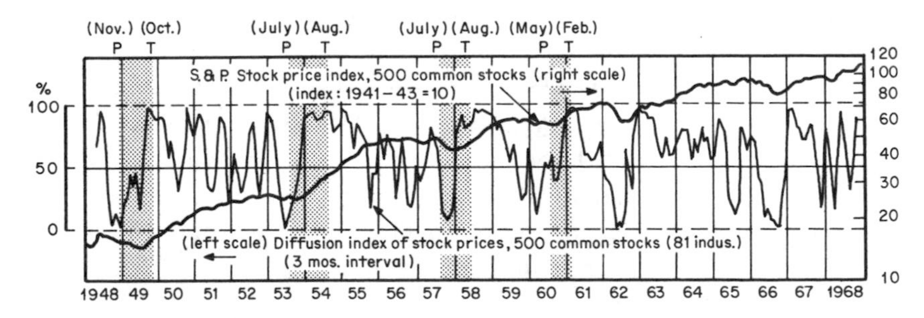


**CHART 11-8*b* Standard & Poor's 500 Stock Index versus *Money Manager* 8-Week Diffusion Index**

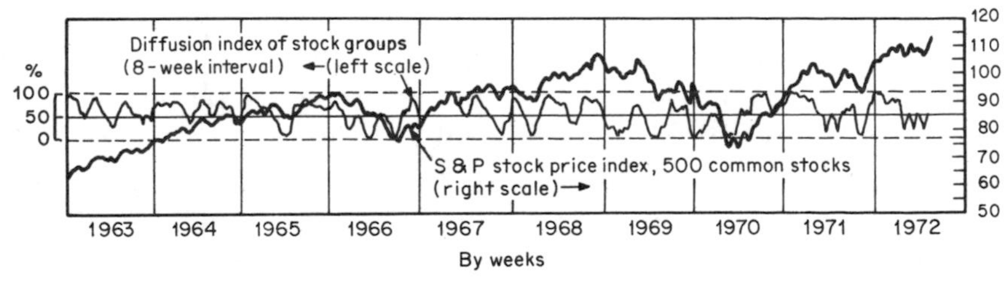


Courtesy of *Money Manager.*

**CHART 11-9 Dow Jones Industrial Average Annual Rate of Change versus the *Bank Credit Analyst* Stock Market Group Momentum Index**

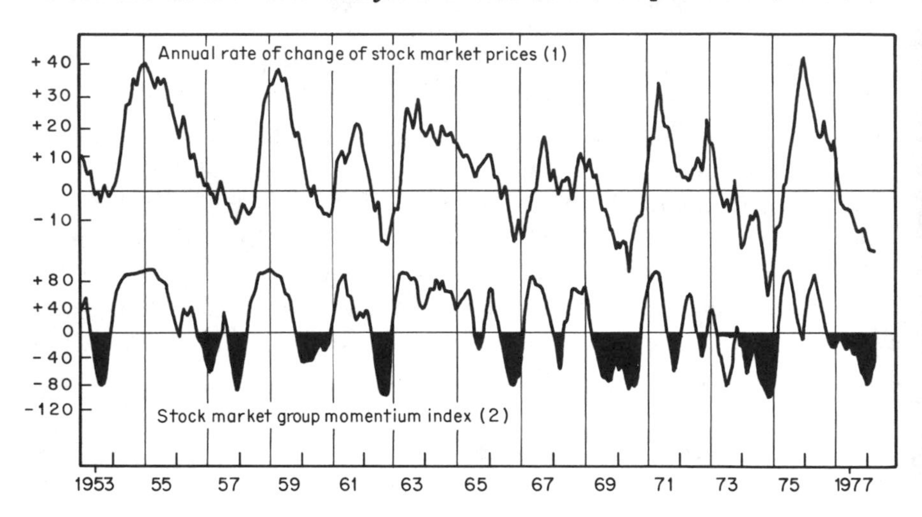


SOURCES: (1) Dow Jones Industrials, (2) 25 S&P groups. Courtesy of *The Bank Credit Analyst*.

Barron's 35 Groups. Indicators calculated in this way are known as *diffusion indexes*. Two such examples are shown in Chart 11-8. These indicators are based on the percentage of Standard & Poor's Industry Groups that are above their level 8 weeks and 3 months ago; they are published by *The Money Manager* (a weekly publication for institutional investors). The shaded areas on Chart 11-8*a* indicate periods of economic recession. Chart 11-9 shows an alternate method of portraying group momentum published by the *Bank Credit Analyst*. These indexes should be interpreted in the same way as any momentum indicator.

## HIGH-LOW FIGURES

The popular press publishes daily and weekly figures for stocks reaching new highs and new lows. These "statistics" relate to the number of stocks making new highs or lows in a calendar year. In order to retain some continuity, the year is not "changed" until the following March, so that January and February figures cover a 13- and 14-month period respectively. More recently, figures covering a 52-week period, as opposed to a calendar year, have been published by the *Wall Street Journal*. In theory, these new figures should offer a much more accurate reflection of the market's

**CHART 11-10 Daily Advance/Decline Line versus the 5-Day High/Low Differential**

A classic signal of improving technical strength was given in December 1974 when the DJIA fell below its October low. The high/low had fallen to its lowest level in September, thereby pointing up that the majority of stocks were in better technical shape than the DJIA. Between March and July the reverse situation occurred as the DJIA made new highs unaccompanied by an expanding high/low differential. This suggested a weakening technical structure, so that once the A/D line had violated the trendline joining the April–May and June bottoms, an intermediate reaction got underway.

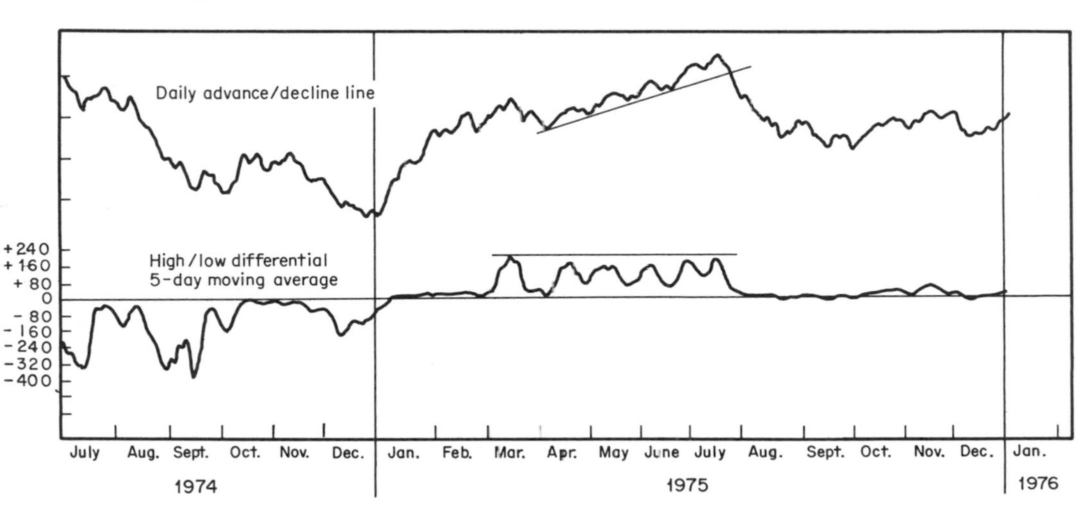

technical structure, but a fair comparison of the two series over a long period is not yet possible, due to the short length of time that the 52-week figures have been available.

There are various methods of measuring the high-low figures. Some technicians prefer to take a moving average of the two series, others a moving average of the net difference between highs and lows. The basic principle is that a rising market over a period of time should be accompanied by a healthy, but not necessarily rising, number of net new highs. When the DJIA continues to make a series of higher peaks following a long advance, and the net number of new highs is tracing out a series of declining peaks, this is a warning of potential trouble. Similarly, in a bear market a new low in the DJIA which is not accompanied by an expanding number of net new lows should be looked at in a favorable light.

Basically, the first example is indicating that while the DJIA is rising, the technical picture is gradually weakening as successive peaks are accompanied by fewer and fewer stocks making breakouts (new highs) from price patterns. Since the net number of new highs also takes into consideration stocks making new lows, the higher peaks in the DJIA may also be accompanied by a larger number of stocks reaching new lows. The reverse, of course, is true at bottoms, where a declining number of stocks reaching new lows implies fewer downside breakouts and therefore a greater number of stocks resisting the downtrend in the major averages, which is clearly a positive sign (see Chart 11-10).

# PART THREE
# INTEREST RATES AND THE STOCK MARKET

# 12

# WHY INTEREST RATES AFFECT THE STOCK MARKET

Changes in interest rates affect the stock market for three basic reasons. First, fluctuations in the price charged for money influence the earnings or profits that companies are able to make and therefore the price which investors are willing to pay for stocks. Second, movements in interest rates alter the relationships between competing financial assets, of which the bond market/stock market relationship is the most important. Third, a substantial number of equities are purchased on borrowed money (known as margin debt), so that changes in the cost of carrying that debt (i.e., the interest rate) influence the desirability of holding such stock. Since significant changes in interest rates usually lead stock prices, it is important to understand and identify reversals in their cyclical trend, but first an examination of the cause-and-effect relationship between interest rates and the stock market is worth close scrutiny.

## THE EFFECT OF INTEREST RATE CHANGES ON CORPORATE PROFITS

The primary motivation for buying or selling equities is the fact that investors as a group consider stocks to be over- or undervalued in relation to the outlook for corporate profits and the future flow of dividends. Since interest rates affect company earnings, fluctuations in their level are clearly of importance to the trend of profits.

Interest rates affect profits in two ways. First, since almost all companies borrow money to finance either plant and equipment or inventory (or both), the cost of money, i.e., the interest rate they pay, is of great importance. Second, since many of the sales which each company produces are in turn financed by borrowing, the level of interest rates has a great deal of influence on the willingness of customers to make additional purchases. One of the most outstanding examples is the automobile business, where both producers and consumers are very heavily financed. Other examples are the utilities and transportation industries, which tend to be large borrowers because of the tremendous capital they require. The construction industry is also highly leveraged, so that the housing market is sensitive to interest rates.

Falling interest rates are normally good for business and profits, while rising interest rates, after a point, have the effect of hurting profits and adversely affecting stock prices.

## INTEREST RATES AND COMPETING FINANCIAL ASSETS

As well as influencing corporate sales and profits, interest rate changes also alter the attractiveness of various investment vehicles among themselves. The most significant relationship is that of stocks to bonds. For example, if at any given point investors judge stocks and bonds to be equally attractive, their investment posture will remain unchanged, other things being equal. However, if interest rates rise, it will be possible to obtain a better rate of return on bonds than was offered before. Consequently there will be a flow of money at the margin out of stocks into bonds. Stocks will then fall in value until the relationship is perceived by investors to be more in balance.

The full effect of interest rate changes on any particular stock group will depend upon the yield being obtained combined with the prospects for profit growth. Most sensitive will be the preferred shares, which are primarily held for their dividends and which do not generally permit their owners to participate in corporate profit growth. Second, while some growth in profits may be expected from utility companies, this growth is substantially influenced by regulation, so dividends are a relatively more important factor rendering the shares sensitive to interest rate changes. On the other hand, companies in a dynamic stage of growth are usually financed from corporate earnings and for this reason pay smaller dividends. These stocks are less affected by fluctuations in the cost of money since they are purchased in anticipation of fast profit growth and future yield rather than an immediate dividend return.

## INTEREST RATES AND MARGIN DEBT

Margin debt is money loaned by brokers on the basis of securities pledged as collateral. Normally this money is used for the acquisition of equities, but sometimes margin debt is used for purchases of consumer items, such as automobiles. The effect of rising interest rates on both forms of margin debt is similar in that rising rates raise the cost of carrying the debt and make it less attractive to carry. There is therefore a reluctance on the part of investors to take on additional debt as its cost rises, and at some point, when the service charges become excessive, stocks are liquidated and the debt is paid off. Thus, rising interest rates have the effect of increasing the supply of stock put up for sale and thereby pushing prices down.

Having looked at the effect of interest rates on the stock market, the relationship between interest rates and bond prices and the calculation of bond yields will now bc examined.

## BOND YIELDS

When a bond is brought to market by a borrower, it is issued at a fixed interest rate (coupon) which is paid over a predetermined period. At the end of this period, the issuer agrees to repay the face amount. Since bonds are normally issued in denominations of $1000 (known as par), this figure usually represents the amount to be repaid at the end of the (loan) period. Bond prices are quoted in percentage terms, so that par ($1000) is expressed as 100. Normally bonds are issued and redeemed at par; however, the issue price is sometimes at a discount, i.e., less than 100, and occasionally at a premium, i.e., over 100.

While it is usual for a bond to be issued and redeemed at 100, over the life of the bond its price can fluctuate quite widely since interest rate levels are continually changing. Assume that a 20-year bond was issued with an 8 percent interest rate (coupon) at par (i.e., 100); if interest rates rose to 9 percent, the bond at 8 percent would be difficult to sell, because investors would be unwilling to purchase the 8 percent bond while their money could earn a superior rate of return at 9 percent. The only way in which the 8 percent bondholder can find a buyer is to accept a lower price that will compensate a prospective purchaser for the 1 percent differential in interest rates. The new owner will then earn 8 percent in interest with some capital appreciation. When spread over the remaining life of the bond, this capital appreciation will be equivalent to the 1 percent loss in interest. This combination of coupon rate and averaged capital appreciation is known as the *yield*. If interest rates decline, the process is reversed, and the 8 percent bond becomes more attractive in relation to prevailing rates, so that its price rises.

The longer the maturity of the bond, the greater will be the fluctuation in its price for any given change in the general level of interest rates. The reason is that a more substantial adjustment in the price of the bond is necessary to compensate for the differential in the prevailing interest rates and the coupon for a longer maturity.

As interest rates rise bond prices fall, and when interest rates fall bond prices rise. Looked at in this way, it is easier to understand the relationship between the bond and stock markets. For this reason, most of the charts in this and subsequent chapters will use interest rates or yields plotted inversely, i.e., upside down, so that a falling line will indicate declining bond prices (i.e., declining debt markets), and vice versa.

## THE STRUCTURE OF THE DEBT MARKETS

The debt markets can be broken down into two main areas known as *the short end* and *the long end*. The short end, more commonly known as the money market, relates to interest rates charged for loans up to one year in maturity. Normally, movements at the short end lead those at the longer end, since short rates are more sensitive to trends in business conditions and changes in Federal Reserve policy. Money-market instruments are issued by governments of all kinds and by all types of corporations.

The "long end" of the market consists of bonds issued for a period of at least 15 years. There are, of course, a myriad of bonds issued for periods of between 1 and 15 years which fall between these definitions. Also, once a 25-year bond has been in existence for 20 years, it trades on the same basis as a bond issued for 5 years.

The bond market (i.e., the long end) has three main sectors, which are classified as to issuer. These are the U.S. government sector; the tax-exempt sector (i.e., state and local governments); and the corporate sector. The financial status of the tax-exempt and corporate sectors varies from issuer to issuer, and the practice of rating each bond for quality of credit has therefore become widespread.

The best possible credit rating is known as AAA; next in order are AA, A, BAA, BA, BB, etc. The better the quality of a bond, the lower the risk that investors undertake, so the lower the interest rate that is needed to compensate them. The credit of the U.S. federal government is higher than that of any other issuer, so it can sell bonds at a relatively low interest rate. The tax-exempt sector (i.e., bonds issued by state and local governments) is able to issue bonds with the lowest rates of all in view of the favored tax treatment assigned to the holders of such issues.

Most of the time, price trends of the various sectors will be similar, but at major cyclical turns some will lag behind others in view of the differing demand and supply conditions in each sector.

## DEBT AND EQUITY MARKETS

At cyclical peaks, debt markets typically top out ahead of the equity market. The lead characteristics and degree of deterioration of bond prices differ from cycle to cycle. Short-term rates usually start to move up first, followed later by long-term rates. There are no hard and fast rules that relate the size of the ensuing equity decline to the time period separating the peaks of bond and equity prices. Thus in 1959 short- and long-term rates peaked 18 and 17 months respectively ahead of the bull market high in the Dow. This compared with 11 months and 1 month for the 1973 bull market peak. While the deterioration in the bond and money markets was sharper and longer in the 1959 period, the Dow, on a monthly average basis, declined only 13 percent compared with 42 percent in the 1973–1974 bear market. The significant point is that every cyclical stock market peak in this century was preceded by, or has coincided with, a peak in both the long and short ends of the bond market.

A further characteristic of cyclical peaks is that higher-quality bonds (such as Treasury or AAA corporate bonds) decline in price ahead of poorer-quality issues such as BAA-rated bonds. This has been true of nearly every cyclical turning point since 1919. This lead characteristic of high-quality bonds results from two factors. First, in the latter stages of an economic expansion, private-sector demand for financing accelerates. Commercial banks, the largest institutional holders of government securities, are the lenders of last resort to private borrowers. As the demand for financing accelerates against a less accommodative central bank posture, banks step up their sales of these and other high-grade investments and reinvest the money in more profitable bank loans. A ripple effect is then set off, both down the yield curve itself and also to lower-quality issues. At the same time these pressures are pushing yields on high-quality bonds upwards. They are also reflecting buoyant business conditions which encourage investors to become less cautious. Consequently, they are willing to overlook the relatively conservative yields on high-quality bonds in favor of the more rewarding lower-rated debt instruments, so, for a temporary period, these bonds are rising while high-quality bonds are falling.

At cyclical bottoms these relationships are similar, in that good-quality bonds lead both debt instruments and equities that are of poorer quality. The lead characteristics of the debt markets are not quite so pronounced as at cyclical peaks, and occasionally bond and stock prices trough out simultaneously.

The trend of interest rates is therefore a useful benchmark for identifying stock market bottoms.

Chart 12-1 shows the 10 cyclical stock market troughs since 1919 that have been associated with tight monetary conditions. The monthly aver-

**CHART 12-1 Interest Rates at Cyclical Stock Market Troughs**

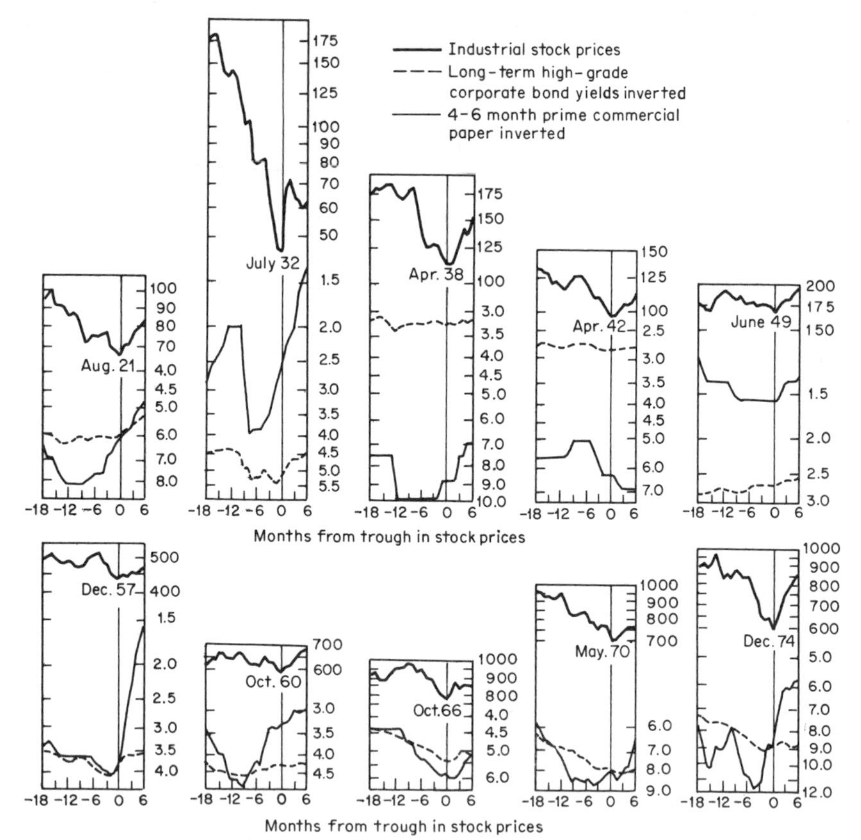


Courtesy of *The Bank Credit Analyst.*

age of stock prices is represented on the top line, with the yields of 4- to 6-month commercial paper and AAA corporate long-term bonds shown underneath. The respective yields have been plotted inversely so that they correspond with the trend of equity and bond prices.

Reference to the chart shows that in every case either one or both yield curves bottomed out ahead of stock prices. This relationship between debt and equity markets may be traced back through the early nineteenth century, and it can be shown that virtually every major cyclical stock market bottom was preceded by strength in one or both areas of the debt market. In most cases where this relationship did not hold, the ensuing rise in stock prices proved to be spurious, and more often than not prices eventually moved well below the previous low.

The fact that interest rates have begun a declining phase is not a sufficient condition in itself to justify the purchase of equities. For example, in the 1919–1921 bear market, bond prices reached their lowest point in June 1920, 14 months ahead or 27 percent above the final stock market bottom in August 1921. An even more dramatic example occurred during the 1929–1932 debacle when money market yields reached their highs in October 1929. Over the next 3 years the discount rate was cut in half, but stock prices lost 85 percent of their October 1929 value. The reason for such excessively long lead times was that these periods were associated with a great deal of debt liquidation and many bankruptcies. Even the sharp reduction in interest rates was not sufficient to encourage consumers and businesses to spend, as is the normal cyclical experience. While falling interest rates are not by themselves a sufficient reason to expect that stock prices will reverse their cyclical decline, they are a necessary one. On the other hand, a continued trend of rising rates has in the past proved to be bearish.

The principles of trend determination appear to apply equally well to the bond market as to the stock market. In fact, trends in bond prices are in many ways easier to identify, since the bulk of the transactions in bonds are made on the basis of money flows caused by a need to finance and an ability to purchase. Consequently, while emotions are still important from the point of view of determining the short-term trends of bond prices, the money flow aspect is responsible for a far smoother cyclical trend than is the case with equities.

# 13

# SHORT-TERM INTEREST RATES

Short-term interest rates are more sensitive to business conditions than long-term rates. This is because decisions to change inventories, for which a substantial amount of short-term money is required, are made much more quickly than decisions to purchase plant and equipment, which form the basis for long-term credit demands. The Federal Reserve, in its management of monetary policy, is also better able to influence short-term rates than bonds at the longer end. For these reasons changes in short-term rates usually precede changes in long-term rates.

The key short-term interest rate that should be monitored is the Federal Funds rate. At any particular point in time, banks are required to hold a certain level of reserves with the Federal Reserve in relation to their deposits. Some banks will find themselves with a surplus of reserves over the required amount, and others will have a deficiency. The practice has evolved of surplus-holding banks lending to deficient ones, so that the latter can meet their reserve requirements. The Federal Funds rate is simply the rate of interest charged for these loans. Through various techniques, the Federal Reserve is able to control the amount of reserves in the banking system, and it is therefore able to influence the Federal Funds rate. The trend in the Federal Funds rate is therefore a good indication of the direction of monetary policy and financial pressures in the system as well as a guide to the future course of short-term rates in general.

Chart 13-1 shows one technique that has proved useful in determining the intermediate movements in the Federal Funds rate. A reversal in the trend of the rate is signaled when the Federal Funds rate crosses over its 10-week moving average (dashed line) and this is confirmed by the mo-

**CHART 13-1 Federal Funds Rate and 13-Week Momentum**

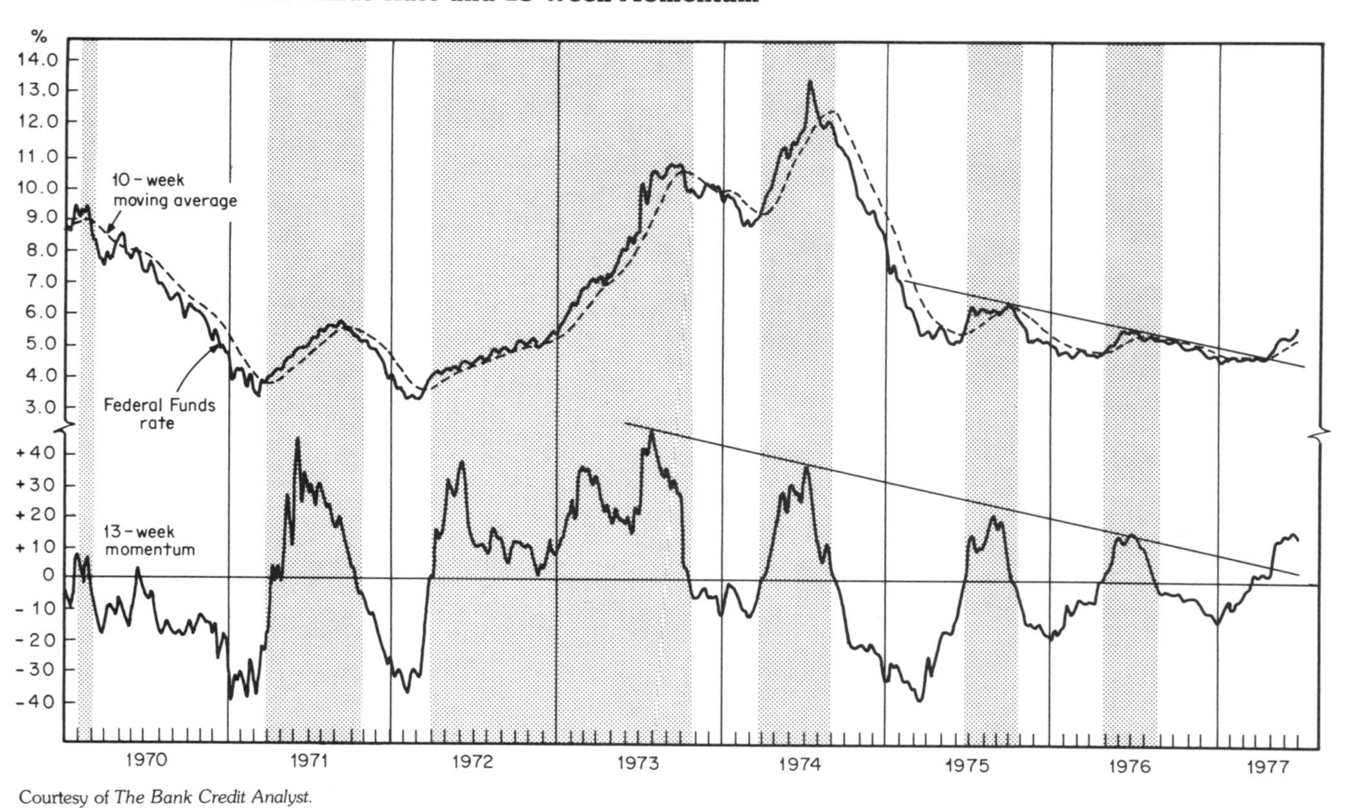


Courtesy of *The Bank Credit Analyst.*

mentum index (13-week rate of change) crossing its zero reference line in the same direction. The trend is assumed to be in existence until both indexes signal a reversal. For instance, at the beginning of 1970 the prevailing trend was down. It appeared to be reversed when the Federal Funds rate crossed above its 10-week average, but this was not confirmed by the momentum index, which remained below its zero reference line. It was not until 1971 that the two indexes confirmed each other and a valid warning of a reversal was given.

Since 1970 moving average crossovers in the Federal Funds rate combined with trend reversals in the 13-week momentum index have been useful in forecasting future trend reversals of an intermediate nature in the DJIA (see Chart 13-2). At this point a word of caution is in order,

**CHART 13-2 Federal Funds Rate versus Dow Jones Industrial Average**

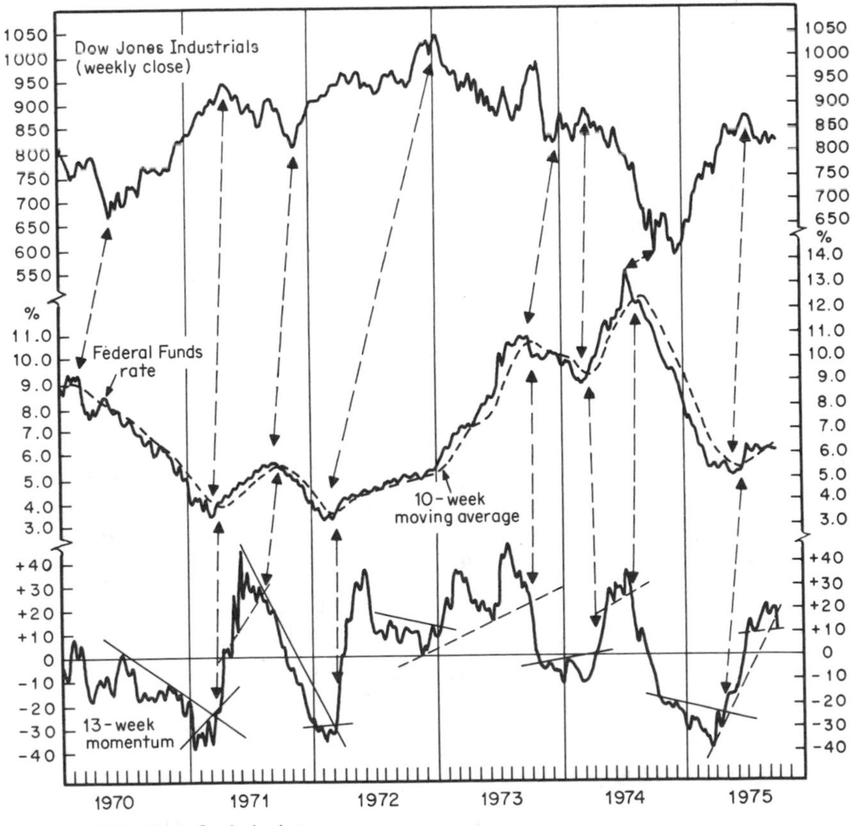


Courtesy of *The Bank Credit Analyst.*

since the relationship between interest rates and the stock market is usually less precise than is indicated in the chart. Consequently it should not be assumed that any change in the Federal Funds rate will necessarily affect the stock market to an equal degree. This relationship should be regarded as an important but *not infallible* indicator of future stock market trends.

## FOUR- TO SIX-MONTH COMMERICAL PAPER

Another key short-term interest rate, one with a long historical record, is that charged for 4- to 6-month prime (i.e., best-quality) commercial paper. From time to time corporations have a substantial amount of surplus cash which will be available for a relatively brief period. Other corporations find themselves in need of money. Normally the two corporations would lend each other money through the intermediacy of a bank. The practice has evolved of bypassing the banks and acting through an investment dealer, who is paid considerably less than the bank for his services. Consequently both borrower and lender are able to obtain more favorable terms. This form of debt is known as commercial paper. The monitoring of commercial paper rates is very useful, since these interest rates give a good indication of financial pressures and trends of short-term interest rates in the private sector.

There are of course many techniques that can be used to determine the change in trend of the commercial paper rate, but one that has been found most useful from the point of view of confirming reversals in cyclical movements is shown in Chart 13-3.

The solid line represents the rate itself and the dashed line an 18-month weighted moving average. Signals are given when the moving average changes direction. An 18-month time span has proved to be useful in determining the cyclical trend of short-term interest rates, since it smooths out whipsaws (such as the 1971 rise) yet retains sufficient sensitivity to respond fairly quickly to changes in longer-term trends. In the 58-year period between 1919 and 1977, a reversal in the downward movement of this moving average, following a cyclical decline in short-term rates, has with only two exceptions confirmed that a cyclical rise in rates was under way. The two exceptions occurred in 1931 and 1933, where the discount rate was temporarily raised and then lowered in response to specific crises.

## TREASURY BILLS

Another money market rate that should be monitored is the yield on 3-month Treasury bills. Treasury bills are short-term debt instruments (i.e.,

**CHART 13-3 Commercial Paper Rate**

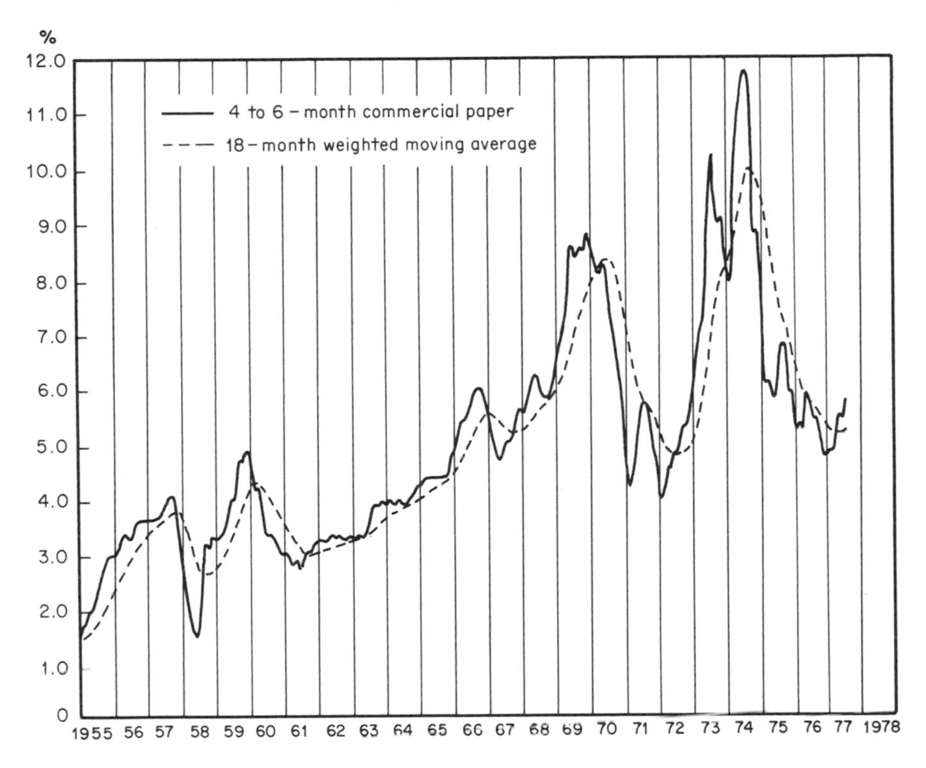


3-month to 1-year) issued by the U.S. government. The bills are redeemed at face value when they mature but pay no interest. The holder obtains his return by purchasing these bills at a discount (i.e., a price below the face value), which compensates for the loss of interest. Unlike the purchaser of commercial paper who earns an income, a Treasury bill holder really receives a guaranteed capital gain from his investment. It is possible to calculate the rate of return on a Treasury bill by relating the capital gain to the time remaining to maturity. Treasury bills are usually quoted on a yield basis, so that their rate of return can be compared with that of other short-term instruments. A Treasury bill yield is shown in Chart 14-1. Most of the time the rate on 90-day Treasury bills will move in the same direction as the commercial paper rate, since both debt instruments are influenced by domestic financial conditions. However, the movement of Treasury bills occasionally diverges for short periods, since they are also bought by foreign investors and foreign central banks. Trends in Treasury bill yields are analyzed by means of essentially the same techniques as the other money market rates discussed above.

## THE IMPORTANCE OF CHANGES IN THE DISCOUNT RATE

The importance of a discount rate change compared with a change in any other interest rate is that it is an indication of official central bank policy. Since the central bank is in a position to influence other interest rates, it reflects the desire of the central bank to raise or lower the general level of interest rates. It is therefore a good indication of the future direction of these market-oriented rates and has a strong psychological influence on both debt and equity markets. Additionally, since the Federal Reserve does not reverse policy decisions on a week-to-week basis, a change in the discount rate implies that the trend in market interest rates is unlikely to be reversed for at least a few months. Just as a corporation does not like to cut dividends shortly after they have been raised, so does the central bank wish to create a feeling of continuity and consistency. A change in the discount rate is therefore helpful in confirming trends in other rates which, when taken by themselves, can sometimes give misleading signals due to temporary technical or psychological factors.

Since the incorporation of the Federal Reserve System, with the exception of the depression and war years of 1937 and 1939 and more recently 1976, every single bull market peak in equities has been preceded by a rise in the discount rate. The leads have varied. In 1973 the discount rate was raised on January 12, 3 days prior to the bull market high, whereas the 1956 peak was preceded by no less than five consecutive hikes. It is worth considering the number of months separating the point where the discount rate reached its cyclical low from the point marking a cyclical high in the equity market. For example, in the 1970–1973 bull cycle, the

**TABLE 13-1**

| Discount rate cyclical low | Stock market cyclical peak | Lead time in months |
|---|---|---|
| August 1924 | September 1929 | 61 |
| February 1934 | March 1937 | 37 |
| August 1937 | September 1939 | 26 |
| October 1942 | May 1946 | 44 |
| April 1954 | April 1956 | 25 |
| April 1958 | December 1959 | 20 |
| August 1960 | February 1966 | 65 |
| April 1967 | December 1968 | 20 |
| December 1971 | January 1973 | 13 |

**CHART 13-4 The Discount Rate and the Stock Market**

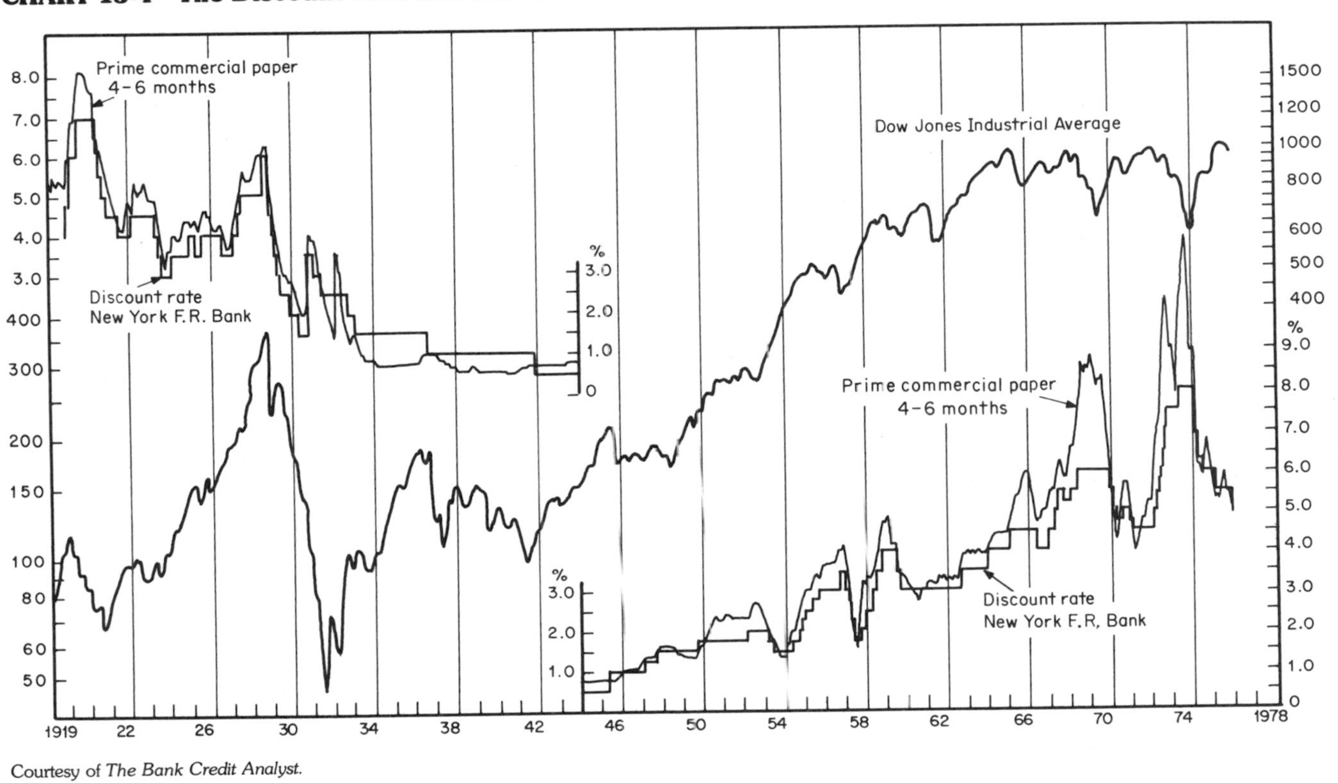


Courtesy of *The Bank Credit Analyst.*

discount rate was cut to its lowest point (4½ percent) in December 1971, 13 months prior to the bull market top in January 1973. Other bull market cycles since 1919 are as shown in Table 13-1.

Looking at the table, it can be seen that historically the lead time can vary from 5½ years, as in the 1960–1966 case, to as little as 13 months in the 1971–1973 period.

As can be seen in Chart 13-4, cuts in the discount rate usually precede stock market bottoms, although the relationship is far less precise than that observed at market tops. Note, for example, how the rate was lowered no less than seven times during the 1929–1932 debacle, while it was not changed at all during the 1946–1949 bear market.

# 14

# LONG-TERM INTEREST RATES

As discussed in Chapter 12, the long-term bond market comprises the U.S. government sector, the corporate sector, and the tax-exempt sector. The yields of a series representative of each are shown in Chart 14-1. During the period under consideration it can be seen that all sectors basically move in the same direction most of the time, as might be expected. At specific points, the supply/demand relationship peculiar to one or other of the sectors sets it apart from the rest. For example, the long-term corporate market reached its low point, i.e., highest yield as shown on Chart 14-1, in October 1974 (along with most bonds of shorter maturity), while the municipal and government indexes did not reach their low point until October 1975. Moreover, the government yield index troughed in January 1977, while its municipal counterpart reached its low many months later. Chart 14-2 shows a composite bond market index constructed from the average of all three sectors. This index is useful since it places movements of all three areas in perspective and serves as a base from which relative-strength analysis can be undertaken.

Technical analysis of bonds is very similar to that of stocks in that trend reversals in yield indexes or price indexes of bonds can be identified by price pattern formations, trendlines, momentum and moving averages, etc. In a sense it is possible to extend a crude form of breadth analysis, for an advance or decline by one sector, if unconfirmed by the other two, is likely to prove spurious since it will result in a change in the competitive spread between the various sectors. One major difference between bonds and stocks is that high-quality bonds generally lead the debt market in both directions, whereas it is generally the poorer-quality stocks which

**CHART 14-1 Key Interest Rates, 1955–1978: AAA, Government Long-Term, and Bond Buyer 20 Bond Yields**

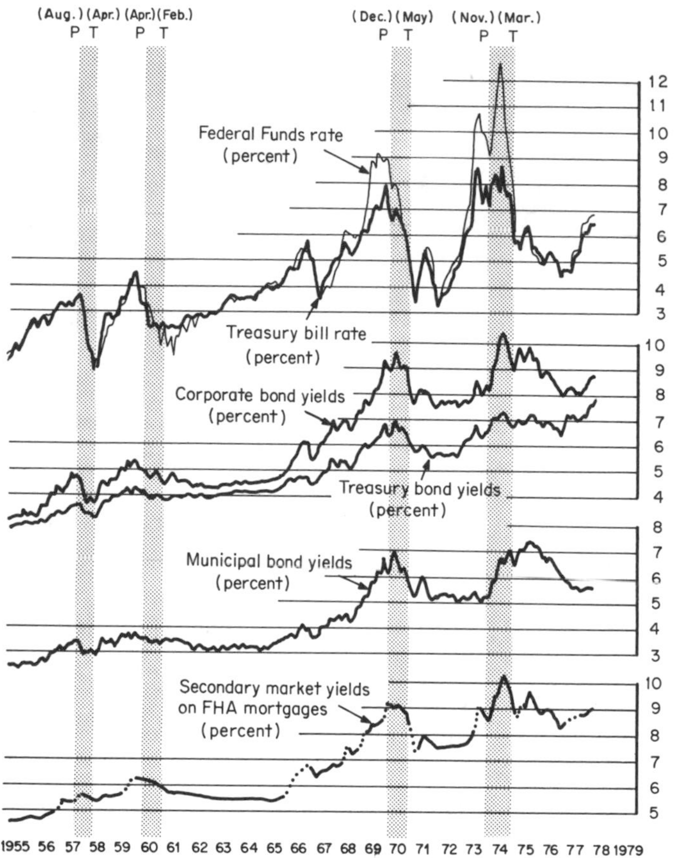


National Bureau of Economic Research.

lead the equity market down; both classes appear to bottom out simultaneously.

## RELATIONSHIP BETWEEN BOND AND STOCK YIELDS

Chart 14-3 shows the ratio of Moody's AAA corporate yields divided by the indicated yield on Standard & Poor's 500 stock index. When the index falls, it shows that the return on stocks based on the indicated dividend is

**CHART 14-2 Composite Long-Term Bond Yield Index, 1974–1978**

This chart of the *Bank Credit Analyst's* composite long-term term bond yield index has been plotted inversely to correspond with bond prices. The dashed lines at +5 and −5 percent are used as a rough benchmark for "overbought" and "oversold" areas. Note the positive divergence between the 13-week momentum index and the index itself between late 1974 and late 1975.

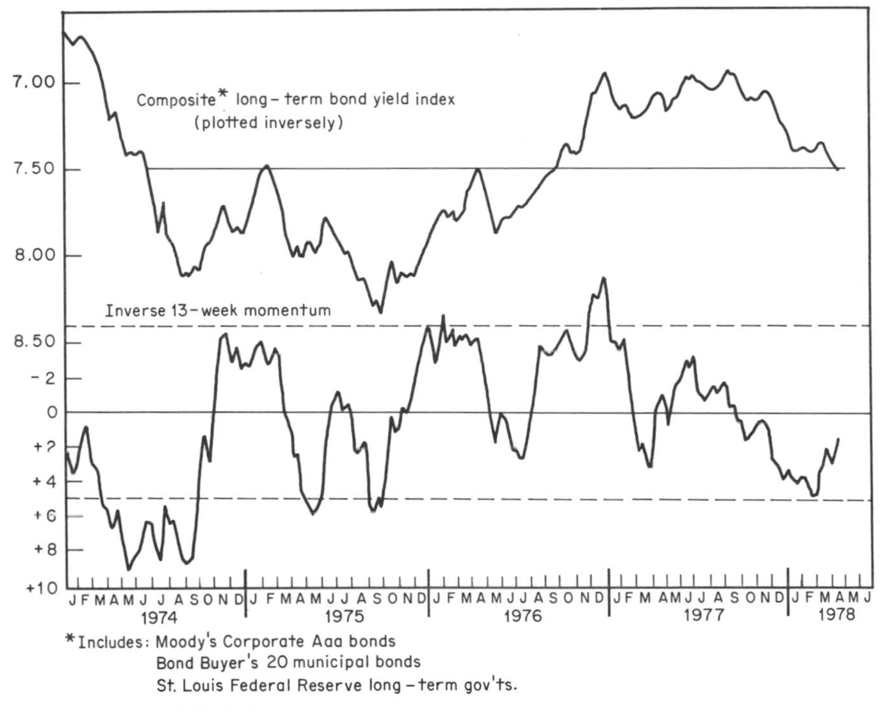


Courtesy of *The Bank Credit Analyst.*

becoming more attractive in relation to bonds. When the line rises, stocks (on a yield basis) are becoming more expensive. When the line has been rising for some time, it is usually indicating a bull market in equities, whereas a declining line is normally associated with a bear market. Quite often the bond/stock index peaks out ahead of the stock market, giving advance warning of a reversal in the primary trend. Usually this lead time is relatively short, as in 1959 and 1972, but in the case of 1936–1937 the line peaked out well ahead of the stock market indicating that as stocks rose in price they were becoming cheaper and cheaper in relation to bonds, since dividends were growing at a faster rate than stock prices.

**CHART 14-3 Moody's AAA/Standard & Poor's 500 Yield, 1926–1977**

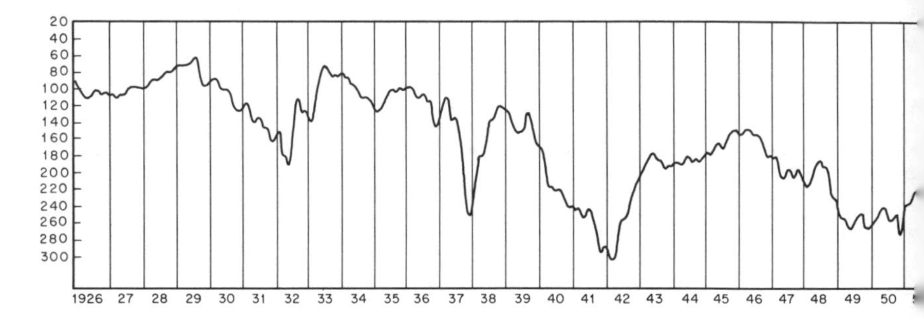


From a longer-term point of view, it is worth noting that in the postwar period stocks at market tops have become successively more and more overvalued in relation to bonds. In 1976 for the first time this series of higher peaks in the index was not achieved, suggesting that the secular uptrend in the index which began in the 1930s may have been reversed.

If this does prove to be the case, there are two important implications. First, the break in the secular trend of higher and higher levels of overvaluation of equities in each stock market cycle would suggest that investors in aggregate are no longer assuming that the postwar period of buoyant growth interrupted by mild recessions will continue. In other words, investors were willing to continue to pay higher prices for stocks in relation to bonds only because they had confidence in the system's ability to produce unlimited profit and dividend growth.

The second implication is more technical in nature. During the secular uptrend, the lead time between the peak in the bond and equity markets could be quite long, i.e., stocks were less sensitive to rising interest rates. On the other hand, the lead between interest rates and equities at market bottoms was fairly short, since stocks were very sensitive to falling interest rates. If in fact a basic change in this relationship has taken place, it means that stocks will in future become more sensitive to rising rates and less sensitive to falling rates, as in the 1930s.

Chart 14-4 helps to show the degree of overvaluation of stocks in an even longer historical context. The index is similar to the one just described, but instead of dividing one yield by another, a subtraction is made. The resulting index is called a *yield gap*, since it measures the difference in yield between bonds and stocks. Normally stock yields are higher than bond yields in view of the greater risk associated with stocks, but sometimes, as in the last 10 to 15 years, bond yields have been higher. During such periods the yield gap becomes negative and is known as the

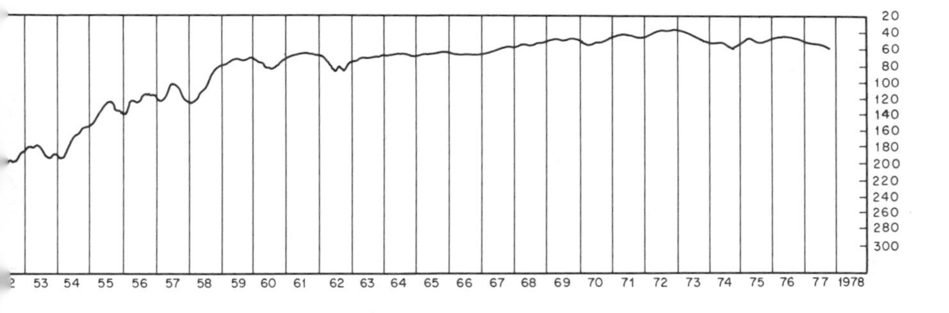


"reverse" yield gap. This negative gap is a direct result of growing investor confidence and expectations arising from the virtually uninterrupted period of postwar prosperity. Under such conditions stocks are obviously less risky than in the more volatile economic conditions of the past.

Chart 14-4 shows that even at the bottom of the 1973–1974 bear market, the most severe in postwar history, stocks were still overvalued on a yield basis in relation to bonds in a long-term historical context, even though they appeared to be inexpensive in relation to the previous 10 years. In this connection, Chart 14-5 shows the postwar period in greater detail.

**CHART 14-4 Long-Term Reverse Yield Gap, 1867–1975**

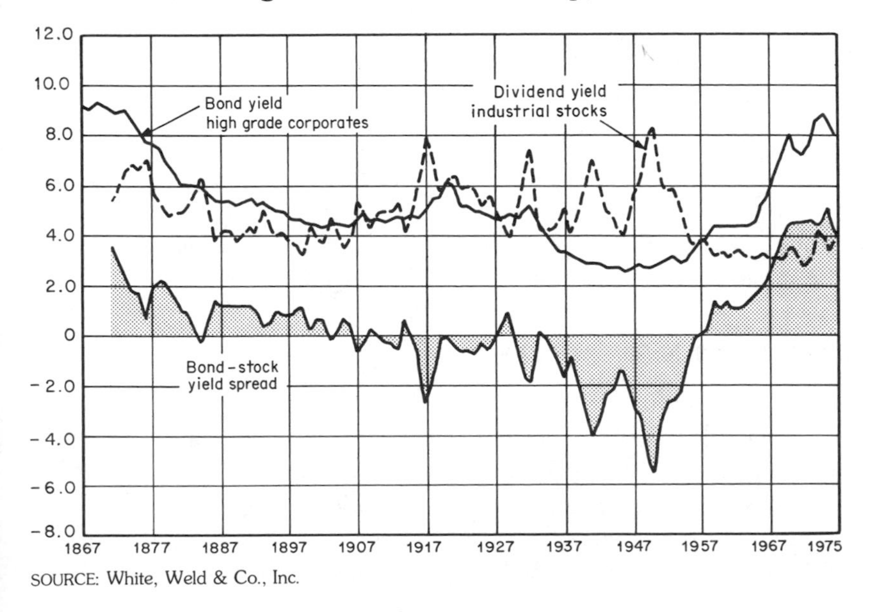


SOURCE: White, Weld & Co., Inc.

**CHART 14-5 Reverse Yield Gap, 1949–1978** [Recent bond/stock yield spread between average quarterly yield on Moody's Composite (average long-term Corporate Bond Series—AAA, AA, A, and BAA) and the indicated quarterly (seasonally adjusted) dividend yield on Standard & Poor's 425 Industrial Stock Price Index]

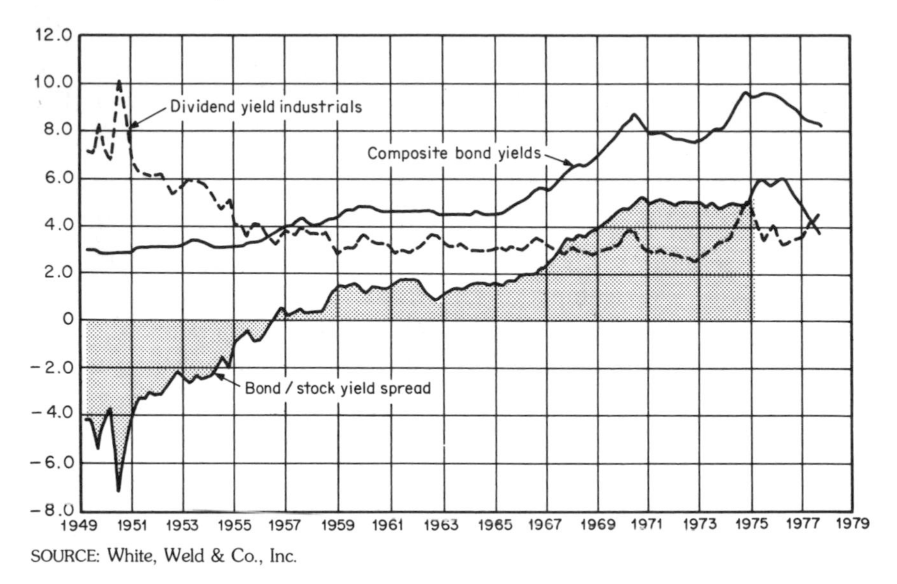


SOURCE: White, Weld & Co., Inc.

## TREND DETERMINATION OF LONG-TERM BONDS

Analysis of the trend in bond prices (or yields) is approached in the same way as analysis of the various equity market indicators. Because the movement of bonds tends to be much smoother than those of stocks, whipsaws resulting from the use of moving averages are usually far less troublesome.

## SHORT-TERM MOVEMENTS

One good method for determining short-term (4- to 12-week) movements in bonds is to plot a price arithmetically (or inverted yield) above a 30-day rate-of-change (momentum) index, as in Chart 14-6. The bond used in the chart is an 8 percent Government National Mortgage Certificate (GNMA), a widely traded U.S. government agency debt instrument which is a good bellwether for the medium-term U.S. government mar-

**CHART 14-6 Government National Mortgage Association Certificates (GNMAs) versus 30-Day Momentum**

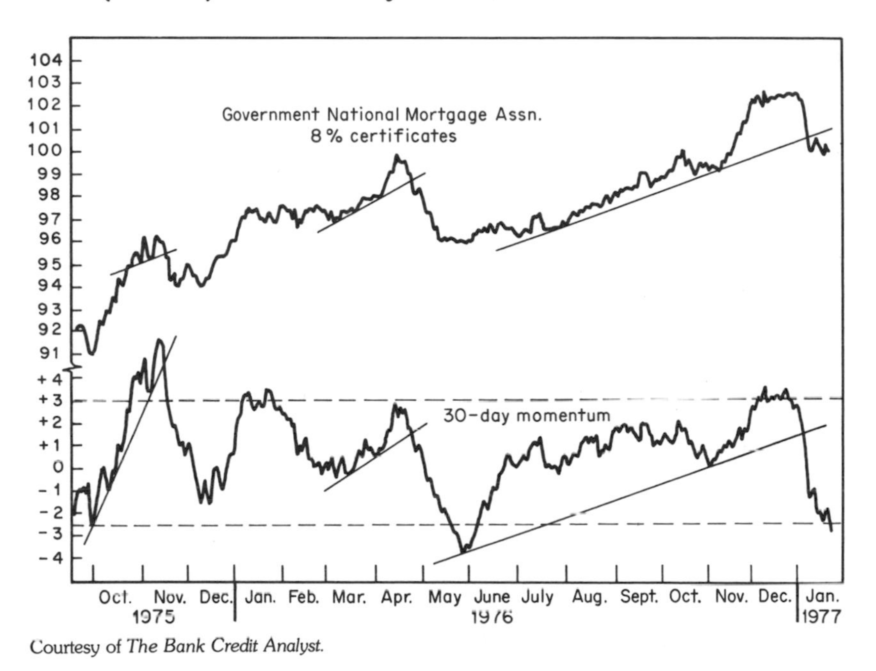


Courtesy of *The Bank Credit Analyst.*

ket. The technique is a very simple one of drawing trendlines for both the GNMA and its 30-day rate of change. When a trendline break is confirmed for both the index and its momentum index, a reliable indication is given of a reversal in the trend of the GNMA itself. The level of the 30-day rate-of-change index is also useful for judging when the price of the GNMA is likely to run into support or resistance. Such "overbought" or "oversold" indications are shown by the two dashed lines.

PART FOUR

# OTHER ASPECTS OF MARKET BEHAVIOR

# 15

# SENTIMENT INDEXES

I find more and more that it is well to be on the right side of the minority since it is always the more intelligent.
GOETHE

During primary bull and bear markets the psychology of all investors moves from pessimism and fear to hope, overconfidence, and greed. For the majority the feeling of confidence is built up over a period of rising prices, so that optimism reaches its peak around the same point that the market is also reaching its high. Conversely, the majority is most pessimistic at market bottoms, at precisely the point when the acquisition of stocks should be taking place.

The better-informed market participants, such as insiders and stock exchange members, act in a contrary manner to the majority by selling at market tops and buying at market bottoms. Both groups go through a complete cycle of emotions, but in completely opposite phases. This is not to suggest that members of the public are always wrong at major market turns and that professionals are always correct, but that in aggregate the opinions of these groups are usually in direct conflict.

Data are available on many of these market participants, so that over a period of years it has been possible to derive parameters that indicate when a particular group has moved to an extreme historically associated with a major market turning point.

Unfortunately there is a strong possibility that many of the indexes that have proved useful historically may now be distorted due to the advent of listed options trading in 1973. This arises because the purchase and sale of options are a substitute for short selling and other speculative activity formerly carried out on the Stock Exchange. If the degree of distortion could be ascertained, it would be possible to make an adjustment to the indexes affected, but the history of options trading is so brief that modifications cannot be made with any degree of reliability. Generally speaking, the data relating to market participants that have not been unduly affected by options trading are unavailable prior to the mid-1960s. Some degree of caution should therefore be exercised in the interpretation of these data in view of their relatively short history. For these reasons, some of the more popular indicators measuring investor expectations which almost certainly have been distorted by options trading, such as the Odd-Lot and Short-Interest Ratios, are not discussed here;[1] but since a description of technical analysis would not be complete without some reference to investor sentiment, some of the more reliable indicators are considered below. Incorporation of three or four indexes measuring sentiment is useful from the point of view of assessing majority opinion, from which a contrary opinion can be taken.

## MEMBERS SHORT INDEX

Each week data are published by the NYSE which break down the total round-lot short sales into two categories, namely, those made by stock exchange members and those made by nonmembers. A ratio of members' short sales to the total short sales is calculated each week and then smoothered by a 10-week moving average (see Chart 15-1). Over the years,

**CHART 15-1 Dow Jones Industrial Average versus Members 10-Week Short Ratio**

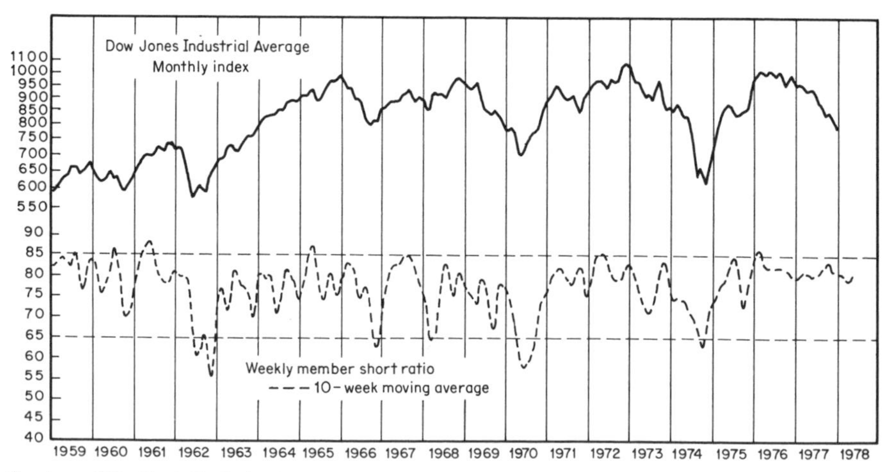


Courtesy of *The Bank Credit Analyst.*

[1] Representative charts of some of these indexes that may be subject to distortion are listed in Appendix 3.

stock exchange members in aggregate have proved to be consistently on the right side of the market. Consequently, when the 10-week Members' Short Ratio reaches the 85 percent level, it indicates that over the previous 10-week period only 15 percent of total NYSE short sales have been made by the public. A 65 percent reading in the ratio, on the other hand, shows that a somewhat higher 35 percent of all short sales have been made by the public. A high level of shorting by the public (30 to 35 percent) is considered bullish, and a low level (15 to 20 percent) bearish. Readings between 70 and 80 percent are considered normal and have little forecasting value. Sometimes the ratio is smoothed by a 2-month moving average, which is shown on Chart 15-2 along with a similar index for the American Stock Exchange. The dashed lines on the chart are representative of the extreme levels of the index, although it can be seen, for example, that in the mid-sixties high readings did not necessarily result immediately in a falling stock market.

In this respect it is worth noting that the index has a habit of achieving

**CHART 15-2 Dow Jones Industrial Average versus Members 4-Week Short Ratio**

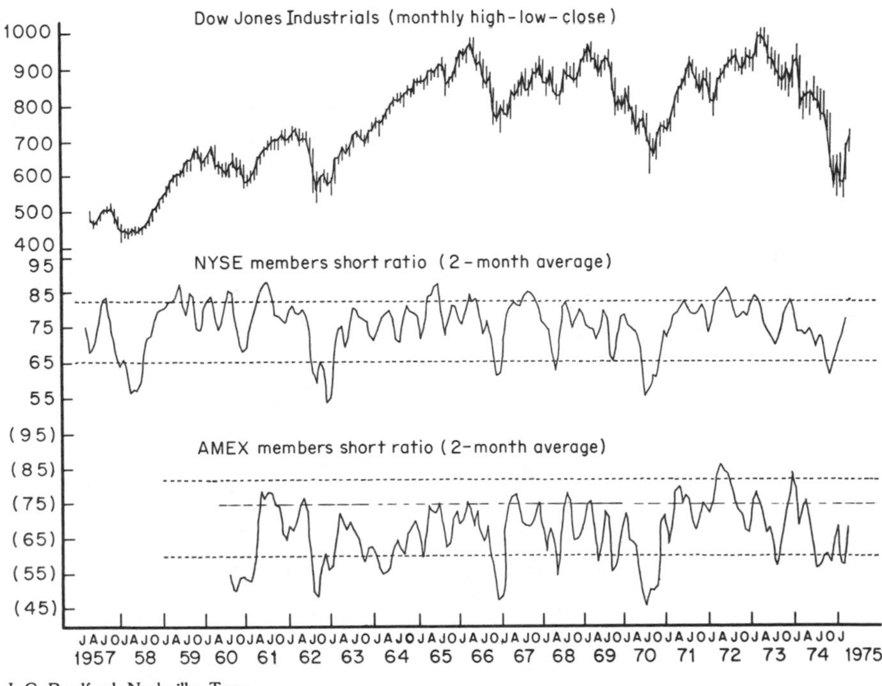


J. C. Bradford, Nashville, Tenn.

a series of declining peaks as the market works its way higher (e.g., 1965 and 1966). These divergences should be viewed as growing technical weakness, just as the two descending troughs in 1962 indicated technical strength in November of that year, since the public was shorting proportionally more stock even though stock prices held above their earlier lows.

The Members' Short Ratio is therefore better employed as a gauge of public confidence in the market rather than as an actual "buy" or "sell" indicator. For example, if the ratio is at a low 65 percent reading, the oppressive nature of public short selling indicates that an extremely pessimistic attitude toward the market has developed. Chances are that most of the discouraging news has already been built into the price level and that the market is likely to advance.

At the other extreme, a high reading in the index indicates a reluctance on the part of the public to go short, since this group is anticipating a further market advance. But chances are that these expectations have already been translated into stock purchases, leaving these participants with little or no cash reserves with which to push stock prices higher.

While it is possible that the Members' Short Ratio may have been distorted by options trading, it is also apparent that the option alternative is open to both members and nonmembers alike, since the ratio relates two categories of short selling to each other. This is not the case for many other indexes constructed from short-sale data, since most of these compare short sales with some other figure, e.g., odd-lot shorts to odd-lot sales or total NYSE short position to total NYSE volume.

## ADVISORY SERVICES

Since 1963 Investors' Intelligence[2] has been compiling data on the opinions of publishers of market letters. It might be expected that this group would be well informed and offer advice of a contrary nature by recommending acquisition of equities at market bottoms and offering selling advice at market tops. The evidence suggests that the advisory services in aggregate act in a completely opposite manner and therefore represent a good proxy for "majority" opinion.

Chart 15-3 shows the percentage of bullish market letter writers in relation to the total of all those expressing an opinion. The data have been smoothed by a 10-week moving average to iron out misleading fluctua-

[2] Larchmont, N.Y.

**CHART 15-3 Advisory Service Sentiment**

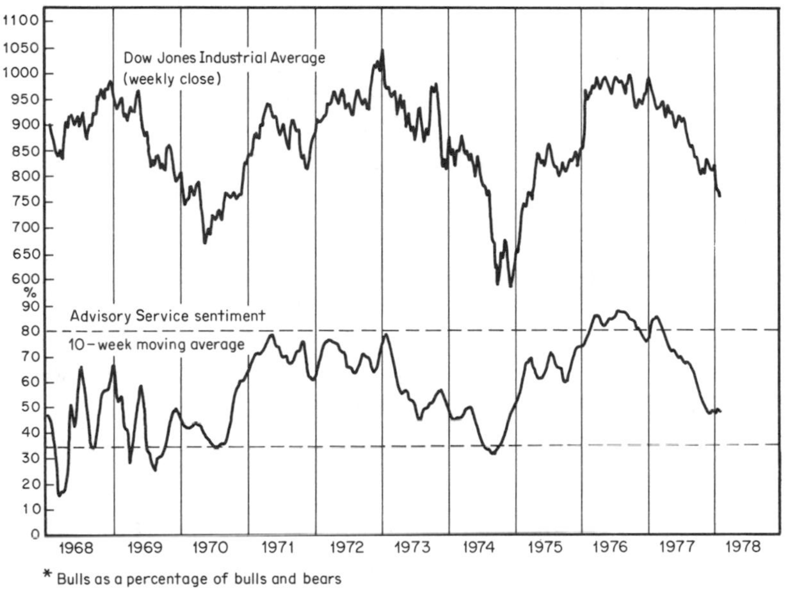


* Bulls as a percentage of bulls and bears

SOURCE: Investors Intelligence, Inc., Larchmont, N.Y.

tions. The resulting index shows that the advisory services follow the trend of equity prices by becoming most bullish near market tops and predominantly pessimistic around market bottoms. Investors would clearly find it more profitable to take a position contrary to that of the advisory service industry.

This index also gives a good indication of how market psychology can swing from outright pessimism to extreme overconfidence. In early 1968, for example, virtually all services were putting out bearish forecasts right at a major low. Then, as prices began to rise, their prognostications became more optimistic and turned to outright bullishness by the time prices reached their highest level. This indicator would have proved very useful during the 1973–1974 bear market, for although the averages experienced two substantial declines in 1973, the index showed that while confidence among the advisory services was low, it was not sufficiently weak to indicate that a major bottom had been reached.

The principles of nonconfirmation and divergence can also be applied to the interpretation of this index. For example, the Dow peak in 1973 was not confirmed by a new high in the Advisory Services Index. During the 1969–1970 bear market prices worked their way lower, but confidence was markedly higher in May 1970 than in July 1969, thereby setting up a positive divergence.

## MUTUAL FUNDS

Data relating to mutual funds are published on a monthly basis by the Investment Company Institute. The statistics are useful because they monitor the actions of both the public and the institutions. In recent years money-market and tax-exempt mutual funds have become widespread, so the data used here have been modified to include only equity funds. There are two useful measurements of mutual fund activity. The first relates to mutual fund cash as a percentage of assets, and the second to mutual fund cash and annual redemptions.

### Mutual Fund Cash/Assets Ratio

At all times mutual funds hold a certain amount of liquid assets, so that investors wishing to cash in or redeem their investment can be accommodated. When the cash position is expressed as a percentage of the total value of mutual funds' portfolios (a figure known as total asset value), a useful indicator is derived (see Chart 15-4). The index moves in the opposite direction to the stock market, since the proportion of cash held by mutual funds rises as prices fall, and vice versa. There are three reasons for this characteristic. First, as the value of a fund's portfolio falls in a declining market, the proportion of cash held will automatically rise even though no new cash is raised. Second, as prices decline the funds become more cautious in their buying policy, since they see fewer opportunities for capital gains. Third, the decision is made to hold more cash reserves as insurance against a rush of redemptions by the public. In a rising market the opposite effect is felt, as advancing prices automatically reduce the proportion of cash, sales increase, and fund managers are under tremendous pressure to capitalize on the bull market by being fully invested.

The cash/assets ratio can be interpreted in two ways. First, as illustrated in Chart 15-4, two parameters are used as benchmarks for buying and selling. In most instances such signals result in an unduly long lead time before peaks and troughs, and an alternative approach is to wait for a reversal in the trend of the index. A useful method of accomplishing this is to construct a 12-month moving average of the ratio, using crossover points as "buy" and "sell" indicators. Most of the time this technique is

**CHART 15-4 Dow Jones Industrial Average versus Mutual Fund Liquidity and 12-Month Moving Average, 1966–1977**

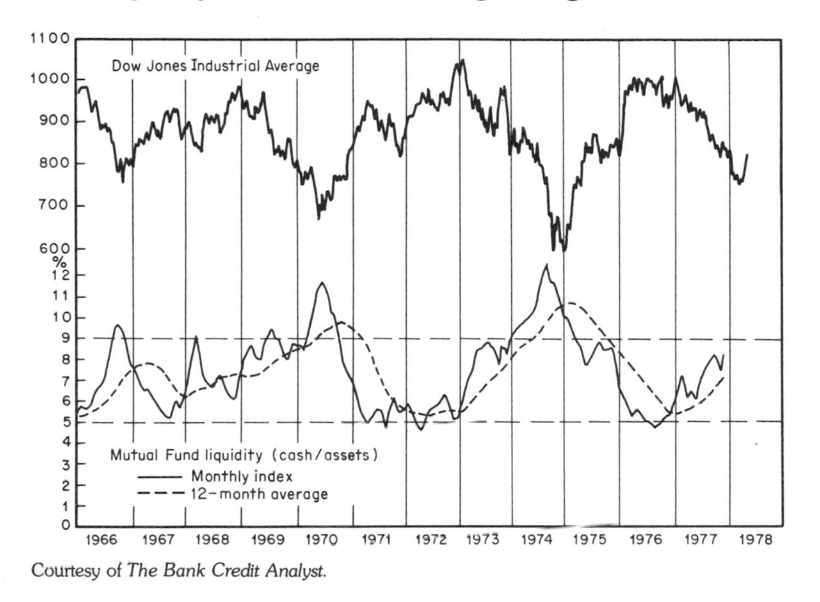


Courtesy of *The Bank Credit Analyst.*

very useful, but occasionally, when the action of the index is indecisive (e.g., the 1971–1972 period), whipsaws can develop. In such instances it is as well to examine the action of other expectational and technical indicators, with the objective of forming an overall opinion rather than trying to fathom the course of one particular indicator.

## Mutual Fund Cash/Redemptions

Another way of looking at the pressure on mutual funds is to compare their cash position with the previous 3 months of redemptions. In this instance the cash reserves are used as a proxy for institutional buying power, and the redemptions as an indication of selling pressure on the market by the public. This ratio is shown in Chart 15-5. When cash on hand is considerably higher than redemptions, i.e., a ratio in excess of 18, selling by the public does not appear to place undue pressure on the institutions. Conversely, when the ratio falls to 8 or below, this combination of low cash reserves (buying power) and strong selling pressure results in declining prices. Quite often a change in the direction of the trend can be as important as the actual level of the index. Conse-

**CHART 15-5 Dow Jones Industrial Average versus Mutual Fund Cash/3-Month Redemption Ratio**

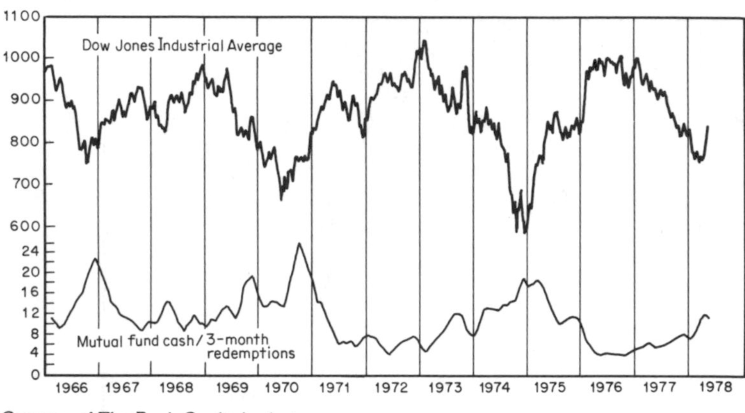


Courtesy of *The Bank Credit Analyst.*

quently, a moving average can prove useful in assessing important juncture points in the index.

## MARGIN DEBT

Trends in margin debt are probably better classified as flow-of-funds indicators, but since the trend and level of margin debt is also a good indication of investor confidence (or lack of confidence), it is discussed in this section.

Margin debt is money borrowed from brokers and bankers using securities as collateral. The credit is normally used for the purchasing of equities. At the beginning of a typical stock market cycle, margin debt is relatively low; it begins to rise very shortly after the final bottom in equity prices. As prices rise, the margin trader becomes more confident, and since he wishes to take maximum advantage of an advancing market, he purchases more stock by taking on additional margin debt.

During a primary uptrend, margin debt is a valuable source of new funds for the stock market. The importance of this factor can be appreciated when it is noted that margin debt increased by $6.3 billion between 1974 and 1977. The essential difference between stock purchased for cash and that bought on margin is that the latter at some point must be sold in order to pay off the debt, whereas stock purchased outright can theoretically be held indefinitely. Consequently, during stock market declines margin debt reverses its positive role and becomes an important source of stock supply.

This occurs for four reasons. First, margin requirements are fairly closely regulated, so the financial position and sophistication of margin-oriented investors is relatively superior to that of other market participants. When this group realizes that the potential for capital gains has greatly diminished, a trend toward liquidation begins. Thus, margin debt has flattened or declined within 3 months of all but one of the 14 stock market peaks since 1932.

Second, primary stock market peaks are invariably preceded by rising interest rates, which in turn increase the carrying cost of margin debt, therefore making it less attractive to maintain.

Third, since 1934 the Federal Reserve Board has been empowered to set and vary margin requirements, which specify the amount that can be lent by a broker or bank to customers for the purpose of holding securities. This measure was considered necessary in view of the substantial expansion of margin debt in the late 1920s, the liquidation of which greatly contributed to the severity of the 1929–1932 bear market. When stock prices have been rising strongly for a period of time, speculation develops, often resulting in a sharp rise in margin debt. Sensing that things could get out of control at this stage, the "Fed" raises the margin requirement, which has the effect of reducing the buying power of the general public from what it might otherwise have been. Normally, it takes several margin-requirement changes to significantly reduce the buying power of margin speculators, since the substantial advance in the price of stocks—which was responsible for the requirements' being raised in the first place—normally creates additional collateral at a rate which is sufficient to offset the rise in reserve requirements.

Fourth, once stock prices have begun their decline, the collateral value of the securities used as a basis for the margin debt falls, and the margin speculator is faced with the option of putting up more money or selling stock in order to pay off the debt. At first, the margin call process is reasonably orderly, as most traders have a sufficient cushion of collateral to protect them from the initial drop in prices. Alternatively, those who are under-margined often choose to put up additional collateral or cash. Toward the end of a bear market prices fall more rapidly, and this unnerving process, combined with the fact that margin customers are either unwilling or unable to come up with additional collateral, triggers a rush of margin calls, which substantially adds to the supply of stock which must be sold regardless of price. This self-feeding downward spiral of

**CHART 15-6 New York Stock Exchange Composite versus Margin Debt Level**

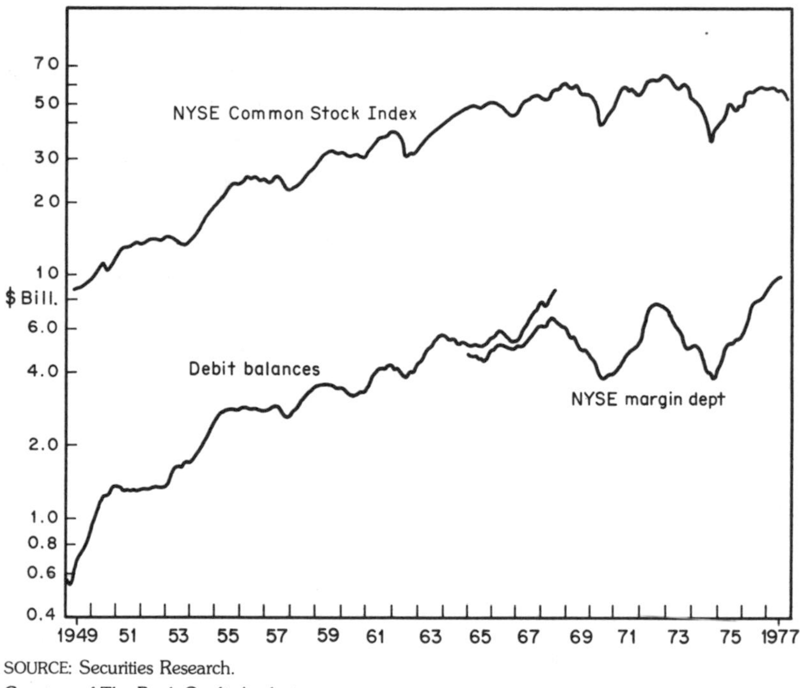


SOURCE: Securities Research.
Courtesy of *The Bank Credit Analyst.*

forced liquidation continues until margin debt has contracted to a more manageable level.

As a potential influence on stock prices, margin debt can be examined with regard to both its trend and its level. Chart 15-6 shows the New York Stock Exchange Composite and the amount of margin debt outstanding with brokers. (The margin debt index has been revised from time to time, and this accounts for the broken nature of this series.) It can be seen that each bear market since 1949 has been accompanied by some form of liquidation in margin debt. Also important is the amount of margin debt in relation to the value of all securities; this measure gives an indication of the presence of the speculative content. This ratio, shown at the bottom of Chart 15-7, uses the market value of stocks listed on the New York Stock Exchange as a proxy for the value of all securities. The scale incorporating the margin debt ratio has been compressed prior to

**CHART 15-7 Margin Debt as a Percentage of Total NYSE Capitalization**

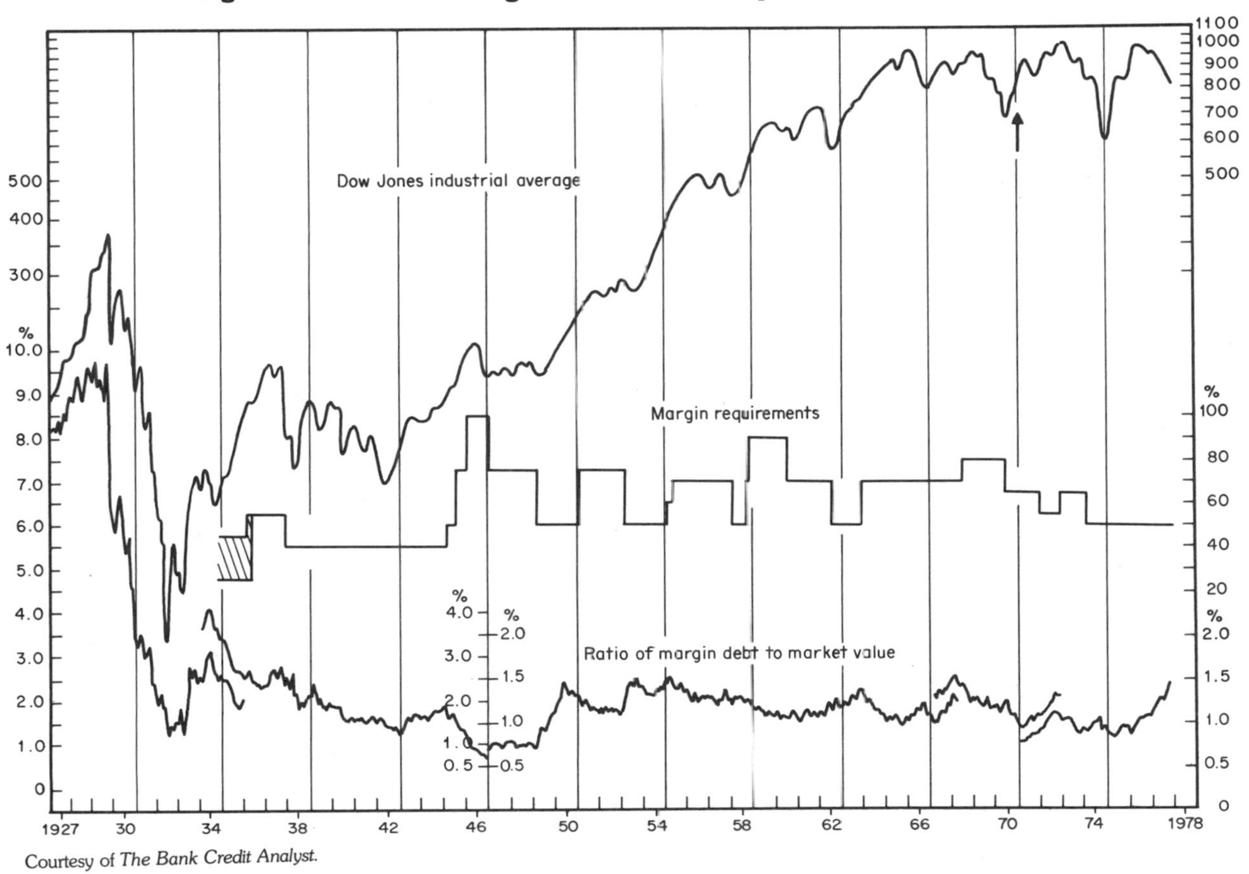


Courtesy of *The Bank Credit Analyst.*

1947, for easier presentation of data. The ratio is affected by three factors: the absolute level of margin debt (both at brokers and banks), the number of shares listed on the New York Stock Exchange, and the price of those shares. Over a period of 12 to 24 months the number of shares outstanding does not fluctuate significantly, so within a particular cycle it is the level of margin debt and stock prices that determines the course of the ratio.

Generally speaking, when the ratio approaches 1.5 percent and then begins to fall, the bull market in stocks is approaching maturity. Moreover, in the postwar period a reading close to 1.5 percent has normally been associated with a hike in the margin requirement.

## CONCLUSION

Sentiment indicators are a useful adjunct to the trend-determining techniques discussed earlier. They should be used for the purpose of assessing a majority or consensus opinion from which a contrary position can be taken. Since it is difficult to derive actual levels from which definitive investment action can be taken, none of these measures of investor psychology should be used in isolation; each should therefore be analyzed from the point of view of confirming the rest. Only in this way can an overall assessment of market sentiment be accurately made.

# 16

# SPECULATIVE ACTIVITY IN THE STOCK MARKET

Speculation has been defined as the undertaking of substantial risk in the hope of gain. Some degree of speculation is always present in market activity, in the sense that there are always individuals willing to bear a risk. However, since virtually all bull markets have in their late stages clearly defined characteristics of speculation, an examination of some indicators that measure this phenomenon of excessive confidence is well worthwhile.

Over the course of the stock market cycle investors move from outright pessimism and sometimes panic at market bottoms to an outlook of overconfidence and avarice at market tops. On the assumption that, once set in motion, a trend in market psychology perpetuates itself, it is important to determine the level of emotion at any point compared with previous periods and also to detect any reversal in its trend at a relatively early stage.

## MEASURING SPECULATION

As discussed in previous chapters, there are four main aspects or dimensions of psychology in the stock market: price, volume, time, and breadth. Price reflects the level of enthusiasm; volume, its intensity; breadth, the extent of participation; and time, the extent of the period. The farther these four factors work in any one direction, the greater the significance of the counterreaction. Thus, by late 1929, stock prices had been rising almost uninterruptedly for 8 years. Prices had increased fivefold, and volume at the time of the peak was exceptionally heavy. The counterreaction was naturally extreme, with an almost total collapse in equity prices. Since speculation is one of the most emotional aspects of the stock market, examination from all four dimensions of price, volume, time, and breadth is relevant.

## Price

One method of evaluating speculation is to compare the prices of low-quality issues with those of high-quality ones. The rationale is that low-quality stocks are more volatile and, in a bull market, will rise proportionately faster than better-quality stocks. The high potential for profit in these equities makes them very popular with speculators. In addition, members of the public who like to buy in 100-share "round" lots but do not have the capital to purchase higher-priced stocks in these quantities find low-priced equities a natural vehicle.

The upper portion of Chart 16-1 shows the Standard & Poor's Low-Priced and High-Grade Stock Indexes. The relatively high volatility of the low-priced stocks is self-evident and is a good reflection of the exaggerated emotions of investors during a market cycle. At market bottoms the Low-Priced Index either lags or coincides with the Dow Jones Industrial Average and is therefore a confirming indicator. At market peaks the index usually leads by anywhere from zero to 15 months. Considered in isolation, the index itself is not too helpful in determining levels of speculation, though obviously a sustained but controlled uptrend points to a good underlying improvement in confidence, while a more exponential trend after a slow but lengthy rise is a good tipoff to a speculative peak.

The lower portion of Chart 16-1 shows what amounts to a relative-strength line comparing the Low-Priced Index to the Standard & Poor's High-Grade Stock Index. It is calculated by dividing the Low-Priced by the High-Grade Index. Whenever this ratio is below 100, the value of the higher-grade index is greater than that of the low-priced. Such a condition reflects a relatively defensive posture by speculators and is normally associated with major bottom areas. With the exception of 1961, the ratio has always been above 100 at major tops.

There is no absolute level that indicates a peak in speculation. However, the high readings in the late 1920s, 1937, 1946, and 1968 (all of which followed lengthy periods of rising stock prices) were clearly pointing out that activity was reaching an extreme from which a major counterreaction could be expected. It is important to distinguish between the high levels which occur in this index after a long and substantial rise in

**CHART 16-1 Low-Priced versus High-Priced Stocks**

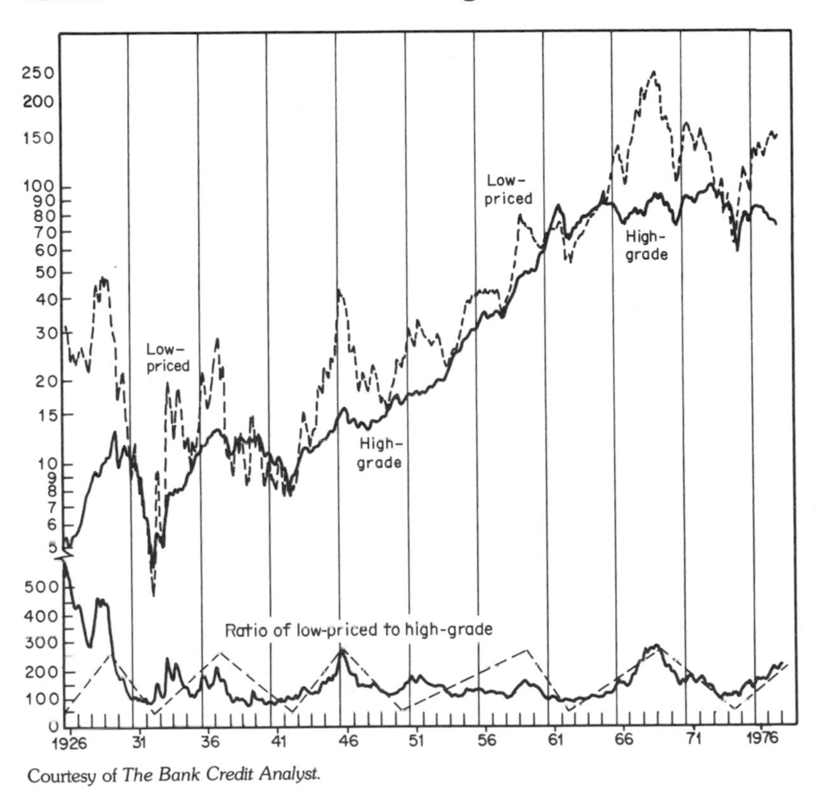


Courtesy of *The Bank Credit Analyst.*

equities and, for example, those that occurred in 1933–1934 and 1951, which were strong upward reactions in low-priced stocks following a long and sharp decline. In retrospect it can be seen that the 1971–1972 rally was really a temporary interruption of what proved to be a long corrective period.

## Volume

Chart 16-2 shows the average monthly volume of the American Stock Exchange expressed as a percentage of that on the New York Stock Exchange. The rationale for using this index is basically the same as the high-grade/low-price comparison. Generally speaking, due to less stringent listing requirements, stocks trading on the ASE are of lower quality than those on the NYSE. Since speculators normally deal in low-quality

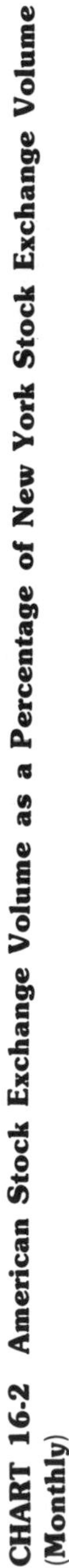

**CHART 16-2 American Stock Exchange Volume as a Percentage of New York Stock Exchange Volume (Monthly)**

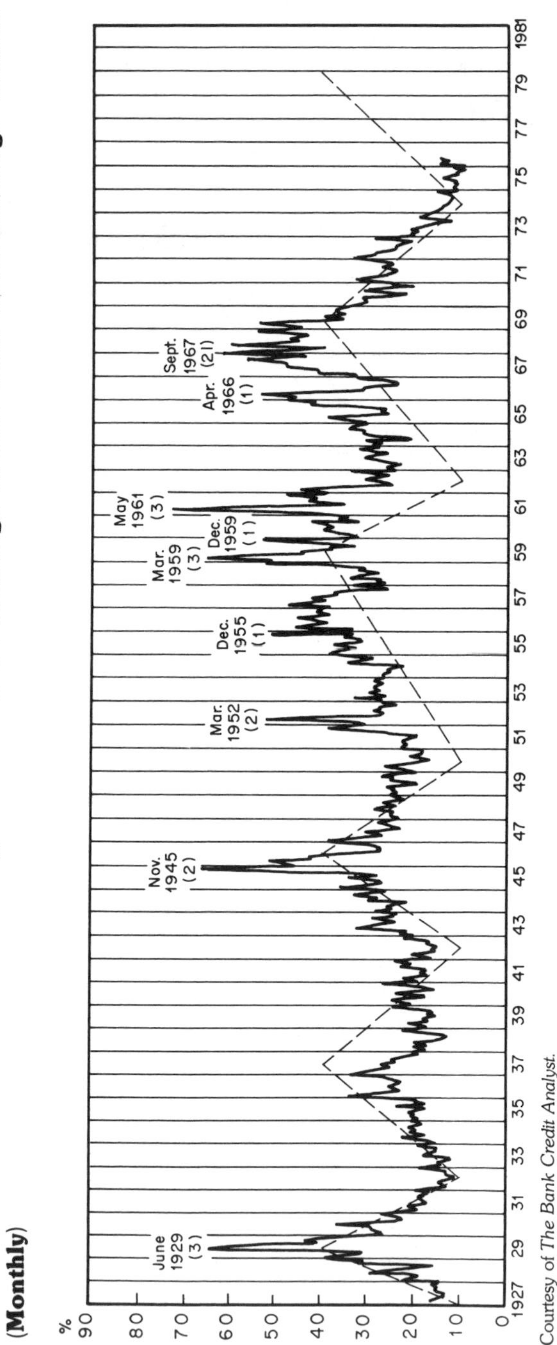


Courtesy of *The Bank Credit Analyst.*

issues, rises and falls in the index therefore reflect volume trends toward more or less speculation. Historically, levels of 50 percent or more have occurred in the area of tops. Normally these high readings lead ultimate peaks in the Dow by several months or more. Conversely, levels of 20 percent or less (when speculation in terms of volume has been low) have often been good buying points, although purchases would have been more profitably made if this somewhat erratic indicator had been used in conjunction with the others.

## Breadth

One prerequisite for a strong surge in speculation is that equities should have been rising on a broad front for a relatively long period of time. For the general level of confidence to mature into a speculative orgy, it is first necessary for the broad spectrum of stocks to advance rather than for the price rise to be confined to a relatively select group.

If only a few investors are making money, overconfidence is unlikely to develop, but if virtually everyone is profiting, the ingredients are in place for a self-feeding spiral resulting in an excess of optimism.

Breadth is portrayed in Chart 11-2 by an advance/decline line constructed on the basis of the $\sqrt{A/u - D/u}$ formula discussed in Chapter 11. Until 1969 every cyclical peak in the Dow since 1931 was confirmed by the A/D line. In 1968 the line made a new cyclical high unaccompanied by the DJIA, while in 1973 the opposite occurred. These examples illustrate the necessity for strong breadth rather than a strong average to presage a speculative flurry. The 1968–1969 rise in the A/D line was a cumulation of the rise in speculation begun in 1962, but it was not reflected in the Dow. On the other hand, the all-time high for the Industrial Average was achieved in January 1973 against an overall declining trend in breadth.

Previous speculative peaks were preceded not only by strong rises in breadth but also by a rise that had extended for a number of years. Only after a long advance was confidence able to develop into excessive optimism.

## Time

For a strong speculative wave to get under way, it is necessary not only to correct the excesses of the previous speculative peak but also to establish a fairly lengthy period of rising stock prices, so that a feeling of gradually improving confidence can erase the fears and caution of the corrective processes. Since many investors suffer staggering losses during the declining period, time is also needed either to restore their financial viability or to develop a new generation of participants who have not undergone the sobering experience of a speculative cleansing. Once confidence resumes, it can become a self-feeding spiral as investors undertake risks as a group which they would never contemplate as independent and clear-thinking individuals.

In observing the patterns of the speculative indicators discussed above, it becomes clear that this repetitive emotional transition from greed to fear and back to greed forms an approximate cycle. Since the 1920s, for example, there have been three outstanding speculative peaks: 1928–1929, 1946, and 1967–1968, each occurring about 20 years apart. In turn, these major peaks were separated by two other peaks at roughly 10-year intervals, namely 1937 and 1959, so that there were really five peaks of speculation in all, each falling in the latter part of their respective decade. Between these periods of excesses, troughs also occurred at 10-year intervals when risk-bearing was at a minimum. These points happened around the years 1932, 1942, 1950, 1962, and 1974. Following these low points there was a period either of bounce-back or of speculative recovery, such as 1934 or 1951–1952, or the beginning of a slow but sustained rise in speculation, as in 1942–1946 or 1962–1968. This idealized cycle has been superimposed on the volume and price indicators discussed above. There are some obvious drawbacks in utilizing this approach, since volume, price, and breadth do not always coincide. For example, in 1936–1937 the high-grade/low-priced ratio was quite high, and breadth had risen sharply for a fairly lengthy 5 years. On the other hand, ASE/NYSE volume was high (34 percent) compared with 1932 (10 percent), but it was still well below other speculative peaks. Conversely, the ASE/NYSE volume reached its highest recorded level in 1961, at around 74 percent, against a relatively short 6-month advance in breadth and a price ratio that showed low-priced stocks only marginally above high-grade issues.

## Yield

As a further approach to the 10-year-cycle concept of speculation it is worth examining stock yields, since yields reflect one form of equity valuation. Consequently, if stocks are overvalued, i.e., possess low yields, some degree of speculation is implied, and vice versa. Chart 16-3 shows the yield of the Barron's 50-Stock Average on which the idealized 10-year cycle has been superimposed. With the obvious exception of 1936–1937, when yields rose as speculation and stock prices advanced, the cycle fits very well.

**CHART 16-3 Dividend Yields (*Barron's* 50-Stock Average) and the 10-Year Cycle in Speculation**

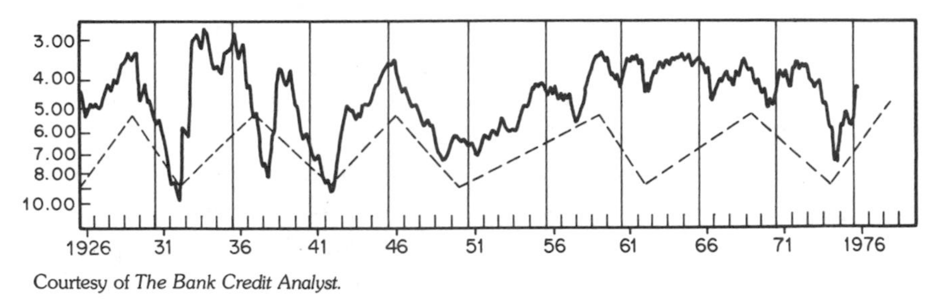


Courtesy of *The Bank Credit Analyst.*

## OPTIONS TRADING

To some extent, traditional methods of measurement of speculation may be distorted by the advent of listed options trading. Unfortunately, data on this new vehicle do not go back far enough to be of any predictive value; indeed, it is not clear what proportion of options trading is due to arbitrage and hedging and what to speculation. While the development of listed options trading may dilute some of the activity associated with lower-quality issues, the concept of a wasting asset is not palatable to all speculators. Consequently, it should be fair to assume that while the indicators discussed above may not reach the fervent levels witnessed on previous occasions, they should continue to give some indication of the general trend and level of speculative activity.

## CONCLUSION

From the point of view of the long-term investor, the establishment of a major cyclical peak in speculation is an important factor, since speculation is normally the last but most dynamic stage of any bull market. Since the 1920s, there have been five such peaks: 1928–1929, 1936–1937, 1946, 1959, and 1967–1969, each separated by about 10 years. A major decline followed these excesses of optimism from which the market, as measured by the A/D line, did not recover to a new high for at least 4 years. In most cases this retracement took much longer.

In addition to these speculative peaks there is also an approximate 10-year cycle of troughs characterized by a dearth of speculative activity and a general disgust with equities. Following these major cyclical lows, there is normally an initial rebound or recovery in speculation.

These recovery periods were not followed by severe and lengthy declines since insufficient time had elapsed from which major distribution could be formed. For example, while the 1951–1952 period showed breadth advancing sharply, accompanied by high readings in the ASE/NYSE volume and high-grade/low-price ratios, the A/D line had merely returned to its 1946 level over a comparatively short period.

These indicators of confidence or speculation are not precise tools like those discussed earlier; rather, they offer a longer-term perspective as to how far the trend of optimism or pessimism has developed. Consequently they are useful background indicators that can lead to a more enlightened interpretation of the more mechanical techniques.

# 17

# TECHNICAL ANALYSIS OF INTERNATIONAL STOCK MARKETS

Since equities are bought and sold throughout the world for essentially the same reasons, the principles of technical analysis can be applied to any stock market. Unfortunately the degree of sophistication in statistical reporting of many countries does not permit the kind of detailed analysis that is available in the United States.

Nevertheless, it is possible to obtain data on price, breadth, and volume for nearly all countries. Information on industry groups and interest rates is also widely available.

Chart 17-1 shows the *International Bank Credit Analyst World Stock Index*, which is constructed from the market averages of 12 individual countries, each in turn weighted by that country's share of the sum total of all their gross national products. The World Stock Index is a good starting point from which to analyze the cyclical trends of the various stock markets, just as a composite index or the Dow Jones Industrial Average would be used as a focal point in assessing the U.S. market. This is because the stock markets of the individual countries tend to move in the same direction, just as the majority of U.S. stocks gravitate in the general direction of the DJIA most of the time. The individual stock markets do not all reach their peaks or troughs concurrently. The United Kingdom market, for example, has a tendency to lead, while the Japanese and Canadian markets are usually late turning down. If the world is treated as one unit comprising many parts (i.e., the various stock markets), it is possible to construct several indicators measuring world breadth, momentum, volume, etc.

In Chart 17-1, the presence of the international 4-year cycle is indi-

**CHART 17-1 World Stock Index versus 12-Month Momentum**

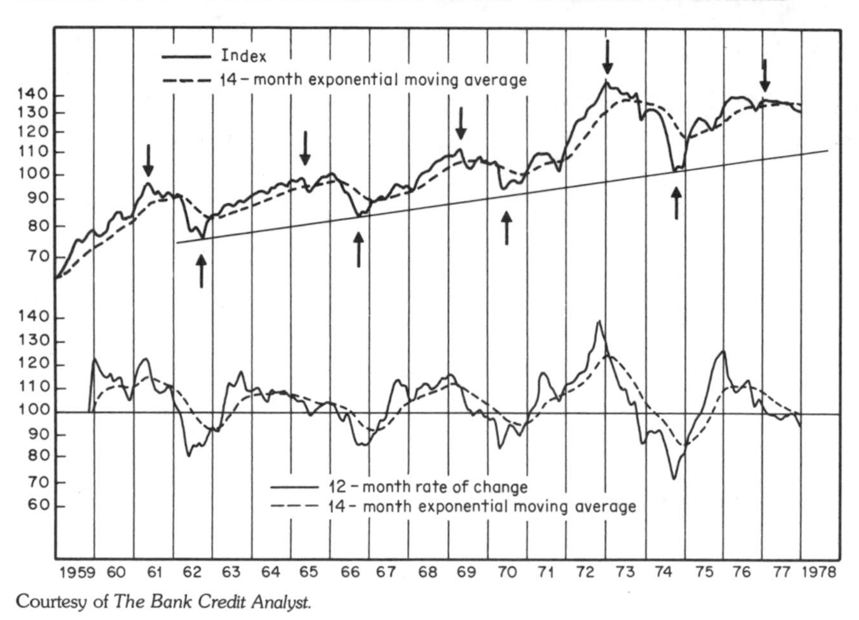


Courtesy of *The Bank Credit Analyst.*

cated by the arrows. It can be seen that the peaks in 1961, 1965, 1969, 1973, and 1977 are separated by approximately 4 years, and the troughs in 1962, 1966, 1970, and 1974 by the same period. The trend in this World Index is therefore the starting point for analyzing the primary trends for individual markets, including that of the United States.

Beneath the World Stock Index is its 12-month rate-of-change index and a 14-month exponential moving average. This moving average, when used in combination with the moving average of the World Stock Index itself, has offered a useful confirmation of primary trend reversals in world equities.

Another way to analyze the global technical position is to calculate the percentage of individual-country indexes that are above their 40-week moving averages (see Chart 17-2). Reference to the chart shows that this measure of momentum is continually moving from the zero (bullish) to the 100 percent (bearish) extremes. By the time the index has reached these outer parameters, the probabilities favor the fact that the larger portion of a price move has already taken place, although confirmation of an actual reversal should be obtained from a bearish signal in the World

**CHART 17-2 World Stock Index versus Percentage of Markets in Positive Trend**

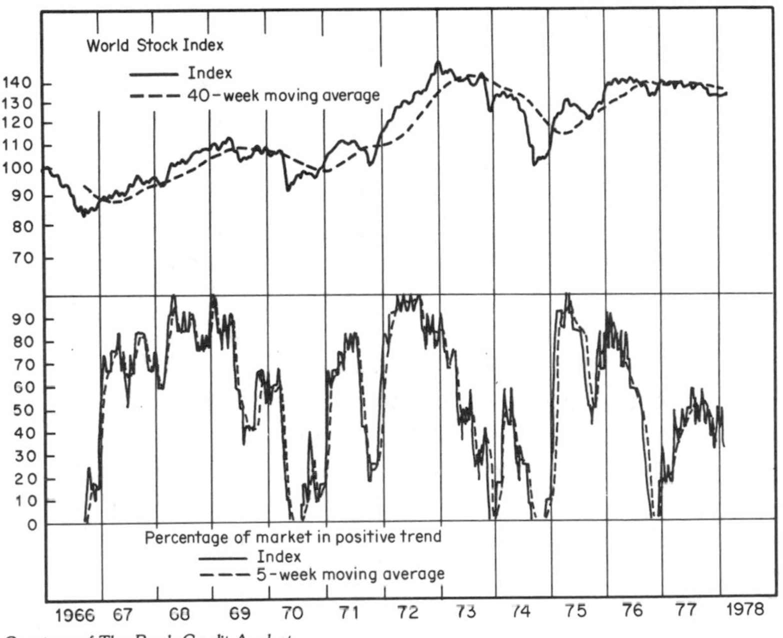


Courtesy of *The Bank Credit Analyst.*

Index itself in the form of a moving-average crossover or other trend reversal technique.

Extreme readings in the index are usually associated with primary reversals, but occasionally they mark the termination of an intermediate move. The action of the index following a decline, especially to a zero reading, can be very useful from the point of view of assessing the overall technical structure of the world equity markets. For example, the 60-point advance in this momentum index from 0 to 60 in early 1974 was accompanied by a relatively small price advance of about 10 points, while a similar rise of 50 percentage points in the index in late 1975 resulted in a proportionately larger price rise of 20 points in the World Index. The first example of strong momentum accompanied by a weak price rise was indicative of technical weakness and was followed by a substantial decline. The second, where strong momentum was accompanied by a healthy price rise, was followed by a more sustainable advance, although prices eventually fell again due to the rapidly maturing position of the cycle.

Chart 17-3 represents the World Stock Index and a composite momentum indicator, the latter constructed by taking a mean average of each country's 12-month rate of change. This series appears to have a more negative bias than the straightforward 12-month rate of change shown in Chart 17-1, but it is still useful as a confirmation of that index.

An alternate measure of world momentum is constructed by means of a diffusion index from the percentage of stock indexes that are recording a rate of change above zero. There are, of course, a whole variety of such measures that could be used. A momentum index constructed in this manner moves from the two extremes of 0 to 100 percent in the same way as the 40-week moving-average indicator discussed above and illustrated in Chart 17-2.

Once the cyclical trend in the World Index has been established, it is possible to determine more accurately the course of each individual stock market within the global equity cycle. Not all countries will rise or decline with the World Stock Index, because the principle of variation (see Chapter 9) applies just as much to countries as to individual stocks or groups. In addition, the long-term economic financial and political situations of countries will vary, so that a good world bull market in equities may be

**CHART 17-3 World Stock Index versus Percentage of Markets**

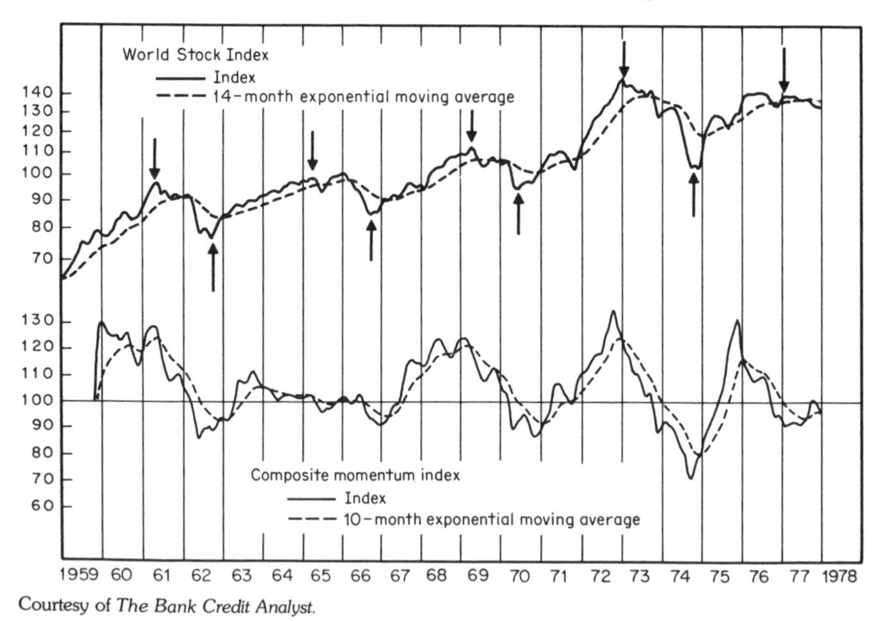


Courtesy of *The Bank Credit Analyst.*

**CHART 17-4 World Stock Index versus World Advance/Decline Line**

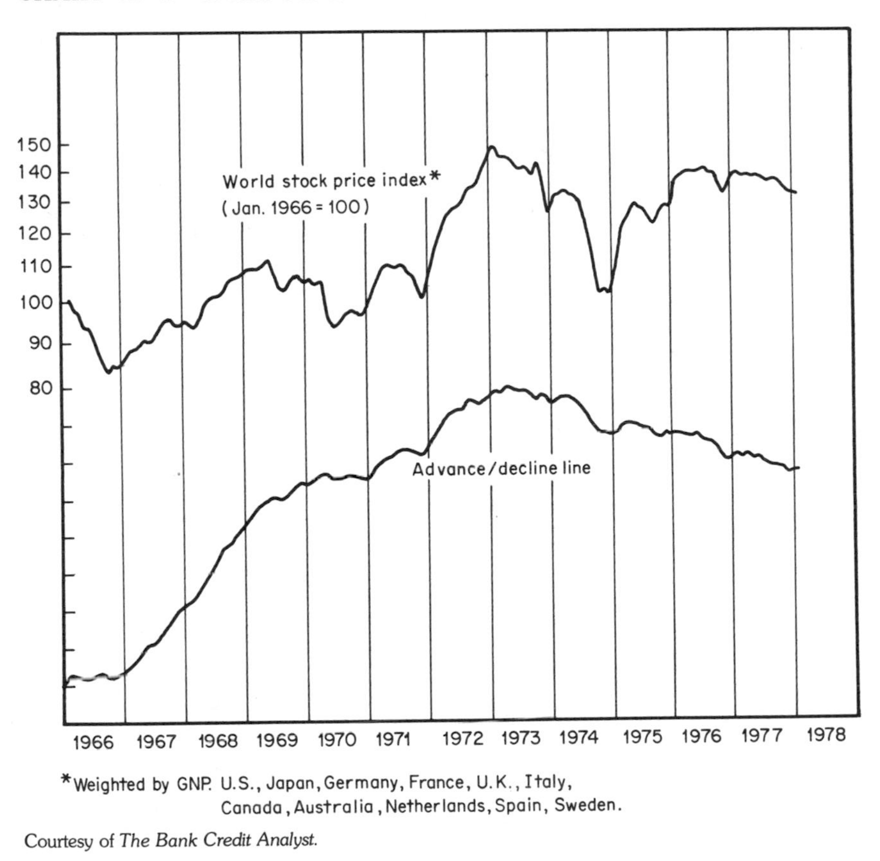


Courtesy of *The Bank Credit Analyst.*

brief or almost nonexistent for a country undergoing severe financial distortions, as was the Italian and French experience in the 1974–1977 global bull market. Generally speaking, as time progresses and improvements in technology and communications break down geographical and trading patterns, countries will become more interdependent, so that their stock market and business cycles are likely to become more closely related. The three stages of the 1974–1977 bull market shown in Charts 17-5 through 17-7 point up this principle of commonality (see Chapter 9) quite well. The arrows above and below the various stock indexes correspond with intermediate "buy" and "sell" signals on the DJIA, as derived from a momentum indicator.[1] Reference to the charts show that the termination

[1] This momentum indicator was developed by I. S. Notley of the Trend and Cycle Department, Dominion Securities, Toronto, Canada.

**CHART 17-5 Commonality Applied to World Stock Indexes, Phase I**

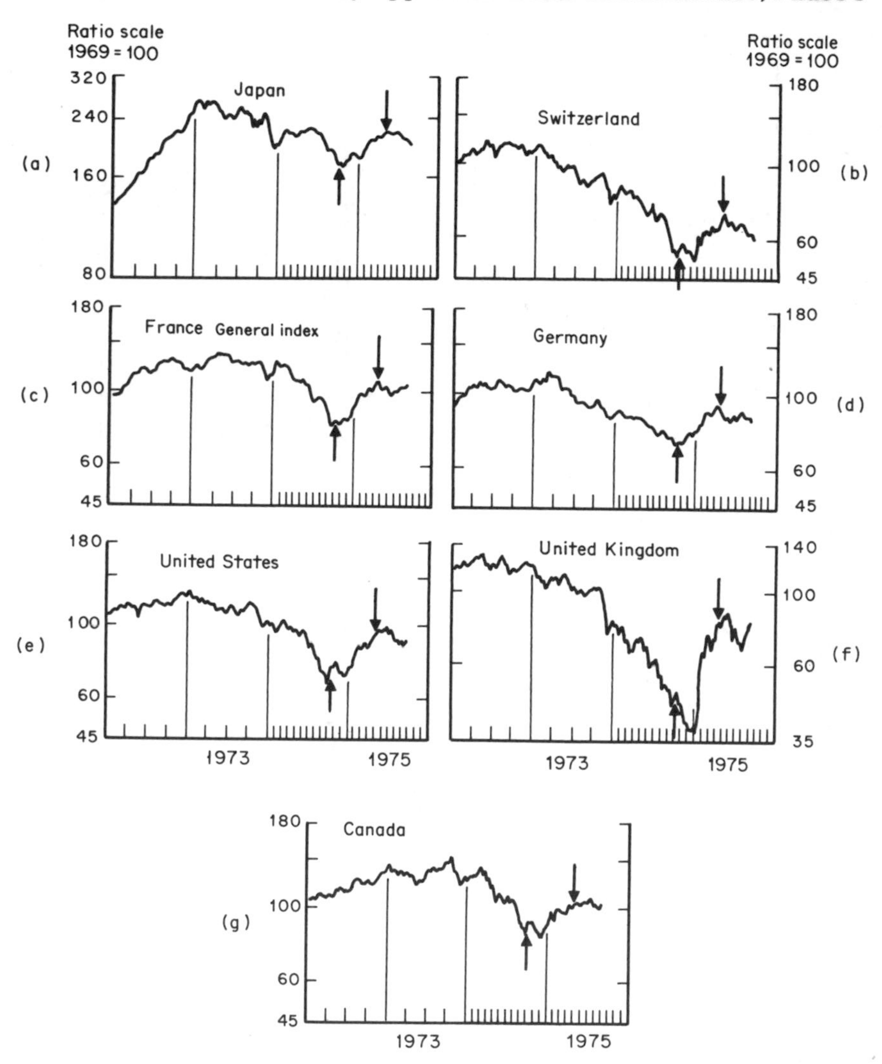


of each up leg was followed sooner or later by a fall in each individual stock index. As the global cycle developed into the third phase, several stock markets (such as the French and Italian) failed to make a new bull market high, resulting in a third up-leg magnitude failure.

A further aid in assessing the technical position of an individual country can be obtained through the construction of a relative-strength index between the stock index of that country and the World Index itself. The

**CHART 17-6 Commonality Applied to World Stock Indexes, Phase II**

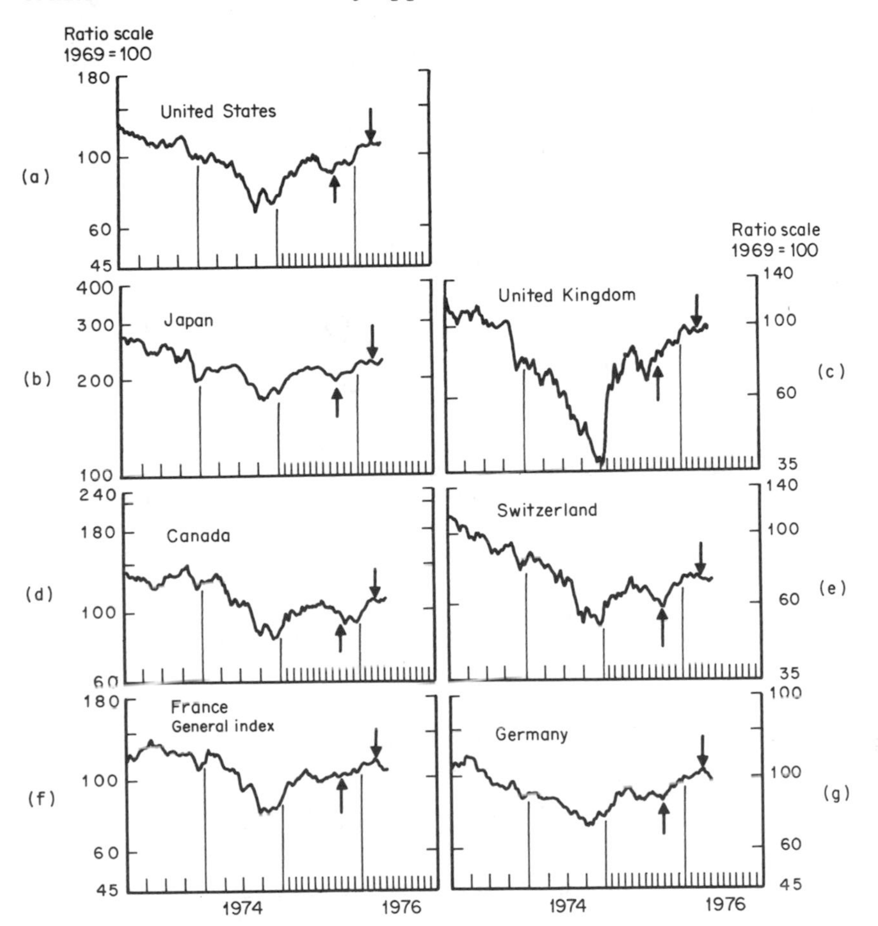


relative-strength line itself will not indicate whether a market will rise or fall in an absolute sense, but only whether it will outperform or underperform the rest of the world. If it has been established that the World Index is in a rising trend, it will be as well to invest in a country with a strong trend of relative strength. A relative-strength index for Japan and the United States in the period from 1959 to 1978 is shown in Charts 17-8 and 17-9. During this period there were isolated points such as 1970–1972 and 1974–1977 when investment in the United States would have given relatively good results, but overall its performance against the rest of the world since 1959 was inferior. In complete contrast, Japan's relative-

**CHART 17-7 Commonality Applied to World Stock Indexes, Phase III**

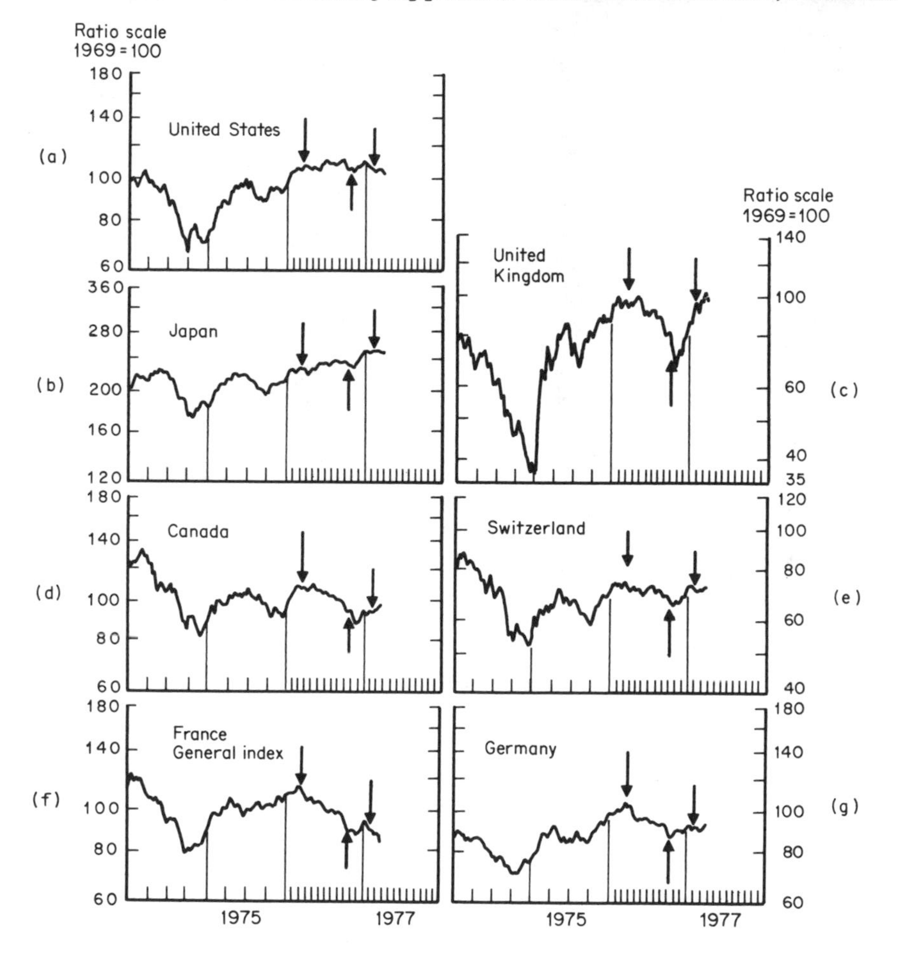


strength line rose strongly during the same period and experienced only minor declines. It is worth noting that the Japanese relative strength in each cycle has a tendency to peak out around the time that the United States stock market is reaching its bear market low. These points are indicated on the chart by arrows.

An analysis of the global technical position should incorporate an assessment of the technical structure of the stock markets in the various countries concerned, as well as the composite measures of momentum, breadth, etc., discussed above. The trend-determining techniques outlined in Part 1 should be used as a basis, but it will be found through ex-

## CHART 17-8 United States Relative Strength versus World Index

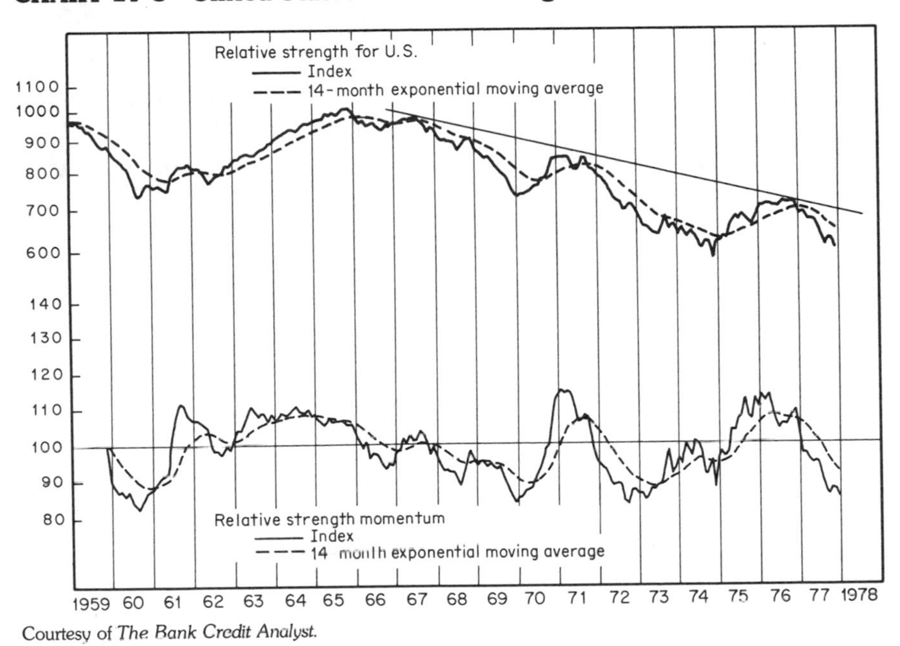


Courtesy of *The Bank Credit Analyst.*

## CHART 17-9 Japanese Relative Strength versus World Index

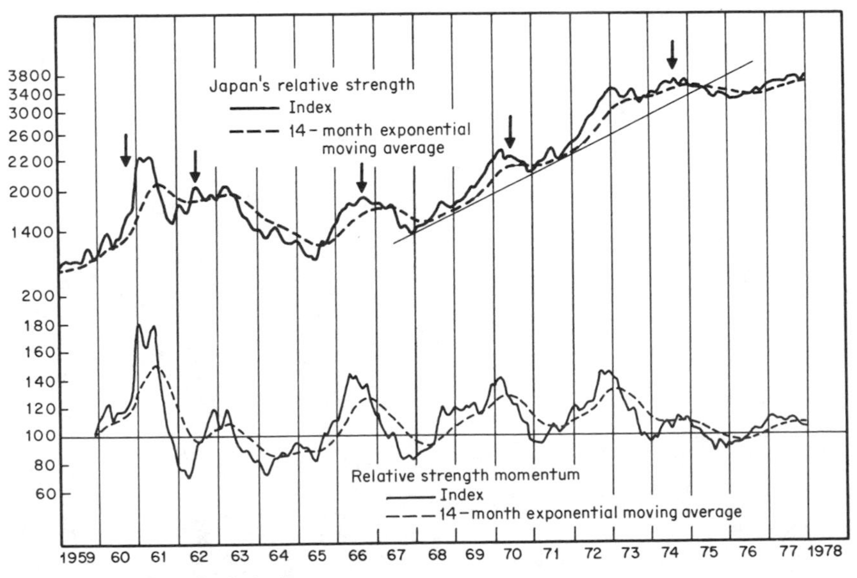


Courtesy of *The Bank Credit Analyst.*

perimentation that different markets respond to different techniques because of the variations in their structures and cycles.

This global technical approach should therefore form an integral part of an assessment of the technical health of any individual stock market, including that of the United States. As time progresses, the economies and financial structures of the countries that make up the world economy are becoming increasingly interwoven, making it extremely difficult for any stock market to move in an opposite direction to the prevailing trend of world markets in general.

# 18

## PUTTING THE INDICATORS TOGETHER

### A Description of the 1970–1974 Stock Market Cycle

It is now time to examine an actual market cycle by putting all the indicators together and observing how they work in practice. The 1970–1974 period has been chosen since that particular cycle offers a substantial variety of technical conditions that are worth studying

Using the Dow Jones Industrial Average, Chart 18-1 illustrates the market action of this period in a simplified wave form, showing the three upward phases of the 1970–1973 bull market and the three downward phases of the 1973–1974 bear market. The description will focus upon these two primary moves and the intervening secondary corrections.

The cycle begins in the dark days of late May 1970. The market had been declining dramatically for 6 weeks on gradually expanding volume, the trend of virtually all the technical indicators was down, and there seemed no end in sight to falling prices. Nevertheless, there were some tentative indications that prices would soon bottom out, since short-term interest rates and long-term government bonds had already reversed their cyclical trends in late 1969.[1]

By the end of May it was obvious that a selling climax had taken place. While this did not necessarily mean that the ultimate low had been reached, it did imply that the May bottom would hold at least temporarily and that prices were unlikely to decline for a while. At the same time, many of the oscillators were recording extreme "oversold" conditions. The annual rate of change of the Dow, for instance (see Chart 8-6), was at a postwar low, while the 10-week moving average of the A/D ratio

[1] See Charts 13-1 and 14-1.

**CHART 18-1 Dow Jones Industrial Average and Major Movements, 1970–1974**

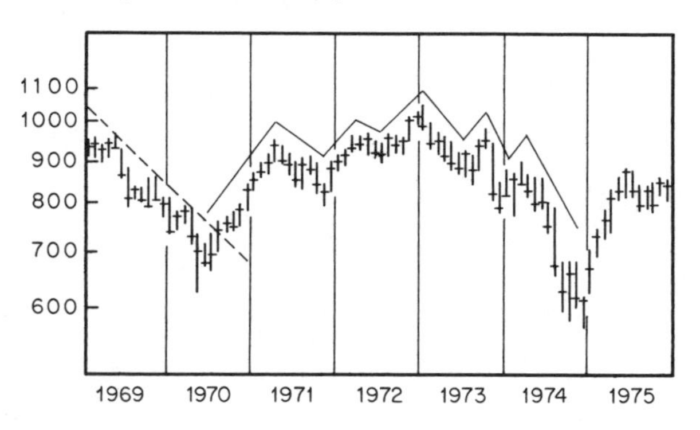


(Chart 11-1) was at a very negative level. In addition, many of the sentiment indicators were pointing up an extreme degree of pessimism. This can be observed from Charts 15-1, 15-3, and 15-4, which represent the Members' Short Ratio, the Advisory Service Index, and the Mutual Funds Cash/Asset Ratio respectively.

At this juncture many of the indicators were pointing out that a new bull market was in the process of being born, but as yet no long-term trends had been broken, no long-term moving averages had been crossed, no bases had been built, and certainly no positive divergences had been observed among the major averages. It was not until the summer months of 1970 that more tangible signs of a cyclical reversal appeared. First, a base was formed by most averages as the early June rally petered out and a successful test of the May lows was achieved. Then as prices rose in July and August, a clearly defined bear market trendline was penetrated on the upside (see Chart 18-1). During this period, the 18-month weighted moving average of the 4- to 6-month commercial paper rate turned down (see Chart 13-3), confirming that a cyclical reversal in short-term rates had taken place. Finally, longer-term indicators of price and volume momentum turned up (see Chart 10-2). These latter indicators offered fairly conclusive proof that a new bull market had begun in May, especially as they were accompanied by a Dow theory "buy" signal in December. Finally, encouragement was given by the World Stock Index, which had crossed above its 40-week moving average (Chart 17-2), accompanied by a rising trend in the smoothed momentum index (Chart 17-1).

Prices generally rose until March 1971, following which an 8-month correction took place. Advance warning of this intermediate decline was given by several interest-sensitive indexes. The Federal Funds rate, for example, had already given a "sell" signal in February as it moved above its 10-week moving average, and its momentum index had crossed above its zero reference line (see Chart 13-1). This signal was validating the action of the Utility Index, which had topped out in January and had refused to confirm the March high in the DJIA. The A/D line had also peaked ahead of the industrial average. Many of the oscillators had also reached extremely overbought conditions earlier in the year, but this was quite a normal and healthy phenomenon for the early stages of a bull market. It is only when a decline in momentum is accompanied by a break in the trend of the index being monitored that a red light is really flashed.

The position at the March peak in the Dow was therefore one in which the interest-sensitive and breadth indexes had begun to diverge, and the oscillators were declining from an overbought condition. In addition, many of the sentiment indicators were also at bearish extremes. Certainly no distribution had formed, the strong uptrend since May 1970 was still intact, and the longer-term momentum indicators were still positive, like the primary trend of the commercial paper rate, as evidenced by the strongly rising 18-month weighted moving average. The situation appeared to be one of cyclical strength but potential intermediate weakness.

Not surprisingly, prices broke and continued falling until August. During this period, the strong July 1970–April 1971 uptrend of many averages was decisively violated on the downside. But since these trends were unsustainably steep, this development merely indicated that the market was adjusting to a more subdued rate of advance. By August the technical picture suggested that the market was oversold and a rally was called for. The 13-week rate of change of the DJIA, for example (see Chart 8-5), reached a low reading of −10 percent, and the percentage of Dow stocks in bullish trends fell to below 30 percent (see Chart 11-5). The rally did take place, and at the same time debt markets began to recover from their decline, which had started earlier than the stock market.

This was a very difficult period, since most long-term factors pointed to a continuation of the bull cycle, and the intermediate trends of interest-sensitive securities (normally a good lead indicator for the stock market) had also turned positive. Given the oversold nature of the market in August and the fact that the April–August downtrend had been violated on the upside, it would have been easy to come to the conclusion that the reaction was over, especially as volume expanded very sharply during late August. But this proved to be one of those times when patience would have paid off, since not all the indicators were in a positive position.

The 30-Day Most Active Indicator was the most notable, since it was unable to cross over its bullish area (+3) despite the surge in volume. In addition, while some of the major averages such as the DJIA were able to better their July highs, others (such as the weekly A/D line) failed to surpass theirs. The Members' Short Index had also crept back toward a very bearish 85 percent reading. Finally, despite the positive trend in the bond and money markets, the Utility Index continued to remain in its intermediate downtrend, while the World Index showed absolutely no indication of a reversal in its intermediate downtrend.

It was not until early December, when the market was much lower, that the Most Active Indicator became positive and the Utility Index was able to reverse its downtrend. By this time there were many more positive signs that the bull market was about to resume. First, the debt markets failed to make new lows in November and had therefore set up some positive divergences with the stock market. Second, the 10-week oscillator of the weekly A/D line had failed to significantly penetrate its August low, thereby creating a further positive divergence between it and the A/D line itself. Third, the indicators of long-term momentum measuring price and volume (Chart 10-2), while falling slightly, had not confirmed a bear market by falling below their zero reference lines. Finally, the Dow Jones Transportation Index had refused to confirm the November low in the Dow, thereby indicating that the Dow theory bull market was still intact.

The rally continued until April 1972, when the DJIA returned to the 950 resistance level, which again served to halt the advance as it had done one year earlier. The Dow did not sell off as it had in 1971, but began a period of consolidation in the 900–950 area that was to last for most of the balance of the year. The action over the ensuing months provided the first indication that the bull market was reaching maturity. First, the up phase of the 4-year cycle was rapidly running out of time, since it was 2 years old by May 1972. This did not necessarily mean that the rise could not continue longer (as shown by the 1962–1966 experience); but in view of renewed weakness of (1) the debt markets during this period, (2) the Utility Average, (3) the A/D line, and (4) the Most Active Index, some ominous rumblings were developing. Indeed, not only was the A/D line falling while the DJIA was marking time, but it had also failed to surpass its old 1971 high (see Chart 11-1). Moreover, even though long-term interest rates held steady until January 1973, the Federal Funds indicator (Chart 13-1) went negative in the late spring of 1972, suggesting that the intermediate trend of short-term interest rates was up. Later on, the 18-month weighted moving average of 4- to 6-month commercial paper (Chart 13-3) started to rise, thereby confirming that the cyclical trend in short-term rates had also been reversed. To make matters worse, the Transportation Average, which had previously outperformed the Dow during the 1971 decline, began to fall off rather badly. Finally, all the sentiment indexes were portraying a picture of widespread confidence and optimism, which, from a contrary-opinion standpoint, suggested that a top was being made.

By October 1972 many of the oscillators measuring intermediate movements began to reach oversold conditions or, in the case of the 10-week A/D ratio, had created a positive divergence. The indications therefore were that a short-term rally would then take place. In the case of the DJIA this proved to be somewhat more of an intermediate rise, as it rallied 15 percent from 900 to 1065, a new all-time high. In doing so, even greater divergences were set up, as this strong performance was not confirmed by many of the other indicators. As if things were not bad enough, short-term interest rates during this period began to rise even more sharply, accompanied by a discount rate hike and a rise in margin requirements in December. By late January, the long end of the bond market (Chart 14-1) also began to fall sharply. The situation in January 1973 was one in which the DJIA had broken out of what appeared to be a 9-year consolidation pattern, but the accompanying technical structure was so weak that the move proved to be spurious. The bear market, which had begun for most issues in mid-1972, had finally reached the blue-chip sector. During the early months of 1973 it also became apparent that the World Stock Index had turned down, although it was not until late spring that the index fell below its 40-week moving average (Chart 17-2). The first leg down of the new bear market was temporarily halted in July. The DJIA, the A/D line, and several other indicators made a new low in August, after a brief rally. But this secondary bottom was unconfirmed by the S&P 500 (Chart 8-1) and the upside-downside volume line (Chart 10-4). Moreover, two important oscillators—the 10-week A/D line ratio (see Chart 11-1) and the Trendline Over-200-Day Indicator (Chart 11-6)—were both well up from their July bottoms at the time of the new Dow low in August. These many divergences and oversold readings suggested that the market was in a position to mount a substantial rally, for the first time since the fall of 1972. However, the fact that some of the longer-term indicators were still in the early stages of deterioration suggested that this rally was likely to be a bear market correction rather than the first leg of a new bull market. For example, both the 18-month weighted moving average of the commercial paper rate and the Federal Funds Indicator were indicating that the trend of short-term interest rates was still adverse for stock prices. In addition, the sentiment indicators were reflecting a fair degree of pessimism, but not enough to be commensurate with a bear market low. This can be observed from the Members' Short Index, Chart 15-1, and the Advisory Service Sentiment Indicator (Chart 15-3).

In 7 weeks the Dow rose 12 percent from 880 right back to the 1000 level on rapidly expanding volume. The speed and strength of the rally confused many into believing that the bear market was over. This deceptive characteristic is most common during the first rally of a bear market, since business conditions are usually still good and negative arguments seem less persuasive under a background of rising stock prices.

However, most of the indicators measuring long-term movements remained in a bearish trend throughout this period. Despite the softening in short-term interest rates, the 18-month weighted moving average of the commercial paper rate was still in a rising trend, and there had been no meaningful rally in long-term bonds. The Dow theory, which had signaled a bear market in April, was still negative, while most of the long-term momentum indicators were below their zero reference lines and falling. Still, some of these more sensitive longer-term indicators began to turn up, and for the first time in many years they gave a misleading signal; generally speaking however, the proof of a new bull market was far from conclusive. In addition to the negative tone of many of the longer-term indicators, the rally had taken place during a period of rising interest rates, whereas falling interest rates normally lead or coincide with bear market bottoms. In any event, by October the market became overbought and was losing momentum in the face of a very major level of resistance around the Dow 1000 level. Whenever the market runs into significant resistance under such conditions, it is always wiser to await the reaction which invariably takes place and then reassess the state of the indicators.

In this case the correction extended for 200 Dow points and left no doubt that the bear market was still very much alive.

The second bear market rally was not difficult to predict. The DJIA had fallen to an extremely significant support area at 800, while the various oscillators had again reached oversold territory. In the case of the 10-week A/D ratio, this was to be its lowest reading of the bear market. Although no major positive divergences were apparent, the Federal Funds Indicator, which had turned positive in November, was still signaling that lower interest rates lay ahead.

The ensuing rally was almost as deceptive as the previous one, but for different reasons. To start with, the oversold condition in December 1973 had been much greater than that of the previous July or August, leading many observers to believe that the December selling climax marked the end of the bear market. Moreover, the A/D line and many of the lower-priced stocks were initially acting much stronger than the Dow itself, as can be seen from a comparison of the two indexes in Chart 11-1.

This strength was short-lived, since the A/D line fell away very sharply following the late March peak in the Dow. Moreover, the 18-month weighted moving average of the commercial paper rate had remained in a negative trend throughout this period and was now joined by the Federal Funds Indicator, which also deteriorated in late March. While the Federal Reserve was obviously easing monetary policy during the December–March period, it did not at any time lower the discount rate. As the rally progressed, the Members' 10-week Short Index and the Advisory Services Indicator both turned over and began to deteriorate.

The ensuing reaction proved to be the third and final leg of the worst bear market since the 1930s. The 1974 bottom was initially signaled by a positive reading in the Federal Funds Indicator in October as well as a major set of divergences between the DJIA, the S&P 500, and the Transportation and Utility Averages during October and December, when most indexes formed classic double bottoms. Several positive divergences were also observed in many of the oscillators. For example, the December reaction in the percentage of stocks above their 200-day moving average was far above the October low, thereby indicating a position of growing technical strength. By October most of the sentiment indexes had reached extreme levels of pessimism, adding further confirmation that a major bottom was being formed. By the end of January all the indicators had formed and broken out of large bases, Dow theory was signaling a new bull market, and most of the long-term momentum indexes had turned up accompanied by an expanding level of volume. These positive technical developments were also confirmed by the debt markets, most of which had made their cyclical bottoms in the fall. Finally, the World Stock Index, confirmed by virtually all the foreign stock indexes, had made a sharp reversal and rose explosively in January 1975. There was little doubt that the 1974–1976 bull market had begun.

# EPILOG

At the outset the suggestion was made that the keys to success in the stock market were knowledge and action. The "knowledge" part of the equation has been discussed as comprehensively as possible, but the final word has been reserved for investor "action," since the way in which knowledge is used is just as important as the understanding process itself.

It remains to point out some common errors which all of us commit more often than we would like to. The most obvious of these can be avoided by applying the following principles.

1. *Perspective*. The interpretation of any indicator should not be based on short-term trading patterns; it should always consider the longer-term implications.
2. *Objectivity*. A conclusion should not be drawn on the basis of one or two "reliable" or "favorite" indicators. The possibility that these indicators could give misleading signals demonstrates the need to form a balanced view derived from *all* available information.
3. *Humility*. One of the hardest lessons in life is learning to admit a mistake. The knowledge of all market participants in the aggregate is, and always will be, greater than that of any one individual or group of individuals. This knowledge is expressed in the action of the market itself as represented by the various indicators. Anyone who "fights the tape" or the verdict of the market will swiftly suffer the consequences. Under such circumstances, it is as well to become humble and let the mar-

ket give its own verdict; a review of the indicators will frequently suggest its future direction. Occasionally the analysis proves to be wrong, and the market fails to act as anticipated. If this unexpected action changes the basis on which the original conclusion was drawn, it is wise to admit the mistake and alter the conclusion.

4. *Tenacity*. If the circumstances outlined above develop but it is considered that the technical position has *not* changed, the original opinion should not be changed either.

5. *Independent Thought*. Should a review of the indicators suggest a position that is not attuned to the majority view, that conclusion is probably well founded. On the other hand, a conclusion should never be drawn simply because it is opposed to that of the majority. In other words, contrariness for its own sake is not valid. Usually the majority conclusion is based on a false assumption, so it is prudent to examine such assumptions to determine their accuracy.

6. *Simplicity*. Most things done well are also done simply. Because the market operates on common sense, the best approaches to it are basically very simple. If one must resort to complex computer programming and model building, the chances are that the basic techniques have not been mastered and therefore an analytical crutch is required.

7. *Discretion*. There is a persistent temptation to call every possible market turn, along with the duration of every move the stock market is likely to make. This deluded belief in one's power to pull off the impossible inevitably results in failure, a loss of confidence, and damage to one's reputation. For this reason, analysis should concentrate on identifying major turning points rather than predicting the duration of a move—for there is no known formula on which consistent and accurate forecasts of this type can be based.

APPENDIX 1

# THE ELLIOT WAVE

The Elliot Wave Principle was established by R. N. Elliot and first published in a series of articles in *Financial World* in 1939. The basis of Elliot Wave theory developed from the observation that rhythmic regularity has been the law of creation since the beginning to time. Elliot noted that all cycles in nature, whether of the tide, the heavenly bodies, the planets, day and night, or even life and death, had the capability for repeating themselves indefinitely. Those cyclical movements were characterized by two forces—one building up, the other tearing down.

This concept of natural law also embraces an extraordinary numerical series discovered by a thirteenth-century mathematician named Fibonacci. The series which carries his name is derived by taking the number 2 and adding to it the previous number in the series. Thus 2 + 1 = 3, then 3 + 2 = 5, 5 + 3 = 8, 8 + 5 = 13, 13 + 8 = 21, 21 + 13 = 34, etc. The series becomes 1, 2, 3, 5, 8, 13, 21, 34, 55, 89, 144, 233, etc. It has a number of fascinating properties, among which are:

1. The sum of any two consecutive numbers forms the number following them. Thus 3 + 5 = 8 and 5 + 8 = 13, etc.
2. The ratio of any number to its next higher is 61.8 to 100, and any number to its next lower is 161.8 to 100.
3. The ratio 1.68 multiplied by the ratio 0.618 equals 1.

The connection between Elliot's observation of repeating cycles of nature and the Fibonacci summation series is that the Fibonacci numbers and proportions are found in many manifestations of nature. For example, a sunflower has 89 curves, of which 55 wind in one direction and 34 in the opposite direction. In music, an octave comprises 13 keys on a piano, with 5 black notes and 8 white. Trees always branch from the base in Fibonacci series, and so on.

Combining his observation of natural cycles with his knowledge of the Fibonacci series, Elliot noted in an 80-year period that the market moved forward in a series of five waves, and then declined in a series of three waves. He concluded that a single cycle comprised eight waves, as shown in Figure A1-1 (3, 5, and 8 are of course Fibonacci numbers).

The longest cycle in the Elliot concept is called the Grand Supercycle. In turn each Grand Supercycle can be subdivided into eight supercycle waves, each of which is then divided into eight cycle waves. The process continues to embrace Primary, Intermediate, Minute, Minuette, and Sub-Minuette waves. The various details are highly intricate, but the general picture is represented in Figures A1-1 and A1-2.

Figures A1-2 and A1-3 show Elliot in historical perspective. Figure A1-2 illustrates the first five waves of the Grand Supercycle, which Elliot deemed to have begun in 1800. As the wave principle is one of form, there is nothing to determine when the three corrective waves are likely to appear. However, the frequent recurrence of Fibonacci numbers representing time spans between peaks and troughs are probably beyond coincidence. These time spans are shown in Table A1-1.

More recently, 8 years occurred between the 1966 and 1974 bottoms and the 1968 and 1976 tops, and 5 years between the 1968 and 1973 tops, for example.

It can readily be seen that the real problem with Elliot is interpretation. Indeed, every wave theorist (including Elliot himself) has at some time or another become entangled with the question of where one wave finished and another started. As far as the Fibonacci time spans are concerned, although these periods recur frequently, it is extremely difficult to use this principle as a basis for forecasting; there are no indications as to whether time spans based on these numbers will produce tops to tops, or bottoms to tops, or something else, and the permutations are infinite.

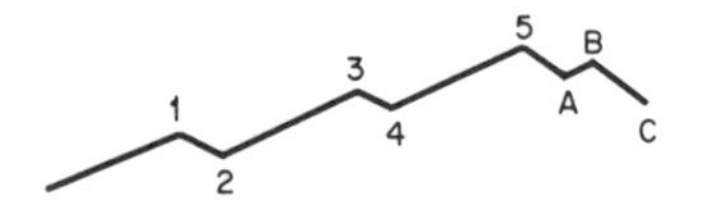

**FIGURE A1-1**

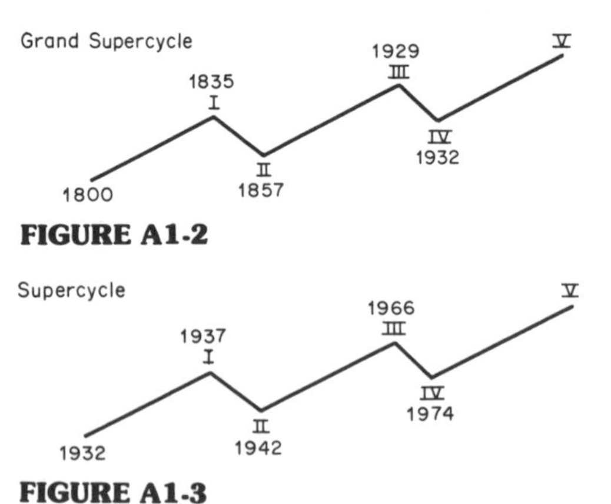


**FIGURE A1-2**

**FIGURE A1-3**

**TABLE A1-1 Time Spans between Stock Market Peaks and Troughs**

| Year started | Position | Year ended | Position | Length of cycle (years) |
|---|---|---|---|---|
| 1916 | Top | 1921 | Bottom | 5 |
| 1919 | Top | 1924 | Bottom | 5 |
| 1924 | Bottom | 1929 | Top | 5 |
| 1932 | Bottom | 1937 | Top | 5 |
| 1937 | Top | 1942 | Bottom | 5 |
| 1956 | Top | 1961 | Top | 5 |
| 1961 | Top | 1966 | Top | 5 |
| 1916 | Top | 1924 | Bottom | 8 |
| 1921 | Bottom | 1929 | Top | 8 |
| 1924 | Bottom | 1932 | Bottom | 8 |
| 1929 | Top | 1937 | Top | 8 |
| 1938 | Bottom | 1946 | Top | 8 |
| 1949 | Bottom | 1957 | Bottom | 8 |
| 1960 | Bottom | 1968 | Top | 8 |
| 1962 | Bottom | 1970 | Bottom | 8 |
| 1916 | Top | 1929 | Top | 13 |
| 1919 | Top | 1932 | Bottom | 13 |
| 1924 | Bottom | 1937 | Top | 13 |
| 1929 | Top | 1942 | Bottom | 13 |
| 1949 | Bottom | 1962 | Bottom | 13 |
| 1953 | Bottom | 1966 | Bottom | 13 |
| 1957 | Bottom | 1970 | Bottom | 13 |
| 1916 | Top | 1937 | Top | 21 |
| 1921 | Bottom | 1942 | Bottom | 21 |
| 1932 | Bottom | 1953 | Bottom | 21 |
| 1949 | Bottom | 1970 | Bottom | 21 |
| 1953 | Bottom | 1974 | Bottom | 21 |
| 1919 | Top | 1953 | Bottom | 34 |
| 1932 | Bottom | 1966 | Top | 34 |
| 1942 | Bottom | 1976 | Top | 34 |
| 1919 | Top | 1974 | Bottom | 55 |
| 1921 | Bottom | 1976 | Top | 55 |

The Elliot Wave is clearly a very subjective tool. This in itself can be dangerous in view of the fact that the market is very subject to emotional influences. For this reason, the weight given to Elliot interpretations should probably be downplayed. The old maxim "a little knowledge is a dangerous thing" applies probably more to Elliot than to any other market theory. The sources listed in the Bibliography will give a fuller understanding of the principles involved, since the theory has been described here only in its barest outline.

# APPENDIX 2

# COMPOSITION OF THE DOW JONES INDUSTRIAL AVERAGE

Allied Chemical
Alcoa
American Brands
American Can
American Telephone and Telegraph
Bethlehem Steel
Chrysler
Du Pont
Eastman Kodak
Esmark
Exxon
General Electric
General Foods
General Motors
Goodyear
International Harvester
Inco
International Paper
Johns Manville
Minnesota Mining
Owens Illinois
Procter and Gamble
Sears Roebuck
Standard Oil of California
Texaco
Union Carbide
U.S. Steel
United Technologies
Westinghouse Electric
Woolworth

# APPENDIX 3

# SELECTED CHARTS

**CHART A3-1 Stock Prices, Bond Prices, and Money Market Rates, 1831–1938**

Due to its construction the stock price series used here does not reflect the long-term uptrend in stock prices that took place between 1832 and 1929. The comparison between stock prices, bond prices, and commercial paper rates should therefore only be used for the purpose of cyclical reversals and not secular trends.

SOURCE: Leonard Ayres, *Stock Prices and Turning Points in Business Cycles.*

140
130
120
110
100
90
80
70
60
50
40
30
140
130
120
110
100
90
80
70
60
50
40
20
15
10
5
86 88 1890 92 94 96 98 1900 02 04 06 08 1910 12 14 16 18 1920 22 24 26 28 1930 32 34 36 38 1940

**CHART A3-2 Dow Jones Industrial Average versus Constant-Dollar Dow Jones Industrial Average, 1897–1976**

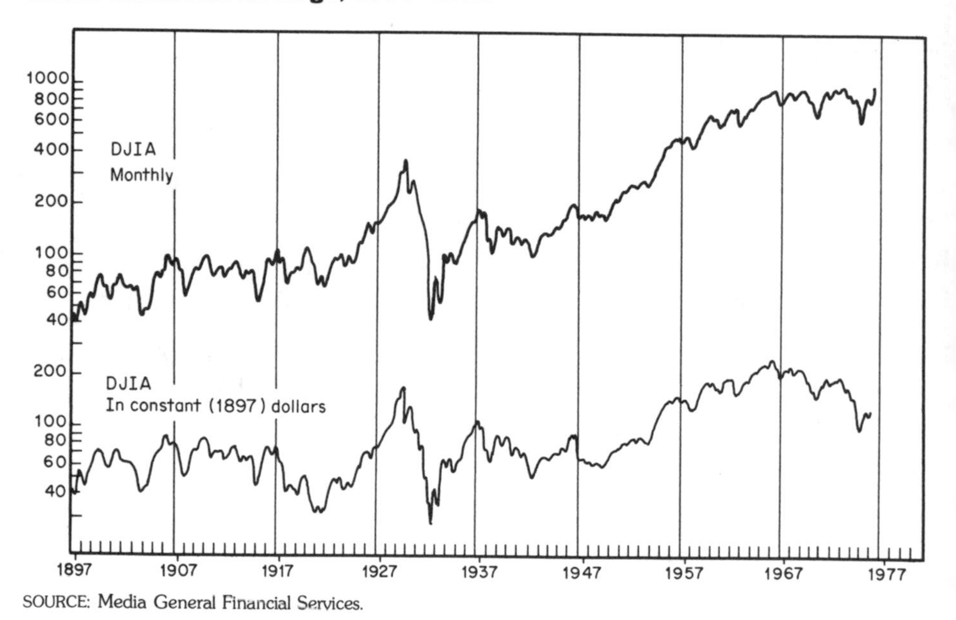


SOURCE: Media General Financial Services.

**CHART A3-3 Dow Jones Industrial Average, 1930–1944**

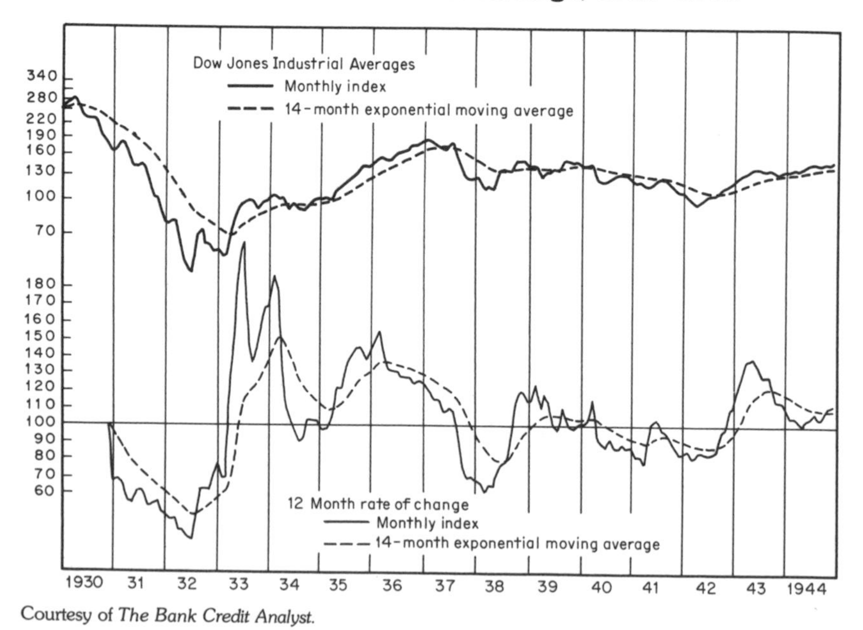


Courtesy of *The Bank Credit Analyst.*

## CHART A3-4 Dow Jones Industrial Average, 1944–1959

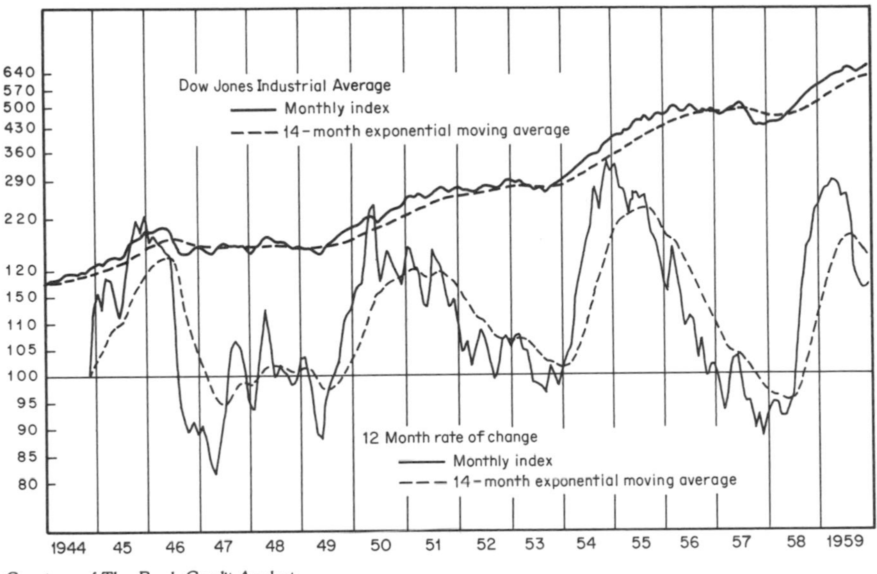


Courtesy of *The Bank Credit Analyst.*

## CHART A3-5 Financial Times Industrial Ordinary Index

(Chart 4-1 reproduced with 13-week momentum)

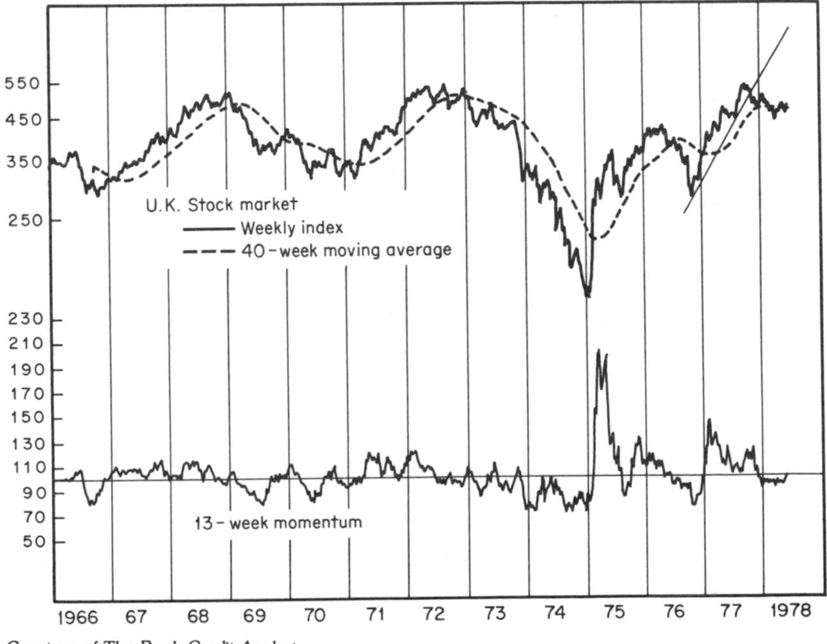


Courtesy of *The Bank Credit Analyst.*

**CHART A3-6 Dow Jones Transportation Average, 1929–1977**

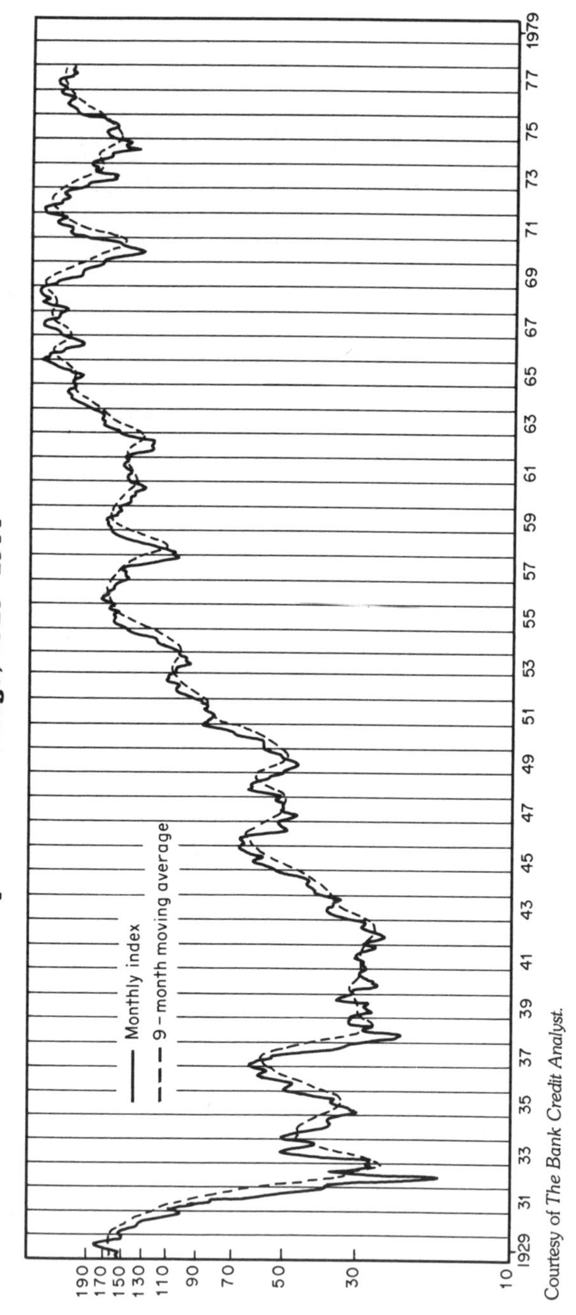


Courtesy of *The Bank Credit Analyst.*

## CHART A3-7 Dow Jones Transportation Average, 1930–1944

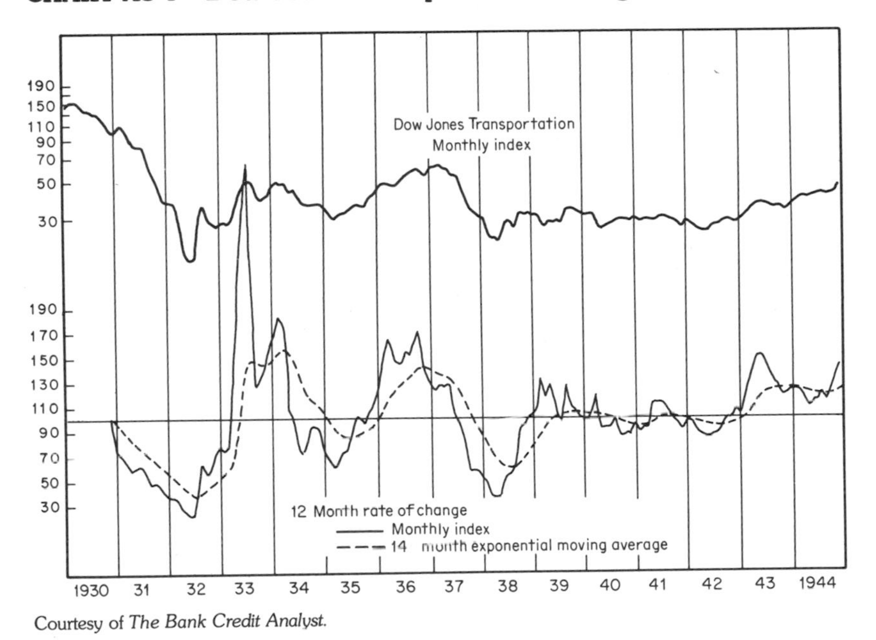


Courtesy of *The Bank Credit Analyst.*

## CHART A3-8 Dow Jones Transportation Average, 1944–1959

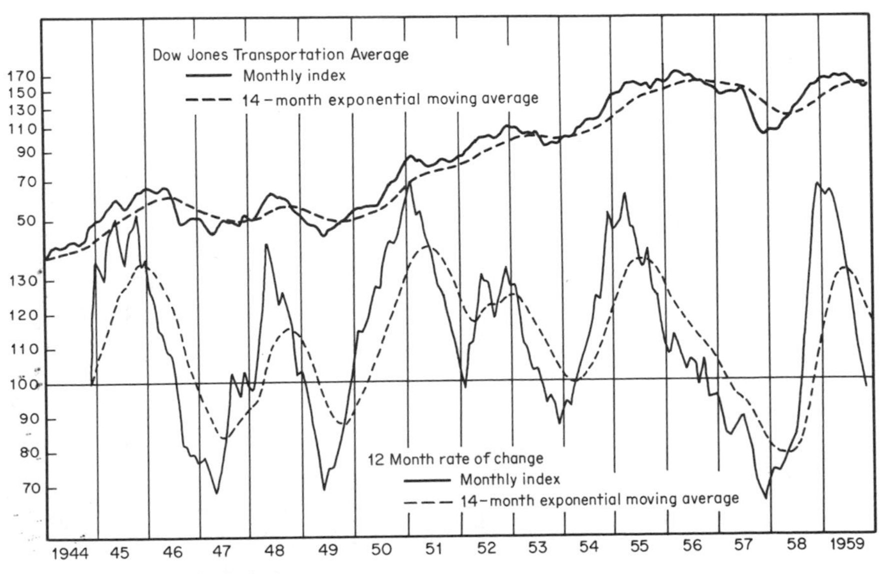


Courtesy of *The Bank Credit Analyst.*

**CHART A3-9 Dow Jones Transportation Average, 1959–1978**

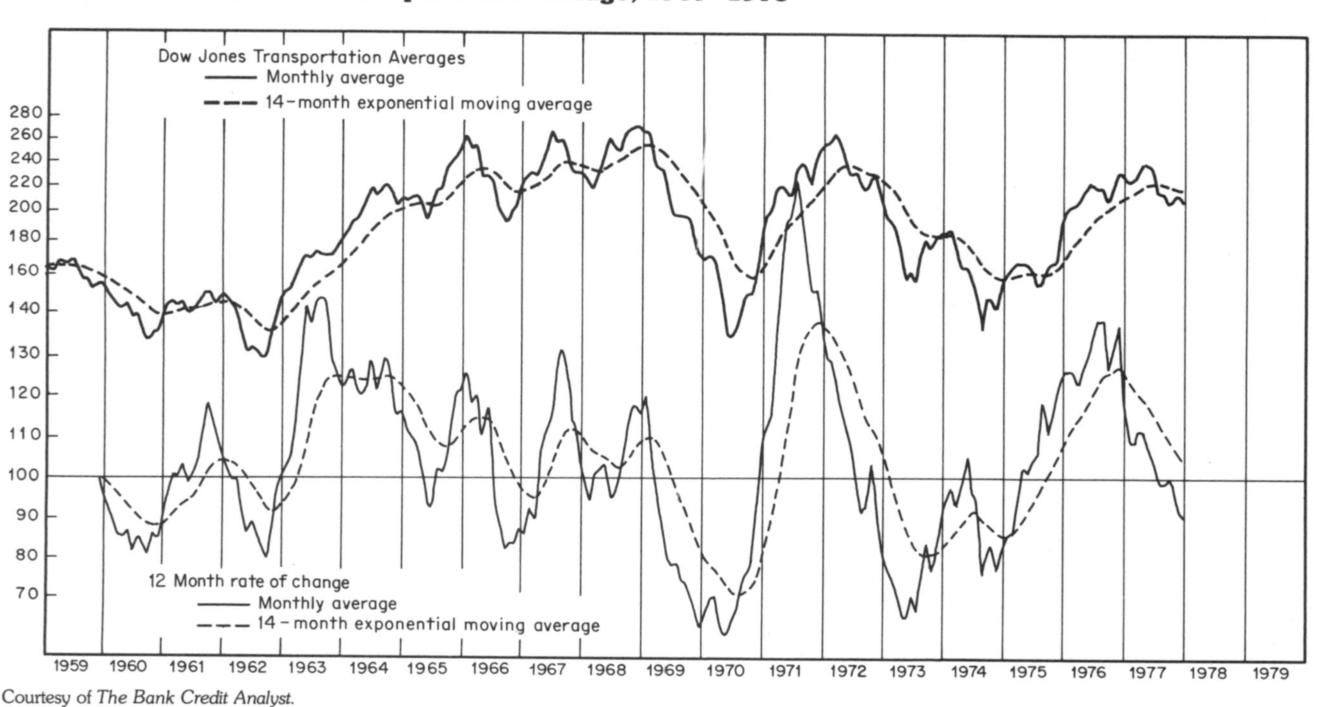


Courtesy of *The Bank Credit Analyst.*

**CHART A3-10 Dow Jones Transportation Average, Relative Strength, 1929–1977**

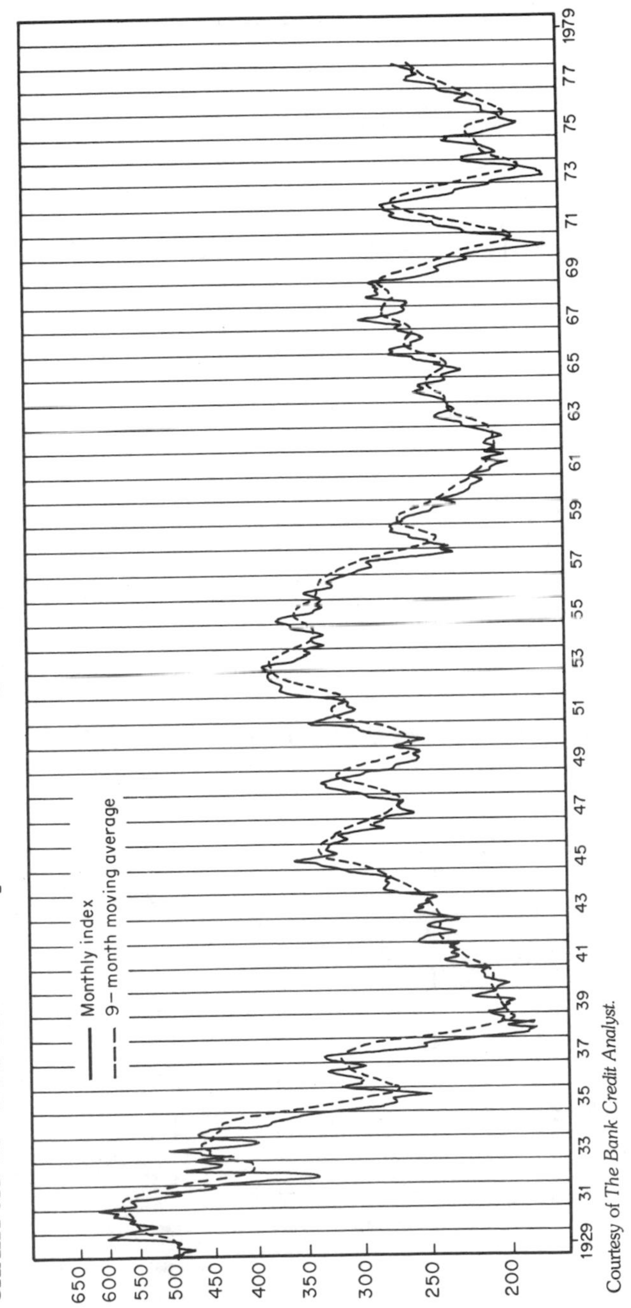


Courtesy of *The Bank Credit Analyst.*

## CHART A3-11 Dow Jones Utility Average, 1930–1944

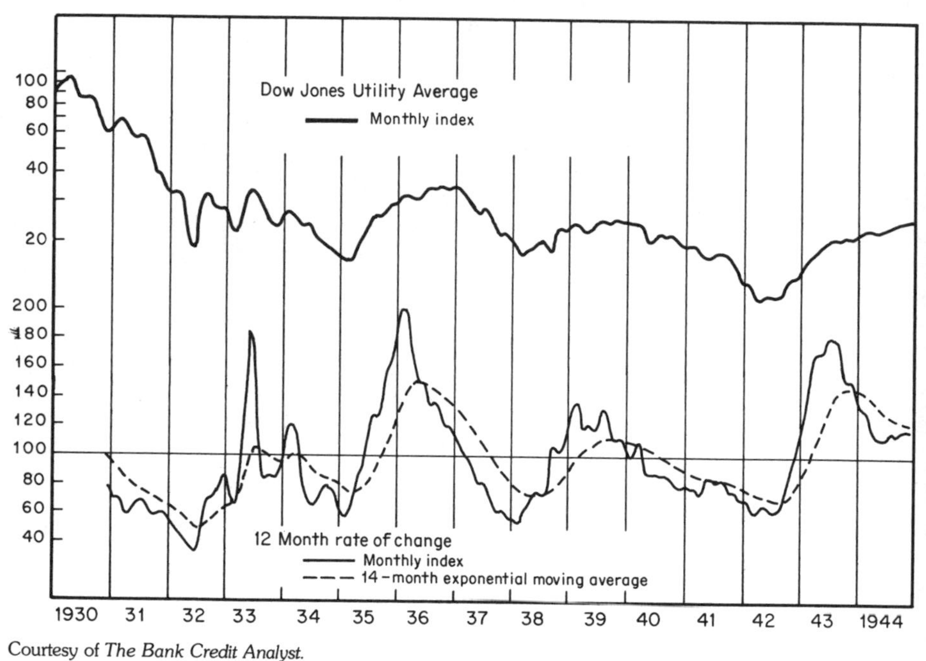


Courtesy of *The Bank Credit Analyst.*

## CHART A3-12 Dow Jones Utility Average, 1944–1959

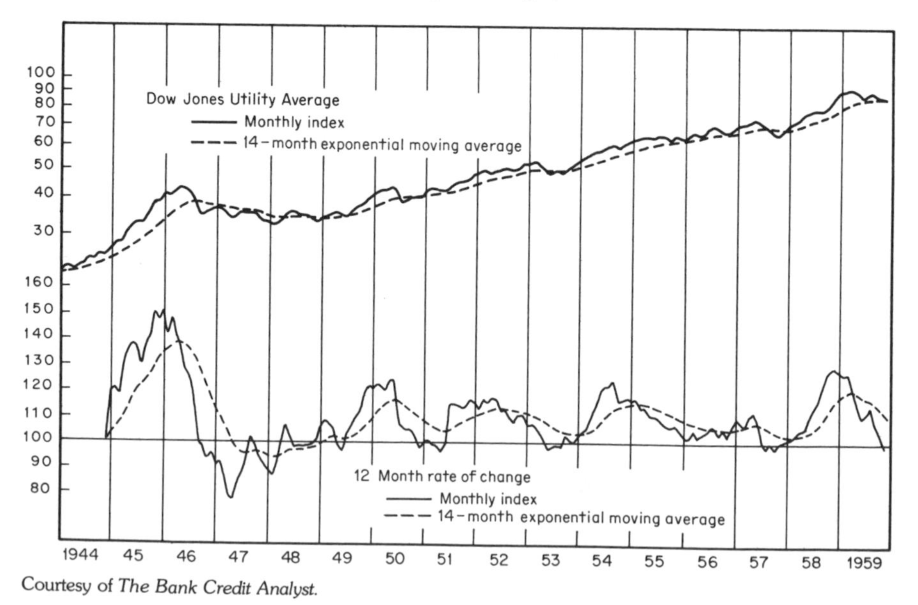


Courtesy of *The Bank Credit Analyst.*

**CHART A3-13 Dow Jones Utility Average, Relative Strength, 1929–1977**

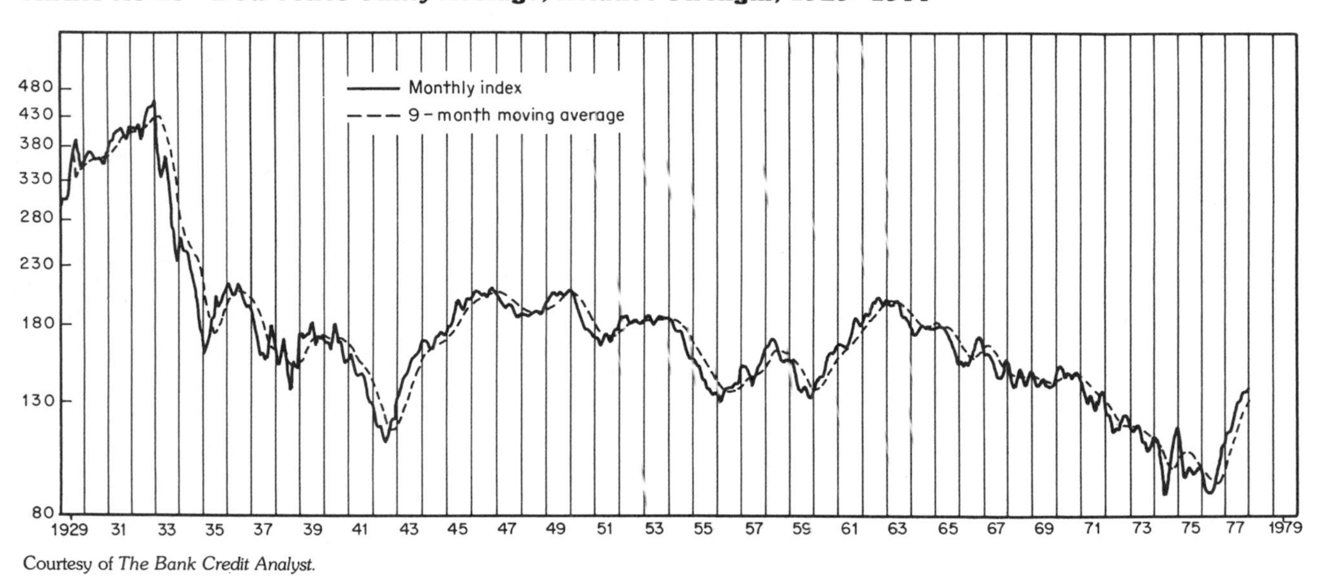


Courtesy of *The Bank Credit Analyst.*

## CHART A3-14 Industry Advance/Decline Line, 1965–1977

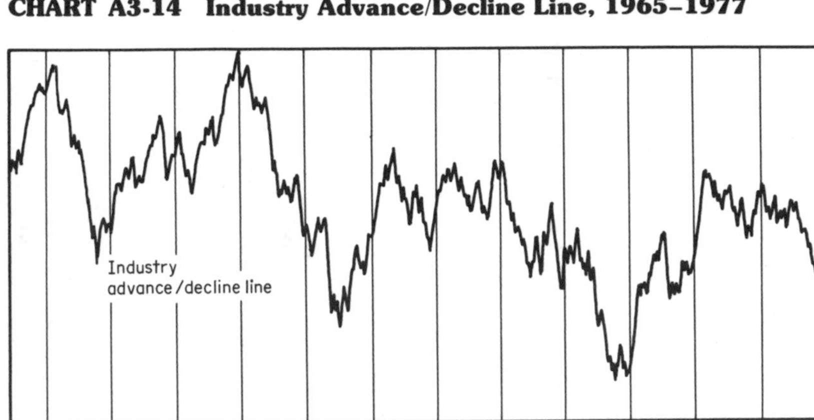


SOURCE: Barron's 35-Industry Groups.
Courtesy of *The Bank Credit Analyst.*

## CHART A3-15 Secondary Stock Offerings versus the Dow Jones Industrial Average, 1958–1976

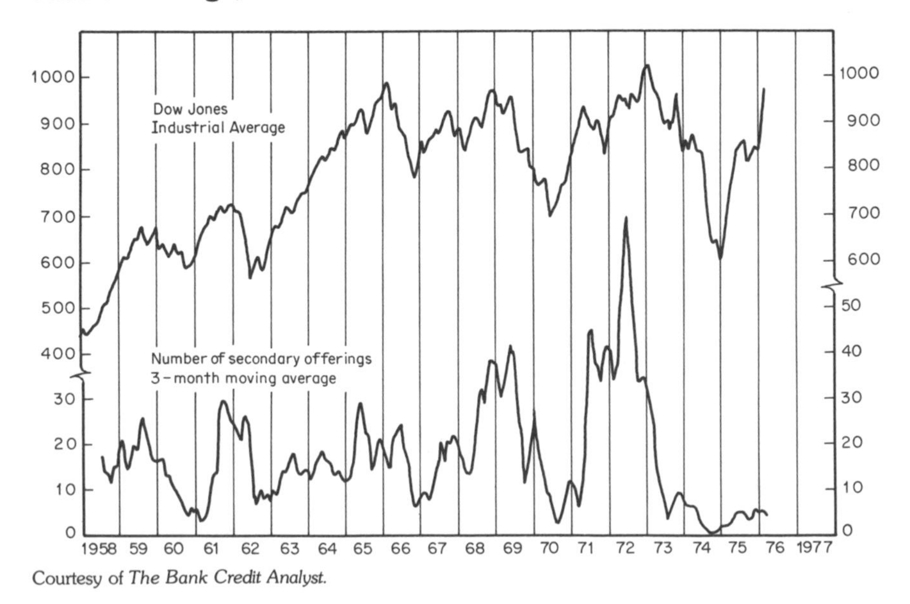


Courtesy of *The Bank Credit Analyst.*

**CHART A3-16 Dow Jones Industrial Average versus Odd-Lot Short-Sale Index**

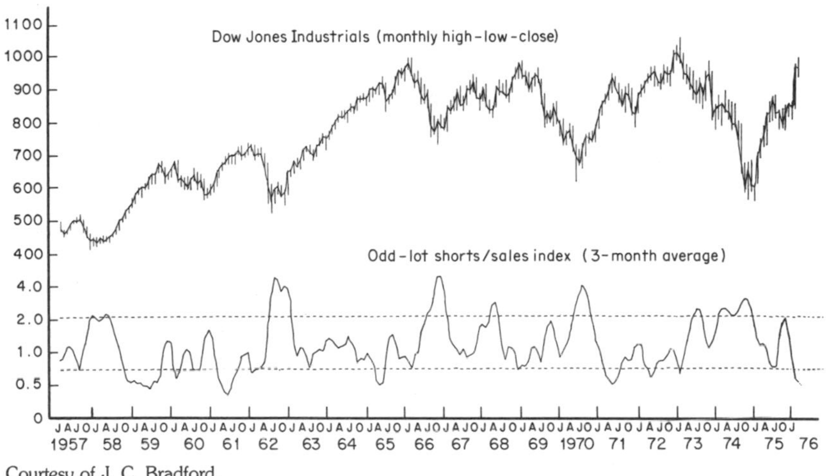


Courtesy of J. C. Bradford.

**CHART A3-17 Annual Rate of Change of Stock Prices versus Short-Interest Ratio**

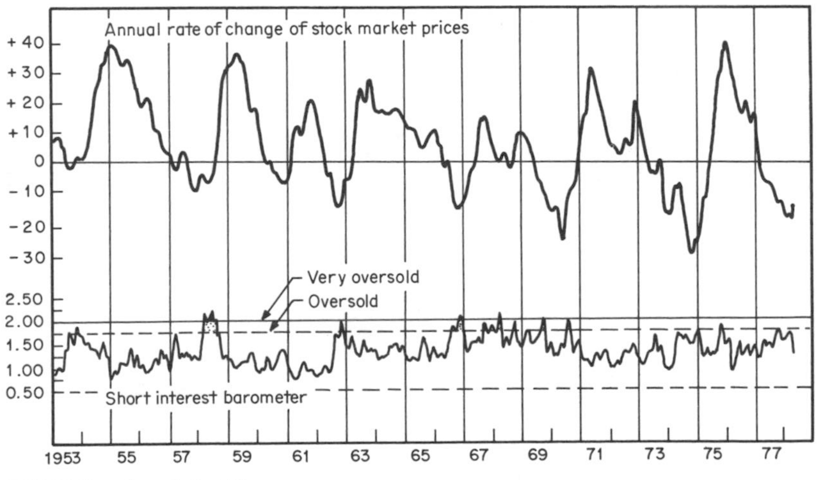


SOURCE: Dow Jones Industrials.

## CHART A3-18 Composite Long-Term Bond Index

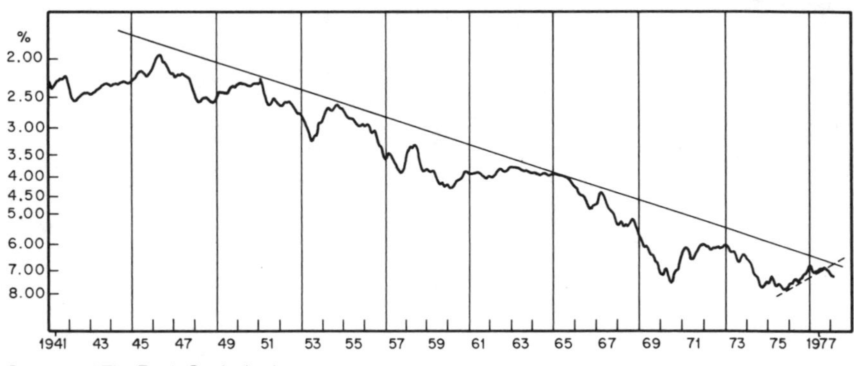

Courtesy of *The Bank Credit Analyst.*

# APPENDIX 4

# A STATISTICAL SUMMARY

**TABLE A4-1 Monthly Values of Dow Jones Industrial Average**

| YEAR | MONTH | DJIA | YEAR | MONTH | DJIA |
|---|---|---|---|---|---|
| 1790 | JAN | 2.85 | 1794 | JAN | 3.62 |
| 1790 | FEB | 2.86 | 1794 | FEB | 3.63 |
| 1790 | MAR | 2.91 | 1794 | MAR | 3.78 |
| 1790 | APR | 2.95 | 1794 | APR | 3.90 |
| 1790 | MAY | 3.00 | 1794 | MAY | 4.05 |
| 1790 | JUN | 3.04 | 1794 | JUN | 3.94 |
| 1790 | JUL | 3.14 | 1794 | JUL | 4.23 |
| 1790 | AUG | 3.11 | 1794 | AUG | 4.43 |
| 1790 | SEP | 3.16 | 1794 | SEP | 4.50 |
| 1790 | OCT | 3.27 | 1794 | OCT | 4.40 |
| 1790 | NOV | 3.31 | 1794 | NOV | 4.24 |
| 1790 | DEC | 3.39 | 1794 | DEC | 4.35 |
| 1791 | JAN | 3.45 | 1795 | JAN | 4.43 |
| 1791 | FEB | 3.47 | 1795 | FEB | 4.49 |
| 1791 | MAR | 3.43 | 1795 | MAR | 4.44 |
| 1791 | APR | 3.40 | 1795 | APR | 4.52 |
| 1791 | MAY | 3.40 | 1795 | MAY | 4.49 |
| 1791 | JUN | 3.46 | 1795 | JUN | 4.38 |
| 1791 | JUL | 3.44 | 1795 | JUL | 4.42 |
| 1791 | AUG | 3.55 | 1795 | AUG | 4.34 |
| 1791 | SEP | 3.53 | 1795 | SEP | 4.34 |
| 1791 | OCT | 3.65 | 1795 | OCT | 4.18 |
| 1791 | NOV | 3.64 | 1795 | NOV | 4.21 |
| 1791 | DEC | 3.66 | 1795 | DEC | 4.11 |
| 1792 | JAN | 3.69 | 1796 | JAN | 4.07 |
| 1792 | FEB | 3.62 | 1796 | FEB | 4.16 |
| 1792 | MAR | 3.48 | 1796 | MAR | 4.05 |
| 1792 | APR | 3.45 | 1796 | APR | 3.96 |
| 1792 | MAY | 3.52 | 1796 | MAY | 3.92 |
| 1792 | JUN | 3.20 | 1796 | JUN | 3.87 |
| 1792 | JUL | 3.08 | 1796 | JUL | 3.84 |
| 1792 | AUG | 3.03 | 1796 | AUG | 3.65 |
| 1792 | SEP | 2.96 | 1796 | SEP | 3.59 |
| 1792 | OCT | 3.06 | 1796 | OCT | 3.66 |
| 1792 | NOV | 3.10 | 1796 | NOV | 3.64 |
| 1792 | DEC | 3.20 | 1796 | DEC | 3.55 |
| 1793 | JAN | 3.26 | 1797 | JAN | 3.62 |
| 1793 | FEB | 3.28 | 1797 | FEB | 3.59 |
| 1793 | MAR | 3.40 | 1797 | MAR | 3.55 |
| 1793 | APR | 3.51 | 1797 | APR | 3.45 |
| 1793 | MAY | 3.50 | 1797 | MAY | 3.59 |
| 1793 | JUN | 3.48 | 1797 | JUN | 3.47 |
| 1793 | JUL | 3.40 | 1797 | JUL | 3.51 |
| 1793 | AUG | 3.49 | 1797 | AUG | 3.35 |
| 1793 | SEP | 3.51 | 1797 | SEP | 3.46 |
| 1793 | OCT | 3.53 | 1797 | OCT | 3.27 |
| 1793 | NOV | 3.53 | 1797 | NOV | 3.19 |
| 1793 | DEC | 3.54 | 1797 | DEC | 3.25 |

| YEAR | MONTH | DJIA | YEAR | MONTH | DJIA |
|---|---|---|---|---|---|
| 1798 | JAN | 3.26 | 1802 | JAN | 3.48 |
| 1798 | FEB | 3.31 | 1802 | FEB | 3.36 |
| 1798 | MAR | 3.41 | 1802 | MAR | 3.43 |
| 1798 | APR | 3.52 | 1802 | APR | 3.51 |
| 1798 | MAY | 3.59 | 1802 | MAY | 3.50 |
| 1798 | JUN | 3.73 | 1802 | JUN | 3.63 |
| 1798 | JUL | 3.73 | 1802 | JUL | 3.63 |
| 1798 | AUG | 3.84 | 1802 | AUG | 3.67 |
| 1798 | SEP | 3.81 | 1802 | SEP | 3.52 |
| 1798 | OCT | 4.01 | 1802 | OCT | 3.68 |
| 1798 | NOV | 3.98 | 1802 | NOV | 3.75 |
| 1798 | DEC | 3.99 | 1802 | DEC | 3.89 |
| 1799 | JAN | 4.05 | 1803 | JAN | 3.95 |
| 1799 | FEB | 4.08 | 1803 | FEB | 4.01 |
| 1799 | MAR | 4.09 | 1803 | MAR | 4.03 |
| 1799 | APR | 4.20 | 1803 | APR | 4.04 |
| 1799 | MAY | 4.21 | 1803 | MAY | 4.07 |
| 1799 | JUN | 4.29 | 1803 | JUN | 4.10 |
| 1799 | JUL | 4.20 | 1803 | JUL | 4.07 |
| 1799 | AUG | 4.34 | 1803 | AUG | 4.10 |
| 1799 | SEP | 4.23 | 1803 | SEP | 4.10 |
| 1799 | OCT | 4.19 | 1803 | OCT | 4.24 |
| 1799 | NOV | 4.43 | 1803 | NOV | 4.09 |
| 1799 | DEC | 4.63 | 1803 | DEC | 4.22 |
| 1800 | JAN | 4.68 | 1804 | JAN | 4.46 |
| 1800 | FEB | 4.96 | 1804 | FEB | 4.68 |
| 1800 | MAR | 4.89 | 1804 | MAR | 4.98 |
| 1800 | APR | 5.14 | 1804 | APR | 5.28 |
| 1800 | MAY | 5.39 | 1804 | MAY | 5.67 |
| 1800 | JUN | 5.60 | 1804 | JUN | 5.81 |
| 1800 | JUL | 5.67 | 1804 | JUL | 5.81 |
| 1800 | AUG | 5.80 | 1804 | AUG | 6.06 |
| 1800 | SEP | 5.80 | 1804 | SEP | 6.30 |
| 1800 | OCT | 5.96 | 1804 | OCT | 6.52 |
| 1800 | NOV | 5.73 | 1804 | NOV | 6.40 |
| 1800 | DEC | 5.63 | 1804 | DEC | 6.26 |
| 1801 | JAN | 5.57 | 1805 | JAN | 6.55 |
| 1801 | FEB | 5.43 | 1805 | FEB | 6.44 |
| 1801 | MAR | 5.18 | 1805 | MAR | 6.38 |
| 1801 | APR | 4.93 | 1805 | APR | 6.23 |
| 1801 | MAY | 4.62 | 1805 | MAY | 6.46 |
| 1801 | JUN | 4.36 | 1805 | JUN | 6.56 |
| 1801 | JUL | 4.07 | 1805 | JUL | 6.38 |
| 1801 | AUG | 3.84 | 1805 | AUG | 6.13 |
| 1801 | SEP | 3.71 | 1805 | SEP | 6.02 |
| 1801 | OCT | 3.57 | 1805 | OCT | 6.10 |
| 1801 | NOV | 3.54 | 1805 | NOV | 6.14 |
| 1801 | DEC | 3.47 | 1805 | DEC | 6.38 |

**TABLE A4-1 Monthly Values of Dow Jones Industrial Average** (cont.)

| YEAR | MONTH | DJIA | YEAR | MONTH | DJIA |
|---|---|---|---|---|---|
| 1806 | JAN | 6.58 | 1810 | JAN | 6.77 |
| 1806 | FEB | 6.61 | 1810 | FEB | 6.84 |
| 1806 | MAR | 6.81 | 1810 | MAR | 6.71 |
| 1806 | APR | 7.20 | 1810 | APR | 6.79 |
| 1806 | MAY | 7.49 | 1810 | MAY | 6.48 |
| 1806 | JUN | 7.49 | 1810 | JUN | 6.46 |
| 1806 | JUL | 7.54 | 1810 | JUL | 6.43 |
| 1806 | AUG | 7.56 | 1810 | AUG | 6.25 |
| 1806 | SEP | 7.56 | 1810 | SEP | 6.00 |
| 1806 | OCT | 7.34 | 1810 | OCT | 5.81 |
| 1806 | NOV | 6.97 | 1810 | NOV | 5.66 |
| 1806 | DEC | 6.71 | 1810 | DEC | 5.61 |
| 1807 | JAN | 6.31 | 1811 | JAN | 5.73 |
| 1807 | FEB | 6.08 | 1811 | FEB | 5.83 |
| 1807 | MAR | 5.63 | 1811 | MAR | 6.04 |
| 1807 | APR | 5.28 | 1811 | APR | 6.12 |
| 1807 | MAY | 4.95 | 1811 | MAY | 6.12 |
| 1807 | JUN | 4.51 | 1811 | JUN | 6.25 |
| 1807 | JUL | 4.18 | 1811 | JUL | 6.14 |
| 1807 | AUG | 3.91 | 1811 | AUG | 6.11 |
| 1807 | SEP | 3.71 | 1811 | SEP | 6.09 |
| 1807 | OCT | 3.57 | 1811 | OCT | 5.79 |
| 1807 | NOV | 3.69 | 1811 | NOV | 5.59 |
| 1807 | DEC | 3.60 | 1811 | DEC | 5.31 |
| 1808 | JAN | 3.65 | 1812 | JAN | 5.61 |
| 1808 | FEB | 3.81 | 1812 | FEB | 5.17 |
| 1808 | MAR | 3.96 | 1812 | MAR | 4.97 |
| 1808 | APR | 4.04 | 1812 | APR | 4.77 |
| 1808 | MAY | 4.17 | 1812 | MAY | 4.66 |
| 1808 | JUN | 4.44 | 1812 | JUN | 4.58 |
| 1808 | JUL | 4.63 | 1812 | JUL | 4.67 |
| 1808 | AUG | 4.76 | 1812 | AUG | 4.59 |
| 1808 | SEP | 5.05 | 1812 | SEP | 4.61 |
| 1808 | OCT | 5.16 | 1812 | OCT | 4.53 |
| 1808 | NOV | 5.33 | 1812 | NOV | 4.53 |
| 1808 | DEC | 5.55 | 1812 | DEC | 4.59 |
| 1809 | JAN | 5.31 | 1813 | JAN | 4.62 |
| 1809 | FEB | 5.54 | 1813 | FEB | 4.61 |
| 1809 | MAR | 5.90 | 1813 | MAR | 4.43 |
| 1809 | APR | 6.12 | 1813 | APR | 4.74 |
| 1809 | MAY | 6.23 | 1813 | MAY | 4.83 |
| 1809 | JUN | 6.45 | 1813 | JUN | 4.92 |
| 1809 | JUL | 6.60 | 1813 | JUL | 4.92 |
| 1809 | AUG | 6.74 | 1813 | AUG | 4.78 |
| 1809 | SEP | 6.73 | 1813 | SEP | 4.58 |
| 1809 | OCT | 6.91 | 1813 | OCT | 4.82 |
| 1809 | NOV | 6.71 | 1813 | NOV | 4.90 |
| 1809 | DEC | 6.69 | 1813 | DEC | 5.06 |

| YEAR | MONTH | DJIA | YEAR | MONTH | DJIA |
|---|---|---|---|---|---|
| 1814 | JAN | 5.40 | 1818 | JAN | 10.46 |
| 1814 | FEB | 5.83 | 1818 | FEB | 10.44 |
| 1814 | MAR | 6.57 | 1818 | MAR | 10.18 |
| 1814 | APR | 7.17 | 1818 | APR | 10.72 |
| 1814 | MAY | 7.96 | 1818 | MAY | 10.81 |
| 1814 | JUN | 8.42 | 1818 | JUN | 10.94 |
| 1814 | JUL | 8.27 | 1818 | JUL | 10.86 |
| 1814 | AUG | 8.26 | 1818 | AUG | 10.72 |
| 1814 | SEP | 8.31 | 1818 | SEP | 10.83 |
| 1814 | OCT | 8.16 | 1818 | OCT | 10.72 |
| 1814 | NOV | 8.18 | 1818 | NOV | 10.60 |
| 1814 | DEC | 8.53 | 1818 | DEC | 11.03 |
| 1815 | JAN | 8.84 | 1819 | JAN | 10.54 |
| 1815 | FEB | 8.66 | 1819 | FEB | 10.55 |
| 1815 | MAR | 8.55 | 1819 | MAR | 10.90 |
| 1815 | APR | 8.78 | 1819 | APR | 10.88 |
| 1815 | MAY | 8.76 | 1819 | MAY | 11.03 |
| 1815 | JUN | 8.76 | 1819 | JUN | 10.42 |
| 1815 | JUL | 8.92 | 1819 | JUL | 10.42 |
| 1815 | AUG | 8.92 | 1819 | AUG | 10.42 |
| 1815 | SEP | 9.05 | 1819 | SEP | 10.00 |
| 1815 | OCT | 9.04 | 1819 | OCT | 10.11 |
| 1815 | NOV | 8.97 | 1819 | NOV | 10.37 |
| 1815 | DEC | 8.83 | 1819 | DEC | 9.92 |
| 1816 | JAN | 8.78 | 1820 | JAN | 9.59 |
| 1816 | FEB | 8.76 | 1820 | FEB | 9.85 |
| 1816 | MAR | 8.86 | 1820 | MAR | 9.98 |
| 1816 | APR | 8.78 | 1820 | APR | 9.97 |
| 1816 | MAY | 8.75 | 1820 | MAY | 10.36 |
| 1816 | JUN | 8.73 | 1820 | JUN | 10.67 |
| 1816 | JUL | 8.65 | 1820 | JUL | 10.46 |
| 1816 | AUG | 8.47 | 1820 | AUG | 10.59 |
| 1816 | SEP | 8.29 | 1820 | SEP | 10.90 |
| 1816 | OCT | 8.57 | 1820 | OCT | 10.94 |
| 1816 | NOV | 8.86 | 1820 | NOV | 10.93 |
| 1816 | DEC | 8.68 | 1820 | DEC | 11.35 |
| 1817 | JAN | 8.60 | 1821 | JAN | 11.01 |
| 1817 | FEB | 8.60 | 1821 | FEB | 11.50 |
| 1817 | MAR | 8.75 | 1821 | MAR | 11.76 |
| 1817 | APR | 8.92 | 1821 | APR | 11.73 |
| 1817 | MAY | 8.73 | 1821 | MAY | 11.87 |
| 1817 | JUN | 8.76 | 1821 | JUN | 11.97 |
| 1817 | JUL | 8.94 | 1821 | JUL | 12.10 |
| 1817 | AUG | 8.94 | 1821 | AUG | 11.84 |
| 1817 | SEP | 9.32 | 1821 | SEP | 11.81 |
| 1817 | OCT | 9.67 | 1821 | OCT | 11.74 |
| 1817 | NOV | 9.85 | 1821 | NOV | 12.15 |
| 1817 | DEC | 10.13 | 1821 | DEC | 12.02 |

**TABLE A4-1 Monthly Values of Dow Jones Industrial Average** (cont.)

| YEAR | MONTH | DJIA | YEAR | MONTH | DJIA |
|---|---|---|---|---|---|
| 1822 | JAN | 11.84 | 1826 | JAN | 11.40 |
| 1822 | FEB | 11.68 | 1826 | FEB | 11.33 |
| 1822 | MAR | 11.69 | 1826 | MAR | 11.43 |
| 1822 | APR | 11.69 | 1826 | APR | 11.76 |
| 1822 | MAY | 11.42 | 1826 | MAY | 11.66 |
| 1822 | JUN | 11.79 | 1826 | JUN | 11.56 |
| 1822 | JUL | 11.64 | 1826 | JUL | 11.68 |
| 1822 | AUG | 11.40 | 1826 | AUG | 11.42 |
| 1822 | SEP | 11.19 | 1826 | SEP | 11.51 |
| 1822 | OCT | 11.40 | 1826 | OCT | 11.33 |
| 1822 | NOV | 11.33 | 1826 | NOV | 11.40 |
| 1822 | DEC | 11.60 | 1826 | DEC | 11.37 |
| 1823 | JAN | 11.79 | 1827 | JAN | 11.25 |
| 1823 | FEB | 11.27 | 1827 | FEB | 11.11 |
| 1823 | MAR | 11.69 | 1827 | MAR | 11.45 |
| 1823 | APR | 11.40 | 1827 | APR | 11.76 |
| 1823 | MAY | 11.74 | 1827 | MAY | 11.90 |
| 1823 | JUN | 11.76 | 1827 | JUN | 12.00 |
| 1823 | JUL | 11.79 | 1827 | JUL | 11.79 |
| 1823 | AUG | 11.94 | 1827 | AUG | 11.94 |
| 1823 | SEP | 11.92 | 1827 | SEP | 12.02 |
| 1823 | OCT | 11.95 | 1827 | OCT | 12.33 |
| 1823 | NOV | 12.17 | 1827 | NOV | 11.79 |
| 1823 | DEC | 12.25 | 1827 | DEC | 11.71 |
| 1824 | JAN | 12.28 | 1828 | JAN | 11.20 |
| 1824 | FEB | 12.41 | 1828 | FEB | 10.93 |
| 1824 | MAR | 12.83 | 1828 | MAR | 10.68 |
| 1824 | APR | 12.75 | 1828 | APR | 10.86 |
| 1824 | MAY | 12.75 | 1828 | MAY | 11.19 |
| 1824 | JUN | 13.01 | 1828 | JUN | 11.22 |
| 1824 | JUL | 13.03 | 1828 | JUL | 11.06 |
| 1824 | AUG | 12.49 | 1828 | AUG | 11.04 |
| 1824 | SEP | 14.38 | 1828 | SEP | 10.90 |
| 1824 | OCT | 13.01 | 1828 | OCT | 11.01 |
| 1824 | NOV | 13.29 | 1828 | NOV | 11.30 |
| 1824 | DEC | 13.04 | 1828 | DEC | 11.07 |
| 1825 | JAN | 12.85 | 1829 | JAN | 10.46 |
| 1825 | FEB | 12.70 | 1829 | FEB | 11.19 |
| 1825 | MAR | 12.82 | 1829 | MAR | 10.83 |
| 1825 | APR | 12.88 | 1829 | APR | 10.67 |
| 1825 | MAY | 12.62 | 1829 | MAY | 11.24 |
| 1825 | JUN | 12.36 | 1829 | JUN | 11.20 |
| 1825 | JUL | 12.52 | 1829 | JUL | 11.50 |
| 1825 | AUG | 12.54 | 1829 | AUG | 11.61 |
| 1825 | SEP | 12.43 | 1829 | SEP | 11.53 |
| 1825 | OCT | 12.43 | 1829 | OCT | 11.58 |
| 1825 | NOV | 11.94 | 1829 | NOV | 11.71 |
| 1825 | DEC | 11.51 | 1829 | DEC | 11.87 |

| YEAR | MONTH | DJIA | YEAR | MONTH | DJIA |
|---|---|---|---|---|---|
| 1830 | JAN | 12.41 | 1834 | JAN | 14.61 |
| 1830 | FEB | 12.62 | 1834 | FEB | 13.31 |
| 1830 | MAR | 12.74 | 1834 | MAR | 13.83 |
| 1830 | APR | 13.16 | 1834 | APR | 15.91 |
| 1830 | MAY | 13.47 | 1834 | MAY | 15.91 |
| 1830 | JUN | 13.81 | 1834 | JUN | 15.39 |
| 1830 | JUL | 14.41 | 1834 | JUL | 15.39 |
| 1830 | AUG | 14.07 | 1834 | AUG | 16.18 |
| 1830 | SEP | 13.70 | 1834 | SEP | 15.91 |
| 1830 | OCT | 13.94 | 1834 | OCT | 16.70 |
| 1830 | NOV | 14.41 | 1834 | NOV | 17.22 |
| 1830 | DEC | 13.76 | 1834 | DEC | 17.74 |
| 1831 | JAN | 13.29 | 1835 | JAN | 19.05 |
| 1831 | FEB | 12.95 | 1835 | FEB | 18.52 |
| 1831 | MAR | 13.01 | 1835 | MAR | 20.09 |
| 1831 | APR | 12.75 | 1835 | APR | 20.61 |
| 1831 | MAY | 12.93 | 1835 | MAY | 23.48 |
| 1831 | JUN | 12.70 | 1835 | JUN | 23.22 |
| 1831 | JUL | 12.52 | 1835 | JUL | 22.70 |
| 1831 | AUG | 12.26 | 1835 | AUG | 23.48 |
| 1831 | SEP | 12.00 | 1835 | SEP | 22.64 |
| 1831 | OCT | 12.00 | 1835 | OCT | 21.39 |
| 1831 | NOV | 11.74 | 1835 | NOV | 19.57 |
| 1831 | DEC | 12.00 | 1835 | DEC | 18.78 |
| 1832 | JAN | 12.26 | 1836 | JAN | 19.05 |
| 1832 | FEB | 12.00 | 1836 | FEB | 21.65 |
| 1832 | MAR | 12.26 | 1836 | MAR | 21.13 |
| 1832 | APR | 12.26 | 1836 | APR | 20.09 |
| 1832 | MAY | 12.26 | 1836 | MAY | 19.83 |
| 1832 | JUN | 12.26 | 1836 | JUN | 20.09 |
| 1832 | JUL | 12.52 | 1836 | JUL | 18.52 |
| 1832 | AUG | 13.04 | 1836 | AUG | 17.48 |
| 1832 | SEP | 13.57 | 1836 | SEP | 16.70 |
| 1832 | OCT | 14.09 | 1836 | OCT | 15.13 |
| 1832 | NOV | 14.87 | 1836 | NOV | 14.33 |
| 1832 | DEC | 15.91 | 1836 | DEC | 14.61 |
| 1833 | JAN | 16.70 | 1837 | JAN | 16.96 |
| 1833 | FEB | 17.22 | 1837 | FEB | 16.70 |
| 1833 | MAR | 17.48 | 1837 | MAR | 14.61 |
| 1833 | APR | 18.00 | 1837 | APR | 13.04 |
| 1833 | MAY | 18.00 | 1837 | MAY | 12.00 |
| 1833 | JUN | 18.00 | 1837 | JUN | 11.22 |
| 1833 | JUL | 17.74 | 1837 | JUL | 13.83 |
| 1833 | AUG | 17.74 | 1837 | AUG | 13.04 |
| 1833 | SEP | 17.22 | 1837 | SEP | 13.04 |
| 1833 | OCT | 16.70 | 1837 | OCT | 13.31 |
| 1833 | NOV | 15.91 | 1837 | NOV | 13.57 |
| 1833 | DEC | 15.13 | 1837 | DEC | 12.78 |

**TABLE A4-1 Monthly Values of Dow Jones Industrial Average** (cont.)

| YEAR | MONTH | DJIA | YEAR | MONTH | DJIA |
|---|---|---|---|---|---|
| 1838 | JAN | 12.26 | 1842 | JAN | 6.78 |
| 1838 | FEB | 11.74 | 1842 | FEB | 6.26 |
| 1838 | MAR | 10.70 | 1842 | MAR | 5.74 |
| 1838 | APR | 10.70 | 1842 | APR | 6.00 |
| 1838 | MAY | 11.48 | 1842 | MAY | 7.04 |
| 1838 | JUN | 12.26 | 1842 | JUN | 7.04 |
| 1838 | JUL | 12.78 | 1842 | JUL | 7.04 |
| 1838 | AUG | 13.31 | 1842 | AUG | 6.52 |
| 1838 | SEP | 13.83 | 1842 | SEP | 7.04 |
| 1838 | OCT | 12.26 | 1842 | OCT | 6.52 |
| 1838 | NOV | 12.26 | 1842 | NOV | 6.52 |
| 1838 | DEC | 11.22 | 1842 | DEC | 6.52 |
| 1839 | JAN | 12.26 | 1843 | JAN | 6.26 |
| 1839 | FEB | 13.04 | 1843 | FEB | 6.26 |
| 1839 | MAR | 12.26 | 1843 | MAR | 6.26 |
| 1839 | APR | 12.26 | 1843 | APR | 6.52 |
| 1839 | MAY | 12.52 | 1843 | MAY | 7.83 |
| 1839 | JUN | 11.74 | 1843 | JUN | 8.61 |
| 1839 | JUL | 10.96 | 1843 | JUL | 8.61 |
| 1839 | AUG | 10.70 | 1843 | AUG | 8.87 |
| 1839 | SEP | 10.17 | 1843 | SEP | 9.13 |
| 1839 | OCT | 9.39 | 1843 | OCT | 8.87 |
| 1839 | NOV | 8.61 | 1843 | NOV | 9.65 |
| 1839 | DEC | 9.13 | 1843 | DEC | 11.22 |
| 1840 | JAN | 9.39 | 1844 | JAN | 11.22 |
| 1840 | FEB | 9.39 | 1844 | FEB | 11.48 |
| 1840 | MAR | 9.13 | 1844 | MAR | 12.26 |
| 1840 | APR | 9.39 | 1844 | APR | 13.31 |
| 1840 | MAY | 9.39 | 1844 | MAY | 14.61 |
| 1840 | JUN | 9.13 | 1844 | JUN | 13.57 |
| 1840 | JUL | 9.39 | 1844 | JUL | 13.57 |
| 1840 | AUG | 9.13 | 1844 | AUG | 13.57 |
| 1840 | SEP | 9.39 | 1844 | SEP | 13.83 |
| 1840 | OCT | 10.70 | 1844 | OCT | 13.83 |
| 1840 | NOV | 10.44 | 1844 | NOV | 12.78 |
| 1840 | DEC | 9.91 | 1844 | DEC | 12.52 |
| 1841 | JAN | 9.39 | 1845 | JAN | 12.26 |
| 1841 | FEB | 9.39 | 1845 | FEB | 13.31 |
| 1841 | MAR | 8.09 | 1845 | MAR | 13.04 |
| 1841 | APR | 8.61 | 1845 | APR | 13.04 |
| 1841 | MAY | 9.39 | 1845 | MAY | 13.04 |
| 1841 | JUN | 9.13 | 1845 | JUN | 12.00 |
| 1841 | JUL | 9.39 | 1845 | JUL | 12.26 |
| 1841 | AUG | 9.39 | 1845 | AUG | 12.00 |
| 1841 | SEP | 8.61 | 1845 | SEP | 12.00 |
| 1841 | OCT | 8.61 | 1845 | OCT | 12.26 |
| 1841 | NOV | 8.61 | 1845 | NOV | 13.57 |
| 1841 | DEC | 8.09 | 1845 | DEC | 13.31 |

| YEAR | MONTH | DJIA | YEAR | MONTH | DJIA |
|---|---|---|---|---|---|
| 1846 | JAN | 12.78 | 1850 | JAN | 13.83 |
| 1846 | FEB | 13.57 | 1850 | FEB | 13.83 |
| 1846 | MAR | 14.35 | 1850 | MAR | 14.61 |
| 1846 | APR | 12.78 | 1850 | APR | 13.83 |
| 1846 | MAY | 13.31 | 1850 | MAY | 14.61 |
| 1846 | JUN | 13.83 | 1850 | JUN | 15.39 |
| 1846 | JUL | 13.83 | 1850 | JUL | 14.87 |
| 1846 | AUG | 13.83 | 1850 | AUG | 15.39 |
| 1846 | SEP | 13.31 | 1850 | SEP | 15.13 |
| 1846 | OCT | 13.04 | 1850 | OCT | 16.96 |
| 1846 | NOV | 13.04 | 1850 | NOV | 16.70 |
| 1846 | DEC | 13.04 | 1850 | DEC | 17.48 |
| 1847 | JAN | 13.04 | 1851 | JAN | 17.48 |
| 1847 | FEB | 13.83 | 1851 | FEB | 17.22 |
| 1847 | MAR | 13.57 | 1851 | MAR | 16.96 |
| 1847 | APR | 13.83 | 1851 | APR | 17.48 |
| 1847 | MAY | 14.35 | 1851 | MAY | 16.96 |
| 1847 | JUN | 15.65 | 1851 | JUN | 17.48 |
| 1847 | JUL | 15.91 | 1851 | JUL | 17.22 |
| 1847 | AUG | 16.18 | 1851 | AUG | 16.18 |
| 1847 | SEP | 15.91 | 1851 | SEP | 16.18 |
| 1847 | OCT | 15.13 | 1851 | OCT | 15.91 |
| 1847 | NOV | 13.83 | 1851 | NOV | 16.70 |
| 1847 | DEC | 13.04 | 1851 | DEC | 16.70 |
| 1848 | JAN | 13.04 | 1852 | JAN | 16.70 |
| 1848 | FEB | 13.83 | 1852 | FEB | 16.70 |
| 1848 | MAR | 13.83 | 1852 | MAR | 17.48 |
| 1848 | APR | 13.83 | 1852 | APR | 18.00 |
| 1848 | MAY | 13.31 | 1852 | MAY | 18.26 |
| 1848 | JUN | 13.31 | 1852 | JUN | 18.78 |
| 1848 | JUL | 13.04 | 1852 | JUL | 19.31 |
| 1848 | AUG | 12.78 | 1852 | AUG | 19.83 |
| 1848 | SEP | 12.78 | 1852 | SEP | 20.09 |
| 1848 | OCT | 12.26 | 1852 | OCT | 20.61 |
| 1848 | NOV | 12.26 | 1852 | NOV | 20.61 |
| 1848 | DEC | 13.04 | 1852 | DEC | 21.13 |
| 1849 | JAN | 13.04 | 1853 | JAN | 21.13 |
| 1849 | FEB | 13.04 | 1853 | FEB | 20.35 |
| 1849 | MAR | 14.09 | 1853 | MAR | 20.35 |
| 1849 | APR | 13.83 | 1853 | APR | 20.09 |
| 1849 | MAY | 13.83 | 1853 | MAY | 20.35 |
| 1849 | JUN | 14.61 | 1853 | JUN | 20.35 |
| 1849 | JUL | 14.61 | 1853 | JUL | 19.83 |
| 1849 | AUG | 13.83 | 1853 | AUG | 18.78 |
| 1849 | SEP | 13.83 | 1853 | SEP | 18.00 |
| 1849 | OCT | 13.83 | 1853 | OCT | 17.74 |
| 1849 | NOV | 13.57 | 1853 | NOV | 16.70 |
| 1849 | DEC | 13.57 | 1853 | DEC | 18.00 |

**TABLE A4-1 Monthly Values of Dow Jones Industrial Average** (cont.)

| YEAR | MONTH | DJIA | YEAR | MONTH | DJIA |
|---|---|---|---|---|---|
| 1854 | JAN | 17.48 | 1858 | JAN | 10.53 |
| 1854 | FEB | 17.48 | 1858 | FEB | 11.34 |
| 1854 | MAR | 18.44 | 1858 | MAR | 10.53 |
| 1854 | APR | 17.67 | 1858 | APR | 10.67 |
| 1854 | MAY | 17.07 | 1858 | MAY | 10.13 |
| 1854 | JUN | 16.58 | 1858 | JUN | 10.07 |
| 1854 | JUL | 15.36 | 1858 | JUL | 9.92 |
| 1854 | AUG | 13.99 | 1858 | AUG | 9.92 |
| 1854 | SEP | 13.89 | 1858 | SEP | 9.88 |
| 1854 | OCT | 14.18 | 1858 | OCT | 10.23 |
| 1854 | NOV | 13.35 | 1858 | NOV | 10.33 |
| 1854 | DEC | 12.87 | 1858 | DEC | 9.90 |
| 1855 | JAN | 13.70 | 1859 | JAN | 9.88 |
| 1855 | FEB | 13.97 | 1859 | FEB | 9.65 |
| 1855 | MAR | 14.26 | 1859 | MAR | 9.65 |
| 1855 | APR | 14.39 | 1859 | APR | 9.18 |
| 1855 | MAY | 14.59 | 1859 | MAY | 8.57 |
| 1855 | JUN | 14.70 | 1859 | JUN | 8.33 |
| 1855 | JUL | 15.03 | 1859 | JUL | 8.60 |
| 1855 | AUG | 15.15 | 1859 | AUG | 8.76 |
| 1855 | SEP | 14.88 | 1859 | SEP | 9.09 |
| 1855 | OCT | 14.42 | 1859 | OCT | 8.95 |
| 1855 | NOV | 13.20 | 1859 | NOV | 8.91 |
| 1855 | DEC | 13.80 | 1859 | DEC | 9.03 |
| 1856 | JAN | 13.95 | 1860 | JAN | 8.76 |
| 1856 | FEB | 14.53 | 1860 | FEB | 8.74 |
| 1856 | MAR | 14.26 | 1860 | MAR | 9.57 |
| 1856 | APR | 14.35 | 1860 | APR | 10.34 |
| 1856 | MAY | 14.15 | 1860 | MAY | 10.84 |
| 1856 | JUN | 14.22 | 1860 | JUN | 11.14 |
| 1856 | JUL | 14.35 | 1860 | JUL | 11.91 |
| 1856 | AUG | 14.55 | 1860 | AUG | 13.20 |
| 1856 | SEP | 14.93 | 1860 | SEP | 13.81 |
| 1856 | OCT | 14.78 | 1860 | OCT | 13.16 |
| 1856 | NOV | 15.55 | 1860 | NOV | 11.29 |
| 1856 | DEC | 15.94 | 1860 | DEC | 10.65 |
| 1857 | JAN | 15.90 | 1861 | JAN | 12.27 |
| 1857 | FEB | 15.36 | 1861 | FEB | 12.00 |
| 1857 | MAR | 15.34 | 1861 | MAR | 12.15 |
| 1857 | APR | 14.55 | 1861 | APR | 10.92 |
| 1857 | MAY | 14.26 | 1861 | MAY | 10.50 |
| 1857 | JUN | 13.16 | 1861 | JUN | 10.73 |
| 1857 | JUL | 12.91 | 1861 | JUL | 11.37 |
| 1857 | AUG | 12.23 | 1861 | AUG | 11.27 |
| 1857 | SEP | 10.07 | 1861 | SEP | 11.58 |
| 1857 | OCT | 7.99 | 1861 | OCT | 12.37 |
| 1857 | NOV | 9.45 | 1861 | NOV | 12.54 |
| 1857 | DEC | 11.92 | 1861 | DEC | 12.02 |

| YEAR | MONTH | DJIA | YEAR | MONTH | DJIA |
|---|---|---|---|---|---|
| 1862 | JAN | 12.99 | 1866 | JAN | 25.04 |
| 1862 | FEB | 13.26 | 1866 | FEB | 24.18 |
| 1862 | MAR | 13.39 | 1866 | MAR | 24.15 |
| 1862 | APR | 13.37 | 1866 | APR | 24.14 |
| 1862 | MAY | 14.55 | 1866 | MAY | 24.19 |
| 1862 | JUN | 14.86 | 1866 | JUN | 24.00 |
| 1862 | JUL | 14.57 | 1866 | JUL | 25.22 |
| 1862 | AUG | 15.61 | 1866 | AUG | 25.95 |
| 1862 | SEP | 16.78 | 1866 | SEP | 26.53 |
| 1862 | OCT | 18.83 | 1866 | OCT | 27.97 |
| 1862 | NOV | 18.64 | 1866 | NOV | 26.89 |
| 1862 | DEC | 17.90 | 1866 | DEC | 25.93 |
| 1863 | JAN | 20.71 | 1867 | JAN | 24.39 |
| 1863 | FEB | 21.48 | 1867 | FEB | 24.35 |
| 1863 | MAR | 22.07 | 1867 | MAR | 24.25 |
| 1863 | APR | 22.26 | 1867 | APR | 23.96 |
| 1863 | MAY | 23.69 | 1867 | MAY | 24.15 |
| 1863 | JUN | 23.46 | 1867 | JUN | 24.89 |
| 1863 | JUL | 23.69 | 1867 | JUL | 26.01 |
| 1863 | AUG | 25.58 | 1867 | AUG | 25.97 |
| 1863 | SEP | 25.27 | 1867 | SEP | 25.74 |
| 1863 | OCT | 25.39 | 1867 | OCT | 26.32 |
| 1863 | NOV | 26.76 | 1867 | NOV | 27.13 |
| 1863 | DEC | 25.34 | 1867 | DEC | 26.97 |
| 1864 | JAN | 25.74 | 1868 | JAN | 26.03 |
| 1864 | FEB | 26.30 | 1868 | FEB | 28.55 |
| 1864 | MAR | 27.23 | 1868 | MAR | 28.36 |
| 1864 | APR | 26.78 | 1868 | APR | 28.09 |
| 1864 | MAY | 25.54 | 1868 | MAY | 28.82 |
| 1864 | JUN | 25.16 | 1868 | JUN | 29.67 |
| 1864 | JUL | 25.51 | 1868 | JUL | 29.79 |
| 1864 | AUG | 25.31 | 1868 | AUG | 29.21 |
| 1864 | SEP | 24.54 | 1868 | SEP | 29.06 |
| 1864 | OCT | 23.07 | 1868 | OCT | 29.13 |
| 1864 | NOV | 24.93 | 1868 | NOV | 28.62 |
| 1864 | DEC | 24.12 | 1868 | DEC | 29.44 |
| 1865 | JAN | 24.77 | 1869 | JAN | 30.98 |
| 1865 | FEB | 23.92 | 1869 | FEB | 30.60 |
| 1865 | MAR | 21.07 | 1869 | MAR | 29.13 |
| 1865 | APR | 21.42 | 1869 | APR | 29.52 |
| 1865 | MAY | 22.13 | 1869 | MAY | 30.91 |
| 1865 | JUN | 22.15 | 1869 | JUN | 31.29 |
| 1865 | JUL | 23.88 | 1869 | JUL | 31.33 |
| 1865 | AUG | 23.58 | 1869 | AUG | 31.33 |
| 1865 | SEP | 24.89 | 1869 | SEP | 28.13 |
| 1865 | OCT | 26.15 | 1869 | OCT | 27.51 |
| 1865 | NOV | 26.22 | 1869 | NOV | 26.74 |
| 1865 | DEC | 26.07 | 1869 | DEC | 25.35 |

**TABLE A4-1 Monthly Values of Dow Jones Industrial Average** (cont.)

| YEAR | MONTH | DJIA | YEAR | MONTH | DJIA |
|---|---|---|---|---|---|
| 1870 | JAN | 25.53 | 1874 | JAN | 33.74 |
| 1870 | FEB | 26.51 | 1874 | FEB | 35.68 |
| 1870 | MAR | 25.53 | 1874 | MAR | 35.55 |
| 1870 | APR | 25.62 | 1874 | APR | 34.57 |
| 1870 | MAY | 26.89 | 1874 | MAY | 33.60 |
| 1870 | JUN | 27.51 | 1874 | JUN | 34.71 |
| 1870 | JUL | 26.74 | 1874 | JUL | 33.74 |
| 1870 | AUG | 25.62 | 1874 | AUG | 32.77 |
| 1870 | SEP | 25.84 | 1874 | SEP | 32.63 |
| 1870 | OCT | 26.39 | 1874 | OCT | 32.91 |
| 1870 | NOV | 26.22 | 1874 | NOV | 34.02 |
| 1870 | DEC | 26.34 | 1874 | DEC | 33.74 |
| 1871 | JAN | 26.39 | 1875 | JAN | 32.49 |
| 1871 | FEB | 26.74 | 1875 | FEB | 32.77 |
| 1871 | MAR | 27.55 | 1875 | MAR | 32.91 |
| 1871 | APR | 28.50 | 1875 | APR | 33.32 |
| 1871 | MAY | 29.04 | 1875 | MAY | 33.32 |
| 1871 | JUN | 29.06 | 1875 | JUN | 31.24 |
| 1871 | JUL | 28.63 | 1875 | JUL | 31.38 |
| 1871 | AUG | 29.44 | 1875 | AUG | 31.24 |
| 1871 | SEP | 30.41 | 1875 | SEP | 31.52 |
| 1871 | OCT | 29.30 | 1875 | OCT | 30.69 |
| 1871 | NOV | 29.57 | 1875 | NOV | 31.52 |
| 1871 | DEC | 29.99 | 1875 | DEC | 31.94 |
| 1872 | JAN | 30.96 | 1876 | JAN | 32.49 |
| 1872 | FEB | 31.94 | 1876 | FEB | 34.43 |
| 1872 | MAR | 33.46 | 1876 | MAR | 33.60 |
| 1872 | APR | 34.57 | 1876 | APR | 33.32 |
| 1872 | MAY | 34.99 | 1876 | MAY | 33.18 |
| 1872 | JUN | 34.85 | 1876 | JUN | 33.05 |
| 1872 | JUL | 34.02 | 1876 | JUL | 32.07 |
| 1872 | AUG | 33.46 | 1876 | AUG | 31.94 |
| 1872 | SEP | 32.35 | 1876 | SEP | 30.13 |
| 1872 | OCT | 33.46 | 1876 | OCT | 30.27 |
| 1872 | NOV | 34.30 | 1876 | NOV | 30.27 |
| 1872 | DEC | 34.85 | 1876 | DEC | 30.27 |
| 1873 | JAN | 35.27 | 1877 | JAN | 29.85 |
| 1873 | FEB | 35.13 | 1877 | FEB | 29.02 |
| 1873 | MAR | 35.27 | 1877 | MAR | 27.63 |
| 1873 | APR | 34.99 | 1877 | APR | 26.24 |
| 1873 | MAY | 35.27 | 1877 | MAY | 25.83 |
| 1873 | JUN | 34.57 | 1877 | JUN | 22.08 |
| 1873 | JUL | 34.43 | 1877 | JUL | 22.77 |
| 1873 | AUG | 34.85 | 1877 | AUG | 23.47 |
| 1873 | SEP | 32.21 | 1877 | SEP | 24.30 |
| 1873 | OCT | 30.13 | 1877 | OCT | 25.13 |
| 1873 | NOV | 28.74 | 1877 | NOV | 24.44 |
| 1873 | DEC | 31.80 | 1877 | DEC | 24.58 |

| YEAR | MONTH | DJIA | YEAR | MONTH | DJIA |
|---|---|---|---|---|---|
| 1878 | JAN | 24.85 | 1882 | JAN | 35.27 |
| 1878 | FEB | 24.72 | 1882 | FEB | 34.43 |
| 1878 | MAR | 24.85 | 1882 | MAR | 34.16 |
| 1878 | APR | 24.72 | 1882 | APR | 34.30 |
| 1878 | MAY | 25.55 | 1882 | MAY | 33.88 |
| 1878 | JUN | 25.41 | 1882 | JUN | 33.32 |
| 1878 | JUL | 25.41 | 1882 | JUL | 34.30 |
| 1878 | AUG | 25.27 | 1882 | AUG | 34.71 |
| 1878 | SEP | 25.83 | 1882 | SEP | 34.85 |
| 1878 | OCT | 25.41 | 1882 | OCT | 34.16 |
| 1878 | NOV | 25.41 | 1882 | NOV | 32.91 |
| 1878 | DEC | 25.13 | 1882 | DEC | 32.63 |
| 1879 | JAN | 24.85 | 1883 | JAN | 32.21 |
| 1879 | FEB | 25.27 | 1883 | FEB | 31.80 |
| 1879 | MAR | 25.27 | 1883 | MAR | 31.38 |
| 1879 | APR | 25.55 | 1883 | APR | 32.07 |
| 1879 | MAY | 25.96 | 1883 | MAY | 32.21 |
| 1879 | JUN | 26.10 | 1883 | JUN | 32.77 |
| 1879 | JUL | 26.10 | 1883 | JUL | 32.77 |
| 1879 | AUG | 25.83 | 1883 | AUG | 32.21 |
| 1879 | SEP | 26.38 | 1883 | SEP | 31.80 |
| 1879 | OCT | 29.85 | 1883 | OCT | 31.10 |
| 1879 | NOV | 30.96 | 1883 | NOV | 31.24 |
| 1879 | DEC | 30.27 | 1883 | DEC | 30.41 |
| 1880 | JAN | 31.24 | 1884 | JAN | 29.85 |
| 1880 | FEB | 30.55 | 1884 | FEB | 30.55 |
| 1880 | MAR | 30.82 | 1884 | MAR | 30.69 |
| 1880 | APR | 29.44 | 1884 | APR | 30.41 |
| 1880 | MAY | 28.60 | 1884 | MAY | 27.91 |
| 1880 | JUN | 28.60 | 1884 | JUN | 27.21 |
| 1880 | JUL | 28.88 | 1884 | JUL | 27.63 |
| 1880 | AUG | 29.44 | 1884 | AUG | 29.16 |
| 1880 | SEP | 29.30 | 1884 | SEP | 29.02 |
| 1880 | OCT | 28.74 | 1884 | OCT | 29.02 |
| 1880 | NOV | 29.44 | 1884 | NOV | 28.88 |
| 1880 | DEC | 30.13 | 1884 | DEC | 28.46 |
| 1881 | JAN | 32.35 | 1885 | JAN | 28.46 |
| 1881 | FEB | 32.91 | 1885 | FEB | 29.30 |
| 1881 | MAR | 33.18 | 1885 | MAR | 29.44 |
| 1881 | APR | 33.60 | 1885 | APR | 29.85 |
| 1881 | MAY | 35.13 | 1885 | MAY | 29.99 |
| 1881 | JUN | 36.24 | 1885 | JUN | 29.99 |
| 1881 | JUL | 35.27 | 1885 | JUL | 30.27 |
| 1881 | AUG | 35.13 | 1885 | AUG | 31.52 |
| 1881 | SEP | 34.99 | 1885 | SEP | 31.94 |
| 1881 | OCT | 35.41 | 1885 | OCT | 33.05 |
| 1881 | NOV | 35.68 | 1885 | NOV | 34.16 |
| 1881 | DEC | 34.99 | 1885 | DEC | 33.88 |

**TABLE A4-1 Monthly Values of Dow Jones Industrial Average** (cont.)

| YEAR | MONTH | DJIA | YEAR | MONTH | DJIA |
|---|---|---|---|---|---|
| 1886 | JAN | 34.16 | 1890 | JAN | 42.35 |
| 1886 | FEB | 34.99 | 1890 | FEB | 41.38 |
| 1886 | MAR | 34.43 | 1890 | MAR | 41.38 |
| 1886 | APR | 34.30 | 1890 | APR | 41.79 |
| 1886 | MAY | 33.60 | 1890 | MAY | 44.43 |
| 1886 | JUN | 34.71 | 1890 | JUN | 44.71 |
| 1886 | JUL | 34.99 | 1890 | JUL | 44.71 |
| 1886 | AUG | 34.99 | 1890 | AUG | 44.43 |
| 1886 | SEP | 35.27 | 1890 | SEP | 44.02 |
| 1886 | OCT | 35.96 | 1890 | OCT | 42.90 |
| 1886 | NOV | 36.79 | 1890 | NOV | 38.46 |
| 1886 | DEC | 36.52 | 1890 | DEC | 36.66 |
| 1887 | JAN | 36.66 | 1891 | JAN | 39.71 |
| 1887 | FEB | 36.79 | 1891 | FEB | 40.68 |
| 1887 | MAR | 37.21 | 1891 | MAR | 40.68 |
| 1887 | APR | 38.04 | 1891 | APR | 41.65 |
| 1887 | MAY | 38.60 | 1891 | MAY | 40.96 |
| 1887 | JUN | 37.91 | 1891 | JUN | 40.13 |
| 1887 | JUL | 36.93 | 1891 | JUL | 39.16 |
| 1887 | AUG | 35.96 | 1891 | AUG | 39.43 |
| 1887 | SEP | 35.55 | 1891 | SEP | 41.38 |
| 1887 | OCT | 35.41 | 1891 | OCT | 40.96 |
| 1887 | NOV | 35.82 | 1891 | NOV | 40.68 |
| 1887 | DEC | 35.82 | 1891 | DEC | 42.07 |
| 1888 | JAN | 36.10 | 1892 | JAN | 42.63 |
| 1888 | FEB | 36.10 | 1892 | FEB | 42.35 |
| 1888 | MAR | 35.13 | 1892 | MAR | 42.90 |
| 1888 | APR | 35.41 | 1892 | APR | 43.04 |
| 1888 | MAY | 35.96 | 1892 | MAY | 43.60 |
| 1888 | JUN | 36.79 | 1892 | JUN | 44.57 |
| 1888 | JUL | 37.77 | 1892 | JUL | 44.85 |
| 1888 | AUG | 38.88 | 1892 | AUG | 46.51 |
| 1888 | SEP | 39.99 | 1892 | SEP | 46.24 |
| 1888 | OCT | 41.10 | 1892 | OCT | 48.18 |
| 1888 | NOV | 41.93 | 1892 | NOV | 48.60 |
| 1888 | DEC | 41.65 | 1892 | DEC | 47.90 |
| 1889 | JAN | 42.49 | 1893 | JAN | 48.18 |
| 1889 | FEB | 44.85 | 1893 | FEB | 47.07 |
| 1889 | MAR | 45.40 | 1893 | MAR | 44.57 |
| 1889 | APR | 45.96 | 1893 | APR | 43.60 |
| 1889 | MAY | 47.35 | 1893 | MAY | 36.66 |
| 1889 | JUN | 49.99 | 1893 | JUN | 34.43 |
| 1889 | JUL | 49.57 | 1893 | JUL | 30.41 |
| 1889 | AUG | 48.04 | 1893 | AUG | 29.44 |
| 1889 | SEP | 47.07 | 1893 | SEP | 33.18 |
| 1889 | OCT | 44.85 | 1893 | OCT | 35.55 |
| 1889 | NOV | 42.49 | 1893 | NOV | 35.13 |
| 1889 | DEC | 41.93 | 1893 | DEC | 33.32 |

| YEAR | MONTH | DJIA | YEAR | MONTH | DJIA |
|---|---|---|---|---|---|
| 1894 | JAN | 32.63 | 1898 | JAN | 35.97 |
| 1894 | FEB | 33.18 | 1898 | FEB | 35.24 |
| 1894 | MAR | 34.71 | 1898 | MAR | 32.56 |
| 1894 | APR | 35.82 | 1898 | APR | 32.72 |
| 1894 | MAY | 34.71 | 1898 | MAY | 36.97 |
| 1894 | JUN | 34.57 | 1898 | JUN | 38.13 |
| 1894 | JUL | 33.46 | 1898 | JUL | 38.62 |
| 1894 | AUG | 35.13 | 1898 | AUG | 42.01 |
| 1894 | SEP | 34.71 | 1898 | SEP | 41.86 |
| 1894 | OCT | 33.05 | 1898 | OCT | 38.53 |
| 1894 | NOV | 33.05 | 1898 | NOV | 40.85 |
| 1894 | DEC | 32.49 | 1898 | DEC | 43.01 |
| 1895 | JAN | 32.35 | 1899 | JAN | 48.28 |
| 1895 | FEB | 31.66 | 1899 | FEB | 45.17 |
| 1895 | MAR | 32.63 | 1899 | MAR | 46.77 |
| 1895 | APR | 34.99 | 1899 | APR | 50.79 |
| 1895 | MAY | 37.77 | 1899 | MAY | 55.02 |
| 1895 | JUN | 38.46 | 1899 | JUN | 52.25 |
| 1895 | JUL | 37.91 | 1899 | JUL | 51.31 |
| 1895 | AUG | 29.30 | 1899 | AUG | 52.12 |
| 1895 | SEP | 37.35 | 1899 | SEP | 54.64 |
| 1895 | OCT | 36.66 | 1899 | OCT | 54.15 |
| 1895 | NOV | 34.57 | 1899 | NOV | 53.06 |
| 1895 | DEC | 31.80 | 1899 | DEC | 54.48 |
| 1896 | JAN | 31.80 | 1900 | JAN | 47.36 |
| 1896 | FEB | 33.46 | 1900 | FEB | 48.19 |
| 1896 | MAR | 33.18 | 1900 | MAR | 45.86 |
| 1896 | APR | 33.88 | 1900 | APR | 45.94 |
| 1896 | MAY | 33.18 | 1900 | MAY | 42.68 |
| 1896 | JUN | 31.80 | 1900 | JUN | 40.56 |
| 1896 | JUL | 28.88 | 1900 | JUL | 41.39 |
| 1896 | AUG | 26.94 | 1900 | AUG | 42.25 |
| 1896 | SEP | 29.16 | 1900 | SEP | 40.77 |
| 1896 | OCT | 29.99 | 1900 | OCT | 41.87 |
| 1896 | NOV | 33.05 | 1900 | NOV | 47.72 |
| 1896 | DEC | 33.38 | 1900 | DEC | 48.65 |
| 1897 | JAN | 31.52 | 1901 | JAN | 44.91 |
| 1897 | FEB | 30.96 | 1901 | FEB | 50.14 |
| 1897 | MAR | 30.82 | 1901 | MAR | 49.52 |
| 1897 | APR | 29.99 | 1901 | APR | 53.46 |
| 1897 | MAY | 29.85 | 1901 | MAY | 53.14 |
| 1897 | JUN | 31.38 | 1901 | JUN | 55.98 |
| 1897 | JUL | 33.39 | 1901 | JUL | 52.62 |
| 1897 | AUG | 37.69 | 1901 | AUG | 56.81 |
| 1897 | SEP | 39.52 | 1901 | SEP | 50.55 |
| 1897 | OCT | 36.34 | 1901 | OCT | 47.24 |
| 1897 | NOV | 34.05 | 1901 | NOV | 47.55 |
| 1897 | DEC | 35.57 | 1901 | DEC | 45.84 |

**TABLE A4-1 Monthly Values of Dow Jones Industrial Average** (cont.)

| YEAR | MONTH | DJIA | YEAR | MONTH | DJIA |
|---|---|---|---|---|---|
| 1902 | JAN | 46.53 | 1906 | JAN | 72.52 |
| 1902 | FEB | 47.21 | 1906 | FEB | 71.49 |
| 1902 | MAR | 47.97 | 1906 | MAR | 69.18 |
| 1902 | APR | 48.65 | 1906 | APR | 69.07 |
| 1902 | MAY | 47.99 | 1906 | MAY | 66.79 |
| 1902 | JUN | 47.36 | 1906 | JUN | 66.78 |
| 1902 | JUL | 47.62 | 1906 | JUL | 63.93 |
| 1902 | AUG | 48.07 | 1906 | AUG | 68.09 |
| 1902 | SEP | 48.28 | 1906 | SEP | 68.99 |
| 1902 | OCT | 47.63 | 1906 | OCT | 69.02 |
| 1902 | NOV | 45.44 | 1906 | NOV | 68.56 |
| 1902 | DEC | 44.89 | 1906 | DEC | 68.67 |
| 1903 | JAN | 47.30 | 1907 | JAN | 68.38 |
| 1903 | FEB | 48.53 | 1907 | FEB | 66.66 |
| 1903 | MAR | 46.87 | 1907 | MAR | 60.12 |
| 1903 | APR | 45.82 | 1907 | APR | 60.66 |
| 1903 | MAY | 45.50 | 1907 | MAY | 59.39 |
| 1903 | JUN | 42.20 | 1907 | JUN | 57.18 |
| 1903 | JUL | 39.31 | 1907 | JUL | 58.88 |
| 1903 | AUG | 37.15 | 1907 | AUG | 52.61 |
| 1903 | SEP | 35.82 | 1907 | SEP | 51.10 |
| 1903 | OCT | 32.78 | 1907 | OCT | 45.13 |
| 1903 | NOV | 31.81 | 1907 | NOV | 40.35 |
| 1903 | DEC | 34.16 | 1907 | DEC | 42.73 |
| 1904 | JAN | 35.26 | 1908 | JAN | 45.57 |
| 1904 | FEB | 34.71 | 1908 | FEB | 43.66 |
| 1904 | MAR | 34.78 | 1908 | MAR | 47.57 |
| 1904 | APR | 35.73 | 1908 | APR | 49.79 |
| 1904 | MAY | 34.95 | 1908 | MAY | 52.66 |
| 1904 | JUN | 35.48 | 1908 | JUN | 52.99 |
| 1904 | JUL | 37.57 | 1908 | JUL | 56.02 |
| 1904 | AUG | 38.94 | 1908 | AUG | 60.34 |
| 1904 | SEP | 40.99 | 1908 | SEP | 58.78 |
| 1904 | OCT | 44.21 | 1908 | OCT | 59.35 |
| 1904 | NOV | 49.85 | 1908 | NOV | 63.17 |
| 1904 | DEC | 50.38 | 1908 | DEC | 62.46 |
| 1905 | JAN | 50.99 | 1909 | JAN | 61.96 |
| 1905 | FEB | 53.45 | 1909 | FEB | 61.14 |
| 1905 | MAR | 56.37 | 1909 | MAR | 60.24 |
| 1905 | APR | 59.17 | 1909 | APR | 63.45 |
| 1905 | MAY | 54.61 | 1909 | MAY | 65.68 |
| 1905 | JUN | 54.23 | 1909 | JUN | 67.24 |
| 1905 | JUL | 57.47 | 1909 | JUL | 68.25 |
| 1905 | AUG | 59.63 | 1909 | AUG | 71.30 |
| 1905 | SEP | 58.50 | 1909 | SEP | 71.46 |
| 1905 | OCT | 59.60 | 1909 | OCT | 71.32 |
| 1905 | NOV | 61.35 | 1909 | NOV | 71.90 |
| 1905 | DEC | 68.04 | 1909 | DEC | 71.51 |

| YEAR | MONTH | DJIA | YEAR | MONTH | DJIA |
|---|---|---|---|---|---|
| 1910 | JAN | 68.78 | 1914 | JAN | 58.68 |
| 1910 | FEB | 64.96 | 1914 | FEB | 59.92 |
| 1910 | MAR | 66.97 | 1914 | MAR | 59.60 |
| 1910 | APR | 65.43 | 1914 | APR | 58.25 |
| 1910 | MAY | 63.89 | 1914 | MAY | 58.58 |
| 1910 | JUN | 61.08 | 1914 | JUN | 58.87 |
| 1910 | JUL | 57.68 | 1914 | JUL | 58.12 |
| 1910 | AUG | 57.49 | 1914 | AUG | 57.44 |
| 1910 | SEP | 57.25 | 1914 | SEP | 56.75 |
| 1910 | OCT | 60.86 | 1914 | OCT | 56.07 |
| 1910 | NOV | 61.53 | 1914 | NOV | 55.38 |
| 1910 | DEC | 58.93 | 1914 | DEC | 54.70 |
| 1911 | JAN | 60.72 | 1915 | JAN | 57.16 |
| 1911 | FEB | 62.00 | 1915 | FEB | 56.06 |
| 1911 | MAR | 60.49 | 1915 | MAR | 57.56 |
| 1911 | APR | 60.03 | 1915 | APR | 66.74 |
| 1911 | MAY | 61.37 | 1915 | MAY | 64.93 |
| 1911 | JUN | 62.78 | 1915 | JUN | 69.76 |
| 1911 | JUL | 62.54 | 1915 | JUL | 71.80 |
| 1911 | AUG | 58.96 | 1915 | AUG | 79.07 |
| 1911 | SEP | 56.01 | 1915 | SEP | 83.82 |
| 1911 | OCT | 55.94 | 1915 | OCT | 93.15 |
| 1911 | NOV | 58.28 | 1915 | NOV | 95.22 |
| 1911 | DEC | 58.93 | 1915 | DEC | 97.49 |
| 1912 | JAN | 59.93 | 1916 | JAN | 95.30 |
| 1912 | FEB | 58.71 | 1916 | FEB | 94.06 |
| 1912 | MAR | 61.98 | 1916 | MAR | 93.49 |
| 1912 | APR | 65.14 | 1916 | APR | 91.11 |
| 1912 | MAY | 64.96 | 1916 | MAY | 90.99 |
| 1912 | JUN | 65.46 | 1916 | JUN | 96.06 |
| 1912 | JUL | 64.97 | 1916 | JUL | 88.55 |
| 1912 | AUG | 66.10 | 1916 | AUG | 91.09 |
| 1912 | SEP | 66.75 | 1916 | SEP | 98.10 |
| 1912 | OCT | 67.32 | 1916 | OCT | 102.98 |
| 1912 | NOV | 65.79 | 1916 | NOV | 107.71 |
| 1912 | DEC | 63.37 | 1916 | DEC | 100.23 |
| 1913 | JAN | 61.91 | 1917 | JAN | 96.64 |
| 1913 | FEB | 59.19 | 1917 | FEB | 91.84 |
| 1913 | MAR | 58.12 | 1917 | MAR | 95.84 |
| 1913 | APR | 59.08 | 1917 | APR | 92.89 |
| 1913 | MAY | 57.49 | 1917 | MAY | 92.75 |
| 1913 | JUN | 54.39 | 1917 | JUN | 96.98 |
| 1913 | JUL | 55.85 | 1917 | JUL | 92.20 |
| 1913 | AUG | 58.04 | 1917 | AUG | 90.36 |
| 1913 | SEP | 59.54 | 1917 | SEP | 83.66 |
| 1913 | OCT | 57.38 | 1917 | OCT | 78.74 |
| 1913 | NOV | 55.83 | 1917 | NOV | 71.57 |
| 1913 | DEC | 55.88 | 1917 | DEC | 70.17 |

**TABLE A4-1 Monthly Values of Dow Jones Industrial Average** (cont.)

| YEAR | MONTH | DJIA | YEAR | MONTH | DJIA |
|---|---|---|---|---|---|
| 1918 | JAN | 75.54 | 1922 | JAN | 80.90 |
| 1918 | FEB | 79.80 | 1922 | FEB | 83.90 |
| 1918 | MAR | 78.37 | 1922 | MAR | 87.00 |
| 1918 | APR | 77.54 | 1922 | APR | 91.70 |
| 1918 | MAY | 80.93 | 1922 | MAY | 93.80 |
| 1918 | JUN | 80.65 | 1922 | JUN | 93.40 |
| 1918 | JUL | 81.72 | 1922 | JUL | 95.10 |
| 1918 | AUG | 81.85 | 1922 | AUG | 98.30 |
| 1918 | SEP | 82.45 | 1922 | SEP | 99.80 |
| 1918 | OCT | 86.06 | 1922 | OCT | 100.40 |
| 1918 | NOV | 84.43 | 1922 | NOV | 95.80 |
| 1918 | DEC | 82.37 | 1922 | DEC | 97.60 |
| 1919 | JAN | 81.30 | 1923 | JAN | 98.00 |
| 1919 | FEB | 82.10 | 1923 | FEB | 101.80 |
| 1919 | MAR | 87.40 | 1923 | MAR | 104.30 |
| 1919 | APR | 91.00 | 1923 | APR | 101.60 |
| 1919 | MAY | 99.30 | 1923 | MAY | 96.20 |
| 1919 | JUN | 105.40 | 1923 | JUN | 93.70 |
| 1919 | JUL | 110.00 | 1923 | JUL | 89.30 |
| 1919 | AUG | 102.60 | 1923 | AUG | 91.20 |
| 1919 | SEP | 107.70 | 1923 | SEP | 90.30 |
| 1919 | OCT | 113.90 | 1923 | OCT | 87.70 |
| 1919 | NOV | 110.80 | 1923 | NOV | 91.00 |
| 1919 | DEC | 105.60 | 1923 | DEC | 94.10 |
| 1920 | JAN | 104.60 | 1924 | JAN | 97.40 |
| 1920 | FEB | 94.90 | 1924 | FEB | 98.80 |
| 1920 | MAR | 99.60 | 1924 | MAR | 96.30 |
| 1920 | APR | 100.80 | 1924 | APR | 91.70 |
| 1920 | MAY | 91.40 | 1924 | MAY | 90.50 |
| 1920 | JUN | 91.40 | 1924 | JUN | 92.60 |
| 1920 | JUL | 90.60 | 1924 | JUL | 98.30 |
| 1920 | AUG | 85.40 | 1924 | AUG | 103.30 |
| 1920 | SEP | 87.00 | 1924 | SEP | 102.90 |
| 1920 | OCT | 84.90 | 1924 | OCT | 102.00 |
| 1920 | NOV | 78.40 | 1924 | NOV | 108.20 |
| 1920 | DEC | 71.90 | 1924 | DEC | 114.20 |
| 1921 | JAN | 75.10 | 1925 | JAN | 122.20 |
| 1921 | FEB | 75.50 | 1925 | FEB | 121.10 |
| 1921 | MAR | 75.40 | 1925 | MAR | 120.90 |
| 1921 | APR | 76.70 | 1925 | APR | 119.80 |
| 1921 | MAY | 77.20 | 1925 | MAY | 126.10 |
| 1921 | JUN | 69.10 | 1925 | JUN | 128.80 |
| 1921 | JUL | 68.50 | 1925 | JUL | 133.60 |
| 1921 | AUG | 66.70 | 1925 | AUG | 139.60 |
| 1921 | SEP | 70.20 | 1925 | SEP | 144.00 |
| 1921 | OCT | 71.30 | 1925 | OCT | 149.90 |
| 1921 | NOV | 76.00 | 1925 | NOV | 153.90 |
| 1921 | DEC | 78.80 | 1925 | DEC | 154.20 |

| YEAR | MONTH | DJIA | YEAR | MONTH | DJIA |
|---|---|---|---|---|---|
| 1926 | JAN | 156.60 | 1930 | JAN | 251.90 |
| 1926 | FEB | 159.20 | 1930 | FEB | 268.60 |
| 1926 | MAR | 146.40 | 1930 | MAR | 277.00 |
| 1926 | APR | 140.50 | 1930 | APR | 288.10 |
| 1926 | MAY | 140.20 | 1930 | MAY | 269.10 |
| 1926 | JUN | 149.20 | 1930 | JUN | 239.30 |
| 1926 | JUL | 156.60 | 1930 | JUL | 231.70 |
| 1926 | AUG | 163.20 | 1930 | AUG | 230.80 |
| 1926 | SEP | 160.10 | 1930 | SEP | 231.50 |
| 1926 | OCT | 151.20 | 1930 | OCT | 196.20 |
| 1926 | NOV | 154.50 | 1930 | NOV | 182.20 |
| 1926 | DEC | 159.30 | 1930 | DEC | 169.90 |
| 1927 | JAN | 155.00 | 1931 | JAN | 168.50 |
| 1927 | FEB | 157.30 | 1931 | FEB | 181.30 |
| 1927 | MAR | 160.20 | 1931 | MAR | 181.80 |
| 1927 | APR | 164.10 | 1931 | APR | 162.00 |
| 1927 | MAY | 169.20 | 1931 | MAY | 142.90 |
| 1927 | JUN | 169.30 | 1931 | JUN | 138.40 |
| 1927 | JUL | 175.50 | 1931 | JUL | 143.50 |
| 1927 | AUG | 184.70 | 1931 | AUG | 138.80 |
| 1927 | SEP | 195.80 | 1931 | SEP | 118.80 |
| 1927 | OCT | 188.60 | 1931 | OCT | 101.80 |
| 1927 | NOV | 193.10 | 1931 | NOV | 104.00 |
| 1927 | DEC | 198.50 | 1931 | DEC | 81.20 |
| 1928 | JAN | 198.90 | 1932 | JAN | 79.40 |
| 1928 | FEB | 195.60 | 1932 | FEB | 80.00 |
| 1928 | MAR | 204.00 | 1932 | MAR | 81.50 |
| 1928 | APR | 211.70 | 1932 | APR | 62.70 |
| 1928 | MAY | 217.20 | 1932 | MAY | 53.30 |
| 1928 | JUN | 209.10 | 1932 | JUN | 46.90 |
| 1928 | JUL | 211.10 | 1932 | JUL | 46.20 |
| 1928 | AUG | 224.00 | 1932 | AUG | 67.50 |
| 1928 | SEP | 239.80 | 1932 | SEP | 72.60 |
| 1928 | OCT | 248.70 | 1932 | OCT | 63.50 |
| 1928 | NOV | 273.80 | 1932 | NOV | 62.10 |
| 1928 | DEC | 280.80 | 1932 | DEC | 59.10 |
| 1929 | JAN | 306.00 | 1933 | JAN | 62.70 |
| 1929 | FEB | 309.90 | 1933 | FEB | 56.10 |
| 1929 | MAR | 311.20 | 1933 | MAR | 57.60 |
| 1929 | APR | 307.90 | 1933 | APR | 65.00 |
| 1929 | MAY | 315.00 | 1933 | MAY | 81.60 |
| 1929 | JUN | 315.40 | 1933 | JUN | 94.10 |
| 1929 | JUL | 343.80 | 1933 | JUL | 100.40 |
| 1929 | AUG | 360.70 | 1933 | AUG | 98.40 |
| 1929 | SEP | 364.90 | 1933 | SEP | 100.30 |
| 1929 | OCT | 320.60 | 1933 | OCT | 92.80 |
| 1929 | NOV | 232.60 | 1933 | NOV | 96.40 |
| 1929 | DEC | 246.90 | 1933 | DEC | 99.30 |

**TABLE A4-1 Monthly Values of Dow Jones Industrial Average** (cont.)

| YEAR | MONTH | DJIA | YEAR | MONTH | DJIA |
|---|---|---|---|---|---|
| 1934 | JAN | 102.70 | 1938 | JAN | 128.40 |
| 1934 | FEB | 107.30 | 1938 | FEB | 126.10 |
| 1934 | MAR | 102.10 | 1938 | MAR | 119.10 |
| 1934 | APR | 104.30 | 1938 | APR | 112.90 |
| 1934 | MAY | 95.30 | 1938 | MAY | 114.20 |
| 1934 | JUN | 96.70 | 1938 | JUN | 118.80 |
| 1934 | JUL | 94.50 | 1938 | JUL | 129.50 |
| 1934 | AUG | 91.60 | 1938 | AUG | 141.00 |
| 1934 | SEP | 90.50 | 1938 | SEP | 137.00 |
| 1934 | OCT | 93.50 | 1938 | OCT | 150.40 |
| 1934 | NOV | 99.30 | 1938 | NOV | 152.00 |
| 1934 | DEC | 101.60 | 1938 | DEC | 150.10 |
| 1935 | JAN | 103.10 | 1939 | JAN | 146.90 |
| 1935 | FEB | 103.00 | 1939 | FEB | 144.60 |
| 1935 | MAR | 99.80 | 1939 | MAR | 145.10 |
| 1935 | APR | 106.00 | 1939 | APR | 127.70 |
| 1935 | MAY | 113.50 | 1939 | MAY | 132.60 |
| 1935 | JUN | 116.90 | 1939 | JUN | 136.50 |
| 1935 | JUL | 122.70 | 1939 | JUL | 139.30 |
| 1935 | AUG | 127.10 | 1939 | AUG | 127.90 |
| 1935 | SEP | 131.50 | 1939 | SEP | 150.70 |
| 1935 | OCT | 130.40 | 1939 | OCT | 152.20 |
| 1935 | NOV | 144.30 | 1939 | NOV | 150.00 |
| 1935 | DEC | 141.80 | 1939 | DEC | 148.50 |
| 1936 | JAN | 145.90 | 1940 | JAN | 147.60 |
| 1936 | FEB | 151.80 | 1940 | FEB | 147.30 |
| 1936 | MAR | 155.90 | 1940 | MAR | 147.10 |
| 1936 | APR | 155.80 | 1940 | APR | 148.90 |
| 1936 | MAY | 149.30 | 1940 | MAY | 130.80 |
| 1936 | JUN | 155.20 | 1940 | JUN | 119.50 |
| 1936 | JUL | 162.30 | 1940 | JUL | 122.20 |
| 1936 | AUG | 165.90 | 1940 | AUG | 125.30 |
| 1936 | SEP | 167.80 | 1940 | SEP | 131.50 |
| 1936 | OCT | 175.00 | 1940 | OCT | 132.40 |
| 1936 | NOV | 182.10 | 1940 | NOV | 133.90 |
| 1936 | DEC | 180.10 | 1940 | DEC | 130.50 |
| 1937 | JAN | 183.50 | 1941 | JAN | 130.20 |
| 1937 | FEB | 188.00 | 1941 | FEB | 121.70 |
| 1937 | MAR | 188.40 | 1941 | MAR | 122.50 |
| 1937 | APR | 179.30 | 1941 | APR | 119.10 |
| 1937 | MAY | 173.10 | 1941 | MAY | 116.40 |
| 1937 | JUN | 170.10 | 1941 | JUN | 121.60 |
| 1937 | JUL | 180.30 | 1941 | JUL | 127.60 |
| 1937 | AUG | 184.40 | 1941 | AUG | 126.70 |
| 1937 | SEP | 160.10 | 1941 | SEP | 127.40 |
| 1937 | OCT | 138.60 | 1941 | OCT | 121.20 |
| 1937 | NOV | 125.10 | 1941 | NOV | 116.90 |
| 1937 | DEC | 125.50 | 1941 | DEC | 110.70 |

| YEAR | MONTH | DJIA | YEAR | MONTH | DJIA |
|---|---|---|---|---|---|
| 1942 | JAN | 111.10 | 1946 | JAN | 199.00 |
| 1942 | FEB | 107.30 | 1946 | FEB | 199.50 |
| 1942 | MAR | 101.60 | 1946 | MAR | 194.40 |
| 1942 | APR | 97.80 | 1946 | APR | 205.80 |
| 1942 | MAY | 98.40 | 1946 | MAY | 206.60 |
| 1942 | JUN | 103.80 | 1946 | JUN | 207.30 |
| 1942 | JUL | 107.00 | 1946 | JUL | 202.30 |
| 1942 | AUG | 106.10 | 1946 | AUG | 199.40 |
| 1942 | SEP | 107.40 | 1946 | SEP | 172.70 |
| 1942 | OCT | 113.50 | 1946 | OCT | 169.50 |
| 1942 | NOV | 115.30 | 1946 | NOV | 168.90 |
| 1942 | DEC | 117.20 | 1946 | DEC | 174.40 |
| 1943 | JAN | 121.50 | 1947 | JAN | 176.10 |
| 1943 | FEB | 127.40 | 1947 | FEB | 181.50 |
| 1943 | MAR | 131.20 | 1947 | MAR | 176.70 |
| 1943 | APR | 134.10 | 1947 | APR | 171.30 |
| 1943 | MAY | 138.60 | 1947 | MAY | 168.70 |
| 1943 | JUN | 141.30 | 1947 | JUN | 173.80 |
| 1943 | JUL | 142.90 | 1947 | JUL | 183.50 |
| 1943 | AUG | 136.30 | 1947 | AUG | 180.00 |
| 1943 | SEP | 138.90 | 1947 | SEP | 176.80 |
| 1943 | OCT | 138.30 | 1947 | OCT | 181.90 |
| 1943 | NOV | 132.70 | 1947 | NOV | 181.40 |
| 1943 | DEC | 134.60 | 1947 | DEC | 179.20 |
| 1944 | JAN | 137.70 | 1948 | JAN | 176.30 |
| 1944 | FEB | 136.00 | 1948 | FEB | 169.50 |
| 1944 | MAR | 139.10 | 1948 | MAR | 169.90 |
| 1944 | APR | 137.20 | 1948 | APR | 180.10 |
| 1944 | MAY | 139.20 | 1948 | MAY | 186.70 |
| 1944 | JUN | 145.50 | 1948 | JUN | 191.00 |
| 1944 | JUL | 148.40 | 1948 | JUL | 187.10 |
| 1944 | AUG | 146.70 | 1948 | AUG | 181.80 |
| 1944 | SEP | 145.20 | 1948 | SEP | 180.30 |
| 1944 | OCT | 147.70 | 1948 | OCT | 185.20 |
| 1944 | NOV | 146.90 | 1948 | NOV | 176.70 |
| 1944 | DEC | 150.40 | 1948 | DEC | 176.30 |
| 1945 | JAN | 154.00 | 1949 | JAN | 179.70 |
| 1945 | FEB | 157.10 | 1949 | FEB | 174.50 |
| 1945 | MAR | 157.20 | 1949 | MAR | 175.90 |
| 1945 | APR | 160.50 | 1949 | APR | 175.70 |
| 1945 | MAY | 165.60 | 1949 | MAY | 174.00 |
| 1945 | JUN | 167.30 | 1949 | JUN | 165.60 |
| 1945 | JUL | 164.00 | 1949 | JUL | 173.30 |
| 1945 | AUG | 166.20 | 1949 | AUG | 179.20 |
| 1945 | SEP | 178.00 | 1949 | SEP | 180.90 |
| 1945 | OCT | 185.10 | 1949 | OCT | 182.50 |
| 1945 | NOV | 190.20 | 1949 | NOV | 191.60 |
| 1945 | DEC | 192.70 | 1949 | DEC | 196.80 |

**TABLE A4-1 Monthly Values of Dow Jones Industrial Average** (cont.)

| YEAR | MONTH | DJIA | YEAR | MONTH | DJIA |
|---|---|---|---|---|---|
| 1950 | JAN | 199.80 | 1954 | JAN | 286.60 |
| 1950 | FEB | 203.40 | 1954 | FEB | 292.10 |
| 1950 | MAR | 206.30 | 1954 | MAR | 299.20 |
| 1950 | APR | 212.70 | 1954 | APR | 310.90 |
| 1950 | MAY | 219.40 | 1954 | MAY | 322.90 |
| 1950 | JUN | 221.00 | 1954 | JUN | 327.90 |
| 1950 | JUL | 205.30 | 1954 | JUL | 341.30 |
| 1950 | AUG | 216.60 | 1954 | AUG | 346.10 |
| 1950 | SEP | 223.20 | 1954 | SEP | 352.70 |
| 1950 | OCT | 229.30 | 1954 | OCT | 358.30 |
| 1950 | NOV | 229.30 | 1954 | NOV | 375.50 |
| 1950 | DEC | 229.30 | 1954 | DEC | 393.80 |
| 1951 | JAN | 244.50 | 1955 | JAN | 398.40 |
| 1951 | FEB | 253.30 | 1955 | FEB | 410.30 |
| 1951 | MAR | 249.50 | 1955 | MAR | 408.90 |
| 1951 | APR | 253.40 | 1955 | APR | 423.00 |
| 1951 | MAY | 254.40 | 1955 | MAY | 421.60 |
| 1951 | JUN | 249.30 | 1955 | JUN | 440.80 |
| 1951 | JUL | 253.60 | 1955 | JUL | 462.20 |
| 1951 | AUG | 264.90 | 1955 | AUG | 457.30 |
| 1951 | SEP | 273.40 | 1955 | SEP | 476.40 |
| 1951 | OCT | 269.70 | 1955 | OCT | 452.70 |
| 1951 | NOV | 259.60 | 1955 | NOV | 476.60 |
| 1951 | DEC | 266.10 | 1955 | DEC | 484.60 |
| 1952 | JAN | 271.70 | 1956 | JAN | 474.80 |
| 1952 | FEB | 265.20 | 1956 | FEB | 475.50 |
| 1952 | MAR | 264.50 | 1956 | MAR | 502.70 |
| 1952 | APR | 262.60 | 1956 | APR | 511.00 |
| 1952 | MAY | 261.60 | 1956 | MAY | 495.20 |
| 1952 | JUN | 268.40 | 1956 | JUN | 485.30 |
| 1952 | JUL | 276.40 | 1956 | JUL | 509.80 |
| 1952 | AUG | 276.70 | 1956 | AUG | 511.70 |
| 1952 | SEP | 272.40 | 1956 | SEP | 495.00 |
| 1952 | OCT | 267.80 | 1956 | OCT | 483.80 |
| 1952 | NOV | 276.40 | 1956 | NOV | 479.30 |
| 1952 | DEC | 286.00 | 1956 | DEC | 492.00 |
| 1953 | JAN | 288.40 | 1957 | JAN | 485.90 |
| 1953 | FEB | 283.90 | 1957 | FEB | 466.80 |
| 1953 | MAR | 286.80 | 1957 | MAR | 472.80 |
| 1953 | APR | 275.30 | 1957 | APR | 485.40 |
| 1953 | MAY | 276.80 | 1957 | MAY | 500.80 |
| 1953 | JUN | 266.90 | 1957 | JUN | 505.30 |
| 1953 | JUL | 270.30 | 1957 | JUL | 514.60 |
| 1953 | AUG | 272.20 | 1957 | AUG | 488.00 |
| 1953 | SEP | 261.90 | 1957 | SEP | 471.80 |
| 1953 | OCT | 270.70 | 1957 | OCT | 443.40 |
| 1953 | NOV | 277.10 | 1957 | NOV | 437.30 |
| 1953 | DEC | 281.20 | 1957 | DEC | 436.90 |

| YEAR | MONTH | DJIA | YEAR | MONTH | DJIA |
|---|---|---|---|---|---|
| 1958 | JAN | 445.70 | 1962 | JAN | 705.20 |
| 1958 | FEB | 444.20 | 1962 | FEB | 712.00 |
| 1958 | MAR | 450.10 | 1962 | MAR | 714.20 |
| 1958 | APR | 446.90 | 1962 | APR | 690.30 |
| 1958 | MAY | 460.00 | 1962 | MAY | 643.70 |
| 1958 | JUN | 472.00 | 1962 | JUN | 572.60 |
| 1958 | JUL | 488.30 | 1962 | JUL | 581.80 |
| 1958 | AUG | 507.60 | 1962 | AUG | 602.50 |
| 1958 | SEP | 521.80 | 1962 | SEP | 597.00 |
| 1958 | OCT | 539.90 | 1962 | OCT | 580.70 |
| 1958 | NOV | 557.10 | 1962 | NOV | 628.80 |
| 1958 | DEC | 566.40 | 1962 | DEC | 648.40 |
| 1959 | JAN | 592.30 | 1963 | JAN | 672.10 |
| 1959 | FEB | 590.70 | 1963 | FEB | 679.80 |
| 1959 | MAR | 609.10 | 1963 | MAR | 674.60 |
| 1959 | APR | 617.00 | 1963 | APR | 707.10 |
| 1959 | MAY | 630.80 | 1963 | MAY | 720.80 |
| 1959 | JUN | 631.50 | 1963 | JUN | 719.10 |
| 1959 | JUL | 662.80 | 1963 | JUL | 700.80 |
| 1959 | AUG | 660.60 | 1963 | AUG | 714.20 |
| 1959 | SEP | 635.50 | 1963 | SEP | 738.50 |
| 1959 | OCT | 637.30 | 1963 | OCT | 747.50 |
| 1959 | NOV | 646.40 | 1963 | NOV | 743.20 |
| 1959 | DEC | 671.40 | 1963 | DEC | 759.90 |
| 1960 | JAN | 655.40 | 1964 | JAN | 776.60 |
| 1960 | FEB | 624.90 | 1964 | FEB | 793.00 |
| 1960 | MAR | 614.70 | 1964 | MAR | 812.20 |
| 1960 | APR | 620.00 | 1964 | APR | 820.90 |
| 1960 | MAY | 615.60 | 1964 | MAY | 823.10 |
| 1960 | JUN | 644.40 | 1964 | JUN | 817.60 |
| 1960 | JUL | 625.80 | 1964 | JUL | 844.20 |
| 1960 | AUG | 624.50 | 1964 | AUG | 835.30 |
| 1960 | SEP | 598.10 | 1964 | SEP | 863.60 |
| 1960 | OCT | 582.50 | 1964 | OCT | 875.30 |
| 1960 | NOV | 601.10 | 1964 | NOV | 880.00 |
| 1960 | DEC | 609.50 | 1964 | DEC | 866.70 |
| 1961 | JAN | 632.20 | 1965 | JAN | 890.20 |
| 1961 | FEB | 650.00 | 1965 | FEB | 894.40 |
| 1961 | MAR | 670.60 | 1965 | MAR | 896.40 |
| 1961 | APR | 684.90 | 1965 | APR | 907.70 |
| 1961 | MAY | 693.00 | 1965 | MAY | 927.50 |
| 1961 | JUN | 691.40 | 1965 | JUN | 878.10 |
| 1961 | JUL | 690.70 | 1965 | JUL | 873.50 |
| 1961 | AUG | 718.60 | 1965 | AUG | 887.60 |
| 1961 | SEP | 711.00 | 1965 | SEP | 922.20 |
| 1961 | OCT | 703.00 | 1965 | OCT | 944.60 |
| 1961 | NOV | 724.70 | 1965 | NOV | 953.30 |
| 1961 | DEC | 728.40 | 1965 | DEC | 955.20 |

**TABLE A4-1 Monthly Values of Dow Jones Industrial Average** (cont.)

| YEAR | MONTH | DJIA | YEAR | MONTH | DJIA |
|---|---|---|---|---|---|
| 1966 | JAN | 985.90 | 1970 | JAN | 782.90 |
| 1966 | FEB | 977.10 | 1970 | FEB | 756.30 |
| 1966 | MAR | 926.40 | 1970 | MAR | 777.60 |
| 1966 | APR | 945.70 | 1970 | APR | 771.70 |
| 1966 | MAY | 890.70 | 1970 | MAY | 691.90 |
| 1966 | JUN | 888.80 | 1970 | JUN | 699.30 |
| 1966 | JUL | 875.90 | 1970 | JUL | 712.80 |
| 1966 | AUG | 817.50 | 1970 | AUG | 731.90 |
| 1966 | SEP | 791.60 | 1970 | SEP | 756.40 |
| 1966 | OCT | 778.10 | 1970 | OCT | 763.70 |
| 1966 | NOV | 806.60 | 1970 | NOV | 769.30 |
| 1966 | DEC | 869.00 | 1970 | DEC | 821.50 |
| 1967 | JAN | 830.60 | 1971 | JAN | 849.00 |
| 1967 | FEB | 851.10 | 1971 | FEB | 879.30 |
| 1967 | MAR | 858.10 | 1971 | MAR | 901.30 |
| 1967 | APR | 868.70 | 1971 | APR | 932.50 |
| 1967 | MAY | 885.70 | 1971 | MAY | 925.50 |
| 1967 | JUN | 872.50 | 1971 | JUN | 900.50 |
| 1967 | JUL | 888.50 | 1971 | JUL | 887.80 |
| 1967 | AUG | 912.50 | 1971 | AUG | 875.40 |
| 1967 | SEP | 923.40 | 1971 | SEP | 901.20 |
| 1967 | OCT | 907.50 | 1971 | OCT | 872.10 |
| 1967 | NOV | 865.40 | 1971 | NOV | 822.10 |
| 1967 | DEC | 887.20 | 1971 | DEC | 869.90 |
| 1968 | JAN | 884.80 | 1972 | JAN | 904.70 |
| 1968 | FEB | 847.20 | 1972 | FEB | 914.40 |
| 1968 | MAR | 864.80 | 1972 | MAR | 939.20 |
| 1968 | APR | 893.40 | 1972 | APR | 958.20 |
| 1968 | MAY | 905.20 | 1972 | MAY | 948.20 |
| 1968 | JUN | 906.80 | 1972 | JUN | 943.90 |
| 1968 | JUL | 905.30 | 1972 | JUL | 925.90 |
| 1968 | AUG | 883.70 | 1972 | AUG | 958.30 |
| 1968 | SEP | 922.80 | 1972 | SEP | 950.70 |
| 1968 | OCT | 954.00 | 1972 | OCT | 944.10 |
| 1968 | NOV | 964.90 | 1972 | NOV | 1001.20 |
| 1968 | DEC | 968.60 | 1972 | DEC | 1020.30 |
| 1969 | JAN | 935.40 | 1973 | JAN | 1026.80 |
| 1969 | FEB | 931.50 | 1973 | FEB | 974.00 |
| 1969 | MAR | 916.50 | 1973 | MAR | 957.40 |
| 1969 | APR | 927.40 | 1973 | APR | 944.10 |
| 1969 | MAY | 954.90 | 1973 | MAY | 922.40 |
| 1969 | JUN | 894.80 | 1973 | JUN | 893.90 |
| 1969 | JUL | 844.00 | 1973 | JUL | 903.30 |
| 1969 | AUG | 825.50 | 1973 | AUG | 883.70 |
| 1969 | SEP | 826.70 | 1973 | SEP | 910.00 |
| 1969 | OCT | 832.50 | 1973 | OCT | 967.60 |
| 1969 | NOV | 841.10 | 1973 | NOV | 879.00 |
| 1969 | DEC | 757.20 | 1973 | DEC | 824.10 |

| YEAR | MONTH | DJIA |
|---|---|---|
| 1974 | JAN | 857.20 |
| 1974 | FEB | 831.30 |
| 1974 | MAR | 874.00 |
| 1974 | APR | 847.80 |
| 1974 | MAY | 830.30 |
| 1974 | JUN | 831.40 |
| 1974 | JUL | 783.00 |
| 1974 | AUG | 729.30 |
| 1974 | SEP | 651.30 |
| 1974 | OCT | 638.60 |
| 1974 | NOV | 642.10 |
| 1974 | DEC | 596.50 |
| 1975 | JAN | 659.10 |
| 1975 | FEB | 724.90 |
| 1975 | MAR | 765.10 |
| 1975 | APR | 790.90 |
| 1975 | MAY | 836.40 |
| 1975 | JUN | 845.70 |
| 1975 | JUL | 856.30 |
| 1975 | AUG | 815.50 |
| 1975 | SEP | 818.30 |
| 1975 | OCT | 831.30 |
| 1975 | NOV | 845.50 |
| 1975 | DEC | 840.80 |
| 1976 | JAN | 929.30 |
| 1976 | FEB | 971.20 |
| 1976 | MAR | 988.60 |
| 1976 | APR | 992.60 |
| 1976 | MAY | 988.80 |
| 1976 | JUN | 985.60 |
| 1976 | JUL | 993.20 |
| 1976 | AUG | 981.60 |
| 1976 | SEP | 994.40 |
| 1976 | OCT | 952.00 |
| 1976 | NOV | 944.60 |
| 1976 | DEC | 976.90 |
| 1977 | JAN | 970.60 |
| 1977 | FEB | 941.80 |
| 1977 | MAR | 946.10 |
| 1977 | APR | 929.10 |
| 1977 | MAY | 926.30 |
| 1977 | JUN | 916.90 |
| 1977 | JUL | 908.70 |
| 1977 | AUG | 872.30 |

**TABLE A4-2 Dow Jones Transportation, Utility, and Industrial Weekly Averages**

| DATE | DJTA | DJUA | DJIA |
|---|---|---|---|
| 5/ 1/1929 | 153.53 | 85.54 | 302.43 |
| 12/ 1/1929 | 152.91 | 86.73 | 301.25 |
| 19/ 1/1929 | 154.41 | 88.15 | 305.96 |
| 26/ 1/1929 | 154.16 | 92.83 | 314.56 |
| 2/ 2/1929 | 161.32 | 97.28 | 319.76 |
| 9/ 2/1929 | 154.23 | 91.20 | 301.53 |
| 16/ 2/1929 | 151.58 | 91.34 | 295.85 |
| 23/ 2/1929 | 154.94 | 94.79 | 310.06 |
| 2/ 3/1929 | 158.46 | 96.89 | 319.12 |
| 9/ 3/1929 | 154.82 | 94.49 | 311.61 |
| 16/ 3/1929 | 153.97 | 95.31 | 320.00 |
| 23/ 3/1929 | 150.63 | 93.67 | 306.21 |
| 30/ 3/1929 | 150.90 | 95.24 | 308.85 |
| 6/ 4/1929 | 152.16 | 92.20 | 302.81 |
| 13/ 4/1929 | 150.13 | 91.71 | 304.41 |
| 20/ 4/1929 | 150.66 | 93.30 | 311.07 |
| 27/ 4/1929 | 152.06 | 95.20 | 315.68 |
| 4/ 5/1929 | 152.87 | 98.69 | 327.08 |
| 11/ 5/1929 | 151.81 | 99.56 | 324.59 |
| 18/ 5/1929 | 149.15 | 98.86 | 321.48 |
| 25/ 5/1929 | 149.73 | 94.33 | 304.33 |
| 1/ 6/1929 | 154.88 | 98.82 | 299.12 |
| 8/ 6/1929 | 154.31 | 100.19 | 305.12 |
| 15/ 6/1929 | 155.88 | 103.54 | 314.26 |
| 22/ 6/1929 | 158.95 | 109.99 | 322.23 |
| 29/ 6/1929 | 161.68 | 118.50 | 333.79 |
| 6/ 7/1929 | 168.00 | 115.96 | 344.66 |
| 13/ 7/1929 | 174.78 | 122.65 | 345.94 |
| 20/ 7/1929 | 179.13 | 125.20 | 345.87 |
| 27/ 7/1929 | 172.53 | 125.04 | 343.73 |
| 3/ 8/1929 | 175.40 | 131.66 | 355.62 |
| 10/ 8/1929 | 172.58 | 125.51 | 344.84 |
| 17/ 8/1929 | 178.90 | 131.90 | 360.70 |
| 24/ 8/1929 | 179.71 | 135.74 | 375.44 |
| 31/ 8/1929 | 188.76 | 140.41 | 380.33 |
| 7/ 9/1929 | 186.61 | 143.58 | 377.56 |
| 14/ 9/1929 | 182.48 | 140.26 | 367.01 |
| 21/ 9/1929 | 181.63 | 144.61 | 361.16 |
| 28/ 9/1929 | 175.88 | 141.71 | 347.17 |
| 5/10/1929 | 171.21 | 133.43 | 341.36 |
| 12/10/1929 | 178.53 | 138.88 | 352.69 |
| 19/10/1929 | 171.72 | 117.13 | 323.87 |
| 26/10/1929 | 166.32 | 104.75 | 298.97 |
| 2/11/1929 | 159.82 | 95.34 | 273.51 |
| 9/11/1929 | 147.52 | 80.46 | 236.53 |
| 16/11/1929 | 141.25 | 76.88 | 228.73 |
| 23/11/1929 | 148.36 | 85.25 | 245.74 |
| 30/11/1929 | 145.89 | 82.63 | 238.95 |
| 7/12/1929 | 151.84 | 93.96 | 263.46 |
| 14/12/1929 | 151.68 | 87.97 | 253.02 |
| 21/12/1929 | 145.42 | 80.88 | 235.42 |
| 28/12/1929 | 143.30 | 82.90 | 238.43 |
| 52 | 52 | | |

| DATE | DJTA | DJUA | DJIA |
|---|---|---|---|
| 4/ 1/1930 | 144.85 | 88.09 | 248.85 |
| 11/ 1/1930 | 145.41 | 88.31 | 248.71 |
| 18/ 1/1930 | 146.54 | 87.06 | 246.84 |
| 25/ 1/1930 | 148.70 | 90.64 | 259.06 |
| 1/ 2/1930 | 149.40 | 93.27 | 268.41 |
| 8/ 2/1930 | 156.07 | 94.31 | 269.78 |
| 15/ 2/1930 | 155.25 | 97.38 | 269.25 |
| 22/ 2/1930 | 154.11 | 98.11 | 265.81 |
| 1/ 3/1930 | 152.52 | 101.64 | 273.24 |
| 8/ 3/1930 | 152.43 | 100.71 | 275.46 |
| 15/ 3/1930 | 151.36 | 98.36 | 270.25 |
| 22/ 3/1930 | 155.11 | 100.22 | 276.43 |
| 29/ 3/1930 | 157.94 | 105.41 | 286.19 |
| 5/ 4/1930 | 156.39 | 107.10 | 289.96 |
| 12/ 4/1930 | 154.04 | 108.62 | 293.43 |
| 19/ 4/1930 | 152.08 | 108.28 | 294.07 |
| 26/ 4/1930 | 149.43 | 107.16 | 285.46 |
| 3/ 5/1930 | 140.24 | 94.61 | 258.31 |
| 10/ 5/1930 | 143.12 | 101.04 | 272.01 |
| 17/ 5/1930 | 144.93 | 102.08 | 272.41 |
| 24/ 5/1930 | 145.11 | 99.92 | 271.33 |
| 31/ 5/1930 | 143.86 | 102.95 | 275.07 |
| 7/ 6/1930 | 139.86 | 96.03 | 257.82 |
| 14/ 6/1930 | 134.46 | 88.91 | 244.25 |
| 21/ 6/1930 | 128.60 | 78.54 | 218.30 |
| 28/ 6/1930 | 126.63 | 80.06 | 219.12 |
| 5/ 7/1930 | 128.36 | 80.85 | 222.46 |
| 12/ 7/1930 | 132.39 | 83.58 | 229.23 |
| 19/ 7/1930 | 135.25 | 86.55 | 236.65 |
| 26/ 7/1930 | 134.06 | 87.24 | 240.31 |
| 2/ 8/1930 | 131.01 | 85.32 | 234.50 |
| 9/ 8/1930 | 127.40 | 80.25 | 222.59 |
| 16/ 8/1930 | 129.03 | 82.85 | 228.02 |
| 23/ 8/1930 | 127.62 | 84.46 | 234.42 |
| 30/ 8/1930 | 131.28 | 86.76 | 240.42 |
| 6/ 9/1930 | 132.16 | 87.19 | 243.64 |
| 13/ 9/1930 | 131.89 | 87.58 | 240.34 |
| 20/ 9/1930 | 129.75 | 84.13 | 229.85 |
| 27/ 9/1930 | 124.19 | 78.80 | 212.52 |
| 4/10/1930 | 122.57 | 78.29 | 211.10 |
| 11/10/1930 | 116.54 | 69.48 | 193.05 |
| 18/10/1930 | 113.67 | 66.76 | 185.29 |
| 25/10/1930 | 113.51 | 70.47 | 193.34 |
| 1/11/1930 | 112.66 | 67.60 | 184.89 |
| 8/11/1930 | 105.70 | 60.96 | 173.14 |
| 15/11/1930 | 108.67 | 66.46 | 186.68 |
| 22/11/1930 | 110.91 | 66.65 | 188.04 |
| 29/11/1930 | 105.54 | 64.10 | 183.39 |
| 6/12/1930 | 103.11 | 61.94 | 178.37 |
| 13/12/1930 | 95.42 | 56.93 | 163.34 |
| 20/12/1930 | 98.96 | 59.03 | 169.42 |
| 27/12/1930 | 94.62 | 56.57 | 160.30 |

52 104

**TABLE A4-2 Dow Jones Transportation, Utility, and Industrial Weekly Averages** (cont.)

| DATE | DJTA | DJUA | DJIA |
|---|---|---|---|
| 3/ 1/1931 | 100.87 | 63.01 | 172.12 |
| 10/ 1/1931 | 105.42 | 62.43 | 171.71 |
| 17/ 1/1931 | 103.41 | 60.17 | 162.89 |
| 24/ 1/1931 | 108.53 | 63.48 | 169.80 |
| 31/ 1/1931 | 108.18 | 62.41 | 167.55 |
| 7/ 2/1931 | 107.99 | 63.91 | 172.90 |
| 14/ 2/1931 | 109.19 | 66.55 | 180.41 |
| 21/ 2/1931 | 111.53 | 70.65 | 191.32 |
| 28/ 2/1931 | 109.49 | 71.61 | 189.66 |
| 7/ 3/1931 | 105.29 | 71.44 | 183.85 |
| 14/ 3/1931 | 101.83 | 71.08 | 180.76 |
| 21/ 3/1931 | 100.65 | 72.61 | 185.24 |
| 28/ 3/1931 | 97.86 | 68.24 | 174.06 |
| 4/ 4/1931 | 95.76 | 66.54 | 172.43 |
| 11/ 4/1931 | 92.34 | 66.14 | 168.03 |
| 18/ 4/1931 | 90.49 | 63.15 | 162.37 |
| 25/ 4/1931 | 86.83 | 59.74 | 151.98 |
| 2/ 5/1931 | 86.31 | 59.52 | 147.49 |
| 9/ 5/1931 | 86.60 | 60.46 | 151.31 |
| 16/ 5/1931 | 79.40 | 58.29 | 142.95 |
| 23/ 5/1931 | 78.81 | 55.67 | 137.90 |
| 30/ 5/1931 | 72.06 | 52.35 | 128.46 |
| 6/ 6/1931 | 73.72 | 52.33 | 129.91 |
| 13/ 6/1931 | 79.65 | 55.37 | 137.03 |
| 20/ 6/1931 | 78.83 | 56.86 | 138.96 |
| 27/ 6/1931 | 88.31 | 62.58 | 156.93 |
| 4/ 7/1931 | 86.13 | 62.24 | 155.26 |
| 11/ 7/1931 | 81.98 | 58.38 | 143.88 |
| 18/ 7/1931 | 79.03 | 57.75 | 142.42 |
| 25/ 7/1931 | 77.79 | 56.44 | 138.24 |
| 1/ 8/1931 | 73.50 | 56.38 | 136.65 |
| 8/ 8/1931 | 70.29 | 55.40 | 134.94 |
| 15/ 8/1931 | 71.67 | 59.05 | 145.80 |
| 22/ 8/1931 | 68.33 | 56.78 | 137.76 |
| 29/ 8/1931 | 69.02 | 58.01 | 142.08 |
| 5/ 9/1931 | 64.66 | 54.47 | 132.62 |
| 12/ 9/1931 | 59.44 | 50.76 | 123.85 |
| 19/ 9/1931 | 53.08 | 44.70 | 111.74 |
| 26/ 9/1931 | 55.85 | 42.33 | 107.36 |
| 3/10/1931 | 49.71 | 37.36 | 92.77 |
| 10/10/1931 | 57.43 | 42.02 | 105.61 |
| 17/10/1931 | 55.94 | 40.56 | 102.28 |
| 24/10/1931 | 54.33 | 43.13 | 109.70 |
| 31/10/1931 | 52.19 | 40.51 | 105.43 |
| 7/11/1931 | 54.43 | 43.77 | 115.60 |
| 14/11/1931 | 50.10 | 40.63 | 106.35 |
| 21/11/1931 | 45.33 | 37.60 | 97.42 |
| 28/11/1931 | 38.96 | 35.47 | 90.02 |
| 5/12/1931 | 40.37 | 36.66 | 90.14 |
| 12/12/1931 | 34.85 | 32.43 | 78.93 |
| 19/12/1931 | 36.72 | 32.71 | 80.75 |
| 26/12/1931 | 33.72 | 31.41 | 75.84 |

52 156

| DATE | DJTA | DJUA | DJIA |
|---|---|---|---|
| 2/ 1/1932 | 33.11 | 30.60 | 74.62 |
| 9/ 1/1932 | 36.70 | 32.68 | 79.98 |
| 16/ 1/1932 | 40.42 | 33.78 | 84.44 |
| 23/ 1/1932 | 38.60 | 31.85 | 77.98 |
| 30/ 1/1932 | 37.02 | 30.61 | 76.19 |
| 6/ 2/1932 | 33.65 | 30.39 | 74.45 |
| 13/ 2/1932 | 39.70 | 34.87 | 85.82 |
| 20/ 2/1932 | 38.16 | 34.28 | 83.59 |
| 27/ 2/1932 | 36.45 | 33.53 | 82.02 |
| 5/ 3/1932 | 38.65 | 35.85 | 88.49 |
| 12/ 3/1932 | 36.21 | 34.40 | 84.52 |
| 19/ 3/1932 | 32.97 | 31.94 | 78.09 |
| 26/ 3/1932 | 31.37 | 30.75 | 75.69 |
| 2/ 4/1932 | 27.53 | 27.68 | 71.30 |
| 9/ 4/1932 | 23.79 | 25.01 | 64.48 |
| 16/ 4/1932 | 22.36 | 25.78 | 63.39 |
| 23/ 4/1932 | 22.90 | 24.56 | 59.22 |
| 30/ 4/1932 | 21.44 | 24.22 | 56.11 |
| 7/ 5/1932 | 21.65 | 24.70 | 58.04 |
| 14/ 5/1932 | 17.83 | 22.60 | 52.48 |
| 21/ 5/1932 | 17.40 | 21.76 | 53.04 |
| 28/ 5/1932 | 15.27 | 19.12 | 47.70 |
| 4/ 6/1932 | 17.64 | 19.83 | 50.88 |
| 11/ 6/1932 | 16.62 | 18.37 | 48.26 |
| 18/ 6/1932 | 16.32 | 18.25 | 47.55 |
| 25/ 6/1932 | 14.74 | 17.66 | 44.76 |
| 2/ 7/1932 | 13.84 | 17.91 | 44.39 |
| 9/ 7/1932 | 13.32 | 16.56 | 41.63 |
| 16/ 7/1932 | 15.73 | 18.33 | 45.29 |
| 23/ 7/1932 | 17.42 | 18.99 | 47.84 |
| 30/ 7/1932 | 21.74 | 22.79 | 54.26 |
| 6/ 8/1932 | 24.72 | 25.48 | 66.56 |
| 13/ 8/1932 | 25.54 | 27.05 | 63.19 |
| 20/ 8/1932 | 30.14 | 29.70 | 67.18 |
| 27/ 8/1932 | 36.25 | 34.21 | 75.61 |
| 3/ 9/1932 | 39.27 | 35.58 | 78.33 |
| 10/ 9/1932 | 37.94 | 34.29 | 76.54 |
| 17/ 9/1932 | 32.04 | 29.11 | 66.44 |
| 24/ 9/1932 | 36.95 | 33.43 | 74.83 |
| 1/10/1932 | 34.80 | 31.88 | 72.09 |
| 8/10/1932 | 25.96 | 27.30 | 61.17 |
| 15/10/1932 | 27.95 | 27.96 | 64.22 |
| 22/10/1932 | 27.26 | 26.94 | 60.85 |
| 29/10/1932 | 28.22 | 27.71 | 62.09 |
| 5/11/1932 | 26.91 | 27.08 | 62.41 |
| 12/11/1932 | 30.61 | 29.90 | 68.04 |
| 19/11/1932 | 27.87 | 28.30 | 64.14 |
| 26/11/1932 | 26.47 | 27.00 | 58.89 |
| 3/12/1932 | 24.33 | 25.40 | 55.83 |
| 10/12/1932 | 27.57 | 27.61 | 61.25 |
| 17/12/1932 | 27.59 | 28.10 | 60.11 |
| 24/12/1932 | 24.48 | 26.65 | 57.98 |
| 31/12/1932 | 25.90 | 27.50 | 59.93 |

53 209

**TABLE A4-2 Dow Jones Transportation, Utility, and Industrial Weekly Averages** (cont.)

| DATE | DJTA | DJUA | DJIA |
|---|---|---|---|
| 7/ 1/1933 | 28.24 | 28.78 | 62.96 |
| 14/ 1/1933 | 28.47 | 28.65 | 63.09 |
| 21/ 1/1933 | 28.38 | 27.75 | 61.79 |
| 28/ 1/1933 | 28.13 | 27.31 | 60.71 |
| 4/ 2/1933 | 27.84 | 24.76 | 57.55 |
| 11/ 2/1933 | 29.40 | 25.50 | 59.43 |
| 18/ 2/1933 | 26.81 | 23.69 | 56.04 |
| 25/ 2/1933 | 23.43 | 21.14 | 50.93 |
| 4/ 3/1933 | 24.76 | 21.95 | 53.84 |
| 11/ 3/1933 | 29.09 | 23.40 | 60.56 |
| 18/ 3/1933 | 27.98 | 21.23 | 57.71 |
| 25/ 3/1933 | 25.06 | 19.38 | 55.66 |
| 1/ 4/1933 | 25.00 | 19.83 | 59.30 |
| 8/ 4/1933 | 26.60 | 20.97 | 62.88 |
| 15/ 4/1933 | 30.66 | 23.17 | 72.24 |
| 22/ 4/1933 | 32.37 | 25.09 | 77.66 |
| 29/ 4/1933 | 34.87 | 26.23 | 77.61 |
| 6/ 5/1933 | 36.88 | 27.90 | 80.85 |
| 13/ 5/1933 | 37.47 | 27.40 | 80.21 |
| 20/ 5/1933 | 42.28 | 29.51 | 89.61 |
| 27/ 5/1933 | 43.27 | 31.51 | 90.02 |
| 3/ 6/1933 | 42.98 | 35.36 | 94.42 |
| 10/ 6/1933 | 41.67 | 33.39 | 90.23 |
| 17/ 6/1933 | 44.48 | 34.56 | 95.67 |
| 24/ 6/1933 | 49.78 | 34.93 | 100.92 |
| 1/ 7/1933 | 55.67 | 36.53 | 105.15 |
| 8/ 7/1933 | 54.69 | 37.19 | 106.10 |
| 15/ 7/1933 | 44.32 | 29.58 | 88.42 |
| 22/ 7/1933 | 47.81 | 31.56 | 94.54 |
| 29/ 7/1933 | 46.77 | 30.73 | 92.62 |
| 5/ 8/1933 | 49.27 | 31.34 | 97.47 |
| 12/ 8/1933 | 48.55 | 30.21 | 98.32 |
| 19/ 8/1933 | 53.37 | 31.18 | 105.07 |
| 26/ 8/1933 | 52.56 | 30.86 | 103.66 |
| 2/ 9/1933 | 49.11 | 29.20 | 99.42 |
| 9/ 9/1933 | 50.38 | 28.38 | 105.32 |
| 16/ 9/1933 | 44.59 | 26.86 | 99.78 |
| 23/ 9/1933 | 40.95 | 25.61 | 94.82 |
| 30/ 9/1933 | 41.84 | 25.77 | 98.20 |
| 7/10/1933 | 40.62 | 26.15 | 95.59 |
| 14/10/1933 | 34.10 | 23.14 | 83.64 |
| 21/10/1933 | 37.84 | 24.36 | 92.01 |
| 28/10/1933 | 38.82 | 23.76 | 93.09 |
| 4/11/1933 | 39.71 | 24.38 | 96.10 |
| 11/11/1933 | 38.29 | 23.03 | 98.67 |
| 18/11/1933 | 39.54 | 24.60 | 99.28 |
| 25/11/1933 | 38.11 | 23.38 | 99.07 |
| 2/12/1933 | 42.07 | 23.67 | 102.92 |
| 9/12/1933 | 40.54 | 23.23 | 98.06 |
| 16/12/1933 | 40.23 | 22.03 | 98.04 |
| 23/12/1933 | 40.80 | 23.29 | 99.90 |
| 30/12/1933 | 39.97 | 22.45 | 96.94 |

52 261

| DATE | DJTA | DJUA | DJIA |
|---|---|---|---|
| 6/ 1/1934 | 41.62 | 24.36 | 98.66 |
| 13/ 1/1934 | 48.02 | 26.95 | 105.52 |
| 20/ 1/1934 | 48.20 | 26.45 | 106.03 |
| 27/ 1/1934 | 51.85 | 28.40 | 109.41 |
| 3/ 2/1934 | 49.66 | 28.55 | 105.47 |
| 10/ 2/1934 | 52.02 | 28.92 | 109.07 |
| 17/ 2/1934 | 49.08 | 27.13 | 104.77 |
| 24/ 2/1934 | 49.22 | 26.92 | 105.56 |
| 3/ 3/1934 | 48.02 | 26.21 | 102.77 |
| 10/ 3/1934 | 48.40 | 26.33 | 101.65 |
| 17/ 3/1934 | 47.92 | 26.37 | 100.92 |
| 24/ 3/1934 | 47.92 | 26.02 | 101.85 |
| 31/ 3/1934 | 49.22 | 26.00 | 103.60 |
| 7/ 4/1934 | 49.57 | 26.21 | 105.04 |
| 14/ 4/1934 | 50.68 | 26.71 | 106.34 |
| 21/ 4/1934 | 48.78 | 25.50 | 102.90 |
| 28/ 4/1934 | 45.68 | 23.16 | 98.20 |
| 5/ 5/1934 | 41.11 | 22.11 | 92.22 |
| 12/ 5/1934 | 43.70 | 23.18 | 91.13 |
| 19/ 5/1934 | 43.44 | 23.30 | 95.05 |
| 26/ 5/1934 | 41.68 | 22.48 | 91.41 |
| 2/ 6/1934 | 45.31 | 24.05 | 98.90 |
| 9/ 6/1934 | 46.25 | 25.08 | 99.85 |
| 16/ 6/1934 | 44.13 | 23.76 | 96.59 |
| 23/ 6/1934 | 43.98 | 23.76 | 95.74 |
| 30/ 6/1934 | 43.57 | 23.77 | 97.15 |
| 7/ 7/1934 | 43.11 | 23.24 | 99.02 |
| 14/ 7/1934 | 40.27 | 21.83 | 94.62 |
| 21/ 7/1934 | 35.47 | 19.86 | 88.72 |
| 28/ 7/1934 | 34.22 | 20.07 | 88.43 |
| 4/ 8/1934 | 33.60 | 20.21 | 89.79 |
| 11/ 8/1934 | 34.36 | 20.13 | 90.86 |
| 18/ 8/1934 | 38.42 | 21.72 | 95.71 |
| 25/ 8/1934 | 36.07 | 20.48 | 92.64 |
| 1/ 9/1934 | 35.14 | 19.96 | 90.83 |
| 8/ 9/1934 | 33.42 | 18.75 | 87.34 |
| 15/ 9/1934 | 35.41 | 19.80 | 91.08 |
| 22/ 9/1934 | 36.33 | 20.40 | 92.63 |
| 29/ 9/1934 | 35.97 | 20.13 | 92.85 |
| 6/10/1934 | 36.39 | 20.21 | 94.90 |
| 13/10/1934 | 36.03 | 19.88 | 95.02 |
| 20/10/1934 | 34.98 | 19.19 | 92.86 |
| 27/10/1934 | 34.76 | 19.36 | 94.95 |
| 3/11/1934 | 36.35 | 19.70 | 99.21 |
| 10/11/1934 | 35.80 | 17.68 | 99.45 |
| 17/11/1934 | 36.44 | 18.35 | 102.40 |
| 24/11/1934 | 36.93 | 18.98 | 102.93 |
| 1/12/1934 | 36.93 | 18.50 | 102.83 |
| 8/12/1934 | 36.36 | 18.01 | 100.84 |
| 15/12/1934 | 35.41 | 17.00 | 99.73 |
| 22/12/1934 | 36.66 | 17.63 | 103.90 |
| 29/12/1934 | 36.82 | 17.68 | 105.56 |

52 313

**TABLE A4-2 Dow Jones Transportation, Utility, and Industrial Weekly Averages** (cont.)

| DATE | DJTA | DJUA | DJIA |
|---|---|---|---|
| 5/ 1/1935 | 35.27 | 17.41 | 102.30 |
| 12/ 1/1935 | 35.14 | 17.35 | 102.96 |
| 19/ 1/1935 | 34.30 | 17.58 | 102.50 |
| 26/ 1/1935 | 33.49 | 17.13 | 102.20 |
| 2/ 2/1935 | 33.18 | 16.80 | 102.66 |
| 9/ 2/1935 | 32.55 | 16.35 | 104.54 |
| 16/ 2/1935 | 31.51 | 15.80 | 103.25 |
| 23/ 2/1935 | 30.91 | 15.88 | 103.22 |
| 2/ 3/1935 | 28.56 | 15.56 | 101.18 |
| 9/ 3/1935 | 27.88 | 14.69 | 97.79 |
| 16/ 3/1935 | 28.69 | 16.25 | 99.84 |
| 23/ 3/1935 | 27.97 | 16.60 | 100.81 |
| 30/ 3/1935 | 29.76 | 17.70 | 103.04 |
| 6/ 4/1935 | 30.80 | 17.94 | 105.42 |
| 13/ 4/1935 | 30.56 | 18.08 | 109.76 |
| 20/ 4/1935 | 31.30 | 18.29 | 109.68 |
| 27/ 4/1935 | 30.82 | 18.80 | 110.83 |
| 4/ 5/1935 | 30.71 | 19.65 | 114.08 |
| 11/ 5/1935 | 31.38 | 19.44 | 114.58 |
| 18/ 5/1935 | 31.65 | 19.15 | 115.90 |
| 25/ 5/1935 | 30.48 | 20.04 | 109.74 |
| 1/ 6/1935 | 31.50 | 20.98 | 114.72 |
| 8/ 6/1935 | 33.54 | 21.14 | 119.17 |
| 15/ 6/1935 | 33.54 | 22.74 | 120.75 |
| 22/ 6/1935 | 32.87 | 21.89 | 118.21 |
| 29/ 6/1935 | 32.48 | 22.81 | 121.02 |
| 6/ 7/1935 | 33.39 | 22.21 | 121.88 |
| 13/ 7/1935 | 33.41 | 22.04 | 122.69 |
| 20/ 7/1935 | 34.22 | 22.95 | 125.27 |
| 27/ 7/1935 | 34.89 | 25.03 | 125.90 |
| 3/ 8/1935 | 35.46 | 26.80 | 127.94 |
| 10/ 8/1935 | 36.98 | 28.18 | 127.96 |
| 17/ 8/1935 | 35.30 | 25.07 | 127.93 |
| 24/ 8/1935 | 35.20 | 25.70 | 127.89 |
| 31/ 8/1935 | 36.94 | 26.71 | 131.86 |
| 7/ 9/1935 | 36.45 | 26.11 | 133.40 |
| 14/ 9/1935 | 35.07 | 24.51 | 128.78 |
| 21/ 9/1935 | 35.32 | 25.18 | 131.75 |
| 28/ 9/1935 | 32.73 | 24.70 | 130.35 |
| 5/10/1935 | 32.80 | 25.81 | 133.56 |
| 12/10/1935 | 33.73 | 26.11 | 137.09 |
| 19/10/1935 | 35.04 | 27.47 | 141.47 |
| 26/10/1935 | 34.87 | 27.78 | 141.20 |
| 2/11/1935 | 35.54 | 28.98 | 144.36 |
| 9/11/1935 | 37.59 | 29.35 | 147.31 |
| 16/11/1935 | 39.17 | 29.60 | 146.12 |
| 23/11/1935 | 39.20 | 28.62 | 142.35 |
| 30/11/1935 | 41.69 | 29.78 | 144.47 |
| 7/12/1935 | 40.05 | 28.37 | 140.38 |
| 14/12/1935 | 39.76 | 28.33 | 140.19 |
| 21/12/1935 | 39.39 | 28.69 | 140.76 |
| 28/12/1935 | 42.14 | 30.03 | 144.08 |

52 365

| DATE | DJTA | DJUA | DJIA |
|---|---|---|---|
| 4/ 1/1936 | 42.68 | 30.36 | 146.73 |
| 11/ 1/1936 | 42.64 | 30.93 | 144.93 |
| 18/ 1/1936 | 44.06 | 31.43 | 147.01 |
| 25/ 1/1936 | 46.10 | 31.83 | 149.58 |
| 1/ 2/1936 | 46.65 | 32.62 | 150.40 |
| 8/ 2/1936 | 48.76 | 33.76 | 152.40 |
| 15/ 2/1936 | 51.07 | 32.50 | 153.74 |
| 22/ 2/1936 | 48.58 | 32.11 | 152.15 |
| 29/ 2/1936 | 49.61 | 32.59 | 157.86 |
| 7/ 3/1936 | 47.13 | 31.84 | 154.07 |
| 14/ 3/1936 | 47.10 | 31.76 | 156.42 |
| 21/ 3/1936 | 47.16 | 31.87 | 155.54 |
| 28/ 3/1936 | 49.10 | 32.66 | 161.50 |
| 4/ 4/1936 | 50.05 | 33.15 | 160.48 |
| 11/ 4/1936 | 47.90 | 31.90 | 156.07 |
| 18/ 4/1936 | 44.93 | 30.80 | 151.93 |
| 25/ 4/1936 | 43.39 | 28.96 | 146.41 |
| 2/ 5/1936 | 43.75 | 29.19 | 147.85 |
| 9/ 5/1936 | 45.11 | 30.53 | 151.42 |
| 16/ 5/1936 | 44.81 | 30.90 | 150.65 |
| 23/ 5/1936 | 46.28 | 31.40 | 152.64 |
| 30/ 5/1936 | 45.40 | 30.96 | 149.84 |
| 6/ 6/1936 | 46.73 | 32.65 | 154.64 |
| 13/ 6/1936 | 47.56 | 32.83 | 157.21 |
| 20/ 6/1936 | 48.11 | 32.48 | 158.46 |
| 27/ 6/1936 | 48.05 | 33.41 | 158.11 |
| 4/ 7/1936 | 50.34 | 34.71 | 160.72 |
| 11/ 7/1936 | 52.87 | 35.01 | 164.42 |
| 18/ 7/1936 | 53.31 | 35.37 | 165.56 |
| 25/ 7/1936 | 53.55 | 34.86 | 165.42 |
| 1/ 8/1936 | 55.74 | 35.83 | 169.10 |
| 8/ 8/1936 | 53.98 | 34.76 | 165.86 |
| 15/ 8/1936 | 52.44 | 33.78 | 162.14 |
| 22/ 8/1936 | 55.01 | 34.70 | 166.91 |
| 29/ 8/1936 | 55.68 | 35.11 | 167.80 |
| 5/ 9/1936 | 55.83 | 34.95 | 168.02 |
| 12/ 9/1936 | 56.36 | 34.50 | 168.93 |
| 19/ 9/1936 | 56.50 | 33.83 | 168.07 |
| 26/ 9/1936 | 57.85 | 34.81 | 172.44 |
| 3/10/1936 | 59.55 | 35.30 | 176.05 |
| 10/10/1936 | 59.85 | 35.09 | 177.63 |
| 17/10/1936 | 58.61 | 35.43 | 175.91 |
| 24/10/1936 | 58.66 | 36.08 | 177.19 |
| 31/10/1936 | 57.92 | 35.33 | 183.38 |
| 7/11/1936 | 56.02 | 33.93 | 181.45 |
| 14/11/1936 | 56.05 | 35.10 | 182.01 |
| 21/11/1936 | 55.73 | 35.88 | 183.32 |
| 28/11/1936 | 54.56 | 35.28 | 181.05 |
| 5/12/1936 | 54.93 | 35.26 | 180.92 |
| 12/12/1936 | 52.70 | 34.44 | 177.61 |
| 19/12/1936 | 52.58 | 34.65 | 178.60 |
| 26/12/1936 | 53.28 | 34.66 | 178.52 |

52 417

**TABLE A4-2 Dow Jones Transportation, Utility, and Industrial Weekly Averages** (cont.)

| DATE | DJTA | DJUA | DJIA |
|---|---|---|---|
| 2/ 1/1937 | 55.13 | 36.38 | 182.75 |
| 9/ 1/1937 | 56.31 | 37.26 | 185.73 |
| 16/ 1/1937 | 56.06 | 37.02 | 186.69 |
| 23/ 1/1937 | 55.00 | 35.83 | 185.74 |
| 30/ 1/1937 | 57.29 | 35.11 | 187.11 |
| 6/ 2/1937 | 58.00 | 35.12 | 190.03 |
| 13/ 2/1937 | 58.73 | 34.76 | 189.37 |
| 20/ 2/1937 | 58.01 | 34.08 | 187.30 |
| 27/ 2/1937 | 62.69 | 34.23 | 194.15 |
| 6/ 3/1937 | 62.06 | 32.88 | 190.58 |
| 13/ 3/1937 | 62.58 | 32.79 | 184.04 |
| 20/ 3/1937 | 61.05 | 32.02 | 184.95 |
| 27/ 3/1937 | 60.46 | 31.56 | 183.54 |
| 3/ 4/1937 | 58.72 | 31.03 | 178.26 |
| 10/ 4/1937 | 60.02 | 31.13 | 180.51 |
| 17/ 4/1937 | 59.54 | 29.94 | 176.98 |
| 24/ 4/1937 | 58.26 | 29.16 | 174.42 |
| 1/ 5/1937 | 60.14 | 29.23 | 175.54 |
| 8/ 5/1937 | 57.90 | 27.53 | 169.60 |
| 15/ 5/1937 | 58.96 | 28.34 | 175.00 |
| 22/ 5/1937 | 56.82 | 27.95 | 174.71 |
| 29/ 5/1937 | 57.13 | 27.75 | 175.00 |
| 5/ 6/1937 | 55.26 | 26.55 | 169.51 |
| 12/ 6/1937 | 53.38 | 26.51 | 168.60 |
| 19/ 6/1937 | 51.06 | 26.06 | 168.45 |
| 26/ 6/1937 | 52.06 | 27.03 | 172.22 |
| 3/ 7/1937 | 54.00 | 28.06 | 176.72 |
| 10/ 7/1937 | 53.28 | 28.17 | 179.72 |
| 17/ 7/1937 | 55.05 | 30.65 | 184.85 |
| 24/ 7/1937 | 52.95 | 30.09 | 185.61 |
| 31/ 7/1937 | 52.95 | 28.92 | 186.41 |
| 7/ 8/1937 | 54.13 | 29.23 | 190.02 |
| 14/ 8/1937 | 52.10 | 27.83 | 183.74 |
| 21/ 8/1937 | 49.46 | 27.26 | 175.93 |
| 28/ 8/1937 | 47.43 | 26.86 | 172.55 |
| 4/ 9/1937 | 42.34 | 24.97 | 159.96 |
| 11/ 9/1937 | 41.80 | 24.50 | 157.83 |
| 18/ 9/1937 | 38.93 | 22.77 | 147.47 |
| 25/ 9/1937 | 41.06 | 24.06 | 154.08 |
| 2/10/1937 | 37.39 | 22.61 | 143.93 |
| 9/10/1937 | 33.33 | 20.96 | 136.30 |
| 16/10/1937 | 32.32 | 20.84 | 127.15 |
| 23/10/1937 | 34.63 | 22.83 | 138.17 |
| 30/10/1937 | 31.67 | 21.21 | 125.25 |
| 6/11/1937 | 34.26 | 23.43 | 133.05 |
| 13/11/1937 | 31.06 | 21.85 | 120.45 |
| 20/11/1937 | 31.71 | 22.96 | 123.71 |
| 27/11/1937 | 32.62 | 22.41 | 127.79 |
| 4/12/1937 | 32.36 | 21.97 | 126.83 |
| 11/12/1937 | 31.91 | 21.56 | 126.63 |
| 18/12/1937 | 31.49 | 21.17 | 127.36 |
| 25/12/1937 | 29.46 | 20.35 | 120.85 |

52 469

| DATE | DJTA | DJUA | DJIA |
|---|---|---|---|
| 1/ 1/1938 | 31.21 | 21.80 | 130.84 |
| 8/ 1/1938 | 32.33 | 21.75 | 134.31 |
| 15/ 1/1938 | 29.79 | 20.58 | 130.00 |
| 22/ 1/1938 | 27.45 | 19.05 | 120.14 |
| 29/ 1/1938 | 27.56 | 18.83 | 122.88 |
| 5/ 2/1938 | 28.96 | 18.74 | 124.94 |
| 12/ 2/1938 | 28.76 | 19.33 | 127.50 |
| 19/ 2/1938 | 30.29 | 20.18 | 131.26 |
| 26/ 2/1938 | 28.75 | 19.35 | 127.67 |
| 5/ 3/1938 | 25.85 | 18.59 | 122.58 |
| 12/ 3/1938 | 23.68 | 17.96 | 120.43 |
| 19/ 3/1938 | 19.68 | 16.09 | 106.63 |
| 26/ 3/1938 | 20.46 | 16.58 | 106.11 |
| 2/ 4/1938 | 22.75 | 18.44 | 115.32 |
| 9/ 4/1938 | 22.00 | 18.33 | 121.00 |
| 16/ 4/1938 | 22.07 | 18.81 | 117.64 |
| 23/ 4/1938 | 21.26 | 17.90 | 111.28 |
| 30/ 4/1938 | 22.64 | 19.85 | 117.21 |
| 7/ 5/1938 | 22.96 | 19.77 | 117.21 |
| 14/ 5/1938 | 21.73 | 19.04 | 113.25 |
| 21/ 5/1938 | 20.58 | 18.12 | 108.90 |
| 28/ 5/1938 | 20.57 | 18.77 | 111.82 |
| 4/ 6/1938 | 20.53 | 19.04 | 114.23 |
| 11/ 6/1938 | 19.73 | 18.54 | 113.23 |
| 18/ 6/1938 | 25.45 | 20.58 | 131.94 |
| 25/ 6/1938 | 27.57 | 22.27 | 138.53 |
| 2/ 7/1938 | 27.31 | 21.70 | 136.20 |
| 9/ 7/1938 | 27.18 | 21.39 | 138.53 |
| 16/ 7/1938 | 30.38 | 21.99 | 144.24 |
| 23/ 7/1938 | 28.45 | 20.63 | 141.25 |
| 30/ 7/1938 | 29.77 | 20.93 | 145.67 |
| 6/ 8/1938 | 27.57 | 19.43 | 136.21 |
| 13/ 8/1938 | 28.45 | 19.79 | 141.20 |
| 20/ 8/1938 | 28.71 | 19.89 | 141.95 |
| 27/ 8/1938 | 27.78 | 19.59 | 142.48 |
| 3/ 9/1938 | 26.63 | 18.60 | 138.29 |
| 10/ 9/1938 | 24.01 | 17.34 | 131.82 |
| 17/ 9/1938 | 24.35 | 17.83 | 133.02 |
| 24/ 9/1938 | 27.43 | 19.93 | 143.13 |
| 1/10/1938 | 30.91 | 21.12 | 149.75 |
| 8/10/1938 | 30.86 | 24.63 | 151.96 |
| 15/10/1938 | 31.59 | 23.69 | 154.11 |
| 22/10/1938 | 31.89 | 24.44 | 151.07 |
| 29/10/1938 | 31.78 | 23.59 | 152.12 |
| 5/11/1938 | 33.17 | 24.69 | 158.14 |
| 12/11/1938 | 30.94 | 22.94 | 150.38 |
| 19/11/1938 | 29.89 | 22.34 | 148.45 |
| 26/11/1938 | 29.23 | 21.83 | 147.50 |
| 3/12/1938 | 29.10 | 21.34 | 148.31 |
| 10/12/1938 | 30.35 | 22.29 | 150.36 |
| 17/12/1938 | 32.02 | 22.04 | 151.38 |
| 24/12/1938 | 33.98 | 23.02 | 154.76 |
| 31/12/1938 | 32.93 | 23.49 | 151.54 |

53 522

**TABLE A4-2 Dow Jones Transportation, Utility, and Industrial Weekly Averages** (cont.)

| DATE | DJTA | DJUA | DJIA |
|---|---|---|---|
| 7/ 1/1939 | 31.95 | 23.49 | 148.26 |
| 14/ 1/1939 | 31.10 | 23.97 | 146.76 |
| 21/ 1/1939 | 28.49 | 22.99 | 138.79 |
| 28/ 1/1939 | 30.38 | 24.60 | 145.07 |
| 4/ 2/1939 | 30.18 | 24.93 | 144.61 |
| 11/ 2/1939 | 30.55 | 25.43 | 145.51 |
| 18/ 2/1939 | 31.25 | 25.60 | 146.82 |
| 25/ 2/1939 | 32.98 | 26.09 | 149.49 |
| 4/ 3/1939 | 33.07 | 26.28 | 151.77 |
| 11/ 3/1939 | 29.58 | 23.92 | 141.68 |
| 18/ 3/1939 | 29.73 | 24.05 | 141.55 |
| 25/ 3/1939 | 27.24 | 22.50 | 132.83 |
| 1/ 4/1939 | 24.14 | 20.17 | 121.44 |
| 8/ 4/1939 | 26.15 | 22.69 | 129.61 |
| 15/ 4/1939 | 25.73 | 22.29 | 128.55 |
| 22/ 4/1939 | 25.86 | 22.06 | 128.45 |
| 29/ 4/1939 | 26.88 | 23.10 | 131.74 |
| 6/ 5/1939 | 27.32 | 23.23 | 132.40 |
| 13/ 5/1939 | 26.49 | 22.69 | 131.22 |
| 20/ 5/1939 | 28.18 | 23.65 | 136.80 |
| 27/ 5/1939 | 27.94 | 23.52 | 137.12 |
| 3/ 6/1939 | 28.33 | 23.91 | 140.14 |
| 10/ 6/1939 | 27.25 | 23.40 | 135.31 |
| 17/ 6/1939 | 27.76 | 24.05 | 137.36 |
| 24/ 6/1939 | 25.93 | 23.09 | 131.73 |
| 1/ 7/1939 | 26.66 | 23.80 | 133.24 |
| 8/ 7/1939 | 27.49 | 24.72 | 137.88 |
| 15/ 7/1939 | 30.20 | 25.91 | 144.71 |
| 22/ 7/1939 | 29.43 | 26.02 | 144.00 |
| 29/ 7/1939 | 29.20 | 26.74 | 142.11 |
| 5/ 8/1939 | 28.35 | 26.36 | 138.42 |
| 12/ 8/1939 | 26.96 | 25.20 | 135.11 |
| 19/ 8/1939 | 26.88 | 25.13 | 136.39 |
| 26/ 8/1939 | 26.18 | 23.41 | 138.09 |
| 2/ 9/1939 | 30.51 | 23.68 | 150.91 |
| 9/ 9/1939 | 32.07 | 25.01 | 152.15 |
| 16/ 9/1939 | 33.34 | 24.63 | 152.99 |
| 23/ 9/1939 | 35.61 | 25.13 | 152.54 |
| 30/ 9/1939 | 33.64 | 24.88 | 149.60 |
| 7/10/1939 | 33.53 | 25.31 | 150.38 |
| 14/10/1939 | 34.78 | 26.05 | 153.86 |
| 21/10/1939 | 34.43 | 26.12 | 153.12 |
| 28/10/1939 | 33.84 | 26.01 | 152.36 |
| 4/11/1939 | 32.79 | 25.71 | 149.09 |
| 11/11/1939 | 33.85 | 26.09 | 151.53 |
| 18/11/1939 | 32.90 | 25.42 | 148.64 |
| 25/11/1939 | 31.76 | 25.11 | 146.62 |
| 2/12/1939 | 31.86 | 25.02 | 147.93 |
| 9/12/1939 | 31.78 | 24.86 | 149.36 |
| 16/12/1939 | 31.58 | 24.91 | 149.85 |
| 23/12/1939 | 31.83 | 25.58 | 150.24 |
| 30/12/1939 | 32.25 | 26.25 | 151.19 |
| 52 | 574 | | |

| DATE | DJTA | DJUA | DJIA |
|---|---|---|---|
| 6/ 1/1940 | 30.60 | 25.21 | 145.19 |
| 13/ 1/1940 | 30.25 | 25.25 | 145.64 |
| 20/ 1/1940 | 30.75 | 24.88 | 146.51 |
| 27/ 1/1940 | 30.65 | 24.74 | 145.59 |
| 3/ 2/1940 | 31.27 | 24.95 | 148.84 |
| 10/ 2/1940 | 30.94 | 24.99 | 148.72 |
| 17/ 2/1940 | 30.63 | 24.90 | 146.72 |
| 24/ 2/1940 | 30.41 | 24.11 | 146.33 |
| 2/ 3/1940 | 30.84 | 24.24 | 148.14 |
| 9/ 3/1940 | 29.78 | 24.02 | 145.76 |
| 16/ 3/1940 | 30.10 | 24.24 | 146.73 |
| 23/ 3/1940 | 30.86 | 25.22 | 147.95 |
| 30/ 3/1940 | 32.08 | 25.92 | 151.10 |
| 6/ 4/1940 | 31.06 | 25.24 | 149.66 |
| 13/ 4/1940 | 30.60 | 24.63 | 147.67 |
| 20/ 4/1940 | 30.71 | 24.81 | 148.12 |
| 27/ 4/1940 | 30.89 | 25.00 | 147.55 |
| 4/ 5/1940 | 29.81 | 23.81 | 144.85 |
| 11/ 5/1940 | 23.65 | 19.20 | 122.43 |
| 18/ 5/1940 | 22.76 | 18.52 | 114.75 |
| 25/ 5/1940 | 23.05 | 18.87 | 115.67 |
| 1/ 6/1940 | 23.79 | 18.57 | 115.36 |
| 8/ 6/1940 | 24.97 | 20.18 | 123.36 |
| 15/ 6/1940 | 25.56 | 21.50 | 122.83 |
| 22/ 6/1940 | 26.18 | 22.67 | 121.87 |
| 29/ 6/1940 | 26.06 | 22.57 | 121.59 |
| 6/ 7/1940 | 26.05 | 22.53 | 121.48 |
| 13/ 7/1940 | 26.26 | 22.30 | 121.87 |
| 20/ 7/1940 | 26.38 | 22.21 | 122.45 |
| 27/ 7/1940 | 26.96 | 22.82 | 126.36 |
| 3/ 8/1940 | 26.86 | 22.60 | 126.99 |
| 10/ 8/1940 | 26.27 | 21.63 | 121.98 |
| 17/ 8/1940 | 26.86 | 22.00 | 125.48 |
| 24/ 8/1940 | 27.90 | 22.45 | 129.42 |
| 31/ 8/1940 | 29.21 | 23.05 | 132.78 |
| 7/ 9/1940 | 27.84 | 21.72 | 128.38 |
| 14/ 9/1940 | 28.45 | 21.95 | 132.45 |
| 21/ 9/1940 | 28.40 | 21.65 | 132.32 |
| 28/ 9/1940 | 29.20 | 21.99 | 133.90 |
| 5/10/1940 | 28.74 | 21.56 | 131.04 |
| 12/10/1940 | 28.89 | 22.20 | 132.18 |
| 19/10/1940 | 28.65 | 22.52 | 132.26 |
| 26/10/1940 | 29.22 | 23.44 | 134.85 |
| 2/11/1940 | 29.96 | 22.05 | 138.12 |
| 9/11/1940 | 29.51 | 20.96 | 134.73 |
| 16/11/1940 | 29.38 | 20.42 | 131.47 |
| 23/11/1940 | 27.97 | 19.97 | 131.00 |
| 30/11/1940 | 27.80 | 20.14 | 131.29 |
| 7/12/1940 | 27.85 | 20.14 | 132.31 |
| 14/12/1940 | 27.17 | 19.59 | 128.89 |
| 21/12/1940 | 27.85 | 19.80 | 130.11 |
| 28/12/1940 | 28.40 | 19.90 | 132.40 |

52 626

**TABLE A4-2 Dow Jones Transportation, Utility, and Industrial Weekly Averages** (cont.)

| DATE | DJTA | DJUA | DJIA |
|---|---|---|---|
| 4/ 1/1941 | 29.65 | 20.53 | 133.49 |
| 11/ 1/1941 | 29.03 | 20.27 | 129.75 |
| 18/ 1/1941 | 29.65 | 20.48 | 128.96 |
| 25/ 1/1941 | 28.10 | 19.56 | 123.28 |
| 1/ 2/1941 | 28.26 | 20.00 | 124.71 |
| 8/ 2/1941 | 26.65 | 18.95 | 118.55 |
| 15/ 2/1941 | 27.15 | 19.00 | 120.24 |
| 22/ 2/1941 | 27.58 | 19.53 | 121.86 |
| 1/ 3/1941 | 27.83 | 19.43 | 121.47 |
| 8/ 3/1941 | 28.09 | 19.74 | 123.40 |
| 15/ 3/1941 | 27.83 | 19.70 | 121.92 |
| 22/ 3/1941 | 28.55 | 19.49 | 122.37 |
| 29/ 3/1941 | 29.57 | 19.58 | 124.32 |
| 5/ 4/1941 | 27.89 | 18.69 | 118.60 |
| 12/ 4/1941 | 27.72 | 18.17 | 116.15 |
| 19/ 4/1941 | 28.42 | 18.12 | 116.43 |
| 26/ 4/1941 | 28.68 | 17.85 | 115.55 |
| 3/ 5/1941 | 29.31 | 17.41 | 117.54 |
| 10/ 5/1941 | 27.82 | 17.22 | 116.11 |
| 17/ 5/1941 | 27.74 | 16.99 | 116.64 |
| 24/ 5/1941 | 27.43 | 16.90 | 115.76 |
| 31/ 5/1941 | 27.69 | 17.32 | 118.89 |
| 7/ 6/1941 | 28.31 | 17.50 | 122.04 |
| 14/ 6/1941 | 27.99 | 17.75 | 122.51 |
| 21/ 6/1941 | 28.49 | 17.86 | 123.40 |
| 28/ 6/1941 | 28.56 | 18.10 | 124.18 |
| 5/ 7/1941 | 29.23 | 18.61 | 127.80 |
| 12/ 7/1941 | 29.51 | 18.51 | 127.98 |
| 19/ 7/1941 | 30.07 | 18.59 | 128.70 |
| 26/ 7/1941 | 30.76 | 18.58 | 128.21 |
| 2/ 8/1941 | 29.85 | 18.58 | 126.40 |
| 9/ 8/1941 | 29.85 | 18.31 | 125.05 |
| 16/ 8/1941 | 30.28 | 18.40 | 125.91 |
| 23/ 8/1941 | 30.19 | 18.73 | 127.70 |
| 30/ 8/1941 | 29.70 | 18.70 | 127.26 |
| 6/ 9/1941 | 29.30 | 18.73 | 127.28 |
| 13/ 9/1941 | 29.02 | 18.70 | 127.54 |
| 20/ 9/1941 | 28.97 | 18.32 | 126.03 |
| 27/ 9/1941 | 29.16 | 18.45 | 126.10 |
| 4/10/1941 | 28.61 | 17.92 | 122.63 |
| 11/10/1941 | 28.19 | 17.50 | 120.10 |
| 18/10/1941 | 28.57 | 17.43 | 120.73 |
| 25/10/1941 | 28.32 | 16.58 | 118.11 |
| 1/11/1941 | 28.27 | 16.21 | 118.26 |
| 8/11/1941 | 27.51 | 15.76 | 116.72 |
| 15/11/1941 | 28.29 | 15.87 | 117.04 |
| 22/11/1941 | 26.96 | 15.63 | 114.23 |
| 29/11/1941 | 27.16 | 16.05 | 116.60 |
| 6/12/1941 | 24.71 | 14.06 | 110.73 |
| 13/12/1941 | 24.66 | 13.52 | 107.81 |
| 20/12/1941 | 24.61 | 13.57 | 107.54 |
| 27/12/1941 | 27.15 | 14.70 | 113.75 |

52 678

| DATE | DJTA | DJUA | DJIA |
| --- | --- | --- | --- |
| 3/ 1/1942 | 27.66 | 14.50 | 110.54 |
| 10/ 1/1942 | 28.01 | 14.58 | 110.68 |
| 17/ 1/1942 | 28.71 | 14.05 | 109.42 |
| 24/ 1/1942 | 28.24 | 14.02 | 109.11 |
| 31/ 1/1942 | 28.73 | 14.10 | 109.56 |
| 7/ 2/1942 | 27.69 | 13.85 | 107.30 |
| 14/ 2/1942 | 27.69 | 13.58 | 105.38 |
| 21/ 2/1942 | 27.52 | 13.60 | 106.79 |
| 28/ 2/1942 | 26.35 | 12.35 | 102.31 |
| 7/ 3/1942 | 26.00 | 11.98 | 99.64 |
| 14/ 3/1942 | 26.05 | 11.90 | 100.82 |
| 21/ 3/1942 | 25.04 | 11.53 | 100.00 |
| 28/ 3/1942 | 25.11 | 11.52 | 101.11 |
| 4/ 4/1942 | 25.11 | 11.20 | 99.45 |
| 11/ 4/1942 | 24.11 | 10.87 | 96.92 |
| 18/ 4/1942 | 23.76 | 10.66 | 94.31 |
| 25/ 4/1942 | 24.67 | 11.85 | 96.44 |
| 2/ 5/1942 | 24.52 | 11.72 | 98.70 |
| 9/ 5/1942 | 24.19 | 11.80 | 98.63 |
| 16/ 5/1942 | 24.01 | 11.75 | 99.25 |
| 23/ 5/1942 | 23.88 | 11.50 | 100.88 |
| 30/ 5/1942 | 23.48 | 12.25 | 104.52 |
| 6/ 6/1942 | 23.45 | 12.15 | 104.08 |
| 13/ 6/1942 | 23.57 | 11.99 | 104.42 |
| 20/ 6/1942 | 23.89 | 11.71 | 102.67 |
| 27/ 6/1942 | 25.02 | 11.77 | 104.49 |
| 4/ 7/1942 | 25.70 | 12.09 | 108.70 |
| 11/ 7/1942 | 25.60 | 11.81 | 107.69 |
| 18/ 7/1942 | 26.19 | 11.47 | 106.53 |
| 25/ 7/1942 | 25.92 | 11.43 | 105.90 |
| 1/ 8/1942 | 25.48 | 11.38 | 104.90 |
| 8/ 8/1942 | 25.88 | 11.45 | 106.38 |
| 15/ 8/1942 | 27.00 | 11.75 | 107.30 |
| 22/ 8/1942 | 26.23 | 11.55 | 106.41 |
| 29/ 8/1942 | 26.51 | 11.57 | 106.68 |
| 5/ 9/1942 | 26.44 | 11.37 | 106.20 |
| 12/ 9/1942 | 26.81 | 11.77 | 107.22 |
| 19/ 9/1942 | 27.28 | 12.12 | 109.32 |
| 26/ 9/1942 | 28.72 | 12.38 | 111.34 |
| 3/10/1942 | 29.02 | 13.40 | 114.93 |
| 10/10/1942 | 28.59 | 13.37 | 113.40 |
| 17/10/1942 | 28.72 | 13.62 | 115.01 |
| 24/10/1942 | 28.85 | 14.16 | 114.07 |
| 31/10/1942 | 29.16 | 14.39 | 116.92 |
| 7/11/1942 | 28.27 | 14.44 | 116.24 |
| 14/11/1942 | 27.95 | 13.96 | 115.38 |
| 21/11/1942 | 27.11 | 13.97 | 114.95 |
| 28/11/1942 | 26.82 | 13.75 | 115.24 |
| 5/12/1942 | 26.16 | 13.78 | 115.82 |
| 12/12/1942 | 27.36 | 14.34 | 118.75 |
| 19/12/1942 | 27.21 | 14.11 | 119.71 |
| 26/12/1942 | 27.59 | 14.69 | 119.93 |

52 730

**TABLE A4-2 Dow Jones Transportation, Utility, and Industrial Weekly Averages** (cont.)

| DATE | DJTA | DJUA | DJIA |
|---|---|---|---|
| 2/ 1/1943 | 28.06 | 15.50 | 119.47 |
| 9/ 1/1943 | 28.96 | 15.58 | 121.60 |
| 16/ 1/1943 | 29.08 | 15.79 | 122.38 |
| 23/ 1/1943 | 29.21 | 16.60 | 125.58 |
| 30/ 1/1943 | 29.08 | 16.40 | 125.81 |
| 6/ 2/1943 | 29.21 | 16.95 | 127.83 |
| 13/ 2/1943 | 29.92 | 17.02 | 127.80 |
| 20/ 2/1943 | 32.06 | 17.48 | 130.11 |
| 27/ 2/1943 | 32.37 | 17.47 | 130.74 |
| 6/ 3/1943 | 32.69 | 17.90 | 130.73 |
| 13/ 3/1943 | 31.71 | 17.30 | 129.13 |
| 20/ 3/1943 | 33.00 | 17.91 | 134.56 |
| 27/ 3/1943 | 34.84 | 18.73 | 135.60 |
| 3/ 4/1943 | 33.70 | 17.90 | 131.63 |
| 10/ 4/1943 | 34.62 | 19.10 | 133.59 |
| 17/ 4/1943 | 35.24 | 19.75 | 134.34 |
| 24/ 4/1943 | 35.84 | 19.70 | 136.20 |
| 1/ 5/1943 | 36.47 | 20.11 | 138.36 |
| 8/ 5/1943 | 35.87 | 19.74 | 137.31 |
| 15/ 5/1943 | 36.45 | 20.25 | 138.78 |
| 22/ 5/1943 | 37.31 | 20.42 | 142.06 |
| 29/ 5/1943 | 36.69 | 20.33 | 143.08 |
| 5/ 6/1943 | 35.56 | 20.22 | 141.32 |
| 12/ 6/1943 | 35.21 | 20.26 | 139.73 |
| 19/ 6/1943 | 35.96 | 20.67 | 142.88 |
| 26/ 6/1943 | 36.50 | 21.54 | 143.70 |
| 3/ 7/1943 | 36.95 | 21.95 | 144.23 |
| 10/ 7/1943 | 37.85 | 22.20 | 144.72 |
| 17/ 7/1943 | 38.30 | 22.17 | 143.99 |
| 24/ 7/1943 | 34.51 | 20.69 | 135.95 |
| 31/ 7/1943 | 34.13 | 20.51 | 135.38 |
| 7/ 8/1943 | 34.81 | 20.92 | 137.23 |
| 14/ 8/1943 | 34.01 | 20.65 | 136.16 |
| 21/ 8/1943 | 34.17 | 20.89 | 135.79 |
| 28/ 8/1943 | 34.37 | 21.35 | 137.33 |
| 4/ 9/1943 | 34.22 | 21.54 | 138.04 |
| 11/ 9/1943 | 35.40 | 21.80 | 140.94 |
| 18/ 9/1943 | 35.17 | 21.74 | 140.18 |
| 25/ 9/1943 | 35.12 | 21.81 | 140.27 |
| 2/10/1943 | 34.84 | 21.38 | 137.10 |
| 9/10/1943 | 35.05 | 21.53 | 138.40 |
| 16/10/1943 | 35.01 | 22.03 | 138.29 |
| 23/10/1943 | 35.24 | 21.95 | 138.27 |
| 30/10/1943 | 33.55 | 21.12 | 135.24 |
| 6/11/1943 | 32.53 | 21.01 | 131.76 |
| 13/11/1943 | 33.32 | 21.26 | 132.94 |
| 20/11/1943 | 31.88 | 20.70 | 131.25 |
| 27/11/1943 | 32.17 | 21.40 | 131.87 |
| 4/12/1943 | 33.29 | 21.85 | 135.28 |
| 11/12/1943 | 33.30 | 21.97 | 135.89 |
| 18/12/1943 | 33.36 | 21.93 | 136.24 |
| 25/12/1943 | 33.56 | 21.87 | 135.89 |

52 782

| DATE | DJTA | DJUA | DJIA |
|---|---|---|---|
| 1/ 1/1944 | 34.67 | 22.32 | 138.09 |
| 8/ 1/1944 | 35.77 | 22.37 | 138.40 |
| 15/ 1/1944 | 36.57 | 22.36 | 138.24 |
| 22/ 1/1944 | 36.03 | 22.55 | 137.15 |
| 29/ 1/1944 | 36.10 | 22.32 | 135.12 |
| 5/ 2/1944 | 37.46 | 22.63 | 135.41 |
| 12/ 2/1944 | 38.22 | 22.89 | 135.91 |
| 19/ 2/1944 | 38.83 | 23.46 | 136.58 |
| 26/ 2/1944 | 38.12 | 23.59 | 136.79 |
| 4/ 3/1944 | 38.99 | 23.80 | 140.44 |
| 11/ 3/1944 | 40.21 | 23.93 | 140.30 |
| 18/ 3/1944 | 40.13 | 23.46 | 139.19 |
| 25/ 3/1944 | 39.61 | 23.08 | 138.84 |
| 1/ 4/1944 | 39.56 | 23.05 | 139.10 |
| 8/ 4/1944 | 39.60 | 23.02 | 138.06 |
| 15/ 4/1944 | 38.48 | 22.52 | 136.19 |
| 22/ 4/1944 | 38.81 | 22.45 | 136.23 |
| 29/ 4/1944 | 39.09 | 22.66 | 138.87 |
| 6/ 5/1944 | 38.62 | 22.50 | 138.60 |
| 13/ 5/1944 | 39.54 | 22.92 | 139.37 |
| 20/ 5/1944 | 40.00 | 23.00 | 141.24 |
| 27/ 5/1944 | 40.19 | 23.10 | 142.34 |
| 3/ 6/1944 | 39.31 | 22.95 | 142.53 |
| 10/ 6/1944 | 41.23 | 23.52 | 147.28 |
| 17/ 6/1944 | 41.56 | 23.86 | 147.48 |
| 24/ 6/1944 | 41.49 | 24.05 | 148.46 |
| 1/ 7/1944 | 42.48 | 24.43 | 150.03 |
| 8/ 7/1944 | 42.46 | 24.21 | 150.49 |
| 15/ 7/1944 | 40.93 | 23.32 | 145.58 |
| 22/ 7/1944 | 41.32 | 23.73 | 146.14 |
| 29/ 7/1944 | 40.86 | 23.98 | 145.07 |
| 5/ 8/1944 | 41.48 | 24.78 | 146.56 |
| 12/ 8/1944 | 41.93 | 24.96 | 148.96 |
| 19/ 8/1944 | 40.65 | 25.20 | 147.02 |
| 26/ 8/1944 | 40.87 | 25.25 | 147.16 |
| 2/ 9/1944 | 39.24 | 24.38 | 143.31 |
| 9/ 9/1944 | 39.19 | 24.53 | 144.36 |
| 16/ 9/1944 | 39.90 | 24.85 | 145.78 |
| 23/ 9/1944 | 40.93 | 24.88 | 146.73 |
| 30/ 9/1944 | 41.93 | 25.68 | 148.92 |
| 7/10/1944 | 41.57 | 25.91 | 148.59 |
| 14/10/1944 | 42.00 | 25.94 | 148.35 |
| 21/10/1944 | 41.45 | 25.45 | 146.50 |
| 28/10/1944 | 42.04 | 25.85 | 147.37 |
| 4/11/1944 | 42.37 | 25.61 | 148.08 |
| 11/11/1944 | 41.64 | 25.10 | 146.02 |
| 18/11/1944 | 42.51 | 25.35 | 146.63 |
| 25/11/1944 | 43.37 | 25.40 | 147.50 |
| 2/12/1944 | 45.62 | 25.59 | 151.31 |
| 9/12/1944 | 46.98 | 26.25 | 152.53 |
| 16/12/1944 | 48.16 | 25.98 | 150.63 |
| 23/12/1944 | 48.40 | 26.37 | 152.32 |
| 30/12/1944 | 49.76 | 26.25 | 153.58 |

53 835

**TABLE A4-2 Dow Jones Transportation, Utility, and Industrial Weekly Averages** (cont.)

| DATE | DJTA | DJUA | DJIA |
|---|---|---|---|
| 6/ 1/1945 | 50.12 | 26.69 | 155.58 |
| 13/ 1/1945 | 47.70 | 26.38 | 152.71 |
| 20/ 1/1945 | 48.17 | 26.65 | 154.13 |
| 27/ 1/1945 | 48.72 | 27.49 | 154.76 |
| 3/ 2/1945 | 49.38 | 27.49 | 154.85 |
| 10/ 2/1945 | 51.31 | 28.20 | 158.23 |
| 17/ 2/1945 | 51.25 | 28.01 | 158.69 |
| 24/ 2/1945 | 52.10 | 28.51 | 159.71 |
| 3/ 3/1945 | 50.31 | 27.75 | 157.21 |
| 10/ 3/1945 | 52.90 | 28.12 | 158.75 |
| 17/ 3/1945 | 51.00 | 27.52 | 154.36 |
| 24/ 3/1945 | 50.71 | 27.64 | 154.41 |
| 31/ 3/1945 | 51.35 | 27.92 | 156.33 |
| 7/ 4/1945 | 53.19 | 29.25 | 159.75 |
| 14/ 4/1945 | 55.19 | 29.84 | 163.20 |
| 21/ 4/1945 | 57.19 | 30.41 | 164.71 |
| 28/ 4/1945 | 56.76 | 30.92 | 166.71 |
| 5/ 5/1945 | 55.84 | 30.73 | 163.96 |
| 12/ 5/1945 | 56.24 | 31.17 | 166.44 |
| 19/ 5/1945 | 57.49 | 30.96 | 166.40 |
| 26/ 5/1945 | 58.90 | 31.82 | 167.96 |
| 2/ 6/1945 | 59.48 | 31.85 | 166.85 |
| 9/ 6/1945 | 61.14 | 32.52 | 167.54 |
| 16/ 6/1945 | 62.29 | 33.08 | 168.24 |
| 23/ 6/1945 | 60.62 | 33.13 | 165.29 |
| 30/ 6/1945 | 59.71 | 33.15 | 164.67 |
| 7/ 7/1945 | 60.40 | 33.43 | 166.67 |
| 14/ 7/1945 | 57.63 | 32.92 | 162.50 |
| 21/ 7/1945 | 56.35 | 32.30 | 160.92 |
| 28/ 7/1945 | 57.41 | 32.42 | 163.06 |
| 4/ 8/1945 | 56.24 | 32.51 | 165.14 |
| 11/ 8/1945 | 53.05 | 32.34 | 164.38 |
| 18/ 8/1945 | 54.61 | 32.36 | 169.89 |
| 25/ 8/1945 | 55.28 | 33.01 | 174.29 |
| 1/ 9/1945 | 55.22 | 33.65 | 176.99 |
| 8/ 9/1945 | 56.02 | 33.50 | 175.65 |
| 15/ 9/1945 | 58.27 | 34.25 | 179.49 |
| 22/ 9/1945 | 59.06 | 34.90 | 181.71 |
| 29/ 9/1945 | 59.31 | 35.01 | 184.77 |
| 6/10/1945 | 60.17 | 35.24 | 185.72 |
| 13/10/1945 | 59.79 | 35.81 | 185.60 |
| 20/10/1945 | 59.76 | 36.15 | 185.39 |
| 27/10/1945 | 62.26 | 37.56 | 188.58 |
| 3/11/1945 | 62.42 | 38.43 | 191.37 |
| 10/11/1945 | 63.97 | 38.36 | 192.27 |
| 17/11/1945 | 62.65 | 37.68 | 186.41 |
| 24/11/1945 | 64.23 | 38.73 | 192.40 |
| 1/12/1945 | 64.39 | 39.02 | 195.18 |
| 8/12/1945 | 64.49 | 38.14 | 193.76 |
| 15/12/1945 | 63.19 | 37.96 | 190.67 |
| 22/12/1945 | 62.90 | 38.22 | 192.84 |
| 29/12/1945 | 62.81 | 38.35 | 191.47 |

52 887

| DATE | DJTA | DJUA | DJIA |
|---|---|---|---|
| 5/ 1/1946 | 64.88 | 39.83 | 199.44 |
| 12/ 1/1946 | 66.28 | 40.07 | 200.21 |
| 19/ 1/1946 | 66.41 | 40.99 | 199.85 |
| 26/ 1/1946 | 68.18 | 41.71 | 206.97 |
| 2/ 2/1946 | 66.20 | 39.91 | 202.30 |
| 9/ 2/1946 | 66.95 | 40.90 | 204.41 |
| 16/ 2/1946 | 63.34 | 39.26 | 195.62 |
| 23/ 2/1946 | 61.86 | 38.80 | 188.73 |
| 2/ 3/1946 | 62.57 | 40.28 | 194.45 |
| 9/ 3/1946 | 62.80 | 40.40 | 193.94 |
| 16/ 3/1946 | 63.49 | 41.17 | 198.09 |
| 23/ 3/1946 | 64.26 | 41.81 | 199.75 |
| 30/ 3/1946 | 64.32 | 43.00 | 204.98 |
| 6/ 4/1946 | 63.43 | 42.54 | 206.02 |
| 13/ 4/1946 | 65.37 | 43.67 | 208.06 |
| 20/ 4/1946 | 63.94 | 43.39 | 206.13 |
| 27/ 4/1946 | 63.15 | 42.75 | 202.52 |
| 4/ 5/1946 | 64.37 | 43.24 | 208.06 |
| 11/ 5/1946 | 63.52 | 42.69 | 205.80 |
| 18/ 5/1946 | 66.39 | 43.32 | 207.69 |
| 25/ 5/1946 | 66.14 | 43.07 | 209.96 |
| 1/ 6/1946 | 68.02 | 42.83 | 210.36 |
| 8/ 6/1946 | 66.50 | 41.83 | 203.09 |
| 15/ 6/1946 | 65.81 | 42.10 | 205.62 |
| 22/ 6/1946 | 65.24 | 41.76 | 206.72 |
| 29/ 6/1946 | 63.87 | 41.33 | 204.20 |
| 6/ 7/1946 | 62.98 | 40.72 | 201.13 |
| 13/ 7/1946 | 61.17 | 40.24 | 197.63 |
| 20/ 7/1946 | 63.63 | 41.50 | 202.82 |
| 27/ 7/1946 | 62.68 | 41.64 | 203.57 |
| 3/ 8/1946 | 62.33 | 41.25 | 200.69 |
| 10/ 8/1946 | 61.35 | 40.75 | 197.75 |
| 17/ 8/1946 | 57.29 | 39.04 | 189.19 |
| 24/ 8/1946 | 53.17 | 36.60 | 179.96 |
| 31/ 8/1946 | 51.03 | 35.66 | 173.39 |
| 7/ 9/1946 | 46.95 | 34.07 | 169.06 |
| 14/ 9/1946 | 48.42 | 34.78 | 174.09 |
| 21/ 9/1946 | 46.79 | 34.38 | 169.00 |
| 28/ 9/1946 | 46.55 | 34.19 | 167.97 |
| 5/10/1946 | 47.50 | 34.85 | 171.34 |
| 12/10/1946 | 47.80 | 34.97 | 168.44 |
| 19/10/1946 | 51.22 | 36.10 | 172.53 |
| 26/10/1946 | 50.73 | 35.95 | 171.80 |
| 2/11/1946 | 49.54 | 35.26 | 169.03 |
| 9/11/1946 | 48.19 | 34.55 | 165.10 |
| 16/11/1946 | 49.58 | 35.35 | 169.80 |
| 23/11/1946 | 50.85 | 36.14 | 171.01 |
| 30/11/1946 | 51.99 | 37.30 | 174.73 |
| 7/12/1946 | 52.40 | 37.52 | 178.32 |
| 14/12/1946 | 51.17 | 36.95 | 175.77 |
| 21/12/1946 | 50.87 | 36.95 | 176.92 |
| 28/12/1946 | 48.59 | 36.15 | 175.25 |

52 939

**TABLE A4-2 Dow Jones Transportation, Utility, and Industrial Weekly Averages** (cont.)

| DATE | DJTA | DJUA | DJIA |
|---|---|---|---|
| 4/ 1/1947 | 49.07 | 36.75 | 175.92 |
| 11/ 1/1947 | 49.22 | 36.42 | 175.35 |
| 18/ 1/1947 | 51.67 | 37.06 | 180.88 |
| 25/ 1/1947 | 53.42 | 37.55 | 184.49 |
| 1/ 2/1947 | 51.86 | 37.21 | 181.36 |
| 8/ 2/1947 | 51.64 | 37.42 | 182.26 |
| 15/ 2/1947 | 50.54 | 36.84 | 179.29 |
| 22/ 2/1947 | 49.32 | 36.10 | 175.84 |
| 1/ 3/1947 | 48.01 | 35.80 | 172.37 |
| 8/ 3/1947 | 49.00 | 35.74 | 177.27 |
| 15/ 3/1947 | 49.15 | 35.99 | 178.36 |
| 22/ 3/1947 | 48.39 | 35.92 | 176.71 |
| 29/ 3/1947 | 46.14 | 34.38 | 171.76 |
| 5/ 4/1947 | 44.32 | 33.96 | 168.44 |
| 12/ 4/1947 | 44.68 | 33.90 | 169.13 |
| 19/ 4/1947 | 45.69 | 34.22 | 174.00 |
| 26/ 4/1947 | 44.77 | 33.86 | 171.67 |
| 3/ 5/1947 | 41.20 | 32.68 | 163.21 |
| 10/ 5/1947 | 42.85 | 33.10 | 166.97 |
| 17/ 5/1947 | 44.24 | 33.37 | 169.25 |
| 24/ 5/1947 | 43.72 | 33.14 | 170.28 |
| 31/ 5/1947 | 45.41 | 34.18 | 175.49 |
| 7/ 6/1947 | 45.82 | 34.55 | 176.44 |
| 14/ 6/1947 | 46.02 | 34.63 | 176.56 |
| 21/ 6/1947 | 48.06 | 35.47 | 181.73 |
| 28/ 6/1947 | 49.21 | 35.95 | 184.77 |
| 5/ 7/1947 | 49.33 | 35.51 | 184.60 |
| 12/ 7/1947 | 51.52 | 35.95 | 186.38 |
| 19/ 7/1947 | 49.83 | 35.78 | 183.81 |
| 26/ 7/1947 | 48.21 | 35.55 | 180.13 |
| 2/ 8/1947 | 49.41 | 35.66 | 181.04 |
| 9/ 8/1947 | 48.94 | 35.78 | 179.74 |
| 16/ 8/1947 | 48.77 | 35.58 | 178.85 |
| 23/ 8/1947 | 47.88 | 35.35 | 177.13 |
| 30/ 8/1947 | 47.51 | 35.24 | 175.92 |
| 6/ 9/1947 | 48.32 | 35.24 | 178.12 |
| 13/ 9/1947 | 47.43 | 35.02 | 174.86 |
| 20/ 9/1947 | 49.39 | 35.51 | 179.44 |
| 27/ 9/1947 | 48.67 | 35.48 | 180.49 |
| 4/10/1947 | 50.28 | 35.73 | 184.25 |
| 11/10/1947 | 49.66 | 35.32 | 182.73 |
| 18/10/1947 | 48.66 | 34.85 | 182.48 |
| 25/10/1947 | 47.59 | 34.73 | 181.49 |
| 1/11/1947 | 46.86 | 34.14 | 180.26 |
| 8/11/1947 | 48.66 | 33.95 | 182.33 |
| 15/11/1947 | 47.12 | 32.94 | 179.40 |
| 22/11/1947 | 46.28 | 32.65 | 175.74 |
| 29/11/1947 | 49.02 | 33.06 | 179.34 |
| 6/12/1947 | 51.70 | 33.25 | 181.06 |
| 13/12/1947 | 51.47 | 33.17 | 179.23 |
| 20/12/1947 | 53.85 | 33.52 | 181.04 |
| 27/12/1947 | 52.87 | 33.86 | 180.20 |

52 991

| DATE | DJTA | DJUA | DJIA |
|---|---|---|---|
| 3/ 1/1948 | 51.02 | 33.29 | 177.24 |
| 10/ 1/1948 | 49.96 | 32.41 | 171.67 |
| 17/ 1/1948 | 51.66 | 32.69 | 175.05 |
| 24/ 1/1948 | 50.05 | 32.21 | 169.79 |
| 31/ 1/1948 | 48.36 | 31.71 | 166.18 |
| 7/ 2/1948 | 48.52 | 31.75 | 167.60 |
| 14/ 2/1948 | 49.27 | 31.70 | 167.30 |
| 21/ 2/1948 | 49.68 | 31.95 | 168.94 |
| 28/ 2/1948 | 50.23 | 31.99 | 167.62 |
| 6/ 3/1948 | 51.78 | 32.62 | 173.12 |
| 13/ 3/1948 | 52.51 | 32.80 | 173.95 |
| 20/ 3/1948 | 53.93 | 33.41 | 177.45 |
| 27/ 3/1948 | 55.27 | 33.71 | 179.48 |
| 3/ 4/1948 | 55.75 | 33.76 | 180.38 |
| 10/ 4/1948 | 58.57 | 34.08 | 183.20 |
| 17/ 4/1948 | 58.05 | 34.02 | 180.28 |
| 24/ 4/1948 | 58.73 | 34.68 | 182.50 |
| 1/ 5/1948 | 62.18 | 35.79 | 190.25 |
| 8/ 5/1948 | 61.65 | 35.77 | 190.00 |
| 15/ 5/1948 | 60.81 | 35.83 | 190.74 |
| 22/ 5/1948 | 59.75 | 35.15 | 190.18 |
| 29/ 5/1948 | 61.64 | 35.98 | 192.96 |
| 5/ 6/1948 | 61.27 | 35.85 | 191.65 |
| 12/ 6/1948 | 62.82 | 35.81 | 190.00 |
| 19/ 6/1948 | 63.94 | 35.64 | 190.06 |
| 26/ 6/1948 | 64.76 | 35.75 | 191.62 |
| 3/ 7/1948 | 62.06 | 35.21 | 185.90 |
| 10/ 7/1948 | 61.15 | 35.30 | 185.31 |
| 17/ 7/1948 | 60.15 | 34.70 | 181.33 |
| 24/ 7/1948 | 61.44 | 34.96 | 183.01 |
| 31/ 7/1948 | 58.79 | 34.15 | 179.63 |
| 7/ 8/1948 | 60.72 | 34.86 | 183.60 |
| 14/ 8/1948 | 61.44 | 34.82 | 183.21 |
| 21/ 8/1948 | 62.77 | 35.39 | 184.35 |
| 28/ 8/1948 | 60.28 | 34.80 | 180.61 |
| 4/ 9/1948 | 60.28 | 34.60 | 180.06 |
| 11/ 9/1948 | 59.48 | 34.71 | 179.28 |
| 18/ 9/1948 | 59.20 | 34.62 | 180.78 |
| 25/ 9/1948 | 59.98 | 34.83 | 182.09 |
| 2/10/1948 | 60.17 | 35.23 | 184.93 |
| 9/10/1948 | 62.24 | 35.75 | 190.19 |
| 16/10/1948 | 61.34 | 35.40 | 188.62 |
| 23/10/1948 | 56.59 | 33.78 | 178.94 |
| 30/10/1948 | 53.72 | 32.88 | 174.32 |
| 6/11/1948 | 55.15 | 33.15 | 177.42 |
| 13/11/1948 | 53.06 | 32.96 | 172.90 |
| 20/11/1948 | 53.76 | 32.95 | 176.22 |
| 27/11/1948 | 53.92 | 33.00 | 177.49 |
| 4/12/1948 | 52.89 | 33.20 | 175.69 |
| 11/12/1948 | 53.53 | 33.20 | 177.42 |
| 18/12/1948 | 52.86 | 33.55 | 177.30 |
| 25/12/1948 | 54.12 | 34.39 | 181.41 |

52 1043

**TABLE A4-2 Dow Jones Transportation, Utility, and Industrial Weekly Averages** (cont.)

| DATE | DJTA | DJUA | DJIA |
|---|---|---|---|
| 1/ 1/1949 | 52.75 | 34.28 | 179.15 |
| 8/ 1/1949 | 53.95 | 35.00 | 181.54 |
| 15/ 1/1949 | 52.69 | 34.70 | 179.35 |
| 22/ 1/1949 | 50.47 | 34.56 | 175.60 |
| 29/ 1/1949 | 48.70 | 34.11 | 171.93 |
| 5/ 2/1949 | 48.99 | 34.55 | 174.53 |
| 12/ 2/1949 | 46.95 | 34.42 | 171.62 |
| 19/ 2/1949 | 48.05 | 34.83 | 174.93 |
| 26/ 2/1949 | 49.28 | 35.19 | 176.96 |
| 5/ 3/1949 | 47.76 | 35.05 | 176.07 |
| 12/ 3/1949 | 47.88 | 35.30 | 175.82 |
| 19/ 3/1949 | 48.72 | 35.45 | 176.88 |
| 26/ 3/1949 | 49.56 | 35.74 | 176.75 |
| 2/ 4/1949 | 48.90 | 36.38 | 177.07 |
| 9/ 4/1949 | 47.45 | 35.55 | 173.76 |
| 16/ 4/1949 | 47.27 | 35.41 | 174.16 |
| 23/ 4/1949 | 47.43 | 35.89 | 175.39 |
| 30/ 4/1949 | 48.13 | 35.95 | 175.20 |
| 7/ 5/1949 | 47.10 | 35.72 | 173.49 |
| 14/ 5/1949 | 46.32 | 35.36 | 171.53 |
| 21/ 5/1949 | 43.76 | 34.87 | 167.24 |
| 28/ 5/1949 | 42.70 | 34.21 | 164.61 |
| 4/ 6/1949 | 42.22 | 33.82 | 163.78 |
| 11/ 6/1949 | 43.25 | 34.37 | 166.99 |
| 18/ 6/1949 | 43.23 | 34.66 | 168.08 |
| 25/ 6/1949 | 43.19 | 35.03 | 170.92 |
| 2/ 7/1949 | 44.32 | 35.39 | 173.48 |
| 9/ 7/1949 | 45.02 | 35.37 | 174.53 |
| 16/ 7/1949 | 44.77 | 35.72 | 175.92 |
| 23/ 7/1949 | 45.92 | 36.69 | 179.07 |
| 30/ 7/1949 | 47.05 | 36.65 | 179.29 |
| 6/ 8/1949 | 46.84 | 36.67 | 181.16 |
| 13/ 8/1949 | 45.55 | 36.63 | 179.24 |
| 20/ 8/1949 | 45.68 | 36.83 | 179.38 |
| 27/ 8/1949 | 45.59 | 37.55 | 180.24 |
| 3/ 9/1949 | 47.00 | 37.90 | 182.32 |
| 10/ 9/1949 | 47.78 | 38.19 | 181.30 |
| 17/ 9/1949 | 47.87 | 37.86 | 182.51 |
| 24/ 9/1949 | 47.63 | 37.75 | 181.98 |
| 1/10/1949 | 48.71 | 38.01 | 185.36 |
| 8/10/1949 | 49.33 | 38.36 | 186.36 |
| 15/10/1949 | 48.49 | 38.34 | 186.20 |
| 22/10/1949 | 48.36 | 38.58 | 190.36 |
| 29/10/1949 | 49.50 | 39.35 | 191.37 |
| 5/11/1949 | 48.23 | 39.15 | 190.46 |
| 12/11/1949 | 49.02 | 39.38 | 193.62 |
| 19/11/1949 | 48.09 | 39.46 | 193.23 |
| 26/11/1949 | 50.76 | 39.58 | 194.43 |
| 3/12/1949 | 50.15 | 40.44 | 194.68 |
| 10/12/1949 | 51.45 | 40.99 | 197.98 |
| 17/12/1949 | 51.53 | 41.02 | 198.88 |
| 24/12/1949 | 52.76 | 41.29 | 200.13 |
| 31/12/1949 | 54.52 | 41.70 | 201.94 |

53 1096

| DATE | DJTA | DJUA | DJIA |
|---|---|---|---|
| 7/ 1/1950 | 53.78 | 40.86 | 196.92 |
| 14/ 1/1950 | 55.18 | 41.86 | 200.97 |
| 21/ 1/1950 | 55.04 | 41.91 | 200.08 |
| 28/ 1/1950 | 55.53 | 42.57 | 205.03 |
| 4/ 2/1950 | 54.78 | 42.65 | 203.36 |
| 11/ 2/1950 | 56.00 | 42.78 | 203.97 |
| 18/ 2/1950 | 55.39 | 42.73 | 204.15 |
| 25/ 2/1950 | 56.60 | 43.01 | 204.71 |
| 4/ 3/1950 | 55.06 | 42.89 | 202.96 |
| 11/ 3/1950 | 55.70 | 43.69 | 208.09 |
| 18/ 3/1950 | 55.68 | 43.37 | 210.62 |
| 25/ 3/1950 | 55.43 | 42.69 | 206.37 |
| 1/ 4/1950 | 56.36 | 43.25 | 212.55 |
| 8/ 4/1950 | 55.16 | 42.96 | 214.48 |
| 15/ 4/1950 | 55.53 | 42.67 | 213.90 |
| 22/ 4/1950 | 56.07 | 42.78 | 214.33 |
| 29/ 4/1950 | 56.60 | 43.32 | 217.03 |
| 6/ 5/1950 | 55.44 | 43.50 | 217.78 |
| 13/ 5/1950 | 56.96 | 44.26 | 222.41 |
| 20/ 5/1950 | 56.29 | 43.67 | 221.71 |
| 27/ 5/1950 | 55.33 | 43.50 | 223.71 |
| 3/ 6/1950 | 56.65 | 43.84 | 226.86 |
| 10/ 6/1950 | 55.95 | 43.34 | 222.71 |
| 17/ 6/1950 | 55.85 | 43.95 | 224.35 |
| 24/ 6/1950 | 52.24 | 40.64 | 209.11 |
| 1/ 7/1950 | 52.29 | 40.73 | 208.59 |
| 8/ 7/1950 | 54.38 | 38.14 | 199.83 |
| 15/ 7/1950 | 59.46 | 38.46 | 207.65 |
| 22/ 7/1950 | 60.71 | 37.80 | 208.21 |
| 29/ 7/1950 | 61.78 | 38.14 | 212.66 |
| 5/ 8/1950 | 62.00 | 38.67 | 215.03 |
| 12/ 8/1950 | 63.39 | 39.62 | 219.23 |
| 19/ 8/1950 | 62.43 | 39.19 | 218.10 |
| 26/ 8/1950 | 63.38 | 38.67 | 218.42 |
| 2/ 9/1950 | 64.39 | 38.87 | 220.03 |
| 9/ 9/1950 | 66.74 | 39.21 | 225.85 |
| 16/ 9/1950 | 67.90 | 39.86 | 226.64 |
| 23/ 9/1950 | 67.64 | 40.46 | 226.36 |
| 30/ 9/1950 | 69.72 | 40.78 | 231.81 |
| 7/10/1950 | 69.36 | 40.52 | 227.63 |
| 14/10/1950 | 69.86 | 40.89 | 230.88 |
| 21/10/1950 | 67.82 | 40.38 | 228.56 |
| 28/10/1950 | 66.22 | 40.38 | 227.42 |
| 4/11/1950 | 67.30 | 40.73 | 229.29 |
| 11/11/1950 | 70.41 | 40.65 | 231.64 |
| 18/11/1950 | 70.80 | 40.77 | 235.06 |
| 25/11/1950 | 69.57 | 39.87 | 227.55 |
| 2/12/1950 | 73.57 | 39.12 | 227.30 |
| 9/12/1950 | 75.61 | 38.89 | 228.34 |
| 16/12/1950 | 76.84 | 40.07 | 231.54 |
| 23/12/1950 | 77.64 | 40.98 | 235.41 |
| 30/12/1950 | 78.91 | 41.91 | 240.68 |

52 1148

**TABLE A4-2 Dow Jones Transportation, Utility, and Industrial Weekly Averages** (cont.)

| DATE | DJTA | DJUA | DJIA |
|---|---|---|---|
| 6/ 1/1951 | 82.08 | 42.08 | 243.61 |
| 13/ 1/1951 | 84.02 | 42.48 | 246.91 |
| 20/ 1/1951 | 83.91 | 42.22 | 247.36 |
| 27/ 1/1951 | 89.27 | 42.44 | 253.92 |
| 3/ 2/1951 | 89.85 | 42.57 | 254.80 |
| 10/ 2/1951 | 88.63 | 42.86 | 254.70 |
| 17/ 2/1951 | 86.58 | 43.91 | 252.93 |
| 24/ 2/1951 | 86.35 | 43.84 | 253.43 |
| 3/ 3/1951 | 84.78 | 43.37 | 252.02 |
| 10/ 3/1951 | 82.55 | 43.10 | 249.03 |
| 17/ 3/1951 | 80.26 | 42.91 | 248.14 |
| 24/ 3/1951 | 80.58 | 42.25 | 247.94 |
| 31/ 3/1951 | 82.82 | 42.49 | 250.28 |
| 7/ 4/1951 | 83.99 | 42.42 | 256.18 |
| 14/ 4/1951 | 82.65 | 42.32 | 255.02 |
| 21/ 4/1951 | 83.52 | 42.33 | 259.08 |
| 28/ 4/1951 | 84.88 | 42.47 | 261.76 |
| 5/ 5/1951 | 82.66 | 42.28 | 257.26 |
| 12/ 5/1951 | 79.54 | 42.22 | 250.63 |
| 19/ 5/1951 | 77.58 | 41.81 | 245.83 |
| 26/ 5/1951 | 79.36 | 42.31 | 249.33 |
| 2/ 6/1951 | 80.03 | 42.69 | 250.39 |
| 9/ 6/1951 | 79.86 | 42.65 | 254.03 |
| 16/ 6/1951 | 77.87 | 42.64 | 247.86 |
| 23/ 6/1951 | 72.39 | 42.08 | 242.64 |
| 30/ 6/1951 | 75.21 | 42.82 | 250.01 |
| 7/ 7/1951 | 76.48 | 43.44 | 254.32 |
| 14/ 7/1951 | 76.83 | 44.37 | 253.73 |
| 21/ 7/1951 | 81.69 | 44.82 | 259.23 |
| 28/ 7/1951 | 81.47 | 45.52 | 262.98 |
| 4/ 8/1951 | 81.00 | 45.24 | 261.92 |
| 11/ 8/1951 | 80.62 | 45.25 | 266.17 |
| 18/ 8/1951 | 79.05 | 44.40 | 266.30 |
| 25/ 8/1951 | 80.33 | 45.00 | 270.25 |
| 1/ 9/1951 | 82.82 | 45.33 | 273.89 |
| 8/ 9/1951 | 84.45 | 45.55 | 276.06 |
| 15/ 9/1951 | 85.34 | 45.35 | 272.11 |
| 22/ 9/1951 | 84.76 | 45.67 | 271.16 |
| 29/ 9/1951 | 87.06 | 46.24 | 275.53 |
| 6/10/1951 | 87.00 | 46.48 | 275.13 |
| 13/10/1951 | 82.80 | 46.01 | 267.42 |
| 20/10/1951 | 78.93 | 45.13 | 258.53 |
| 27/10/1951 | 79.05 | 46.00 | 259.57 |
| 3/11/1951 | 79.78 | 46.45 | 261.29 |
| 10/11/1951 | 81.25 | 46.42 | 260.82 |
| 17/11/1951 | 77.91 | 46.01 | 255.95 |
| 24/11/1951 | 81.75 | 46.08 | 262.29 |
| 1/12/1951 | 83.94 | 46.55 | 266.90 |
| 8/12/1951 | 82.11 | 46.86 | 265.48 |
| 15/12/1951 | 82.05 | 46.97 | 265.94 |
| 22/12/1951 | 81.89 | 47.16 | 268.52 |
| 29/12/1951 | 83.77 | 47.90 | 271.26 |

52 1200

| DATE | DJTA | DJUA | DJIA |
|---|---|---|---|
| 5/ 1/1952 | 84.12 | 48.94 | 270.73 |
| 12/ 1/1952 | 86.51 | 48.74 | 272.93 |
| 19/ 1/1952 | 86.10 | 49.03 | 273.69 |
| 26/ 1/1952 | 86.56 | 49.18 | 272.51 |
| 2/ 2/1952 | 86.09 | 49.17 | 269.83 |
| 9/ 2/1952 | 85.35 | 49.21 | 266.30 |
| 16/ 2/1952 | 84.76 | 48.43 | 261.40 |
| 23/ 2/1952 | 84.94 | 48.56 | 260.27 |
| 1/ 3/1952 | 89.71 | 49.93 | 264.14 |
| 8/ 3/1952 | 90.28 | 49.81 | 264.43 |
| 15/ 3/1952 | 89.99 | 50.44 | 265.69 |
| 22/ 3/1952 | 93.58 | 50.11 | 269.00 |
| 29/ 3/1952 | 92.38 | 49.71 | 265.44 |
| 5/ 4/1952 | 93.40 | 49.53 | 266.29 |
| 12/ 4/1952 | 91.25 | 48.85 | 260.14 |
| 19/ 4/1952 | 94.22 | 48.69 | 260.27 |
| 26/ 4/1952 | 93.50 | 48.55 | 260.55 |
| 3/ 5/1952 | 94.58 | 49.26 | 262.50 |
| 10/ 5/1952 | 93.59 | 49.11 | 259.88 |
| 17/ 5/1952 | 96.17 | 49.82 | 263.23 |
| 24/ 5/1952 | 97.29 | 49.94 | 262.94 |
| 31/ 5/1952 | 100.36 | 50.12 | 268.03 |
| 7/ 6/1952 | 100.43 | 49.80 | 268.56 |
| 14/ 6/1952 | 100.44 | 49.77 | 270.19 |
| 21/ 6/1952 | 102.39 | 49.69 | 272.44 |
| 28/ 6/1952 | 102.36 | 49.72 | 274.95 |
| 5/ 7/1952 | 101.55 | 49.54 | 274.22 |
| 12/ 7/1952 | 100.60 | 49.56 | 273.90 |
| 19/ 7/1952 | 102.29 | 50.20 | 277.71 |
| 26/ 7/1952 | 103.81 | 50.51 | 279.80 |
| 2/ 8/1952 | 104.78 | 51.12 | 279.84 |
| 9/ 8/1952 | 103.50 | 50.66 | 277.37 |
| 16/ 8/1952 | 101.94 | 50.75 | 274.43 |
| 23/ 8/1952 | 103.31 | 50.79 | 275.04 |
| 30/ 8/1952 | 102.48 | 50.99 | 276.50 |
| 6/ 9/1952 | 98.50 | 50.17 | 271.02 |
| 13/ 9/1952 | 99.15 | 49.87 | 270.55 |
| 20/ 9/1952 | 101.59 | 50.29 | 271.95 |
| 27/ 9/1952 | 99.98 | 50.36 | 270.55 |
| 4/10/1952 | 101.25 | 49.99 | 270.61 |
| 11/10/1952 | 100.43 | 49.09 | 267.30 |
| 18/10/1952 | 99.67 | 49.00 | 265.46 |
| 25/10/1952 | 100.77 | 49.94 | 269.23 |
| 1/11/1952 | 101.89 | 50.45 | 273.47 |
| 8/11/1952 | 101.51 | 50.86 | 273.27 |
| 15/11/1952 | 104.36 | 51.48 | 279.32 |
| 22/11/1952 | 108.16 | 51.60 | 283.66 |
| 29/11/1952 | 107.94 | 51.51 | 282.06 |
| 6/12/1952 | 109.37 | 52.11 | 285.20 |
| 13/12/1952 | 111.46 | 52.33 | 286.52 |
| 20/12/1952 | 111.28 | 52.33 | 288.23 |
| 27/12/1952 | 111.18 | 52.35 | 292.14 |

52 1252

**TABLE A4-2 Dow Jones Transportation, Utility, and Industrial Weekly Averages** (cont.)

| DATE | DJTA | DJUA | DJIA |
|---|---|---|---|
| 3/ 1/1953 | 109.47 | 51.95 | 287.52 |
| 10/ 1/1953 | 108.64 | 52.06 | 287.17 |
| 17/ 1/1953 | 109.11 | 51.98 | 286.89 |
| 24/ 1/1953 | 112.21 | 52.68 | 289.77 |
| 31/ 1/1953 | 107.85 | 52.45 | 282.85 |
| 7/ 2/1953 | 108.79 | 52.61 | 283.11 |
| 14/ 2/1953 | 109.22 | 52.34 | 281.89 |
| 21/ 2/1953 | 110.05 | 52.50 | 284.27 |
| 28/ 2/1953 | 108.93 | 52.57 | 284.82 |
| 7/ 3/1953 | 110.45 | 53.88 | 289.04 |
| 14/ 3/1953 | 111.40 | 53.65 | 289.69 |
| 21/ 3/1953 | 110.94 | 53.50 | 287.33 |
| 28/ 3/1953 | 106.11 | 52.42 | 280.03 |
| 4/ 4/1953 | 104.27 | 51.95 | 275.50 |
| 11/ 4/1953 | 103.47 | 51.73 | 274.41 |
| 18/ 4/1953 | 101.88 | 50.74 | 271.26 |
| 25/ 4/1953 | 103.37 | 51.22 | 275.66 |
| 2/ 5/1953 | 104.85 | 50.97 | 278.22 |
| 9/ 5/1953 | 105.22 | 50.97 | 277.90 |
| 16/ 5/1953 | 108.30 | 51.21 | 278.16 |
| 23/ 5/1953 | 105.42 | 50.83 | 272.28 |
| 30/ 5/1953 | 103.28 | 49.40 | 268.32 |
| 6/ 6/1953 | 101.98 | 48.48 | 265.78 |
| 13/ 6/1953 | 103.18 | 47.88 | 265.80 |
| 20/ 6/1953 | 105.33 | 48.52 | 269.05 |
| 27/ 6/1953 | 107.15 | 49.15 | 270.53 |
| 4/ 7/1953 | 106.81 | 49.39 | 271.06 |
| 11/ 7/1953 | 106.03 | 49.08 | 270.96 |
| 18/ 7/1953 | 105.22 | 48.85 | 269.76 |
| 25/ 7/1953 | 105.86 | 49.45 | 275.38 |
| 1/ 8/1953 | 106.63 | 50.19 | 275.54 |
| 8/ 8/1953 | 104.70 | 50.88 | 275.71 |
| 15/ 8/1953 | 102.78 | 50.79 | 271.93 |
| 22/ 8/1953 | 98.37 | 50.05 | 265.74 |
| 29/ 8/1953 | 96.93 | 49.71 | 264.34 |
| 5/ 9/1953 | 92.97 | 49.07 | 259.71 |
| 12/ 9/1953 | 92.01 | 48.42 | 258.78 |
| 19/ 9/1953 | 93.99 | 49.35 | 263.31 |
| 26/ 9/1953 | 93.91 | 49.78 | 266.70 |
| 3/10/1953 | 93.68 | 50.23 | 267.04 |
| 10/10/1953 | 97.05 | 50.77 | 272.80 |
| 17/10/1953 | 96.92 | 50.94 | 275.34 |
| 24/10/1953 | 97.26 | 51.14 | 275.81 |
| 31/10/1953 | 98.25 | 51.50 | 278.83 |
| 7/11/1953 | 97.37 | 51.57 | 277.53 |
| 14/11/1953 | 97.00 | 51.52 | 276.05 |
| 21/11/1953 | 97.68 | 51.94 | 280.23 |
| 28/11/1953 | 98.33 | 53.01 | 282.71 |
| 5/12/1953 | 96.22 | 52.86 | 279.91 |
| 12/12/1953 | 97.02 | 52.58 | 283.54 |
| 19/12/1953 | 95.22 | 52.18 | 280.92 |
| 26/12/1953 | 94.03 | 52.04 | 280.90 |

52 1304

| DATE | DJTA | DJUA | DJIA |
|---|---|---|---|
| 2/ 1/1954 | 95.14 | 52.55 | 281.51 |
| 9/ 1/1954 | 98.08 | 53.54 | 286.72 |
| 16/ 1/1954 | 100.32 | 53.88 | 289.65 |
| 23/ 1/1954 | 101.84 | 54.09 | 292.39 |
| 30/ 1/1954 | 103.35 | 54.49 | 293.97 |
| 6/ 2/1954 | 103.49 | 54.58 | 293.99 |
| 13/ 2/1954 | 102.38 | 54.39 | 291.07 |
| 20/ 2/1954 | 102.20 | 54.67 | 294.54 |
| 27/ 2/1954 | 102.99 | 54.92 | 299.45 |
| 6/ 3/1954 | 101.52 | 55.91 | 299.71 |
| 13/ 3/1954 | 102.28 | 56.35 | 301.44 |
| 20/ 3/1954 | 99.47 | 55.95 | 299.08 |
| 27/ 3/1954 | 101.49 | 56.26 | 306.67 |
| 3/ 4/1954 | 101.84 | 56.45 | 309.39 |
| 10/ 4/1954 | 103.09 | 56.78 | 313.77 |
| 17/ 4/1954 | 102.09 | 56.34 | 313.37 |
| 24/ 4/1954 | 104.31 | 56.49 | 319.33 |
| 1/ 5/1954 | 108.52 | 56.71 | 321.30 |
| 8/ 5/1954 | 108.60 | 57.45 | 322.50 |
| 15/ 5/1954 | 110.24 | 58.11 | 326.09 |
| 22/ 5/1954 | 110.60 | 58.07 | 327.49 |
| 29/ 5/1954 | 110.24 | 58.16 | 327.63 |
| 5/ 6/1954 | 108.61 | 57.59 | 322.09 |
| 12/ 6/1954 | 112.02 | 58.00 | 327.91 |
| 19/ 6/1954 | 113.76 | 58.16 | 332.53 |
| 26/ 6/1954 | 112.87 | 58.60 | 337.66 |
| 3/ 7/1954 | 115.22 | 59.19 | 341.25 |
| 10/ 7/1954 | 117.16 | 59.57 | 339.96 |
| 17/ 7/1954 | 117.95 | 59.54 | 343.48 |
| 24/ 7/1954 | 119.56 | 60.10 | 347.92 |
| 31/ 7/1954 | 117.25 | 60.59 | 343.06 |
| 7/ 8/1954 | 119.47 | 61.21 | 346.64 |
| 14/ 8/1954 | 120.25 | 61.58 | 350.38 |
| 21/ 8/1954 | 116.01 | 61.06 | 344.48 |
| 28/ 8/1954 | 114.91 | 60.57 | 343.10 |
| 4/ 9/1954 | 115.68 | 60.87 | 347.83 |
| 11/ 9/1954 | 116.33 | 61.29 | 355.32 |
| 18/ 9/1954 | 118.44 | 61.45 | 361.67 |
| 25/ 9/1954 | 117.00 | 61.16 | 359.88 |
| 2/10/1954 | 117.81 | 60.57 | 363.77 |
| 9/10/1954 | 118.36 | 58.81 | 353.20 |
| 16/10/1954 | 120.55 | 58.79 | 358.61 |
| 23/10/1954 | 117.69 | 57.81 | 352.14 |
| 30/10/1954 | 121.59 | 59.35 | 366.00 |
| 6/11/1954 | 127.65 | 60.60 | 377.10 |
| 13/11/1954 | 129.72 | 60.57 | 378.01 |
| 20/11/1954 | 132.27 | 60.75 | 387.79 |
| 27/11/1954 | 133.30 | 61.20 | 389.60 |
| 4/12/1954 | 135.87 | 61.32 | 390.08 |
| 11/12/1954 | 142.47 | 61.62 | 394.94 |
| 18/12/1954 | 144.76 | 61.51 | 397.15 |
| 25/12/1954 | 145.86 | 62.47 | 404.39 |

52 1356

**TABLE A4-2 Dow Jones Transportation, Utility, and Industrial Weekly Averages** (cont.)

| DATE | DJTA | DJUA | DJIA |
|---|---|---|---|
| 1/ 1/1955 | 144.34 | 62.10 | 395.60 |
| 8/ 1/1955 | 142.02 | 63.00 | 396.54 |
| 15/ 1/1955 | 140.91 | 62.86 | 395.90 |
| 22/ 1/1955 | 144.20 | 61.88 | 404.68 |
| 29/ 1/1955 | 143.37 | 62.81 | 409.76 |
| 5/ 2/1955 | 145.52 | 63.66 | 413.99 |
| 12/ 2/1955 | 146.61 | 63.65 | 411.63 |
| 19/ 2/1955 | 147.75 | 63.78 | 409.50 |
| 26/ 2/1955 | 153.52 | 65.52 | 419.68 |
| 5/ 3/1955 | 146.79 | 63.23 | 401.08 |
| 12/ 3/1955 | 146.44 | 63.69 | 404.75 |
| 19/ 3/1955 | 150.26 | 64.21 | 414.77 |
| 26/ 3/1955 | 151.07 | 63.91 | 413.84 |
| 2/ 4/1955 | 154.42 | 63.78 | 418.20 |
| 9/ 4/1955 | 158.30 | 64.72 | 425.45 |
| 16/ 4/1955 | 159.11 | 64.62 | 425.52 |
| 23/ 4/1955 | 160.52 | 64.79 | 425.65 |
| 30/ 4/1955 | 160.83 | 64.70 | 423.84 |
| 7/ 5/1955 | 157.01 | 63.89 | 419.57 |
| 14/ 5/1955 | 157.41 | 63.91 | 422.89 |
| 21/ 5/1955 | 158.84 | 63.85 | 425.66 |
| 28/ 5/1955 | 161.31 | 64.41 | 428.53 |
| 4/ 6/1955 | 160.56 | 64.21 | 437.72 |
| 11/ 6/1955 | 161.16 | 64.22 | 444.08 |
| 18/ 6/1955 | 162.20 | 64.27 | 448.93 |
| 25/ 6/1955 | 161.42 | 64.48 | 453.82 |
| 2/ 7/1955 | 157.65 | 64.76 | 461.18 |
| 9/ 7/1955 | 159.17 | 65.15 | 460.23 |
| 16/ 7/1955 | 159.98 | 66.45 | 464.69 |
| 23/ 7/1955 | 158.19 | 66.59 | 465.85 |
| 30/ 7/1955 | 155.00 | 66.23 | 456.40 |
| 6/ 8/1955 | 154.09 | 65.40 | 457.01 |
| 13/ 8/1955 | 154.99 | 65.34 | 453.57 |
| 20/ 8/1955 | 157.11 | 66.07 | 463.70 |
| 27/ 8/1955 | 157.40 | 66.30 | 472.53 |
| 3/ 9/1955 | 162.27 | 66.35 | 474.59 |
| 10/ 9/1955 | 164.29 | 65.79 | 483.67 |
| 17/ 9/1955 | 164.28 | 65.50 | 487.45 |
| 24/ 9/1955 | 155.05 | 63.14 | 466.62 |
| 1/10/1955 | 151.00 | 62.54 | 454.41 |
| 8/10/1955 | 148.47 | 61.48 | 444.68 |
| 15/10/1955 | 151.45 | 62.51 | 458.47 |
| 22/10/1955 | 150.10 | 63.25 | 454.85 |
| 29/10/1955 | 152.42 | 64.17 | 467.35 |
| 5/11/1955 | 159.60 | 64.68 | 476.54 |
| 12/11/1955 | 160.05 | 64.89 | 482.91 |
| 19/11/1955 | 167.83 | 65.31 | 482.88 |
| 26/11/1955 | 165.45 | 65.87 | 482.72 |
| 3/12/1955 | 164.74 | 65.82 | 437.64 |
| 10/12/1955 | 161.36 | 64.37 | 482.08 |
| 17/12/1955 | 162.50 | 64.33 | 486.59 |
| 24/12/1955 | 163.29 | 64.16 | 488.40 |
| 31/12/1955 | 161.13 | 64.04 | 485.68 |

53 1409

| DATE | DJTA | DJUA | DJIA |
|---|---|---|---|
| 7/ 1/1956 | 160.68 | 64.01 | 481.80 |
| 14/ 1/1956 | 154.52 | 63.07 | 464.40 |
| 21/ 1/1956 | 155.19 | 63.44 | 466.56 |
| 28/ 1/1956 | 159.22 | 64.68 | 477.44 |
| 4/ 2/1956 | 155.38 | 65.09 | 467.66 |
| 11/ 2/1956 | 158.83 | 64.90 | 477.05 |
| 18/ 2/1956 | 160.36 | 65.27 | 485.66 |
| 25/ 2/1956 | 162.79 | 65.90 | 488.84 |
| 3/ 3/1956 | 165.99 | 67.06 | 497.84 |
| 10/ 3/1956 | 169.04 | 67.47 | 507.60 |
| 17/ 3/1956 | 171.75 | 67.32 | 513.03 |
| 24/ 3/1956 | 171.82 | 67.39 | 511.79 |
| 31/ 3/1956 | 171.58 | 67.14 | 521.05 |
| 7/ 4/1956 | 171.96 | 66.29 | 509.99 |
| 14/ 4/1956 | 174.24 | 65.76 | 507.20 |
| 21/ 4/1956 | 176.96 | 65.01 | 512.03 |
| 28/ 4/1956 | 178.23 | 65.66 | 516.44 |
| 5/ 5/1956 | 177.97 | 66.38 | 501.25 |
| 12/ 5/1956 | 175.17 | 66.17 | 496.39 |
| 19/ 5/1956 | 165.56 | 65.13 | 472.49 |
| 26/ 5/1956 | 164.86 | 65.48 | 480.63 |
| 2/ 6/1956 | 162.03 | 65.21 | 475.29 |
| 9/ 6/1956 | 166.80 | 66.32 | 485.91 |
| 16/ 6/1956 | 167.54 | 66.49 | 487.95 |
| 23/ 6/1956 | 166.69 | 67.38 | 492.78 |
| 30/ 6/1956 | 167.16 | 68.71 | 504.14 |
| 7/ 7/1956 | 168.02 | 69.38 | 511.10 |
| 14/ 7/1956 | 167.96 | 69.83 | 514.57 |
| 21/ 7/1956 | 169.07 | 70.85 | 512.30 |
| 28/ 7/1956 | 170.23 | 71.11 | 520.27 |
| 4/ 8/1956 | 168.57 | 70.84 | 517.38 |
| 11/ 8/1956 | 165.65 | 70.57 | 515.79 |
| 18/ 8/1956 | 162.66 | 68.70 | 507.91 |
| 25/ 8/1956 | 160.65 | 68.63 | 502.04 |
| 1/ 9/1956 | 159.62 | 68.72 | 506.76 |
| 8/ 9/1956 | 160.19 | 68.03 | 500.32 |
| 15/ 9/1956 | 159.43 | 67.39 | 490.33 |
| 22/ 9/1956 | 154.01 | 65.57 | 475.25 |
| 29/ 9/1956 | 158.31 | 66.55 | 482.39 |
| 6/10/1956 | 160.79 | 66.49 | 490.19 |
| 13/10/1956 | 162.38 | 65.83 | 486.12 |
| 20/10/1956 | 160.19 | 66.11 | 486.06 |
| 27/10/1956 | 159.54 | 67.08 | 490.47 |
| 3/11/1956 | 157.22 | 67.08 | 485.35 |
| 10/11/1956 | 156.51 | 66.81 | 480.67 |
| 17/11/1956 | 154.16 | 66.35 | 472.56 |
| 24/11/1956 | 151.69 | 66.42 | 472.78 |
| 1/12/1956 | 158.38 | 67.95 | 494.79 |
| 8/12/1956 | 155.24 | 67.59 | 492.08 |
| 15/12/1956 | 153.70 | 67.66 | 494.38 |
| 22/12/1956 | 153.56 | 68.33 | 496.41 |
| 29/12/1956 | 156.42 | 69.07 | 498.22 |

52 1461

**TABLE A4-2 Dow Jones Transportation, Utility, and Industrial Weekly Averages** (cont.)

| DATE | DJTA | DJUA | DJIA |
|---|---|---|---|
| 5/ 1/1957 | 157.33 | 69.62 | 493.81 |
| 12/ 1/1957 | 151.13 | 69.22 | 477.46 |
| 19/ 1/1957 | 148.96 | 70.68 | 478.34 |
| 26/ 1/1957 | 148.47 | 71.31 | 477.22 |
| 2/ 2/1957 | 144.10 | 71.16 | 466.29 |
| 9/ 2/1957 | 143.09 | 69.89 | 468.07 |
| 16/ 2/1957 | 142.22 | 70.34 | 466.93 |
| 23/ 2/1957 | 142.42 | 70.79 | 468.91 |
| 2/ 3/1957 | 142.72 | 70.74 | 471.63 |
| 9/ 3/1957 | 142.87 | 71.58 | 474.28 |
| 16/ 3/1957 | 143.52 | 70.78 | 472.94 |
| 23/ 3/1957 | 144.05 | 71.47 | 474.81 |
| 30/ 3/1957 | 145.44 | 71.61 | 477.61 |
| 6/ 4/1957 | 146.51 | 71.61 | 486.72 |
| 13/ 4/1957 | 145.26 | 72.22 | 488.03 |
| 20/ 4/1957 | 146.98 | 72.99 | 491.50 |
| 27/ 4/1957 | 146.84 | 73.67 | 497.54 |
| 4/ 5/1957 | 147.20 | 73.46 | 498.30 |
| 11/ 5/1957 | 147.85 | 74.28 | 505.60 |
| 18/ 5/1957 | 145.86 | 74.36 | 504.02 |
| 25/ 5/1957 | 145.55 | 74.03 | 504.93 |
| 1/ 6/1957 | 145.01 | 73.54 | 505.63 |
| 8/ 6/1957 | 147.27 | 73.47 | 511.79 |
| 15/ 6/1957 | 144.21 | 70.41 | 500.00 |
| 22/ 6/1957 | 146.46 | 69.84 | 503.29 |
| 29/ 6/1957 | 148.36 | 71.32 | 516.89 |
| 6/ 7/1957 | 152.51 | 71.70 | 520.77 |
| 13/ 7/1957 | 151.84 | 70.67 | 515.73 |
| 20/ 7/1957 | 152.33 | 70.03 | 514.59 |
| 27/ 7/1957 | 150.55 | 69.69 | 505.10 |
| 3/ 8/1957 | 146.02 | 69.21 | 496.78 |
| 10/ 8/1957 | 142.74 | 68.47 | 488.20 |
| 17/ 8/1957 | 138.21 | 67.51 | 475.74 |
| 24/ 8/1957 | 137.49 | 67.84 | 484.35 |
| 31/ 8/1957 | 133.94 | 67.73 | 478.63 |
| 7/ 9/1957 | 132.72 | 67.58 | 481.02 |
| 14/ 9/1957 | 128.48 | 67.64 | 468.42 |
| 21/ 9/1957 | 124.12 | 66.88 | 456.89 |
| 28/ 9/1957 | 125.66 | 66.69 | 461.70 |
| 5/10/1957 | 116.47 | 65.00 | 441.16 |
| 12/10/1957 | 113.93 | 64.55 | 433.83 |
| 19/10/1957 | 112.41 | 63.98 | 435.15 |
| 26/10/1957 | 108.03 | 65.54 | 434.71 |
| 2/11/1957 | 107.19 | 64.85 | 434.12 |
| 9/11/1957 | 106.67 | 65.89 | 439.35 |
| 16/11/1957 | 103.46 | 67.10 | 442.68 |
| 23/11/1957 | 103.97 | 67.73 | 449.87 |
| 30/11/1957 | 99.20 | 67.80 | 447.20 |
| 7/12/1957 | 101.12 | 68.39 | 440.48 |
| 14/12/1957 | 96.92 | 67.95 | 427.20 |
| 21/12/1957 | 97.39 | 68.65 | 432.90 |
| 28/12/1957 | 103.53 | 69.46 | 444.56 |

52 1513

| DATE | DJTA | DJUA | DJIA |
|---|---|---|---|
| 4/ 1/1958 | 100.16 | 69.94 | 438.68 |
| 11/ 1/1958 | 107.10 | 71.38 | 444.12 |
| 18/ 1/1958 | 107.63 | 71.93 | 450.66 |
| 25/ 1/1958 | 109.04 | 72.27 | 450.02 |
| 1/ 2/1958 | 108.70 | 72.53 | 448.76 |
| 8/ 2/1958 | 107.70 | 72.05 | 444.44 |
| 15/ 2/1958 | 105.31 | 71.93 | 439.62 |
| 22/ 2/1958 | 102.95 | 72.49 | 439.92 |
| 1/ 3/1958 | 104.69 | 73.05 | 451.49 |
| 8/ 3/1958 | 106.27 | 73.12 | 453.04 |
| 15/ 3/1958 | 105.69 | 73.44 | 452.49 |
| 22/ 3/1958 | 104.60 | 74.03 | 448.61 |
| 29/ 3/1958 | 101.43 | 74.06 | 440.50 |
| 5/ 4/1958 | 104.95 | 75.13 | 441.24 |
| 12/ 4/1958 | 110.18 | 76.36 | 449.31 |
| 19/ 4/1958 | 111.52 | 77.38 | 454.92 |
| 26/ 4/1958 | 112.27 | 77.44 | 459.56 |
| 3/ 5/1958 | 114.86 | 77.84 | 462.56 |
| 10/ 5/1958 | 111.98 | 77.62 | 457.10 |
| 17/ 5/1958 | 115.15 | 78.12 | 461.03 |
| 24/ 5/1958 | 116.00 | 78.19 | 462.70 |
| 31/ 5/1958 | 117.27 | 78.88 | 469.60 |
| 7/ 6/1958 | 119.21 | 78.70 | 474.77 |
| 14/ 6/1958 | 119.17 | 78.59 | 473.60 |
| 21/ 6/1958 | 118.95 | 78.83 | 475.42 |
| 28/ 6/1958 | 119.42 | 79.57 | 480.17 |
| 5/ 7/1958 | 122.34 | 80.55 | 482.85 |
| 12/ 7/1958 | 125.31 | 79.36 | 486.55 |
| 19/ 7/1958 | 130.84 | 79.74 | 501.76 |
| 26/ 7/1958 | 132.47 | 79.77 | 505.43 |
| 2/ 8/1958 | 133.61 | 79.41 | 510.13 |
| 9/ 8/1958 | 130.22 | 78.57 | 506.13 |
| 16/ 8/1958 | 132.96 | 78.03 | 508.28 |
| 23/ 8/1958 | 132.52 | 77.97 | 508.63 |
| 30/ 8/1958 | 132.43 | 79.27 | 512.77 |
| 6/ 9/1958 | 132.34 | 80.36 | 519.43 |
| 13/ 9/1958 | 140.39 | 80.96 | 526.48 |
| 20/ 9/1958 | 141.80 | 80.23 | 526.83 |
| 27/ 9/1958 | 145.71 | 80.72 | 533.73 |
| 4/10/1958 | 147.36 | 82.18 | 543.36 |
| 11/10/1958 | 147.29 | 82.64 | 546.36 |
| 18/10/1958 | 147.76 | 82.28 | 539.52 |
| 25/10/1958 | 148.56 | 83.22 | 543.22 |
| 1/11/1958 | 151.80 | 84.05 | 554.26 |
| 8/11/1958 | 154.70 | 86.88 | 564.68 |
| 15/11/1958 | 154.99 | 86.81 | 559.57 |
| 22/11/1958 | 155.68 | 85.25 | 557.46 |
| 29/11/1958 | 154.07 | 86.47 | 556.75 |
| 6/12/1958 | 154.70 | 87.95 | 562.27 |
| 13/12/1958 | 154.68 | 89.00 | 573.17 |
| 20/12/1958 | 157.00 | 89.22 | 572.73 |
| 27/12/1958 | 159.72 | 91.22 | 587.59 |

52 1565

**TABLE A4.2 Dow Jones Transportation, Utility, and Industrial Weekly Averages** (cont.)

| DATE | DJTA | DJUA | DJIA |
|---|---|---|---|
| 3/ 1/1959 | 163.58 | 91.77 | 592.72 |
| 10/ 1/1959 | 167.17 | 92.18 | 595.75 |
| 17/ 1/1959 | 165.66 | 91.99 | 596.07 |
| 24/ 1/1959 | 161.91 | 90.88 | 593.96 |
| 31/ 1/1959 | 160.35 | 90.40 | 582.33 |
| 7/ 2/1959 | 160.77 | 90.81 | 587.97 |
| 14/ 2/1959 | 164.39 | 91.74 | 602.21 |
| 21/ 2/1959 | 162.20 | 92.05 | 603.50 |
| 28/ 2/1959 | 163.74 | 93.05 | 609.52 |
| 7/ 3/1959 | 164.25 | 94.28 | 614.69 |
| 14/ 3/1959 | 162.62 | 94.41 | 610.37 |
| 21/ 3/1959 | 159.74 | 93.80 | 606.58 |
| 28/ 3/1959 | 162.22 | 93.88 | 611.93 |
| 4/ 4/1959 | 163.11 | 92.59 | 605.97 |
| 11/ 4/1959 | 168.92 | 93.22 | 624.06 |
| 18/ 4/1959 | 168.00 | 91.66 | 627.39 |
| 25/ 4/1959 | 167.67 | 91.44 | 625.06 |
| 2/ 5/1959 | 163.85 | 91.60 | 621.36 |
| 9/ 5/1959 | 165.90 | 92.10 | 634.53 |
| 16/ 5/1959 | 169.67 | 91.26 | 634.74 |
| 23/ 5/1959 | 167.33 | 89.80 | 643.79 |
| 30/ 5/1959 | 163.98 | 87.51 | 629.98 |
| 6/ 6/1959 | 163.02 | 86.54 | 627.42 |
| 13/ 6/1959 | 164.21 | 85.78 | 629.76 |
| 20/ 6/1959 | 167.17 | 86.10 | 639.25 |
| 27/ 6/1959 | 168.92 | 88.10 | 654.76 |
| 4/ 7/1959 | 172.22 | 88.78 | 663.56 |
| 11/ 7/1959 | 166.95 | 88.95 | 657.13 |
| 18/ 7/1959 | 167.69 | 89.63 | 663.72 |
| 25/ 7/1959 | 167.80 | 89.99 | 674.88 |
| 1/ 8/1959 | 164.45 | 91.11 | 668.57 |
| 8/ 8/1959 | 162.35 | 91.80 | 658.74 |
| 15/ 8/1959 | 163.20 | 91.28 | 655.39 |
| 22/ 8/1959 | 163.49 | 91.20 | 663.06 |
| 29/ 8/1959 | 158.61 | 90.07 | 652.18 |
| 5/ 9/1959 | 156.42 | 88.24 | 637.36 |
| 12/ 9/1959 | 152.45 | 85.71 | 625.78 |
| 19/ 9/1959 | 154.25 | 86.89 | 632.59 |
| 26/ 9/1959 | 158.85 | 88.86 | 636.57 |
| 3/10/1959 | 158.67 | 87.93 | 636.98 |
| 10/10/1959 | 159.99 | 88.15 | 643.22 |
| 17/10/1959 | 156.95 | 87.45 | 633.07 |
| 24/10/1959 | 154.50 | 87.47 | 646.60 |
| 31/10/1959 | 154.10 | 87.06 | 650.92 |
| 7/11/1959 | 149.45 | 86.67 | 641.71 |
| 14/11/1959 | 149.36 | 86.04 | 645.46 |
| 21/11/1959 | 148.60 | 86.40 | 652.52 |
| 28/11/1959 | 152.73 | 86.75 | 664.00 |
| 5/12/1959 | 153.65 | 87.08 | 670.50 |
| 12/12/1959 | 154.78 | 87.00 | 676.65 |
| 19/12/1959 | 154.28 | 87.01 | 670.69 |
| 26/12/1959 | 154.05 | 87.83 | 679.36 |

52 1617

| DATE | DJTA | DJUA | DJIA |
|---|---|---|---|
| 2/ 1/1960 | 158.10 | 87.69 | 675.73 |
| 9/ 1/1960 | 157.98 | 87.13 | 659.68 |
| 16/ 1/1960 | 155.63 | 86.38 | 645.85 |
| 23/ 1/1960 | 151.60 | 85.56 | 622.62 |
| 30/ 1/1960 | 151.50 | 85.75 | 626.77 |
| 6/ 2/1960 | 151.20 | 85.47 | 622.23 |
| 13/ 2/1960 | 151.90 | 85.99 | 628.45 |
| 20/ 2/1960 | 150.86 | 86.57 | 632.00 |
| 27/ 2/1960 | 141.83 | 86.53 | 609.79 |
| 5/ 3/1960 | 143.17 | 86.28 | 605.83 |
| 12/ 3/1960 | 145.44 | 87.44 | 616.42 |
| 19/ 3/1960 | 146.44 | 88.15 | 622.47 |
| 26/ 3/1960 | 143.43 | 88.42 | 615.98 |
| 2/ 4/1960 | 144.96 | 89.12 | 628.10 |
| 9/ 4/1960 | 143.91 | 89.36 | 630.12 |
| 16/ 4/1960 | 142.98 | 89.65 | 616.32 |
| 23/ 4/1960 | 139.83 | 88.71 | 601.70 |
| 30/ 4/1960 | 140.85 | 88.98 | 607.62 |
| 7/ 5/1960 | 139.47 | 89.19 | 616.03 |
| 14/ 5/1960 | 143.91 | 89.18 | 625.24 |
| 21/ 5/1960 | 141.12 | 88.43 | 624.78 |
| 28/ 5/1960 | 139.66 | 89.09 | 628.98 |
| 4/ 6/1960 | 146.01 | 90.89 | 654.88 |
| 11/ 6/1960 | 142.72 | 92.23 | 650.89 |
| 18/ 6/1960 | 143.79 | 93.30 | 647.01 |
| 25/ 6/1960 | 142.76 | 93.49 | 641.30 |
| 2/ 7/1960 | 142.29 | 94.87 | 646.91 |
| 9/ 7/1960 | 138.89 | 93.99 | 630.24 |
| 16/ 7/1960 | 135.84 | 93.12 | 609.87 |
| 23/ 7/1960 | 135.26 | 92.83 | 616.73 |
| 30/ 7/1960 | 134.64 | 92.83 | 614.29 |
| 6/ 8/1960 | 138.35 | 94.05 | 626.18 |
| 13/ 8/1960 | 139.73 | 95.53 | 629.27 |
| 20/ 8/1960 | 139.92 | 96.02 | 636.13 |
| 27/ 8/1960 | 136.15 | 96.10 | 625.22 |
| 3/ 9/1960 | 134.76 | 96.45 | 614.12 |
| 10/ 9/1960 | 132.42 | 94.97 | 602.18 |
| 17/ 9/1960 | 129.25 | 93.62 | 585.20 |
| 24/ 9/1960 | 125.42 | 91.29 | 580.14 |
| 1/10/1960 | 126.21 | 93.34 | 586.42 |
| 8/10/1960 | 127.62 | 94.19 | 596.48 |
| 15/10/1960 | 124.71 | 92.92 | 577.55 |
| 22/10/1960 | 125.71 | 92.52 | 577.92 |
| 29/10/1960 | 128.22 | 93.75 | 596.07 |
| 5/11/1960 | 129.46 | 93.58 | 608.61 |
| 12/11/1960 | 128.98 | 94.13 | 603.62 |
| 19/11/1960 | 130.13 | 95.45 | 606.47 |
| 26/11/1960 | 128.12 | 95.38 | 596.00 |
| 3/12/1960 | 127.77 | 97.02 | 610.90 |
| 10/12/1960 | 127.86 | 98.22 | 617.78 |
| 17/12/1960 | 129.65 | 98.71 | 613.23 |
| 24/12/1960 | 130.85 | 100.02 | 615.89 |
| 31/12/1960 | 135.65 | 100.83 | 621.64 |

53 1670

**TABLE A4-2 Dow Jones Transportation, Utility, and Industrial Weekly Averages** (cont.)

| DATE | DJTA | DJUA | DJIA |
|---|---|---|---|
| 7/ 1/1961 | 139.63 | 102.62 | 633.65 |
| 14/ 1/1961 | 142.84 | 103.45 | 634.37 |
| 21/ 1/1961 | 140.54 | 105.14 | 643.59 |
| 28/ 1/1961 | 143.10 | 107.79 | 652.97 |
| 4/ 2/1961 | 140.64 | 107.72 | 639.67 |
| 11/ 2/1961 | 144.32 | 107.56 | 651.67 |
| 18/ 2/1961 | 145.49 | 107.89 | 655.60 |
| 25/ 2/1961 | 144.84 | 108.74 | 671.57 |
| 4/ 3/1961 | 143.00 | 108.38 | 663.56 |
| 11/ 3/1961 | 144.67 | 110.40 | 676.48 |
| 18/ 3/1961 | 148.18 | 111.75 | 672.48 |
| 25/ 3/1961 | 146.20 | 111.91 | 676.63 |
| 1/ 4/1961 | 144.41 | 112.23 | 683.68 |
| 8/ 4/1961 | 142.31 | 111.97 | 693.72 |
| 15/ 4/1961 | 140.88 | 112.16 | 685.26 |
| 22/ 4/1961 | 141.07 | 111.72 | 678.71 |
| 29/ 4/1961 | 143.86 | 111.81 | 690.67 |
| 6/ 5/1961 | 144.77 | 113.02 | 687.91 |
| 13/ 5/1961 | 147.56 | 113.69 | 705.96 |
| 20/ 5/1961 | 145.27 | 113.30 | 696.28 |
| 27/ 5/1961 | 143.89 | 113.01 | 697.70 |
| 3/ 6/1961 | 143.93 | 113.77 | 700.90 |
| 10/ 6/1961 | 140.64 | 112.60 | 685.50 |
| 17/ 6/1961 | 139.90 | 111.81 | 688.66 |
| 24/ 6/1961 | 139.47 | 111.74 | 683.96 |
| 1/ 7/1961 | 141.36 | 113.93 | 692.73 |
| 8/ 7/1961 | 137.05 | 114.37 | 690.95 |
| 15/ 7/1961 | 134.69 | 113.99 | 682.81 |
| 22/ 7/1961 | 139.06 | 115.39 | 705.13 |
| 29/ 7/1961 | 140.37 | 117.78 | 720.69 |
| 5/ 8/1961 | 139.81 | 119.40 | 722.61 |
| 12/ 8/1961 | 144.52 | 120.00 | 723.54 |
| 19/ 8/1961 | 143.02 | 119.96 | 716.70 |
| 26/ 8/1961 | 144.19 | 120.80 | 721.19 |
| 2/ 9/1961 | 142.76 | 121.32 | 720.91 |
| 9/ 9/1961 | 143.79 | 121.77 | 716.30 |
| 16/ 9/1961 | 144.28 | 121.11 | 701.57 |
| 23/ 9/1961 | 143.96 | 122.44 | 701.21 |
| 30/ 9/1961 | 150.74 | 127.13 | 708.25 |
| 7/10/1961 | 151.77 | 127.17 | 703.31 |
| 14/10/1961 | 150.14 | 130.14 | 705.62 |
| 21/10/1961 | 148.08 | 129.52 | 698.74 |
| 28/10/1961 | 149.46 | 130.84 | 709.26 |
| 4/11/1961 | 149.85 | 133.10 | 724.83 |
| 11/11/1961 | 151.02 | 135.79 | 729.53 |
| 18/11/1961 | 147.75 | 135.39 | 732.60 |
| 25/11/1961 | 146.39 | 134.77 | 728.80 |
| 2/12/1961 | 145.71 | 133.84 | 728.23 |
| 9/12/1961 | 143.91 | 132.93 | 729.40 |
| 16/12/1961 | 141.61 | 128.12 | 720.87 |
| 23/12/1961 | 143.84 | 129.16 | 731.14 |
| 30/12/1961 | 146.60 | 124.46 | 714.84 |

52 1722

| DATE | DJTA | DJUA | DJIA |
|---|---|---|---|
| 6/ 1/1962 | 148.38 | 124.81 | 711.73 |
| 13/ 1/1962 | 148.26 | 123.63 | 700.72 |
| 20/ 1/1962 | 146.86 | 123.60 | 692.19 |
| 27/ 1/1962 | 149.83 | 124.94 | 706.55 |
| 3/ 2/1962 | 148.64 | 126.73 | 714.27 |
| 10/ 2/1962 | 149.04 | 128.77 | 716.46 |
| 17/ 2/1962 | 147.14 | 128.75 | 709.54 |
| 24/ 2/1962 | 146.25 | 128.54 | 711.00 |
| 3/ 3/1962 | 145.71 | 129.90 | 714.44 |
| 10/ 3/1962 | 145.83 | 130.51 | 722.27 |
| 17/ 3/1962 | 144.40 | 130.55 | 716.46 |
| 24/ 3/1962 | 144.28 | 130.01 | 706.95 |
| 31/ 3/1962 | 142.86 | 130.15 | 699.63 |
| 7/ 4/1962 | 142.18 | 128.22 | 687.90 |
| 14/ 4/1962 | 143.86 | 129.56 | 694.25 |
| 21/ 4/1962 | 138.76 | 128.77 | 672.20 |
| 28/ 4/1962 | 140.68 | 124.77 | 671.20 |
| 5/ 5/1962 | 134.24 | 119.61 | 640.63 |
| 12/ 5/1962 | 136.32 | 122.29 | 650.70 |
| 19/ 5/1962 | 129.23 | 112.57 | 611.88 |
| 26/ 5/1962 | 128.90 | 113.96 | 611.05 |
| 2/ 6/1962 | 126.52 | 113.95 | 601.61 |
| 9/ 6/1962 | 121.48 | 109.42 | 578.18 |
| 16/ 6/1962 | 117.22 | 104.67 | 539.19 |
| 23/ 6/1962 | 118.63 | 108.28 | 561.28 |
| 30/ 6/1962 | 121.46 | 112.36 | 576.17 |
| 7/ 7/1962 | 125.28 | 114.69 | 590.19 |
| 14/ 7/1962 | 123.05 | 113.56 | 577.18 |
| 21/ 7/1962 | 121.83 | 115.61 | 585.00 |
| 28/ 7/1962 | 122.26 | 117.38 | 596.38 |
| 4/ 8/1962 | 119.54 | 117.06 | 592.32 |
| 11/ 8/1962 | 121.44 | 119.17 | 610.02 |
| 18/ 8/1962 | 124.34 | 121.27 | 613.74 |
| 25/ 8/1962 | 123.75 | 120.83 | 609.18 |
| 1/ 9/1962 | 122.72 | 121.39 | 600.86 |
| 8/ 9/1962 | 121.23 | 121.48 | 605.84 |
| 15/ 9/1962 | 117.79 | 120.80 | 591.78 |
| 22/ 9/1962 | 115.68 | 117.61 | 578.98 |
| 29/ 9/1962 | 116.36 | 118.91 | 586.59 |
| 6/10/1962 | 118.04 | 120.59 | 586.47 |
| 13/10/1962 | 116.17 | 117.80 | 573.29 |
| 20/10/1962 | 118.93 | 113.12 | 569.02 |
| 27/10/1962 | 122.51 | 117.67 | 604.58 |
| 3/11/1962 | 126.05 | 120.06 | 616.13 |
| 10/11/1962 | 131.03 | 122.79 | 630.98 |
| 17/11/1962 | 135.15 | 125.05 | 644.87 |
| 24/11/1962 | 138.97 | 125.27 | 649.30 |
| 1/12/1962 | 140.27 | 126.87 | 652.10 |
| 8/12/1962 | 137.64 | 127.56 | 648.09 |
| 15/12/1962 | 138.96 | 127.96 | 646.41 |
| 22/12/1962 | 140.00 | 129.08 | 651.43 |
| 29/12/1962 | 147.51 | 131.01 | 662.23 |

52 1774

**TABLE A4-2 Dow Jones Transportation, Utility, and Industrial Weekly Averages** (cont.)

| DATE | DJTA | DJUA | DJIA |
|---|---|---|---|
| 5/ 1/1963 | 148.68 | 133.39 | 671.60 |
| 12/ 1/1963 | 146.25 | 133.85 | 672.52 |
| 19/ 1/1963 | 149.97 | 135.12 | 679.71 |
| 26/ 1/1963 | 149.45 | 135.82 | 683.19 |
| 2/ 2/1963 | 151.41 | 135.72 | 679.92 |
| 9/ 2/1963 | 154.96 | 137.33 | 686.07 |
| 16/ 2/1963 | 153.12 | 135.72 | 681.64 |
| 23/ 2/1963 | 149.67 | 132.28 | 659.72 |
| 2/ 3/1963 | 151.04 | 133.85 | 672.43 |
| 9/ 3/1963 | 151.71 | 135.65 | 676.33 |
| 16/ 3/1963 | 151.58 | 135.67 | 677.83 |
| 23/ 3/1963 | 152.92 | 136.19 | 682.52 |
| 30/ 3/1963 | 155.03 | 137.45 | 702.43 |
| 6/ 4/1963 | 156.87 | 137.58 | 708.45 |
| 13/ 4/1963 | 159.57 | 138.67 | 711.68 |
| 20/ 4/1963 | 163.24 | 138.78 | 717.16 |
| 27/ 4/1963 | 164.33 | 139.61 | 718.08 |
| 4/ 5/1963 | 164.60 | 139.90 | 723.30 |
| 11/ 5/1963 | 167.88 | 140.68 | 724.81 |
| 18/ 5/1963 | 170.93 | 141.27 | 720.53 |
| 25/ 5/1963 | 173.38 | 140.33 | 726.96 |
| 1/ 6/1963 | 170.98 | 139.96 | 722.41 |
| 8/ 6/1963 | 169.79 | 139.63 | 722.03 |
| 15/ 6/1963 | 174.00 | 140.24 | 720.78 |
| 22/ 6/1963 | 173.66 | 139.08 | 706.88 |
| 29/ 6/1963 | 174.75 | 139.35 | 716.45 |
| 6/ 7/1963 | 174.00 | 139.61 | 707.70 |
| 13/ 7/1963 | 169.29 | 137.95 | 693.89 |
| 20/ 7/1963 | 165.79 | 138.87 | 689.38 |
| 27/ 7/1963 | 168.00 | 140.16 | 697.83 |
| 3/ 8/1963 | 170.61 | 142.09 | 708.39 |
| 10/ 8/1963 | 176.31 | 144.03 | 719.32 |
| 17/ 8/1963 | 175.81 | 144.37 | 723.14 |
| 24/ 8/1963 | 176.86 | 143.96 | 729.32 |
| 31/ 8/1963 | 173.48 | 144.06 | 735.37 |
| 7/ 9/1963 | 172.79 | 143.46 | 740.13 |
| 14/ 9/1963 | 173.11 | 142.71 | 743.60 |
| 21/ 9/1963 | 170.65 | 140.43 | 737.98 |
| 28/ 9/1963 | 170.93 | 139.35 | 745.06 |
| 5/10/1963 | 169.39 | 139.13 | 741.76 |
| 12/10/1963 | 172.17 | 138.65 | 750.60 |
| 19/10/1963 | 171.50 | 137.71 | 755.61 |
| 26/10/1963 | 170.56 | 138.99 | 753.73 |
| 2/11/1963 | 171.80 | 138.12 | 750.81 |
| 9/11/1963 | 172.34 | 138.36 | 740.00 |
| 16/11/1963 | 166.41 | 134.97 | 711.49 |
| 23/11/1963 | 171.85 | 136.44 | 750.52 |
| 30/11/1963 | 173.43 | 136.80 | 760.25 |
| 7/12/1963 | 178.19 | 137.83 | 760.17 |
| 14/12/1963 | 177.35 | 138.60 | 762.08 |
| 21/12/1963 | 177.28 | 138.36 | 762.95 |
| 28/12/1963 | 178.81 | 138.48 | 767.68 |

52 1826

| DATE | DJTA | DJUA | DJIA |
|---|---|---|---|
| 4/ 1/1964 | 180.15 | 140.38 | 774.33 |
| 11/ 1/1964 | 181.87 | 140.72 | 775.69 |
| 18/ 1/1964 | 182.53 | 140.88 | 783.04 |
| 25/ 1/1964 | 181.39 | 139.49 | 785.34 |
| 1/ 2/1964 | 182.06 | 140.14 | 791.59 |
| 8/ 2/1964 | 183.75 | 140.07 | 794.56 |
| 15/ 2/1964 | 186.83 | 139.80 | 796.99 |
| 22/ 2/1964 | 190.74 | 140.50 | 800.14 |
| 29/ 2/1964 | 191.98 | 140.45 | 806.03 |
| 7/ 3/1964 | 192.60 | 139.75 | 816.22 |
| 14/ 3/1964 | 193.47 | 138.43 | 814.93 |
| 21/ 3/1964 | 192.16 | 137.76 | 815.91 |
| 28/ 3/1964 | 195.46 | 137.71 | 822.99 |
| 4/ 4/1964 | 195.28 | 138.58 | 821.75 |
| 11/ 4/1964 | 197.07 | 140.00 | 827.33 |
| 18/ 4/1964 | 196.18 | 139.95 | 814.89 |
| 25/ 4/1964 | 196.50 | 139.63 | 817.10 |
| 2/ 5/1964 | 199.35 | 140.98 | 828.57 |
| 9/ 5/1964 | 201.58 | 141.44 | 826.23 |
| 16/ 5/1964 | 207.29 | 141.04 | 820.87 |
| 23/ 5/1964 | 205.95 | 140.06 | 820.56 |
| 30/ 5/1964 | 202.50 | 139.98 | 806.03 |
| 6/ 6/1964 | 203.74 | 141.60 | 809.39 |
| 13/ 6/1964 | 206.82 | 141.75 | 825.25 |
| 20/ 6/1964 | 212.25 | 143.81 | 830.99 |
| 27/ 6/1964 | 217.80 | 143.83 | 841.47 |
| 4/ 7/1964 | 218.20 | 146.57 | 847.51 |
| 11/ 7/1964 | 220.98 | 148.25 | 851.35 |
| 18/ 7/1964 | 220.48 | 149.36 | 845.64 |
| 25/ 7/1964 | 217.80 | 149.89 | 841.10 |
| 1/ 8/1964 | 213.02 | 148.98 | 829.16 |
| 8/ 8/1964 | 213.59 | 149.11 | 838.81 |
| 15/ 8/1964 | 211.38 | 149.31 | 838.62 |
| 22/ 8/1964 | 207.49 | 149.51 | 839.09 |
| 29/ 8/1964 | 211.25 | 150.60 | 848.31 |
| 5/ 9/1964 | 214.73 | 152.12 | 867.13 |
| 12/ 9/1964 | 215.30 | 151.90 | 865.12 |
| 19/ 9/1964 | 218.03 | 152.20 | 874.71 |
| 26/ 9/1964 | 218.10 | 153.37 | 872.65 |
| 3/10/1964 | 222.12 | 154.20 | 878.08 |
| 10/10/1964 | 221.84 | 154.71 | 873.54 |
| 17/10/1964 | 224.28 | 153.93 | 877.62 |
| 24/10/1964 | 223.54 | 152.96 | 873.08 |
| 31/10/1964 | 219.10 | 153.52 | 876.87 |
| 7/11/1964 | 216.82 | 154.55 | 874.11 |
| 14/11/1964 | 218.47 | 155.71 | 890.72 |
| 21/11/1964 | 215.24 | 154.19 | 882.12 |
| 28/11/1964 | 210.55 | 154.50 | 870.93 |
| 5/12/1964 | 205.52 | 154.94 | 864.34 |
| 12/12/1964 | 206.15 | 154.76 | 868.73 |
| 19/12/1964 | 204.48 | 153.91 | 868.16 |
| 26/12/1964 | 205.34 | 155.17 | 874.13 |

52 1878

**TABLE A4-2 Dow Jones Transportation, Utility, and Industrial Weekly Averages** (cont.)

| DATE | DJTA | DJUA | DJIA |
|---|---|---|---|
| 2/ 1/1965 | 208.06 | 155.97 | 882.60 |
| 9/ 1/1965 | 212.93 | 158.05 | 891.15 |
| 16/ 1/1965 | 212.09 | 159.91 | 893.59 |
| 23/ 1/1965 | 212.78 | 160.68 | 902.86 |
| 30/ 1/1965 | 210.98 | 163.07 | 901.57 |
| 6/ 2/1965 | 208.18 | 161.07 | 888.47 |
| 13/ 2/1965 | 210.42 | 160.68 | 885.61 |
| 20/ 2/1965 | 211.38 | 161.22 | 903.48 |
| 27/ 2/1965 | 209.30 | 162.09 | 895.98 |
| 6/ 3/1965 | 213.08 | 161.48 | 900.33 |
| 13/ 3/1965 | 213.87 | 160.94 | 895.79 |
| 20/ 3/1965 | 213.69 | 162.09 | 891.66 |
| 27/ 3/1965 | 211.08 | 162.00 | 893.38 |
| 3/ 4/1965 | 213.69 | 162.22 | 901.29 |
| 10/ 4/1965 | 213.16 | 162.85 | 911.91 |
| 17/ 4/1965 | 212.12 | 162.36 | 916.41 |
| 24/ 4/1965 | 212.63 | 161.76 | 922.31 |
| 1/ 5/1965 | 213.39 | 161.26 | 932.52 |
| 8/ 5/1965 | 209.50 | 161.87 | 939.62 |
| 15/ 5/1965 | 206.26 | 161.60 | 922.01 |
| 22/ 5/1965 | 205.04 | 160.17 | 918.04 |
| 29/ 5/1965 | 200.47 | 158.12 | 900.87 |
| 5/ 6/1965 | 195.80 | 154.33 | 881.70 |
| 12/ 6/1965 | 196.56 | 154.81 | 879.17 |
| 19/ 6/1965 | 190.74 | 152.00 | 854.36 |
| 26/ 6/1965 | 197.70 | 155.18 | 875.16 |
| 3/ 7/1965 | 200.24 | 157.48 | 879.49 |
| 10/ 7/1965 | 201.79 | 156.78 | 880.43 |
| 17/ 7/1965 | 197.12 | 155.30 | 863.97 |
| 24/ 7/1965 | 207.73 | 155.36 | 881.74 |
| 31/ 7/1965 | 210.14 | 155.69 | 882.51 |
| 7/ 8/1965 | 215.32 | 155.24 | 888.82 |
| 14/ 8/1965 | 215.67 | 155.51 | 889.92 |
| 21/ 8/1965 | 219.18 | 155.30 | 895.96 |
| 28/ 8/1965 | 217.75 | 156.66 | 907.97 |
| 4/ 9/1965 | 216.21 | 157.78 | 918.95 |
| 11/ 9/1965 | 216.79 | 157.84 | 928.99 |
| 18/ 9/1965 | 222.38 | 157.33 | 929.54 |
| 25/ 9/1965 | 222.93 | 157.60 | 929.65 |
| 2/10/1965 | 228.87 | 157.09 | 938.32 |
| 9/10/1965 | 232.95 | 157.15 | 940.68 |
| 16/10/1965 | 235.61 | 156.66 | 952.42 |
| 23/10/1965 | 235.86 | 157.96 | 960.82 |
| 30/10/1965 | 238.53 | 158.72 | 959.46 |
| 6/11/1965 | 238.55 | 157.78 | 956.29 |
| 13/11/1965 | 237.08 | 156.81 | 952.72 |
| 20/11/1965 | 242.33 | 155.24 | 948.16 |
| 27/11/1965 | 243.70 | 153.51 | 946.10 |
| 4/12/1965 | 244.72 | 152.57 | 952.72 |
| 11/12/1965 | 249.55 | 150.57 | 957.85 |
| 18/12/1965 | 246.84 | 151.78 | 966.36 |
| 25/12/1965 | 247.48 | 152.63 | 969.26 |

52 1930

| DATE | DJTA | DJUA | DJIA |
|---|---|---|---|
| 1/ 1/1966 | 248.20 | 152.30 | 986.13 |
| 8/ 1/1966 | 257.73 | 151.60 | 987.30 |
| 15/ 1/1966 | 257.29 | 151.45 | 988.14 |
| 22/ 1/1966 | 263.46 | 149.12 | 985.35 |
| 29/ 1/1966 | 259.70 | 148.48 | 986.35 |
| 5/ 2/1966 | 269.40 | 147.63 | 989.03 |
| 12/ 2/1966 | 267.73 | 144.51 | 975.22 |
| 19/ 2/1966 | 264.23 | 141.78 | 953.00 |
| 26/ 2/1966 | 259.90 | 139.93 | 932.34 |
| 5/ 3/1966 | 254.40 | 142.21 | 927.95 |
| 12/ 3/1966 | 249.62 | 142.93 | 922.88 |
| 19/ 3/1966 | 252.28 | 142.90 | 929.95 |
| 26/ 3/1966 | 251.66 | 141.21 | 931.29 |
| 2/ 4/1966 | 262.56 | 140.96 | 945.76 |
| 9/ 4/1966 | 263.16 | 140.54 | 947.77 |
| 16/ 4/1966 | 263.98 | 139.06 | 949.83 |
| 23/ 4/1966 | 253.68 | 140.21 | 933.68 |
| 30/ 4/1966 | 240.54 | 138.09 | 902.83 |
| 7/ 5/1966 | 228.50 | 136.63 | 876.11 |
| 14/ 5/1966 | 226.08 | 137.03 | 876.89 |
| 21/ 5/1966 | 230.89 | 136.60 | 897.04 |
| 28/ 5/1966 | 228.00 | 135.66 | 887.86 |
| 4/ 6/1966 | 231.24 | 135.27 | 891.75 |
| 11/ 6/1966 | 231.96 | 134.03 | 894.26 |
| 18/ 6/1966 | 231.56 | 132.48 | 897.16 |
| 25/ 6/1966 | 227.25 | 131.66 | 877.06 |
| 2/ 7/1966 | 233.30 | 134.63 | 894.04 |
| 9/ 7/1966 | 231.24 | 135.03 | 889.36 |
| 16/ 7/1966 | 224.77 | 133.96 | 869.15 |
| 23/ 7/1966 | 220.19 | 132.42 | 847.38 |
| 30/ 7/1966 | 219.62 | 131.33 | 852.39 |
| 6/ 8/1966 | 214.54 | 129.57 | 840.53 |
| 13/ 8/1966 | 202.55 | 125.57 | 804.62 |
| 20/ 8/1966 | 195.03 | 120.90 | 780.56 |
| 27/ 8/1966 | 195.18 | 124.36 | 787.69 |
| 3/ 9/1966 | 194.46 | 124.30 | 775.55 |
| 10/ 9/1966 | 201.68 | 129.33 | 814.30 |
| 17/ 9/1966 | 196.78 | 126.60 | 790.97 |
| 24/ 9/1966 | 193.49 | 124.72 | 774.22 |
| 1/10/1966 | 184.34 | 122.72 | 744.32 |
| 8/10/1966 | 191.92 | 128.48 | 771.71 |
| 15/10/1966 | 193.07 | 133.03 | 787.30 |
| 22/10/1966 | 199.51 | 136.87 | 807.96 |
| 29/10/1966 | 199.66 | 137.54 | 805.06 |
| 5/11/1966 | 200.53 | 137.54 | 819.09 |
| 12/11/1966 | 203.77 | 136.84 | 809.40 |
| 19/11/1966 | 203.44 | 134.81 | 803.34 |
| 26/11/1966 | 201.50 | 134.12 | 789.47 |
| 3/12/1966 | 206.68 | 135.03 | 813.02 |
| 10/12/1966 | 206.80 | 136.75 | 807.18 |
| 17/12/1966 | 207.92 | 136.72 | 799.10 |
| 24/12/1966 | 202.97 | 136.18 | 785.69 |
| 31/12/1966 | 210.98 | 138.61 | 808.74 |

53 1983

**TABLE A4-2 Dow Jones Transportation, Utility, and Industrial Weekly Averages** (cont.)

| DATE | DJTA | DJUA | DJIA |
|---|---|---|---|
| 7/ 1/1967 | 220.41 | 138.93 | 835.13 |
| 14/ 1/1967 | 226.84 | 139.79 | 847.16 |
| 21/ 1/1967 | 226.71 | 138.80 | 844.04 |
| 28/ 1/1967 | 228.03 | 138.90 | 857.46 |
| 4/ 2/1967 | 227.93 | 138.55 | 855.73 |
| 11/ 2/1967 | 230.34 | 138.13 | 850.84 |
| 18/ 2/1967 | 229.15 | 136.79 | 847.33 |
| 25/ 2/1967 | 229.08 | 136.28 | 846.60 |
| 4/ 3/1967 | 234.18 | 135.22 | 848.50 |
| 11/ 3/1967 | 235.17 | 135.19 | 869.77 |
| 18/ 3/1967 | 233.33 | 136.18 | 876.67 |
| 25/ 3/1967 | 230.59 | 138.55 | 865.98 |
| 1/ 4/1967 | 227.48 | 138.90 | 853.34 |
| 8/ 4/1967 | 228.85 | 139.70 | 859.74 |
| 15/ 4/1967 | 230.52 | 139.95 | 883.18 |
| 22/ 4/1967 | 231.91 | 139.35 | 897.05 |
| 29/ 4/1967 | 235.87 | 138.61 | 905.96 |
| 6/ 5/1967 | 237.68 | 137.91 | 890.03 |
| 13/ 5/1967 | 239.81 | 137.33 | 874.55 |
| 20/ 5/1967 | 247.33 | 135.03 | 870.32 |
| 27/ 5/1967 | 247.46 | 133.05 | 863.31 |
| 3/ 6/1967 | 254.55 | 132.76 | 874.89 |
| 10/ 6/1967 | 256.47 | 131.65 | 880.61 |
| 17/ 6/1967 | 255.05 | 131.87 | 877.37 |
| 24/ 6/1967 | 254.84 | 131.39 | 860.26 |
| 1/ 7/1967 | 260.75 | 132.25 | 869.05 |
| 8/ 7/1967 | 267.27 | 132.51 | 882.05 |
| 15/ 7/1967 | 272.99 | 132.57 | 909.56 |
| 22/ 7/1967 | 272.38 | 133.79 | 901.53 |
| 29/ 7/1967 | 274.49 | 134.23 | 923.77 |
| 5/ 8/1967 | 262.04 | 133.79 | 920.65 |
| 12/ 8/1967 | 258.07 | 132.09 | 919.04 |
| 19/ 8/1967 | 256.96 | 130.21 | 894.07 |
| 26/ 8/1967 | 262.37 | 130.34 | 901.18 |
| 2/ 9/1967 | 260.88 | 131.04 | 907.54 |
| 9/ 9/1967 | 261.42 | 132.03 | 933.48 |
| 16/ 9/1967 | 262.37 | 131.49 | 934.35 |
| 23/ 9/1967 | 261.83 | 130.34 | 926.66 |
| 30/ 9/1967 | 258.74 | 128.99 | 928.74 |
| 7/10/1967 | 251.55 | 125.96 | 918.17 |
| 14/10/1967 | 247.50 | 124.65 | 896.73 |
| 21/10/1967 | 243.07 | 124.04 | 888.18 |
| 28/10/1967 | 229.74 | 122.70 | 856.62 |
| 4/11/1967 | 231.70 | 121.42 | 862.81 |
| 11/11/1967 | 230.92 | 123.82 | 862.11 |
| 18/11/1967 | 231.31 | 123.98 | 877.60 |
| 25/11/1967 | 235.38 | 124.74 | 879.16 |
| 2/12/1967 | 234.51 | 124.90 | 887.25 |
| 9/12/1967 | 234.35 | 124.36 | 880.61 |
| 16/12/1967 | 230.54 | 125.61 | 887.37 |
| 23/12/1967 | 233.24 | 127.91 | 905.11 |
| 30/12/1967 | 235.62 | 133.37 | 901.24 |

52 2035

| DATE | DJTA | DJUA | DJIA |
|---|---|---|---|
| 6/ 1/1968 | 237.06 | 134.84 | 898.98 |
| 13/ 1/1968 | 233.09 | 133.53 | 880.32 |
| 20/ 1/1968 | 231.75 | 130.24 | 865.06 |
| 27/ 1/1968 | 228.31 | 129.54 | 863.56 |
| 3/ 2/1968 | 223.63 | 128.90 | 840.04 |
| 10/ 2/1968 | 224.66 | 128.04 | 836.34 |
| 17/ 2/1968 | 225.84 | 128.48 | 849.80 |
| 24/ 2/1968 | 217.41 | 128.36 | 840.44 |
| 2/ 3/1968 | 215.14 | 126.02 | 835.24 |
| 9/ 3/1968 | 217.95 | 123.11 | 837.55 |
| 16/ 3/1968 | 218.54 | 120.91 | 826.05 |
| 23/ 3/1968 | 218.99 | 121.58 | 840.67 |
| 30/ 3/1968 | 223.90 | 123.56 | 865.81 |
| 6/ 4/1968 | 229.40 | 124.27 | 905.69 |
| 13/ 4/1968 | 236.31 | 124.36 | 897.65 |
| 20/ 4/1968 | 234.67 | 122.41 | 906.03 |
| 27/ 4/1968 | 240.35 | 122.48 | 919.21 |
| 4/ 5/1968 | 241.71 | 123.31 | 912.91 |
| 11/ 5/1968 | 246.42 | 122.51 | 898.98 |
| 18/ 5/1968 | 253.71 | 123.02 | 893.15 |
| 25/ 5/1968 | 255.66 | 123.02 | 899.00 |
| 1/ 6/1968 | 266.17 | 124.04 | 914.88 |
| 8/ 6/1968 | 265.58 | 125.35 | 913.62 |
| 15/ 6/1968 | 264.15 | 133.44 | 900.93 |
| 22/ 6/1968 | 261.77 | 132.60 | 897.80 |
| 29/ 6/1968 | 266.88 | 133.82 | 903.51 |
| 6/ 7/1968 | 265.82 | 134.71 | 922.46 |
| 13/ 7/1968 | 257.80 | 133.28 | 913.92 |
| 20/ 7/1968 | 250.86 | 131.81 | 888.47 |
| 27/ 7/1968 | 246.42 | 130.85 | 871.27 |
| 3/ 8/1968 | 245.76 | 131.52 | 869.65 |
| 10/ 8/1968 | 250.45 | 131.52 | 877.17 |
| 17/ 8/1968 | 251.11 | 131.55 | 890.17 |
| 24/ 8/1968 | 251.11 | 130.53 | 896.01 |
| 31/ 8/1968 | 255.65 | 131.93 | 921.25 |
| 7/ 9/1968 | 256.08 | 131.23 | 916.21 |
| 14/ 9/1968 | 261.13 | 129.95 | 924.42 |
| 21/ 9/1968 | 266.08 | 130.24 | 933.80 |
| 28/ 9/1968 | 273.04 | 129.86 | 949.40 |
| 5/10/1968 | 269.46 | 130.18 | 949.65 |
| 12/10/1968 | 272.46 | 130.85 | 955.06 |
| 19/10/1968 | 268.40 | 130.62 | 955.93 |
| 26/10/1968 | 265.37 | 131.33 | 946.42 |
| 2/11/1968 | 266.76 | 133.56 | 958.98 |
| 9/11/1968 | 271.83 | 139.86 | 965.88 |
| 16/11/1968 | 272.46 | 140.34 | 965.13 |
| 23/11/1968 | 279.28 | 140.34 | 985.08 |
| 30/11/1968 | 278.06 | 139.06 | 978.24 |
| 7/12/1968 | 278.64 | 138.52 | 976.32 |
| 14/12/1968 | 273.62 | 138.20 | 966.99 |
| 21/12/1968 | 272.36 | 137.59 | 957.51 |
| 28/12/1968 | 272.61 | 136.98 | 951.89 |

52 2087

**TABLE A4-2 Dow Jones Transportation, Utility, and Industrial Weekly Averages** (cont.)

| DATE | DJTA | DJUA | DJIA |
|---|---|---|---|
| 4/ 1/1969 | 263.10 | 134.07 | 925.53 |
| 11/ 1/1969 | 267.82 | 134.39 | 938.84 |
| 18/ 1/1969 | 272.36 | 135.77 | 938.59 |
| 25/ 1/1969 | 274.88 | 139.95 | 946.05 |
| 1/ 2/1969 | 279.88 | 139.28 | 947.85 |
| 8/ 2/1969 | 275.72 | 138.10 | 951.95 |
| 15/ 2/1969 | 263.55 | 135.32 | 916.65 |
| 22/ 2/1969 | 253.68 | 132.57 | 905.21 |
| 1/ 3/1969 | 246.26 | 131.45 | 911.18 |
| 8/ 3/1969 | 241.92 | 130.72 | 904.28 |
| 15/ 3/1969 | 243.97 | 130.34 | 920.00 |
| 22/ 3/1969 | 243.69 | 129.67 | 935.48 |
| 29/ 3/1969 | 241.52 | 129.06 | 927.30 |
| 5/ 4/1969 | 239.48 | 128.32 | 932.89 |
| 12/ 4/1969 | 236.40 | 129.35 | 924.12 |
| 19/ 4/1969 | 236.02 | 130.72 | 921.20 |
| 26/ 4/1969 | 237.36 | 130.08 | 957.17 |
| 3/ 5/1969 | 238.85 | 131.42 | 961.61 |
| 10/ 5/1969 | 241.41 | 132.54 | 967.30 |
| 17/ 5/1969 | 238.30 | 130.56 | 947.45 |
| 24/ 5/1969 | 233.40 | 129.15 | 937.56 |
| 31/ 5/1969 | 230.39 | 128.42 | 924.77 |
| 7/ 6/1969 | 222.69 | 124.27 | 894.84 |
| 14/ 6/1969 | 216.13 | 121.61 | 876.16 |
| 21/ 6/1969 | 212.62 | 120.94 | 869.76 |
| 28/ 6/1969 | 212.30 | 123.79 | 886.12 |
| 5/ 7/1969 | 205.58 | 121.80 | 847.79 |
| 12/ 7/1969 | 201.52 | 120.97 | 853.09 |
| 19/ 7/1969 | 196.86 | 118.58 | 826.53 |
| 26/ 7/1969 | 199.31 | 117.62 | 826.59 |
| 2/ 8/1969 | 197.81 | 116.09 | 824.46 |
| 9/ 8/1969 | 198.12 | 114.04 | 820.88 |
| 16/ 8/1969 | 202.02 | 116.28 | 837.18 |
| 23/ 8/1969 | 201.18 | 116.31 | 836.72 |
| 30/ 8/1969 | 197.88 | 114.30 | 819.50 |
| 6/ 9/1969 | 198.45 | 113.91 | 824.25 |
| 13/ 9/1969 | 200.35 | 112.92 | 830.39 |
| 20/ 9/1969 | 199.54 | 111.39 | 824.18 |
| 27/ 9/1969 | 196.07 | 111.36 | 808.41 |
| 4/10/1969 | 196.09 | 110.78 | 806.93 |
| 11/10/1969 | 199.56 | 116.72 | 836.06 |
| 18/10/1969 | 201.23 | 119.15 | 862.26 |
| 25/10/1969 | 200.20 | 119.02 | 855.99 |
| 1/11/1969 | 199.16 | 119.09 | 860.48 |
| 8/11/1969 | 196.22 | 117.36 | 849.26 |
| 15/11/1969 | 192.91 | 112.99 | 823.13 |
| 22/11/1969 | 186.64 | 111.39 | 812.30 |
| 29/11/1969 | 179.76 | 108.74 | 793.03 |
| 6/12/1969 | 173.06 | 107.75 | 786.69 |
| 13/12/1969 | 172.50 | 108.77 | 789.86 |
| 20/12/1969 | 176.90 | 109.82 | 797.65 |
| 27/12/1969 | 181.07 | 112.25 | 809.20 |

52 2139

| DATE | DJTA | DJUA | DJIA |
|---|---|---|---|
| 3/ 1/1970 | 177.77 | 111.58 | 798.11 |
| 10/ 1/1970 | 173.39 | 109.53 | 782.60 |
| 17/ 1/1970 | 170.24 | 107.71 | 775.54 |
| 24/ 1/1970 | 163.72 | 105.19 | 744.06 |
| 31/ 1/1970 | 166.96 | 106.63 | 752.77 |
| 7/ 2/1970 | 170.82 | 107.49 | 753.30 |
| 14/ 2/1970 | 170.76 | 110.17 | 757.46 |
| 21/ 2/1970 | 177.58 | 115.25 | 779.59 |
| 28/ 2/1970 | 177.86 | 118.51 | 787.12 |
| 7/ 3/1970 | 173.21 | 115.73 | 772.11 |
| 14/ 3/1970 | 170.78 | 113.98 | 763.66 |
| 21/ 3/1970 | 173.28 | 117.94 | 791.05 |
| 28/ 3/1970 | 174.35 | 118.26 | 791.84 |
| 4/ 4/1970 | 172.38 | 117.65 | 790.46 |
| 11/ 4/1970 | 167.40 | 114.33 | 775.94 |
| 18/ 4/1970 | 161.82 | 110.05 | 747.29 |
| 25/ 4/1970 | 156.53 | 108.29 | 733.63 |
| 2/ 5/1970 | 154.34 | 106.08 | 717.73 |
| 9/ 5/1970 | 147.66 | 103.18 | 702.22 |
| 16/ 5/1970 | 139.25 | 100.52 | 662.17 |
| 23/ 5/1970 | 144.46 | 102.25 | 700.44 |
| 30/ 5/1970 | 142.21 | 101.61 | 695.03 |
| 6/ 6/1970 | 138.27 | 98.13 | 684.21 |
| 13/ 6/1970 | 139.87 | 98.54 | 720.43 |
| 20/ 6/1970 | 126.75 | 96.59 | 687.84 |
| 27/ 6/1970 | 120.47 | 96.88 | 689.14 |
| 4/ 7/1970 | 123.80 | 103.53 | 700.10 |
| 11/ 7/1970 | 128.73 | 104.93 | 735.08 |
| 18/ 7/1970 | 130.60 | 104.93 | 730.22 |
| 25/ 7/1970 | 130.73 | 104.93 | 734.12 |
| 1/ 8/1970 | 129.83 | 104.36 | 725.70 |
| 8/ 8/1970 | 129.59 | 103.24 | 710.84 |
| 15/ 8/1970 | 130.60 | 106.82 | 745.41 |
| 22/ 8/1970 | 138.12 | 110.30 | 765.81 |
| 29/ 8/1970 | 137.65 | 110.27 | 771.15 |
| 5/ 9/1970 | 140.02 | 108.70 | 761.84 |
| 12/ 9/1970 | 142.55 | 107.97 | 758.49 |
| 19/ 9/1970 | 145.23 | 108.86 | 761.77 |
| 26/ 9/1970 | 158.71 | 107.84 | 766.16 |
| 3/10/1970 | 154.83 | 106.47 | 768.69 |
| 10/10/1970 | 154.06 | 106.60 | 763.35 |
| 17/10/1970 | 150.25 | 106.56 | 759.38 |
| 24/10/1970 | 145.72 | 106.37 | 755.61 |
| 31/10/1970 | 148.92 | 109.41 | 771.97 |
| 7/11/1970 | 147.51 | 110.88 | 759.79 |
| 14/11/1970 | 146.74 | 110.75 | 761.57 |
| 21/11/1970 | 149.60 | 114.68 | 781.35 |
| 28/11/1970 | 158.59 | 118.48 | 816.06 |
| 5/12/1970 | 158.16 | 118.99 | 825.92 |
| 12/12/1970 | 159.02 | 118.71 | 822.11 |
| 19/12/1970 | 162.59 | 119.92 | 828.38 |
| 26/12/1970 | 171.52 | 121.84 | 838.92 |

52 2191

**TABLE A4-2 Dow Jones Transportation, Utility, and Industrial Weekly Averages** (cont.)

| DATE | DJTA | DJUA | DJIA |
|---|---|---|---|
| 2/ 1/1971 | 176.75 | 122.48 | 837.83 |
| 9/ 1/1971 | 179.53 | 126.69 | 845.70 |
| 16/ 1/1971 | 186.95 | 126.95 | 861.31 |
| 23/ 1/1971 | 192.06 | 124.30 | 868.50 |
| 30/ 1/1971 | 192.58 | 123.79 | 876.57 |
| 6/ 2/1971 | 195.01 | 125.32 | 888.83 |
| 13/ 2/1971 | 193.69 | 124.20 | 878.56 |
| 20/ 2/1971 | 196.40 | 121.42 | 878.83 |
| 27/ 2/1971 | 203.20 | 122.12 | 898.00 |
| 6/ 3/1971 | 202.03 | 123.05 | 898.34 |
| 13/ 3/1971 | 202.28 | 125.42 | 912.12 |
| 20/ 3/1971 | 198.77 | 123.69 | 903.48 |
| 27/ 3/1971 | 201.72 | 122.92 | 903.88 |
| 3/ 4/1971 | 208.87 | 123.69 | 920.39 |
| 10/ 4/1971 | 218.23 | 124.01 | 938.17 |
| 17/ 4/1971 | 225.40 | 122.28 | 940.63 |
| 24/ 4/1971 | 225.89 | 119.79 | 948.15 |
| 1/ 5/1971 | 225.34 | 118.39 | 936.97 |
| 8/ 5/1971 | 222.60 | 118.80 | 936.06 |
| 15/ 5/1971 | 216.56 | 117.40 | 921.87 |
| 22/ 5/1971 | 217.40 | 114.42 | 907.81 |
| 29/ 5/1971 | 225.69 | 114.33 | 922.15 |
| 5/ 6/1971 | 223.08 | 113.94 | 916.47 |
| 12/ 6/1971 | 213.77 | 114.39 | 889.16 |
| 19/ 6/1971 | 210.33 | 114.78 | 876.68 |
| 26/ 6/1971 | 215.70 | 118.32 | 890.19 |
| 3/ 7/1971 | 219.98 | 119.92 | 901.80 |
| 10/ 7/1971 | 217.17 | 118.45 | 888.51 |
| 17/ 7/1971 | 214.83 | 117.46 | 887.78 |
| 24/ 7/1971 | 206.39 | 115.09 | 858.43 |
| 31/ 7/1971 | 205.98 | 112.83 | 850.61 |
| 7/ 8/1971 | 214.23 | 113.02 | 856.02 |
| 14/ 8/1971 | 232.60 | 113.31 | 880.91 |
| 21/ 8/1971 | 241.82 | 113.05 | 908.15 |
| 23/ 8/1971 | 245.99 | 112.70 | 912.75 |
| 4/ 9/1971 | 245.52 | 112.73 | 911.00 |
| 11/ 9/1971 | 242.62 | 111.32 | 908.22 |
| 18/ 9/1971 | 237.01 | 109.50 | 889.31 |
| 25/ 9/1971 | 237.86 | 110.91 | 893.98 |
| 2/10/1971 | 241.89 | 114.39 | 893.91 |
| 9/10/1971 | 237.54 | 115.86 | 874.85 |
| 16/10/1971 | 232.64 | 113.78 | 852.37 |
| 23/10/1971 | 229.21 | 111.90 | 839.00 |
| 30/10/1971 | 229.32 | 112.76 | 840.39 |
| 6/11/1971 | 223.04 | 111.42 | 812.94 |
| 13/11/1971 | 215.82 | 110.91 | 810.67 |
| 20/11/1971 | 214.18 | 108.86 | 816.59 |
| 27/11/1971 | 231.71 | 111.80 | 859.59 |
| 4/12/1971 | 238.81 | 111.74 | 856.75 |
| 11/12/1971 | 239.39 | 111.48 | 873.80 |
| 18/12/1971 | 238.88 | 112.38 | 881.17 |
| 25/12/1971 | 243.72 | 117.75 | 890.20 |

52 2243

| DATE | DJTA | DJUA | DJIA |
|---|---|---|---|
| 1/ 1/1972 | 245.63 | 119.34 | 910.37 |
| 8/ 1/1972 | 248.50 | 120.69 | 906.68 |
| 15/ 1/1972 | 251.71 | 118.03 | 907.44 |
| 22/ 1/1972 | 256.76 | 117.17 | 906.38 |
| 29/ 1/1972 | 256.01 | 114.87 | 906.68 |
| 5/ 2/1972 | 255.70 | 112.57 | 917.59 |
| 12/ 2/1972 | 254.73 | 112.25 | 917.52 |
| 19/ 2/1972 | 256.05 | 113.59 | 922.79 |
| 26/ 2/1972 | 258.71 | 114.23 | 942.43 |
| 4/ 3/1972 | 260.63 | 115.41 | 939.87 |
| 11/ 3/1972 | 259.82 | 115.70 | 942.88 |
| 18/ 3/1972 | 259.92 | 113.24 | 942.28 |
| 25/ 3/1972 | 258.93 | 112.47 | 940.70 |
| 1/ 4/1972 | 275.71 | 112.19 | 962.60 |
| 8/ 4/1972 | 274.08 | 111.00 | 967.72 |
| 15/ 4/1972 | 271.66 | 109.38 | 963.80 |
| 22/ 4/1972 | 258.78 | 109.76 | 954.17 |
| 29/ 4/1972 | 254.30 | 109.79 | 941.23 |
| 6/ 5/1972 | 256.58 | 108.86 | 941.83 |
| 13/ 5/1972 | 261.06 | 108.03 | 961.54 |
| 20/ 5/1972 | 261.06 | 108.16 | 971.25 |
| 27/ 5/1972 | 253.84 | 107.87 | 961.39 |
| 3/ 6/1972 | 244.77 | 106.63 | 934.45 |
| 10/ 6/1972 | 243.35 | 105.38 | 945.06 |
| 17/ 6/1972 | 241.85 | 105.67 | 944.69 |
| 24/ 6/1972 | 233.30 | 106.63 | 929.03 |
| 1/ 7/1972 | 235.17 | 108.45 | 938.06 |
| 8/ 7/1972 | 228.26 | 106.82 | 916.99 |
| 15/ 7/1972 | 229.30 | 106.56 | 910.45 |
| 22/ 7/1972 | 227.56 | 106.66 | 926.85 |
| 29/ 7/1972 | 233.94 | 106.95 | 951.76 |
| 5/ 8/1972 | 237.62 | 107.78 | 964.18 |
| 12/ 8/1972 | 232.09 | 110.25 | 965.83 |
| 19/ 8/1972 | 232.78 | 111.42 | 959.36 |
| 26/ 8/1972 | 233.91 | 110.75 | 970.05 |
| 2/ 9/1972 | 227.87 | 110.46 | 961.24 |
| 9/ 9/1972 | 221.22 | 109.25 | 947.32 |
| 16/ 9/1972 | 219.05 | 109.44 | 943.03 |
| 23/ 9/1972 | 217.70 | 110.56 | 953.27 |
| 30/ 9/1972 | 215.84 | 111.90 | 945.36 |
| 7/10/1972 | 214.83 | 111.36 | 930.46 |
| 14/10/1972 | 212.24 | 113.15 | 942.81 |
| 21/10/1972 | 218.55 | 116.40 | 946.42 |
| 28/10/1972 | 224.43 | 119.28 | 984.12 |
| 4/11/1972 | 224.35 | 119.73 | 995.26 |
| 11/11/1972 | 227.60 | 121.96 | 1005.57 |
| 18/11/1972 | 232.63 | 124.14 | 1025.21 |
| 25/11/1972 | 237.19 | 123.11 | 1023.93 |
| 2/12/1972 | 238.66 | 122.67 | 1033.19 |
| 9/12/1972 | 232.71 | 122.00 | 1027.24 |
| 16/12/1972 | 224.81 | 118.93 | 1004.21 |
| 23/12/1972 | 227.17 | 119.50 | 1020.02 |
| 30/12/1972 | 225.20 | 120.49 | 1047.49 |

53 2296

**TABLE A4-2 Dow Jones Transportation, Utility, and Industrial Weekly Averages** (cont.)

| DATE | DJTA | DJUA | DJIA |
|---|---|---|---|
| 6/ 1/1973 | 217.35 | 120.05 | 1039.36 |
| 13/ 1/1973 | 214.06 | 117.33 | 1026.19 |
| 20/ 1/1973 | 207.71 | 114.58 | 1003.54 |
| 27/ 1/1973 | 203.38 | 113.75 | 980.81 |
| 3/ 2/1973 | 206.17 | 113.34 | 979.46 |
| 10/ 2/1973 | 203.30 | 113.15 | 979.23 |
| 17/ 2/1973 | 198.20 | 112.44 | 959.89 |
| 24/ 2/1973 | 191.62 | 111.48 | 961.32 |
| 3/ 3/1973 | 195.61 | 110.88 | 972.23 |
| 10/ 3/1973 | 193.83 | 110.14 | 963.05 |
| 17/ 3/1973 | 189.22 | 106.60 | 922.71 |
| 24/ 3/1973 | 200.13 | 108.00 | 951.01 |
| 31/ 3/1973 | 195.30 | 106.50 | 931.01 |
| 7/ 4/1973 | 199.82 | 109.15 | 959.36 |
| 14/ 4/1973 | 196.61 | 109.25 | 963.20 |
| 21/ 4/1973 | 183.23 | 106.85 | 922.19 |
| 28/ 4/1973 | 188.06 | 108.61 | 953.87 |
| 5/ 5/1973 | 179.51 | 108.29 | 927.98 |
| 12/ 5/1973 | 165.43 | 105.96 | 895.17 |
| 19/ 5/1973 | 170.30 | 107.52 | 930.84 |
| 26/ 5/1973 | 161.52 | 106.85 | 893.96 |
| 2/ 6/1973 | 162.41 | 107.33 | 920.00 |
| 9/ 6/1973 | 162.88 | 106.12 | 888.55 |
| 16/ 6/1973 | 155.91 | 103.82 | 879.82 |
| 23/ 6/1973 | 156.18 | 102.12 | 891.71 |
| 30/ 6/1973 | 155.83 | 100.62 | 870.11 |
| 7/ 7/1973 | 161.75 | 101.55 | 885.99 |
| 14/ 7/1973 | 164.19 | 101.39 | 910.90 |
| 21/ 7/1973 | 166.74 | 100.56 | 936.71 |
| 28/ 7/1973 | 162.29 | 97.39 | 908.87 |
| 4/ 8/1973 | 158.35 | 96.08 | 892.38 |
| 11/ 8/1973 | 155.37 | 95.12 | 871.84 |
| 18/ 8/1973 | 154.44 | 95.16 | 863.49 |
| 25/ 8/1973 | 159.35 | 96.02 | 887.57 |
| 1/ 9/1973 | 162.88 | 100.84 | 893.63 |
| 8/ 9/1973 | 162.57 | 98.45 | 886.36 |
| 15/ 9/1973 | 172.47 | 100.65 | 927.90 |
| 22/ 9/1973 | 176.96 | 103.40 | 947.10 |
| 29/ 9/1973 | 183.34 | 103.02 | 971.25 |
| 6/10/1973 | 183.11 | 103.08 | 978.63 |
| 13/10/1973 | 183.38 | 99.37 | 963.73 |
| 20/10/1973 | 184.85 | 101.16 | 987.06 |
| 27/10/1973 | 180.01 | 98.38 | 935.28 |
| 3/11/1973 | 185.39 | 96.11 | 908.41 |
| 10/11/1973 | 175.68 | 93.27 | 891.33 |
| 17/11/1973 | 172.90 | 91.13 | 854.00 |
| 24/11/1973 | 175.18 | 87.93 | 822.25 |
| 1/12/1973 | 174.60 | 86.98 | 838.05 |
| 8/12/1973 | 172.39 | 87.17 | 815.65 |
| 15/12/1973 | 177.46 | 87.39 | 818.73 |
| 22/12/1973 | 194.33 | 89.21 | 848.02 |
| 29/12/1973 | 201.06 | 94.52 | 880.23 |

52 2348

| DATE | DJTA | DJUA | DJIA |
|---|---|---|---|
| 5/ 1/1974 | 188.10 | 91.86 | 841.48 |
| 12/ 1/1974 | 189.61 | 92.89 | 855.47 |
| 19/ 1/1974 | 188.95 | 93.46 | 859.39 |
| 26/ 1/1974 | 184.15 | 93.27 | 843.94 |
| 2/ 2/1974 | 182.68 | 93.49 | 820.40 |
| 9/ 2/1974 | 184.77 | 93.24 | 820.32 |
| 16/ 2/1974 | 190.96 | 94.10 | 855.99 |
| 23/ 2/1974 | 195.68 | 93.24 | 851.92 |
| 2/ 3/1974 | 195.45 | 93.69 | 878.05 |
| 9/ 3/1974 | 195.41 | 93.33 | 887.83 |
| 16/ 3/1974 | 193.17 | 91.90 | 878.13 |
| 23/ 3/1974 | 185.08 | 90.75 | 846.68 |
| 30/ 3/1974 | 183.50 | 89.24 | 847.54 |
| 6/ 4/1974 | 181.21 | 87.36 | 844.81 |
| 13/ 4/1974 | 183.92 | 86.69 | 859.90 |
| 20/ 4/1974 | 172.51 | 78.51 | 834.64 |
| 27/ 4/1974 | 174.48 | 78.28 | 845.90 |
| 4/ 5/1974 | 171.12 | 78.03 | 850.44 |
| 11/ 5/1974 | 166.36 | 75.57 | 818.84 |
| 18/ 5/1974 | 162.14 | 74.39 | 816.65 |
| 25/ 5/1974 | 160.09 | 73.36 | 802.17 |
| 1/ 6/1974 | 175.68 | 75.12 | 853.72 |
| 8/ 6/1974 | 173.63 | 72.47 | 843.09 |
| 15/ 6/1974 | 165.89 | 68.16 | 815.39 |
| 22/ 6/1974 | 162.18 | 68.22 | 802.41 |
| 29/ 6/1974 | 156.76 | 69.15 | 791.77 |
| 6/ 7/1974 | 155.33 | 67.23 | 787.23 |
| 13/ 7/1974 | 161.99 | 68.51 | 787.94 |
| 20/ 7/1974 | 163.30 | 70.62 | 787.57 |
| 27/ 7/1974 | 157.77 | 67.68 | 752.58 |
| 3/ 8/1974 | 161.52 | 69.72 | 777.30 |
| 10/ 8/1974 | 152.12 | 66.97 | 731.54 |
| 17/ 8/1974 | 143.07 | 64.16 | 686.80 |
| 24/ 8/1974 | 140.94 | 60.71 | 678.58 |
| 31/ 8/1974 | 140.24 | 61.19 | 677.88 |
| 7/ 9/1974 | 127.21 | 57.93 | 627.19 |
| 14/ 9/1974 | 137.61 | 63.52 | 670.76 |
| 21/ 9/1974 | 132.20 | 61.92 | 621.95 |
| 28/ 9/1974 | 127.71 | 61.54 | 584.56 |
| 5/10/1974 | 148.14 | 68.92 | 658.17 |
| 12/10/1974 | 149.03 | 70.20 | 654.88 |
| 19/10/1974 | 146.28 | 67.29 | 636.19 |
| 26/10/1974 | 153.55 | 68.60 | 665.21 |
| 2/11/1974 | 154.91 | 71.19 | 667.16 |
| 9/11/1974 | 151.66 | 69.24 | 647.61 |
| 16/11/1974 | 146.78 | 66.49 | 615.30 |
| 23/11/1974 | 148.25 | 67.39 | 618.66 |
| 30/11/1974 | 138.70 | 65.89 | 577.60 |
| 7/12/1974 | 139.20 | 67.90 | 592.77 |
| 14/12/1974 | 140.94 | 66.43 | 598.48 |
| 21/12/1974 | 140.90 | 66.53 | 602.16 |
| 28/12/1974 | 146.84 | 74.39 | 634.54 |

52 2400

**TABLE A4-2 Dow Jones Transportation, Utility, and Industrial Weekly Averages** (cont.)

| DATE | DJTA | DJUA | DJIA |
|---|---|---|---|
| 4/ 1/1975 | 153.19 | 77.90 | 658.79 |
| 11/ 1/1975 | 154.62 | 77.55 | 644.63 |
| 18/ 1/1975 | 153.19 | 78.96 | 666.61 |
| 25/ 1/1975 | 159.62 | 80.27 | 703.69 |
| 1/ 2/1975 | 157.90 | 81.64 | 711.91 |
| 8/ 2/1975 | 161.18 | 81.70 | 734.20 |
| 15/ 2/1975 | 162.53 | 81.96 | 749.77 |
| 22/ 2/1975 | 163.80 | 79.34 | 739.05 |
| 1/ 3/1975 | 166.46 | 80.39 | 770.10 |
| 8/ 3/1975 | 166.13 | 79.37 | 773.47 |
| 15/ 3/1975 | 162.98 | 76.97 | 763.06 |
| 22/ 3/1975 | 166.13 | 76.94 | 770.26 |
| 29/ 3/1975 | 161.63 | 76.18 | 747.26 |
| 5/ 4/1975 | 164.50 | 75.98 | 789.50 |
| 12/ 4/1975 | 169.37 | 76.56 | 808.43 |
| 19/ 4/1975 | 169.45 | 75.03 | 811.80 |
| 26/ 4/1975 | 171.99 | 74.64 | 848.48 |
| 3/ 5/1975 | 171.91 | 77.90 | 850.13 |
| 10/ 5/1975 | 168.80 | 78.16 | 837.61 |
| 17/ 5/1975 | 167.98 | 77.17 | 831.90 |
| 24/ 5/1975 | 167.85 | 79.82 | 832.29 |
| 31/ 5/1975 | 170.52 | 82.82 | 839.64 |
| 7/ 6/1975 | 167.53 | 81.90 | 824.47 |
| 14/ 6/1975 | 166.75 | 86.24 | 855.44 |
| 21/ 6/1975 | 171.34 | 86.02 | 873.12 |
| 28/ 6/1975 | 169.78 | 84.45 | 871.79 |
| 5/ 7/1975 | 172.69 | 84.48 | 871.09 |
| 12/ 7/1975 | 170.80 | 83.30 | 862.41 |
| 19/ 7/1975 | 161.22 | 80.11 | 834.09 |
| 26/ 7/1975 | 160.40 | 79.66 | 826.50 |
| 2/ 8/1975 | 156.63 | 78.67 | 817.74 |
| 9/ 8/1975 | 158.23 | 77.68 | 825.64 |
| 16/ 8/1975 | 154.58 | 76.43 | 804.76 |
| 23/ 8/1975 | 157.24 | 79.24 | 835.34 |
| 30/ 8/1975 | 155.32 | 77.74 | 835.97 |
| 6/ 9/1975 | 151.43 | 77.23 | 809.29 |
| 13/ 9/1975 | 156.83 | 76.97 | 829.79 |
| 20/ 9/1975 | 159.58 | 78.00 | 818.60 |
| 27/ 9/1975 | 158.39 | 78.19 | 813.21 |
| 4/10/1975 | 163.06 | 80.39 | 823.91 |
| 11/10/1975 | 164.86 | 82.31 | 832.18 |
| 18/10/1975 | 166.58 | 82.47 | 840.52 |
| 25/10/1975 | 166.38 | 82.63 | 836.04 |
| 1/11/1975 | 169.99 | 82.73 | 835.80 |
| 8/11/1975 | 173.43 | 83.56 | 853.67 |
| 15/11/1975 | 170.60 | 82.66 | 840.76 |
| 22/11/1975 | 169.29 | 83.27 | 860.67 |
| 29/11/1975 | 163.84 | 80.81 | 818.80 |
| 6/12/1975 | 164.78 | 81.00 | 832.81 |
| 13/12/1975 | 166.95 | 81.16 | 844.38 |
| 20/12/1975 | 170.68 | 82.73 | 859.81 |
| 27/12/1975 | 175.69 | 84.84 | 858.71 |

52 2452

| DATE | DJTA | DJUA | DJIA |
|---|---|---|---|
| 3/ 1/1976 | 185.81 | 87.97 | 911.13 |
| 10/ 1/1976 | 191.32 | 89.95 | 929.63 |
| 17/ 1/1976 | 197.38 | 90.62 | 953.95 |
| 24/ 1/1976 | 199.35 | 90.87 | 975.28 |
| 31/ 1/1976 | 198.36 | 89.82 | 954.90 |
| 7/ 2/1976 | 200.96 | 88.03 | 958.36 |
| 14/ 2/1976 | 206.94 | 89.37 | 987.80 |
| 21/ 2/1976 | 205.57 | 87.58 | 972.61 |
| 28/ 2/1976 | 205.21 | 85.79 | 972.92 |
| 6/ 3/1976 | 209.62 | 86.88 | 987.64 |
| 13/ 3/1976 | 206.35 | 86.85 | 979.85 |
| 20/ 3/1976 | 209.62 | 87.30 | 1003.46 |
| 27/ 3/1976 | 206.98 | 87.10 | 991.58 |
| 3/ 4/1976 | 202.34 | 85.79 | 968.28 |
| 10/ 4/1976 | 204.23 | 87.17 | 980.48 |
| 17/ 4/1976 | 212.14 | 87.87 | 1001.71 |
| 24/ 4/1976 | 212.77 | 87.74 | 996.85 |
| 1/ 5/1976 | 214.15 | 87.87 | 996.22 |
| 8/ 5/1976 | 219.54 | 86.98 | 992.60 |
| 15/ 5/1976 | 218.75 | 86.34 | 990.75 |
| 22/ 5/1976 | 212.96 | 85.28 | 975.23 |
| 29/ 5/1976 | 214.38 | 85.63 | 963.90 |
| 5/ 6/1976 | 214.74 | 85.89 | 978.80 |
| 12/ 6/1976 | 220.48 | 86.15 | 1001.88 |
| 19/ 6/1976 | 222.21 | 87.52 | 999.84 |
| 26/ 6/1976 | 224.26 | 88.54 | 999.84 |
| 3/ 7/1976 | 228.35 | 89.53 | 1003.11 |
| 10/ 7/1976 | 228.67 | 90.55 | 993.21 |
| 17/ 7/1976 | 226.70 | 91.10 | 990.91 |
| 24/ 7/1976 | 221.54 | 91.55 | 984.64 |
| 31/ 7/1976 | 222.52 | 93.56 | 986.00 |
| 7/ 8/1976 | 222.28 | 92.98 | 990.19 |
| 14/ 8/1976 | 217.51 | 92.66 | 974.07 |
| 21/ 8/1976 | 216.08 | 92.22 | 963.93 |
| 28/ 8/1976 | 220.40 | 94.52 | 989.11 |
| 4/ 9/1976 | 218.73 | 96.15 | 988.36 |
| 11/ 9/1976 | 218.77 | 97.36 | 995.10 |
| 18/ 9/1976 | 221.34 | 97.81 | 1009.31 |
| 25/ 9/1976 | 215.10 | 97.74 | 979.89 |
| 2/10/1976 | 209.56 | 98.16 | 952.38 |
| 9/10/1976 | 204.70 | 96.69 | 937.00 |
| 16/10/1976 | 205.23 | 96.18 | 938.75 |
| 23/10/1976 | 210.37 | 98.03 | 964.93 |
| 30/10/1976 | 214.23 | 98.00 | 943.07 |
| 6/11/1976 | 211.31 | 98.67 | 927.69 |
| 13/11/1976 | 221.90 | 100.24 | 948.80 |
| 20/11/1976 | 226.26 | 102.76 | 956.62 |
| 27/11/1976 | 228.59 | 103.14 | 950.55 |
| 4/12/1976 | 230.88 | 105.70 | 973.15 |
| 11/12/1976 | 234.87 | 105.13 | 979.06 |
| 18/12/1976 | 231.87 | 104.39 | 985.62 |
| 25/12/1976 | 237.03 | 108.38 | 999.09 |

52 2504

**TABLE A4-2 Dow Jones Transportation, Utility, and Industrial Weekly Averages** (cont.)

| DATE | DJTA | DJUA | DJIA |
|---|---|---|---|
| 1/ 1/1977 | 236.13 | 107.81 | 983.13 |
| 8/ 1/1977 | 233.66 | 108.22 | 972.16 |
| 15/ 1/1977 | 230.61 | 109.57 | 962.43 |
| 22/ 1/1977 | 226.97 | 110.08 | 957.53 |
| 29/ 1/1977 | 227.78 | 109.18 | 947.89 |
| 5/ 2/1977 | 224.24 | 106.08 | 931.52 |
| 12/ 2/1977 | 224.92 | 106.69 | 940.24 |
| 19/ 2/1977 | 221.81 | 104.97 | 933.43 |
| 26/ 2/1977 | 224.14 | 107.23 | 953.46 |
| 5/ 3/1977 | 224.48 | 106.69 | 947.72 |
| 12/ 3/1977 | 231.35 | 107.52 | 961.02 |
| 19/ 3/1977 | 226.29 | 105.22 | 928.86 |
| 26/ 3/1977 | 223.61 | 106.88 | 927.36 |
| 2/ 4/1977 | 224.00 | 107.07 | 918.88 |
| 9/ 4/1977 | 234.42 | 108.83 | 947.76 |
| 16/ 4/1977 | 233.59 | 107.97 | 927.07 |
| 23/ 4/1977 | 234.51 | 108.67 | 926.90 |
| 30/ 4/1977 | 238.36 | 110.11 | 936.74 |
| 7/ 5/1977 | 240.26 | 110.46 | 928.34 |
| 14/ 5/1977 | 245.03 | 111.74 | 930.46 |
| 21/ 5/1977 | 238.02 | 110.17 | 898.83 |
| 28/ 5/1977 | 237.14 | 111.77 | 912.23 |
| 4/ 6/1977 | 237.29 | 112.79 | 910.79 |
| 11/ 6/1977 | 238.80 | 113.88 | 920.45 |
| 18/ 6/1977 | 238.41 | 115.73 | 929.70 |
| 25/ 6/1977 | 237.83 | 115.06 | 912.65 |
| 2/ 7/1977 | 237.29 | 116.09 | 907.99 |
| 9/ 7/1977 | 235.93 | 117.84 | 905.95 |
| 16/ 7/1977 | 239.34 | 118.67 | 923.42 |
| 23/ 7/1977 | 229.30 | 116.37 | 890.07 |
| 30/ 7/1977 | 225.51 | 115.64 | 888.69 |
| 6/ 8/1977 | 218.88 | 114.68 | 871.10 |
| 13/ 8/1977 | 214.70 | 110.81 | 863.48 |
| 20/ 8/1977 | 214.55 | 109.95 | 855.42 |
| 27/ 8/1977 | 218.06 | 112.28 | 872.31 |
| 3/ 9/1977 | 216.64 | 112.50 | 857.07 |
| 10/ 9/1977 | 215.18 | 112.63 | 856.81 |
| 17/ 9/1977 | 213.48 | 112.50 | 839.14 |
| 24/ 9/1977 | 215.48 | 113.25 | 847.11 |
| 1/10/1977 | 216.89 | 114.04 | 840.35 |
| 8/10/1977 | 210.61 | 112.17 | 821.64 |
| 15/10/1977 | 204.81 | 110.52 | 808.30 |
| 22/10/1977 | 205.78 | 109.40 | 822.68 |
| 29/10/1977 | 204.37 | 107.79 | 809.94 |
| 5/11/1977 | 215.96 | 112.50 | 845.89 |
| 12/11/1977 | 215.18 | 112.10 | 835.76 |
| 19/11/1977 | 219.76 | 113.06 | 844.42 |
| 26/11/1977 | 215.57 | 112.92 | 823.98 |
| 3/12/1977 | 212.12 | 112.37 | 815.23 |
| 10/12/1977 | 214.26 | 111.28 | 815.32 |
| 17/12/1977 | 217.13 | 110.69 | 829.87 |
| 24/12/1977 | 217.18 | 111.28 | 831.17 |
| 31/12/1977 | 210.17 | 109.24 | 793.49 |

53 2557

**TABLE A4-3 New York Stock Exchange Volume**

| YEAR | MONTH | NYSTXCH* | 6MTH MA† | 12PC.6MA‡ |
|---|---|---|---|---|
| 1881 | JAN | 13.00 | 13.00 | 100.00 |
| 1881 | FEB | 12.13 | 12.13 | 100.00 |
| 1881 | MAR | 10.84 | 10.84 | 100.00 |
| 1881 | APR | 8.19 | 8.19 | 100.00 |
| 1881 | MAY | 12.38 | 12.38 | 100.00 |
| 1881 | JUN | 8.63 | 10.86 | 100.00 |
| 1881 | JUL | 8.61 | 10.13 | 100.00 |
| 1881 | AUG | 6.87 | 9.25 | 100.00 |
| 1881 | SEP | 6.35 | 8.50 | 100.00 |
| 1881 | OCT | 9.61 | 8.74 | 100.00 |
| 1881 | NOV | 7.98 | 8.01 | 100.00 |
| 1881 | DEC | 9.42 | 8.14 | 100.00 |
| 1882 | JAN | 11.31 | 8.59 | 66.08 |
| 1882 | FEB | 8.30 | 8.83 | 72.78 |
| 1882 | MAR | 12.82 | 9.91 | 91.39 |
| 1882 | APR | 9.89 | 9.95 | 121.53 |
| 1882 | MAY | 5.96 | 9.62 | 77.68 |
| 1882 | JUN | 9.25 | 9.59 | 88.28 |
| 1882 | JUL | 8.28 | 9.08 | 89.67 |
| 1882 | AUG | 9.10 | 9.22 | 99.60 |
| 1882 | SEP | 8.39 | 8.48 | 99.69 |
| 1882 | OCT | 10.26 | 8.54 | 97.69 |
| 1882 | NOV | 13.61 | 9.81 | 122.56 |
| 1882 | DEC | 9.13 | 9.79 | 120.33 |
| 1883 | JAN | 8.18 | 9.78 | 113.83 |
| 1883 | FEB | 8.62 | 9.70 | 109.85 |
| 1883 | MAR | 5.97 | 9.29 | 93.83 |
| 1883 | APR | 8.58 | 9.01 | 90.57 |
| 1883 | MAY | 7.30 | 7.96 | 82.81 |
| 1883 | JUN | 7.62 | 7.71 | 80.43 |
| 1883 | JUL | 5.54 | 7.27 | 80.05 |
| 1883 | AUG | 9.05 | 7.34 | 79.67 |
| 1883 | SEP | 9.07 | 7.86 | 92.71 |
| 1883 | OCT | 11.56 | 8.36 | 97.85 |
| 1883 | NOV | 6.37 | 8.20 | 83.56 |
| 1883 | DEC | 9.19 | 8.46 | 86.40 |
| 1884 | JAN | 10.09 | 9.22 | 94.31 |
| 1884 | FEB | 9.12 | 9.23 | 95.21 |
| 1884 | MAR | 6.29 | 8.77 | 94.35 |
| 1884 | APR | 9.16 | 8.37 | 92.34 |
| 1884 | MAY | 12.01 | 9.31 | 116.91 |
| 1884 | JUN | 9.19 | 9.31 | 120.73 |
| 1884 | JUL | 7.25 | 8.84 | 121.52 |
| 1884 | AUG | 7.50 | 8.57 | 116.66 |
| 1884 | SEP | 6.02 | 8.52 | 108.42 |
| 1884 | OCT | 6.80 | 8.13 | 97.27 |
| 1884 | NOV | 5.44 | 7.03 | 85.75 |
| 1884 | DEC | 7.92 | 6.82 | 80.60 |

* Monthly volume.

† 6-month moving average of monthly volume.

‡ 12-month rate of change of 6-month moving average.

| YEAR | MONTH | NYSTXCH | 6MTH MA | 12FC.6MA |
|---|---|---|---|---|
| 1885 | JAN | 7.13 | 6.80 | 73.76 |
| 1885 | FEB | 7.41 | 6.79 | 73.50 |
| 1885 | MAR | 6.54 | 6.87 | 78.37 |
| 1885 | APR | 4.49 | 6.49 | 77.52 |
| 1885 | MAY | 5.07 | 6.43 | 69.03 |
| 1885 | JUN | 4.44 | 5.85 | 62.80 |
| 1885 | JUL | 7.93 | 5.98 | 67.67 |
| 1885 | AUG | 6.87 | 5.89 | 68.75 |
| 1885 | SEP | 5.90 | 5.78 | 67.87 |
| 1885 | OCT | 12.66 | 7.14 | 87.90 |
| 1885 | NOV | 13.27 | 8.51 | 121.02 |
| 1885 | DEC | 10.84 | 9.58 | 140.41 |
| 1886 | JAN | 8.67 | 9.70 | 142.64 |
| 1886 | FEB | 9.41 | 10.12 | 149.19 |
| 1886 | MAR | 10.15 | 10.83 | 157.61 |
| 1886 | APR | 6.41 | 9.79 | 150.91 |
| 1886 | MAY | 6.64 | 8.69 | 135.17 |
| 1886 | JUN | 7.11 | 8.06 | 137.94 |
| 1886 | JUL | 5.07 | 7.46 | 124.83 |
| 1886 | AUG | 5.05 | 6.74 | 114.40 |
| 1886 | SEP | 8.54 | 6.47 | 111.87 |
| 1886 | OCT | 10.74 | 7.19 | 100.65 |
| 1886 | NOV | 10.88 | 7.90 | 92.79 |
| 1886 | DEC | 12.14 | 8.74 | 91.21 |
| 1887 | JAN | 8.15 | 9.25 | 95.34 |
| 1887 | FEB | 7.21 | 9.61 | 94.91 |
| 1887 | MAR | 7.15 | 9.38 | 86.57 |
| 1887 | APR | 9.47 | 9.17 | 93.62 |
| 1887 | MAY | 6.56 | 8.45 | 97.24 |
| 1887 | JUN | 6.99 | 7.59 | 94.09 |
| 1887 | JUL | 4.66 | 7.01 | 93.86 |
| 1887 | AUG | 6.39 | 6.87 | 101.95 |
| 1887 | SEP | 7.38 | 6.91 | 106.77 |
| 1887 | OCT | 8.30 | 6.71 | 93.35 |
| 1887 | NOV | 7.50 | 6.87 | 86.98 |
| 1887 | DEC | 5.16 | 6.56 | 75.14 |
| 1888 | JAN | 3.93 | 6.44 | 69.66 |
| 1888 | FEB | 3.15 | 5.90 | 61.43 |
| 1888 | MAR | 5.25 | 5.55 | 59.16 |
| 1888 | APR | 7.61 | 5.43 | 59.27 |
| 1888 | MAY | 6.21 | 5.22 | 61.78 |
| 1888 | JUN | 3.83 | 5.00 | 65.85 |
| 1888 | JUL | 4.68 | 5.12 | 73.10 |
| 1888 | AUG | 4.74 | 5.39 | 78.41 |
| 1888 | SEP | 7.32 | 5.73 | 82.97 |
| 1888 | OCT | 6.74 | 5.59 | 83.22 |
| 1888 | NOV | 5.34 | 5.44 | 79.21 |
| 1888 | DEC | 6.38 | 5.87 | 89.36 |

**TABLE A4-3 New York Stock Exchange Volume** (cont.)

| YEAR | MONTH | NYSTXCH | 6MTH MA | 12RC.6MA |
|---|---|---|---|---|
| 1889 | JAN | 4.87 | 5.90 | 91.54 |
| 1889 | FEB | 5.93 | 6.10 | 103.27 |
| 1889 | MAR | 6.15 | 5.90 | 106.37 |
| 1889 | APR | 4.82 | 5.58 | 102.73 |
| 1889 | MAY | 7.16 | 5.88 | 112.78 |
| 1889 | JUN | 6.78 | 5.95 | 119.11 |
| 1889 | JUL | 5.63 | 6.08 | 118.68 |
| 1889 | AUG | 5.06 | 5.93 | 110.15 |
| 1889 | SEP | 5.64 | 5.85 | 102.04 |
| 1889 | OCT | 7.58 | 6.31 | 112.92 |
| 1889 | NOV | 6.98 | 6.28 | 115.38 |
| 1889 | DEC | 5.64 | 6.09 | 103.78 |
| 1890 | JAN | 6.35 | 6.21 | 105.26 |
| 1890 | FEB | 5.20 | 6.23 | 102.21 |
| 1890 | MAR | 4.50 | 6.04 | 102.37 |
| 1890 | APR | 5.08 | 5.62 | 100.78 |
| 1890 | MAY | 11.05 | 6.30 | 107.11 |
| 1890 | JUN | 5.44 | 6.27 | 105.35 |
| 1890 | JUL | 3.01 | 5.71 | 93.99 |
| 1890 | AUG | 4.14 | 5.54 | 93.31 |
| 1890 | SEP | 5.14 | 5.64 | 96.49 |
| 1890 | OCT | 7.26 | 6.01 | 95.22 |
| 1890 | NOV | 8.97 | 5.66 | 90.15 |
| 1890 | DEC | 5.14 | 5.61 | 92.14 |
| 1891 | JAN | 5.62 | 6.04 | 97.37 |
| 1891 | FEB | 3.28 | 5.90 | 94.70 |
| 1891 | MAR | 3.65 | 5.65 | 93.57 |
| 1891 | APR | 7.18 | 5.64 | 100.27 |
| 1891 | MAY | 6.29 | 5.19 | 82.39 |
| 1891 | JUN | 3.98 | 5.00 | 79.74 |
| 1891 | JUL | 3.15 | 4.59 | 80.31 |
| 1891 | AUG | 5.85 | 5.02 | 90.61 |
| 1891 | SEP | 11.18 | 6.27 | 111.13 |
| 1891 | OCT | 6.74 | 6.20 | 103.19 |
| 1891 | NOV | 5.35 | 6.04 | 106.74 |
| 1891 | DEC | 6.78 | 6.51 | 116.01 |
| 1892 | JAN | 9.83 | 7.62 | 126.08 |
| 1892 | FEB | 11.43 | 8.55 | 144.90 |
| 1892 | MAR | 8.90 | 8.17 | 144.55 |
| 1892 | APR | 6.83 | 8.19 | 145.16 |
| 1892 | MAY | 6.15 | 8.32 | 160.21 |
| 1892 | JUN | 4.66 | 7.97 | 159.34 |
| 1892 | JUL | 3.47 | 6.91 | 150.53 |
| 1892 | AUG | 5.47 | 5.91 | 117.87 |
| 1892 | SEP | 6.85 | 5.57 | 88.84 |
| 1892 | OCT | 7.02 | 5.60 | 90.40 |
| 1892 | NOV | 5.82 | 5.55 | 91.83 |
| 1892 | DEC | 8.31 | 6.16 | 94.60 |

| YEAR | MONTH | NYSTXCH | 6MTH MA | 12PC.6MA |
|---|---|---|---|---|
| 1893 | JAN | 10.56 | 7.34 | 96.28 |
| 1893 | FEB | 10.82 | 8.23 | 96.24 |
| 1893 | MAR | 7.54 | 8.34 | 102.12 |
| 1893 | APR | 6.24 | 8.21 | 100.35 |
| 1893 | MAY | 8.81 | 8.71 | 104.73 |
| 1893 | JUN | 4.82 | 8.13 | 102.07 |
| 1893 | JUL | 5.89 | 7.35 | 106.47 |
| 1893 | AUG | 4.90 | 6.37 | 107.67 |
| 1893 | SEP | 4.71 | 5.89 | 105.80 |
| 1893 | OCT | 6.06 | 5.86 | 104.67 |
| 1893 | NOV | 5.45 | 5.30 | 95.61 |
| 1893 | DEC | 4.89 | 5.32 | 86.36 |
| 1894 | JAN | 4.59 | 5.10 | 69.50 |
| 1894 | FEB | 3.17 | 4.81 | 58.46 |
| 1894 | MAR | 4.73 | 4.81 | 57.70 |
| 1894 | APR | 4.01 | 4.47 | 54.45 |
| 1894 | MAY | 5.07 | 4.41 | 50.61 |
| 1894 | JUN | 3.38 | 4.16 | 51.14 |
| 1894 | JUL | 2.82 | 3.86 | 52.54 |
| 1894 | AUG | 5.04 | 4.17 | 65.57 |
| 1894 | SEP | 4.07 | 4.06 | 68.95 |
| 1894 | OCT | 3.84 | 4.04 | 68.82 |
| 1894 | NOV | 4.53 | 3.95 | 74.39 |
| 1894 | DEC | 4.07 | 4.06 | 76.39 |
| 1895 | JAN | 3.24 | 4.13 | 81.01 |
| 1895 | FEB | 3.02 | 3.79 | 78.87 |
| 1895 | MAR | 5.11 | 3.97 | 82.41 |
| 1895 | APR | 5.00 | 4.16 | 93.03 |
| 1895 | MAY | 8.89 | 4.89 | 110.85 |
| 1895 | JUN | 6.00 | 5.21 | 125.29 |
| 1895 | JUL | 5.85 | 5.64 | 146.12 |
| 1895 | AUG | 5.28 | 6.02 | 144.24 |
| 1895 | SEP | 6.81 | 6.30 | 155.11 |
| 1895 | OCT | 5.21 | 6.34 | 157.07 |
| 1895 | NOV | 4.99 | 5.69 | 144.18 |
| 1895 | DEC | 6.75 | 5.81 | 143.17 |
| 1896 | JAN | 4.40 | 5.57 | 134.90 |
| 1896 | FEB | 5.19 | 5.56 | 146.47 |
| 1896 | MAR | 4.58 | 5.19 | 130.70 |
| 1896 | APR | 4.06 | 4.99 | 120.03 |
| 1896 | MAY | 2.78 | 4.63 | 94.65 |
| 1896 | JUN | 4.36 | 4.23 | 81.16 |
| 1896 | JUL | 5.54 | 4.42 | 78.27 |
| 1896 | AUG | 4.27 | 4.26 | 70.93 |
| 1896 | SEP | 4.57 | 4.26 | 67.62 |
| 1896 | OCT | 4.90 | 4.40 | 69.45 |
| 1896 | NOV | 5.87 | 4.92 | 86.44 |
| 1896 | DEC | 2.92 | 4.68 | 80.45 |

**TABLE A4-3 New York Stock Exchange Volume** (cont.)

| YEAR | MONTH | NYSTXCH | 6MTH MA | 12PC.6MA |
|---|---|---|---|---|
| 1897 | JAN | 3.37 | 4.32 | 77.45 |
| 1897 | FEB | 2.82 | 4.07 | 73.31 |
| 1897 | MAR | 5.07 | 4.16 | 80.17 |
| 1897 | APR | 3.54 | 3.93 | 78.71 |
| 1897 | MAY | 4.21 | 3.65 | 79.00 |
| 1897 | JUN | 6.42 | 4.24 | 100.24 |
| 1897 | JUL | 7.01 | 4.84 | 109.66 |
| 1897 | AUG | 11.46 | 6.28 | 147.37 |
| 1897 | SEP | 13.09 | 7.62 | 178.78 |
| 1897 | OCT | 8.01 | 8.37 | 190.02 |
| 1897 | NOV | 5.76 | 8.62 | 175.37 |
| 1897 | DEC | 7.44 | 8.79 | 188.00 |
| 1898 | JAN | 9.22 | 9.16 | 212.29 |
| 1898 | FEB | 8.98 | 8.75 | 214.74 |
| 1898 | MAR | 9.95 | 8.23 | 197.85 |
| 1898 | APR | 6.00 | 7.89 | 200.73 |
| 1898 | MAY | 9.17 | 8.46 | 231.48 |
| 1898 | JUN | 9.10 | 8.74 | 206.15 |
| 1898 | JUL | 4.78 | 8.00 | 165.06 |
| 1898 | AUG | 12.01 | 8.50 | 135.27 |
| 1898 | SEP | 9.37 | 8.40 | 110.28 |
| 1898 | OCT | 7.42 | 8.64 | 103.29 |
| 1898 | NOV | 10.94 | 8.94 | 103.61 |
| 1898 | DEC | 15.22 | 9.96 | 113.21 |
| 1899 | JAN | 24.14 | 13.18 | 143.87 |
| 1899 | FEB | 15.98 | 13.84 | 158.23 |
| 1899 | MAR | 17.68 | 15.23 | 185.13 |
| 1899 | APR | 16.98 | 16.82 | 213.19 |
| 1899 | MAY | 14.79 | 17.46 | 206.45 |
| 1899 | JUN | 10.88 | 16.74 | 191.63 |
| 1899 | JUL | 8.02 | 14.05 | 175.77 |
| 1899 | AUG | 12.81 | 13.53 | 159.11 |
| 1899 | SEP | 12.35 | 12.64 | 150.37 |
| 1899 | OCT | 10.80 | 11.61 | 134.33 |
| 1899 | NOV | 13.58 | 11.41 | 127.64 |
| 1899 | DEC | 17.05 | 12.43 | 124.89 |
| 1900 | JAN | 9.86 | 12.74 | 96.65 |
| 1900 | FEB | 10.21 | 12.31 | 88.90 |
| 1900 | MAR | 14.45 | 12.66 | 83.11 |
| 1900 | APR | 14.65 | 13.30 | 79.06 |
| 1900 | MAY | 9.49 | 12.62 | 72.25 |
| 1900 | JUN | 7.29 | 10.99 | 65.65 |
| 1900 | JUL | 6.27 | 10.39 | 73.95 |
| 1900 | AUG | 4.01 | 9.36 | 69.20 |
| 1900 | SEP | 5.16 | 7.81 | 61.81 |
| 1900 | OCT | 10.90 | 7.19 | 61.91 |
| 1900 | NOV | 22.65 | 9.38 | 82.23 |
| 1900 | DEC | 23.38 | 12.06 | 97.00 |

| YEAR | MONTH | NYSTXCH' | 6MTH MA | 12RC.6MA |
|---|---|---|---|---|
| 1901 | JAN | 30.21 | 16.05 | 125.98 |
| 1901 | FEB | 21.88 | 19.03 | 154.61 |
| 1901 | MAR | 27.00 | 22.67 | 179.09 |
| 1901 | APR | 41.69 | 27.80 | 209.04 |
| 1901 | MAY | 35.20 | 29.89 | 236.91 |
| 1901 | JUN | 19.82 | 29.30 | 266.57 |
| 1901 | JUL | 15.92 | 26.92 | 259.00 |
| 1901 | AUG | 10.77 | 25.07 | 267.82 |
| 1901 | SEP | 14.03 | 22.90 | 293.23 |
| 1901 | OCT | 14.02 | 18.29 | 254.56 |
| 1901 | NOV | 18.36 | 15.49 | 165.11 |
| 1901 | DEC | 16.67 | 14.96 | 124.04 |
| 1902 | JAN | 14.76 | 14.77 | 92.00 |
| 1902 | FEB | 12.95 | 15.13 | 79.51 |
| 1902 | MAR | 11.95 | 14.78 | 65.22 |
| 1902 | APR | 26.58 | 16.88 | 60.71 |
| 1902 | MAY | 13.49 | 16.07 | 53.75 |
| 1902 | JUN | 7.81 | 14.59 | 49.79 |
| 1902 | JUL | 16.32 | 14.85 | 55.17 |
| 1902 | AUG | 14.32 | 15.08 | 60.15 |
| 1902 | SEP | 20.95 | 16.58 | 72.38 |
| 1902 | OCT | 16.35 | 14.87 | 81.30 |
| 1902 | NOV | 17.12 | 15.48 | 99.95 |
| 1902 | DEC | 15.72 | 16.80 | 112.26 |
| 1903 | JAN | 16.01 | 16.74 | 113.38 |
| 1903 | FEB | 10.93 | 16.18 | 106.93 |
| 1903 | MAR | 15.02 | 15.19 | 102.75 |
| 1903 | APR | 12.24 | 14.51 | 85.95 |
| 1903 | MAY | 12.46 | 13.73 | 85.46 |
| 1903 | JUN | 15.54 | 13.70 | 93.90 |
| 1903 | JUL | 14.78 | 13.49 | 90.87 |
| 1903 | AUG | 14.46 | 14.08 | 93.40 |
| 1903 | SEP | 10.71 | 13.36 | 80.62 |
| 1903 | OCT | 12.67 | 13.44 | 90.34 |
| 1903 | NOV | 10.74 | 13.15 | 84.96 |
| 1903 | DEC | 15.18 | 13.09 | 77.93 |
| 1904 | JAN | 12.24 | 12.67 | 75.64 |
| 1904 | FEB | 8.57 | 11.68 | 72.22 |
| 1904 | MAR | 11.42 | 11.80 | 77.70 |
| 1904 | APR | 8.16 | 11.05 | 76.18 |
| 1904 | MAY | 5.26 | 10.14 | 73.84 |
| 1904 | JUN | 4.99 | 8.44 | 61.60 |
| 1904 | JUL | 12.13 | 8.42 | 62.40 |
| 1904 | AUG | 12.44 | 9.07 | 64.38 |
| 1904 | SEP | 18.70 | 10.28 | 76.92 |
| 1904 | OCT | 32.48 | 14.33 | 106.67 |
| 1904 | NOV | 31.96 | 18.78 | 142.84 |
| 1904 | DEC | 28.18 | 22.65 | 173.02 |

**TABLE A4-3 New York Stock Exchange Volume** (cont.)

| YEAR | MONTH | NYSTXCH | 6MTH MA | 12RC.6MA |
|---|---|---|---|---|
| 1905 | JAN | 20.77 | 24.09 | 190.18 |
| 1905 | FEB | 25.36 | 26.24 | 224.58 |
| 1905 | MAR | 29.06 | 27.97 | 236.96 |
| 1905 | APR | 29.37 | 27.45 | 248.39 |
| 1905 | MAY | 20.54 | 25.55 | 251.99 |
| 1905 | JUN | 12.54 | 22.94 | 271.81 |
| 1905 | JUL | 13.02 | 21.65 | 257.07 |
| 1905 | AUG | 20.25 | 20.80 | 229.38 |
| 1905 | SEP | 16.09 | 18.63 | 181.28 |
| 1905 | OCT | 17.74 | 16.70 | 116.49 |
| 1905 | NOV | 26.88 | 17.75 | 94.52 |
| 1905 | DEC | 31.41 | 20.90 | 92.27 |
| 1906 | JAN | 38.55 | 25.15 | 104.42 |
| 1906 | FEB | 21.69 | 25.39 | 96.77 |
| 1906 | MAR | 19.33 | 25.93 | 92.72 |
| 1906 | APR | 24.30 | 27.03 | 98.46 |
| 1906 | MAY | 23.95 | 26.54 | 103.88 |
| 1906 | JUN | 20.28 | 24.68 | 107.60 |
| 1906 | JUL | 16.30 | 20.97 | 96.89 |
| 1906 | AUG | 31.72 | 22.65 | 108.90 |
| 1906 | SEP | 26.12 | 23.78 | 127.60 |
| 1906 | OCT | 21.80 | 23.36 | 139.92 |
| 1906 | NOV | 19.41 | 22.60 | 127.33 |
| 1906 | DEC | 20.26 | 22.60 | 108.15 |
| 1907 | JAN | 22.89 | 23.70 | 94.22 |
| 1907 | FEB | 16.48 | 21.16 | 83.33 |
| 1907 | MAR | 32.25 | 22.18 | 85.53 |
| 1907 | APR | 19.22 | 21.75 | 80.48 |
| 1907 | MAY | 15.76 | 21.14 | 79.67 |
| 1907 | JUN | 9.73 | 19.39 | 78.55 |
| 1907 | JUL | 12.80 | 17.71 | 84.42 |
| 1907 | AUG | 14.50 | 17.38 | 76.73 |
| 1907 | SEP | 12.14 | 14.02 | 58.98 |
| 1907 | OCT | 17.31 | 13.71 | 58.67 |
| 1907 | NOV | 9.65 | 12.69 | 56.13 |
| 1907 | DEC | 12.54 | 13.16 | 58.21 |
| 1908 | JAN | 16.62 | 13.79 | 58.20 |
| 1908 | FEB | 9.92 | 13.03 | 61.58 |
| 1908 | MAR | 15.80 | 13.64 | 61.49 |
| 1908 | APR | 11.61 | 12.69 | 58.34 |
| 1908 | MAY | 20.92 | 14.57 | 68.90 |
| 1908 | JUN | 9.54 | 14.07 | 72.56 |
| 1908 | JUL | 13.87 | 13.61 | 76.86 |
| 1908 | AUG | 18.85 | 15.10 | 86.89 |
| 1908 | SEP | 17.50 | 15.38 | 109.67 |
| 1908 | OCT | 14.27 | 15.82 | 115.46 |
| 1908 | NOV | 24.88 | 16.48 | 129.92 |
| 1908 | DEC | 22.96 | 18.72 | 142.30 |

| YEAR | MONTH | NYSTXCH | 6MTH MA | 12RC.6MA |
|---|---|---|---|---|
| 1909 | JAN | 17.27 | 19.29 | 139.84 |
| 1909 | FEB | 12.74 | 18.20 | 139.71 |
| 1909 | MAR | 13.65 | 17.56 | 128.75 |
| 1909 | APR | 18.97 | 18.34 | 144.56 |
| 1909 | MAY | 16.51 | 16.95 | 116.35 |
| 1909 | JUN | 20.36 | 16.52 | 117.40 |
| 1909 | JUL | 12.81 | 15.77 | 115.90 |
| 1909 | AUG | 24.51 | 17.80 | 117.91 |
| 1909 | SEP | 20.05 | 18.87 | 122.67 |
| 1909 | OCT | 21.71 | 19.32 | 122.12 |
| 1909 | NOV | 18.74 | 19.70 | 119.48 |
| 1909 | DEC | 17.49 | 19.22 | 102.65 |
| 1910 | JAN | 24.12 | 21.10 | 109.41 |
| 1910 | FEB | 15.99 | 19.68 | 108.13 |
| 1910 | MAR | 14.99 | 18.84 | 107.28 |
| 1910 | APR | 14.07 | 17.57 | 95.76 |
| 1910 | MAY | 11.95 | 16.43 | 96.96 |
| 1910 | JUN | 16.28 | 16.23 | 98.28 |
| 1910 | JUL | 14.30 | 14.60 | 92.54 |
| 1910 | AUG | 10.22 | 13.63 | 76.59 |
| 1910 | SEP | 7.68 | 12.42 | 65.81 |
| 1910 | OCT | 13.43 | 12.31 | 63.70 |
| 1910 | NOV | 10.81 | 12.12 | 51.53 |
| 1910 | DEC | 9.89 | 11.05 | 57.52 |
| 1911 | JAN | 10.38 | 10.40 | 49.29 |
| 1911 | FEB | 10.17 | 10.39 | 52.80 |
| 1911 | MAR | 6.92 | 10.27 | 54.49 |
| 1911 | APR | 5.04 | 8.87 | 50.48 |
| 1911 | MAY | 10.69 | 8.85 | 53.84 |
| 1911 | JUN | 10.43 | 8.94 | 55.06 |
| 1911 | JUL | 5.44 | 8.11 | 55.59 |
| 1911 | AUG | 15.04 | 8.93 | 65.47 |
| 1911 | SEP | 17.37 | 10.67 | 85.92 |
| 1911 | OCT | 11.05 | 11.67 | 94.80 |
| 1911 | NOV | 14.90 | 12.37 | 102.08 |
| 1911 | DEC | 9.07 | 12.14 | 109.86 |
| 1912 | JAN | 10.91 | 13.06 | 125.53 |
| 1912 | FEB | 7.09 | 11.73 | 112.88 |
| 1912 | MAR | 14.55 | 11.26 | 109.69 |
| 1912 | APR | 15.99 | 12.08 | 136.27 |
| 1912 | MAY | 13.66 | 11.88 | 134.25 |
| 1912 | JUN | 7.20 | 11.57 | 129.41 |
| 1912 | JUL | 7.17 | 10.94 | 134.86 |
| 1912 | AUG | 8.97 | 11.26 | 126.10 |
| 1912 | SEP | 10.06 | 10.51 | 98.50 |
| 1912 | OCT | 14.15 | 10.20 | 87.42 |
| 1912 | NOV | 8.71 | 9.38 | 75.79 |
| 1912 | DEC | 12.60 | 10.28 | 84.62 |

**TABLE A4-3 New York Stock Exchange Volume** (cont.)

| YEAR | MONTH | NYSTXCH | 6MTH MA | 12RC.6MA |
|---|---|---|---|---|
| 1913 | JAN | 8.73 | 10.54 | 80.70 |
| 1913 | FEB | 6.64 | 10.15 | 86.50 |
| 1913 | MAR | 7.18 | 9.67 | 85.85 |
| 1913 | APR | 8.46 | 8.72 | 72.15 |
| 1913 | MAY | 5.46 | 8.18 | 68.85 |
| 1913 | JUN | 9.59 | 7.68 | 66.37 |
| 1913 | JUL | 5.12 | 7.07 | 64.65 |
| 1913 | AUG | 6.08 | 6.98 | 62.02 |
| 1913 | SEP | 7.68 | 7.06 | 67.23 |
| 1913 | OCT | 7.41 | 6.89 | 67.54 |
| 1913 | NOV | 3.77 | 6.61 | 70.47 |
| 1913 | DEC | 7.15 | 6.20 | 60.34 |
| 1914 | JAN | 10.11 | 7.03 | 66.75 |
| 1914 | FEB | 6.23 | 7.06 | 69.55 |
| 1914 | MAR | 5.86 | 6.75 | 69.86 |
| 1914 | APR | 7.14 | 6.71 | 76.95 |
| 1914 | MAY | 4.76 | 6.87 | 84.06 |
| 1914 | JUN | 4.00 | 6.35 | 82.72 |
| 1914 | JUL | 7.89 | 5.98 | 84.52 |
| 1914 | AUG | 6.50 | 6.02 | 86.30 |
| 1914 | SEP | 5.50 | 5.96 | 84.43 |
| 1914 | OCT | 3.50 | 5.36 | 77.77 |
| 1914 | NOV | 2.50 | 4.98 | 75.38 |
| 1914 | DEC | 1.91 | 4.63 | 74.71 |
| 1915 | JAN | 5.08 | 4.16 | 59.21 |
| 1915 | FEB | 4.38 | 3.81 | 54.00 |
| 1915 | MAR | 7.88 | 4.21 | 62.29 |
| 1915 | APR | 21.05 | 7.13 | 106.31 |
| 1915 | MAY | 12.67 | 8.83 | 128.42 |
| 1915 | JUN | 11.21 | 10.38 | 163.45 |
| 1915 | JUL | 14.33 | 11.92 | 199.34 |
| 1915 | AUG | 20.42 | 14.59 | 242.23 |
| 1915 | SEP | 18.50 | 16.36 | 274.35 |
| 1915 | OCT | 26.64 | 17.29 | 322.80 |
| 1915 | NOV | 17.56 | 18.11 | 363.58 |
| 1915 | DEC | 13.68 | 18.52 | 399.80 |
| 1916 | JAN | 15.94 | 18.79 | 451.21 |
| 1916 | FEB | 12.20 | 17.42 | 457.09 |
| 1916 | MAR | 15.13 | 16.86 | 400.65 |
| 1916 | APR | 12.53 | 14.51 | 203.38 |
| 1916 | MAY | 16.40 | 14.31 | 162.14 |
| 1916 | JUN | 12.79 | 14.16 | 136.49 |
| 1916 | JUL | 9.18 | 13.04 | 109.38 |
| 1916 | AUG | 14.60 | 13.44 | 92.08 |
| 1916 | SEP | 29.85 | 15.89 | 97.12 |
| 1916 | OCT | 27.98 | 18.47 | 106.77 |
| 1916 | NOV | 34.51 | 21.48 | 118.64 |
| 1916 | DEC | 31.71 | 24.64 | 133.03 |

| YEAR | MONTH | NYSTXCH | 6MTH MA | 12RC.6MA |
|---|---|---|---|---|
| 1917 | JAN | 16.42 | 25.84 | 137.55 |
| 1917 | FEB | 13.63 | 25.68 | 147.44 |
| 1917 | MAR | 18.42 | 23.78 | 141.05 |
| 1917 | APR | 14.28 | 21.49 | 148.13 |
| 1917 | MAY | 19.54 | 19.00 | 132.75 |
| 1917 | JUN | 18.99 | 16.88 | 119.17 |
| 1917 | JUL | 12.79 | 16.27 | 124.83 |
| 1917 | AUG | 11.51 | 15.92 | 118.48 |
| 1917 | SEP | 13.70 | 15.13 | 95.24 |
| 1917 | OCT | 17.43 | 15.65 | 84.80 |
| 1917 | NOV | 14.71 | 14.85 | 69.14 |
| 1917 | DEC | 12.78 | 13.82 | 56.09 |
| 1918 | JAN | 12.60 | 13.79 | 53.35 |
| 1918 | FEB | 11.32 | 13.76 | 53.56 |
| 1918 | MAR | 8.21 | 12.84 | 54.00 |
| 1918 | APR | 7.44 | 11.18 | 51.99 |
| 1918 | MAY | 21.10 | 12.24 | 64.43 |
| 1918 | JUN | 11.60 | 12.04 | 71.36 |
| 1918 | JUL | 8.35 | 11.34 | 69.66 |
| 1918 | AUG | 6.86 | 10.59 | 66.53 |
| 1918 | SEP | 8.07 | 10.57 | 69.84 |
| 1918 | OCT | 20.29 | 12.71 | 81.17 |
| 1918 | NOV | 14.67 | 11.64 | 78.36 |
| 1918 | DEC | 11.96 | 11.70 | 84.66 |
| 1919 | JAN | 11.63 | 12.25 | 88.82 |
| 1919 | FEB | 11.75 | 13.06 | 94.95 |
| 1919 | MAR | 21.17 | 15.24 | 118.72 |
| 1919 | APR | 28.64 | 16.64 | 148.86 |
| 1919 | MAY | 34.18 | 19.89 | 162.47 |
| 1919 | JUN | 32.83 | 23.37 | 194.00 |
| 1919 | JUL | 34.10 | 27.11 | 239.16 |
| 1919 | AUG | 24.14 | 29.18 | 275.44 |
| 1919 | SEP | 24.24 | 29.69 | 280.89 |
| 1919 | OCT | 36.89 | 31.06 | 244.33 |
| 1919 | NOV | 29.75 | 30.32 | 260.54 |
| 1919 | DEC | 24.39 | 28.92 | 247.13 |
| 1920 | JAN | 19.65 | 26.51 | 216.48 |
| 1920 | FEB | 21.73 | 26.11 | 199.89 |
| 1920 | MAR | 28.80 | 26.87 | 176.25 |
| 1920 | APR | 27.98 | 25.38 | 152.58 |
| 1920 | MAY | 16.37 | 23.15 | 116.42 |
| 1920 | JUN | 9.20 | 20.62 | 88.25 |
| 1920 | JUL | 12.40 | 19.41 | 71.60 |
| 1920 | AUG | 13.70 | 18.07 | 61.95 |
| 1920 | SEP | 15.32 | 15.83 | 53.31 |
| 1920 | OCT | 13.61 | 13.43 | 43.24 |
| 1920 | NOV | 22.16 | 14.40 | 47.48 |
| 1920 | DEC | 23.83 | 16.84 | 58.22 |

**TABLE A4-3 New York Stock Exchange Volume** (cont.)

| YEAR | MONTH | NYSTXCH | 6MTH MA | 12ROC.6MA |
|---|---|---|---|---|
| 1921 | JAN | 15.98 | 17.43 | 65.76 |
| 1921 | FEB | 10.15 | 16.84 | 64.51 |
| 1921 | MAR | 15.91 | 16.94 | 63.05 |
| 1921 | APR | 15.27 | 17.22 | 67.83 |
| 1921 | MAY | 17.03 | 16.36 | 70.67 |
| 1921 | JUN | 18.17 | 15.42 | 74.77 |
| 1921 | JUL | 9.30 | 14.30 | 73.69 |
| 1921 | AUG | 10.99 | 14.44 | 79.92 |
| 1921 | SEP | 12.81 | 13.93 | 88.00 |
| 1921 | OCT | 12.88 | 13.53 | 100.72 |
| 1921 | NOV | 15.33 | 13.25 | 92.00 |
| 1921 | DEC | 17.62 | 13.15 | 78.13 |
| 1922 | JAN | 15.39 | 14.17 | 81.28 |
| 1922 | FEB | 16.19 | 15.04 | 89.28 |
| 1922 | MAR | 22.73 | 16.69 | 98.52 |
| 1922 | APR | 30.47 | 19.62 | 113.97 |
| 1922 | MAY | 28.91 | 21.88 | 133.76 |
| 1922 | JUN | 24.04 | 22.95 | 148.88 |
| 1922 | JUL | 15.15 | 22.91 | 160.19 |
| 1922 | AUG | 17.85 | 23.19 | 160.56 |
| 1922 | SEP | 21.78 | 23.03 | 165.37 |
| 1922 | OCT | 25.68 | 22.23 | 164.34 |
| 1922 | NOV | 20.78 | 20.88 | 157.63 |
| 1922 | DEC | 19.66 | 20.15 | 153.18 |
| 1923 | JAN | 20.21 | 20.99 | 148.16 |
| 1923 | FEB | 22.69 | 21.80 | 144.98 |
| 1923 | MAR | 25.86 | 22.48 | 134.69 |
| 1923 | APR | 20.04 | 21.54 | 109.78 |
| 1923 | MAY | 23.11 | 21.93 | 100.20 |
| 1923 | JUN | 19.65 | 21.93 | 95.52 |
| 1923 | JUL | 12.67 | 20.67 | 90.20 |
| 1923 | AUG | 13.12 | 19.07 | 82.25 |
| 1923 | SEP | 14.61 | 17.20 | 74.67 |
| 1923 | OCT | 15.82 | 16.50 | 74.19 |
| 1923 | NOV | 22.57 | 16.41 | 78.57 |
| 1923 | DEC | 24.59 | 17.23 | 85.51 |
| 1924 | JAN | 26.73 | 19.57 | 93.24 |
| 1924 | FEB | 20.64 | 20.83 | 95.53 |
| 1924 | MAR | 18.21 | 21.43 | 95.31 |
| 1924 | APR | 17.79 | 21.75 | 101.00 |
| 1924 | MAY | 13.44 | 20.23 | 92.27 |
| 1924 | JUN | 16.80 | 18.93 | 86.36 |
| 1924 | JUL | 24.23 | 18.52 | 89.59 |
| 1924 | AUG | 22.43 | 18.82 | 98.65 |
| 1924 | SEP | 18.15 | 18.81 | 109.34 |
| 1924 | OCT | 17.83 | 18.81 | 114.04 |
| 1924 | NOV | 41.37 | 23.47 | 143.04 |
| 1924 | DEC | 42.88 | 27.81 | 161.44 |

| YEAR | MONTH | NYSTXCH | 6MTH MA | 12PC.6MA |
|---|---|---|---|---|
| 1925 | JAN | 41.43 | 30.68 | 156.76 |
| 1925 | FEB | 32.75 | 32.40 | 155.58 |
| 1925 | MAR | 38.57 | 35.80 | 167.11 |
| 1925 | APR | 24.84 | 36.97 | 169.96 |
| 1925 | MAY | 36.46 | 36.15 | 178.69 |
| 1925 | JUN | 30.86 | 34.15 | 180.37 |
| 1925 | JUL | 32.28 | 32.63 | 176.19 |
| 1925 | AUG | 32.87 | 32.65 | 173.50 |
| 1925 | SEP | 36.89 | 32.37 | 172.11 |
| 1925 | OCT | 53.42 | 37.13 | 197.36 |
| 1925 | NOV | 48.98 | 39.22 | 167.11 |
| 1925 | DEC | 42.88 | 41.22 | 148.20 |
| 1926 | JAN | 39.09 | 42.35 | 138.05 |
| 1926 | FEB | 35.46 | 42.79 | 132.05 |
| 1926 | MAR | 52.04 | 45.31 | 126.55 |
| 1926 | APR | 30.22 | 41.44 | 112.09 |
| 1926 | MAY | 23.19 | 37.15 | 102.74 |
| 1926 | JUN | 37.99 | 36.33 | 106.38 |
| 1926 | JUL | 36.73 | 35.94 | 110.15 |
| 1926 | AUG | 44.19 | 37.39 | 114.54 |
| 1926 | SEP | 36.90 | 34.87 | 107.73 |
| 1926 | OCT | 40.21 | 36.53 | 98.40 |
| 1926 | NOV | 31.18 | 37.87 | 96.56 |
| 1926 | DEC | 41.89 | 38.52 | 93.44 |
| 1927 | JAN | 34.26 | 38.10 | 89.97 |
| 1927 | FEB | 44.16 | 38.10 | 89.05 |
| 1927 | MAR | 49.06 | 40.13 | 88.56 |
| 1927 | APR | 49.64 | 41.70 | 100.61 |
| 1927 | MAY | 46.60 | 44.27 | 119.17 |
| 1927 | JUN | 47.63 | 45.22 | 124.48 |
| 1927 | JUL | 38.49 | 45.93 | 127.80 |
| 1927 | AUG | 51.06 | 47.08 | 125.91 |
| 1927 | SEP | 51.92 | 47.56 | 136.38 |
| 1927 | OCT | 50.46 | 47.69 | 130.54 |
| 1927 | NOV | 51.36 | 48.49 | 128.05 |
| 1927 | DEC | 62.31 | 50.93 | 132.24 |
| 1928 | JAN | 56.96 | 54.01 | 141.74 |
| 1928 | FEB | 47.17 | 53.36 | 140.06 |
| 1928 | MAR | 84.99 | 58.87 | 146.72 |
| 1928 | APR | 80.57 | 63.89 | 153.23 |
| 1928 | MAY | 82.16 | 69.03 | 155.93 |
| 1928 | JUN | 63.74 | 69.26 | 153.16 |
| 1928 | JUL | 39.00 | 66.27 | 144.29 |
| 1928 | AUG | 67.65 | 69.68 | 148.01 |
| 1928 | SEP | 90.91 | 70.67 | 148.61 |
| 1928 | OCT | 99.08 | 73.76 | 154.65 |
| 1928 | NOV | 115.40 | 79.30 | 163.54 |
| 1928 | DEC | 92.84 | 84.15 | 165.21 |

**TABLE A4-3 New York Stock Exchange Volume** (cont.)

| YEAR | MONTH | NYSTXCH | 6MTH MA | 12PC.6MA |
|---|---|---|---|---|
| 1929 | JAN | 110.80 | 96.11 | 177.95 |
| 1929 | FEB | 77.97 | 97.83 | 183.34 |
| 1929 | MAR | 105.60 | 100.28 | 170.33 |
| 1929 | APR | 82.60 | 97.53 | 152.65 |
| 1929 | MAY | 91.28 | 93.51 | 135.48 |
| 1929 | JUN | 69.55 | 89.63 | 129.41 |
| 1929 | JUL | 93.38 | 86.73 | 130.87 |
| 1929 | AUG | 95.70 | 89.68 | 128.70 |
| 1929 | SEP | 100.10 | 88.77 | 125.61 |
| 1929 | OCT | 141.70 | 98.62 | 133.71 |
| 1929 | NOV | 72.46 | 95.48 | 120.41 |
| 1929 | DEC | 83.86 | 97.86 | 116.30 |
| 1930 | JAN | 62.31 | 92.69 | 96.44 |
| 1930 | FEB | 68.72 | 88.19 | 90.14 |
| 1930 | MAR | 96.55 | 87.60 | 87.35 |
| 1930 | APR | 111.00 | 82.48 | 84.57 |
| 1930 | MAY | 78.04 | 83.41 | 89.20 |
| 1930 | JUN | 76.59 | 82.20 | 91.71 |
| 1930 | JUL | 47.75 | 79.77 | 91.98 |
| 1930 | AUG | 39.87 | 74.96 | 83.59 |
| 1930 | SEP | 53.55 | 67.80 | 76.38 |
| 1930 | OCT | 65.50 | 60.21 | 61.06 |
| 1930 | NOV | 51.95 | 55.87 | 58.51 |
| 1930 | DEC | 58.76 | 52.89 | 54.05 |
| 1931 | JAN | 42.63 | 52.04 | 56.15 |
| 1931 | FEB | 64.15 | 56.09 | 63.60 |
| 1931 | MAR | 65.49 | 58.08 | 66.30 |
| 1931 | APR | 54.33 | 56.22 | 68.16 |
| 1931 | MAY | 46.66 | 55.33 | 66.34 |
| 1931 | JUN | 58.72 | 55.33 | 67.31 |
| 1931 | JUL | 33.54 | 53.81 | 67.46 |
| 1931 | AUG | 24.89 | 47.27 | 63.06 |
| 1931 | SEP | 51.14 | 44.88 | 66.19 |
| 1931 | OCT | 47.89 | 43.80 | 72.75 |
| 1931 | NOV | 37.37 | 42.26 | 75.64 |
| 1931 | DEC | 50.19 | 40.83 | 77.20 |
| 1932 | JAN | 34.34 | 40.97 | 78.72 |
| 1932 | FEB | 31.72 | 42.11 | 75.07 |
| 1932 | MAR | 33.06 | 39.09 | 67.31 |
| 1932 | APR | 31.40 | 36.34 | 64.65 |
| 1932 | MAY | 23.15 | 33.97 | 61.40 |
| 1932 | JUN | 23.00 | 29.44 | 53.21 |
| 1932 | JUL | 23.06 | 27.56 | 51.22 |
| 1932 | AUG | 82.54 | 36.03 | 76.23 |
| 1932 | SEP | 67.42 | 41.76 | 93.05 |
| 1932 | OCT | 29.19 | 41.39 | 94.49 |
| 1932 | NOV | 23.04 | 41.37 | 97.91 |
| 1932 | DEC | 23.21 | 41.41 | 101.40 |

| YEAR | MONTH | NYSTXCH | 6MTH MA | 12PC.6MA |
|---|---|---|---|---|
| 1933 | JAN | 18.72 | 40.68 | 99.31 |
| 1933 | FEB | 19.32 | 30.15 | 71.60 |
| 1933 | MAR | 20.09 | 22.26 | 56.94 |
| 1933 | APR | 52.90 | 26.21 | 72.12 |
| 1933 | MAY | 104.20 | 39.74 | 116.96 |
| 1933 | JUN | 125.60 | 56.80 | 192.93 |
| 1933 | JUL | 120.30 | 73.73 | 267.51 |
| 1933 | AUG | 42.47 | 77.59 | 215.34 |
| 1933 | SEP | 43.32 | 81.46 | 195.08 |
| 1933 | OCT | 39.38 | 79.21 | 191.37 |
| 1933 | NOV | 33.65 | 67.45 | 163.03 |
| 1933 | DEC | 34.88 | 52.33 | 126.38 |
| 1934 | JAN | 54.57 | 41.38 | 101.70 |
| 1934 | FEB | 56.83 | 43.77 | 145.18 |
| 1934 | MAR | 29.92 | 41.54 | 186.60 |
| 1934 | APR | 29.85 | 39.95 | 152.41 |
| 1934 | MAY | 25.34 | 38.56 | 97.04 |
| 1934 | JUN | 16.80 | 35.55 | 62.58 |
| 1934 | JUL | 21.12 | 29.97 | 40.65 |
| 1934 | AUG | 16.69 | 23.28 | 30.01 |
| 1934 | SEP | 12.64 | 20.40 | 25.05 |
| 1934 | OCT | 15.66 | 18.04 | 22.77 |
| 1934 | NOV | 20.87 | 17.29 | 25.64 |
| 1934 | DEC | 23.59 | 18.43 | 35.21 |
| 1935 | JAN | 19.41 | 18.14 | 43.84 |
| 1935 | FEB | 14.40 | 17.76 | 40.57 |
| 1935 | MAR | 15.85 | 18.29 | 44.04 |
| 1935 | APR | 22.41 | 19.42 | 48.61 |
| 1935 | MAY | 30.44 | 21.01 | 54.49 |
| 1935 | JUN | 22.34 | 20.81 | 58.53 |
| 1935 | JUL | 29.43 | 22.48 | 74.98 |
| 1935 | AUG | 42.92 | 27.23 | 116.94 |
| 1935 | SEP | 34.75 | 30.38 | 148.89 |
| 1935 | OCT | 46.66 | 34.42 | 190.81 |
| 1935 | NOV | 57.46 | 38.92 | 225.08 |
| 1935 | DEC | 45.59 | 42.80 | 232.28 |
| 1936 | JAN | 67.21 | 49.10 | 270.64 |
| 1936 | FEB | 60.87 | 52.09 | 293.30 |
| 1936 | MAR | 51.03 | 54.80 | 299.56 |
| 1936 | APR | 39.62 | 53.63 | 276.16 |
| 1936 | MAY | 20.61 | 47.49 | 225.97 |
| 1936 | JUN | 21.43 | 43.46 | 208.88 |
| 1936 | JUL | 34.79 | 38.05 | 169.32 |
| 1936 | AUG | 26.56 | 32.34 | 118.76 |
| 1936 | SEP | 30.87 | 28.98 | 95.39 |
| 1936 | OCT | 44.00 | 29.71 | 86.31 |
| 1936 | NOV | 50.47 | 34.68 | 89.11 |
| 1936 | DEC | 48.61 | 39.21 | 91.62 |

**TABLE A4-3 New York Stock Exchange Volume** (cont.)

| YEAR | MONTH | NYSTXCH | 6MTH MA | 12PC.6MA |
|---|---|---|---|---|
| 1937 | JAN | 58.68 | 43.19 | 87.98 |
| 1937 | FEB | 50.26 | 47.14 | 90.51 |
| 1937 | MAR | 50.34 | 50.39 | 91.95 |
| 1937 | APR | 34.61 | 48.82 | 91.05 |
| 1937 | MAY | 18.56 | 43.51 | 91.62 |
| 1937 | JUN | 16.44 | 38.14 | 87.77 |
| 1937 | JUL | 20.72 | 31.82 | 83.61 |
| 1937 | AUG | 17.22 | 26.31 | 81.37 |
| 1937 | SEP | 33.86 | 23.56 | 81.32 |
| 1937 | OCT | 51.09 | 26.31 | 88.57 |
| 1937 | NOV | 29.26 | 28.09 | 81.00 |
| 1937 | DEC | 28.42 | 30.09 | 76.74 |
| 1938 | JAN | 24.15 | 30.66 | 70.99 |
| 1938 | FEB | 14.52 | 30.21 | 64.09 |
| 1938 | MAR | 23.00 | 28.40 | 56.37 |
| 1938 | APR | 17.12 | 22.74 | 46.58 |
| 1938 | MAY | 14.01 | 20.20 | 46.43 |
| 1938 | JUN | 24.36 | 19.52 | 51.18 |
| 1938 | JUL | 38.76 | 21.96 | 69.01 |
| 1938 | AUG | 20.72 | 22.99 | 87.38 |
| 1938 | SEP | 23.83 | 23.13 | 98.15 |
| 1938 | OCT | 41.56 | 27.20 | 103.39 |
| 1938 | NOV | 27.92 | 29.52 | 105.08 |
| 1938 | DEC | 27.49 | 30.04 | 99.84 |
| 1939 | JAN | 25.19 | 27.78 | 90.60 |
| 1939 | FEB | 13.88 | 26.64 | 88.18 |
| 1939 | MAR | 24.57 | 26.76 | 94.23 |
| 1939 | APR | 20.25 | 23.21 | 102.07 |
| 1939 | MAY | 12.93 | 20.71 | 102.55 |
| 1939 | JUN | 11.97 | 18.13 | 92.85 |
| 1939 | JUL | 18.07 | 16.94 | 77.15 |
| 1939 | AUG | 17.37 | 17.52 | 76.22 |
| 1939 | SEP | 57.09 | 22.94 | 99.19 |
| 1939 | OCT | 23.73 | 23.52 | 86.47 |
| 1939 | NOV | 19.23 | 24.57 | 83.24 |
| 1939 | DEC | 17.77 | 25.54 | 85.01 |
| 1940 | JAN | 15.99 | 25.19 | 90.68 |
| 1940 | FEB | 13.47 | 24.54 | 92.12 |
| 1940 | MAR | 16.27 | 17.74 | 66.28 |
| 1940 | APR | 26.70 | 18.23 | 78.55 |
| 1940 | MAY | 38.97 | 21.52 | 103.91 |
| 1940 | JUN | 15.57 | 21.16 | 116.71 |
| 1940 | JUL | 7.31 | 19.71 | 116.35 |
| 1940 | AUG | 7.62 | 18.74 | 106.92 |
| 1940 | SEP | 11.94 | 18.01 | 78.52 |
| 1940 | OCT | 14.49 | 15.98 | 67.93 |
| 1940 | NOV | 20.89 | 12.97 | 52.77 |
| 1940 | DEC | 18.40 | 13.44 | 52.62 |

| YEAR | MONTH | NYSTXCH | 6MTH MA | 12RC.6MA |
|---|---|---|---|---|
| 1941 | JAN | 13.29 | 14.43 | 57.30 |
| 1941 | FEB | 8.97 | 14.66 | 59.73 |
| 1941 | MAR | 10.11 | 14.35 | 80.92 |
| 1941 | APR | 11.18 | 13.80 | 75.70 |
| 1941 | MAY | 9.66 | 11.93 | 55.43 |
| 1941 | JUN | 10.45 | 10.61 | 50.13 |
| 1941 | JUL | 17.87 | 11.37 | 57.68 |
| 1941 | AUG | 10.88 | 11.69 | 62.38 |
| 1941 | SEP | 13.54 | 12.26 | 68.05 |
| 1941 | OCT | 13.14 | 12.59 | 78.76 |
| 1941 | NOV | 15.05 | 13.48 | 104.00 |
| 1941 | DEC | 36.39 | 17.81 | 132.52 |
| 1942 | JAN | 12.99 | 16.99 | 117.73 |
| 1942 | FEB | 7.93 | 16.50 | 112.57 |
| 1942 | MAR | 8.58 | 15.68 | 109.21 |
| 1942 | APR | 7.59 | 14.75 | 106.87 |
| 1942 | MAY | 7.23 | 13.45 | 112.71 |
| 1942 | JUN | 7.47 | 8.63 | 81.35 |
| 1942 | JUL | 8.37 | 7.86 | 69.11 |
| 1942 | AUG | 7.39 | 7.77 | 66.46 |
| 1942 | SEP | 9.45 | 7.91 | 64.55 |
| 1942 | OCT | 15.93 | 9.30 | 73.91 |
| 1942 | NOV | 13.44 | 10.34 | 76.67 |
| 1942 | DEC | 19.31 | 12.31 | 69.13 |
| 1943 | JAN | 18.03 | 13.92 | 81.92 |
| 1943 | FEB | 24.43 | 16.76 | 101.57 |
| 1943 | MAR | 37.00 | 21.35 | 136.21 |
| 1943 | APR | 33.55 | 24.29 | 164.66 |
| 1943 | MAY | 35.05 | 27.89 | 207.40 |
| 1943 | JUN | 23.42 | 28.58 | 331.20 |
| 1943 | JUL | 26.32 | 29.96 | 381.23 |
| 1943 | AUG | 14.25 | 28.26 | 363.81 |
| 1943 | SEP | 14.99 | 24.59 | 310.79 |
| 1943 | OCT | 13.92 | 21.32 | 229.18 |
| 1943 | NOV | 18.25 | 18.52 | 179.16 |
| 1943 | DEC | 19.53 | 17.87 | 145.17 |
| 1944 | JAN | 17.81 | 16.45 | 118.20 |
| 1944 | FEB | 17.10 | 16.93 | 101.00 |
| 1944 | MAR | 27.64 | 19.04 | 89.16 |
| 1944 | APR | 13.85 | 19.03 | 78.33 |
| 1944 | MAY | 17.23 | 18.86 | 67.61 |
| 1944 | JUN | 37.71 | 21.89 | 76.59 |
| 1944 | JUL | 28.22 | 23.62 | 78.85 |
| 1944 | AUG | 20.75 | 24.23 | 85.73 |
| 1944 | SEP | 15.95 | 22.28 | 90.60 |
| 1944 | OCT | 17.53 | 22.89 | 107.38 |
| 1944 | NOV | 18.02 | 23.03 | 124.32 |
| 1944 | DEC | 31.26 | 21.95 | 122.82 |

**TABLE A4-3 New York Stock Exchange Volume** (cont.)

| YEAR | MONTH | NYSTXCH | 6MTH MA | 12RC.6MA |
|---|---|---|---|---|
| 1945 | JAN | 39.00 | 23.75 | 144.32 |
| 1945 | FEB | 32.61 | 25.72 | 151.95 |
| 1945 | MAR | 27.49 | 27.65 | 145.22 |
| 1945 | APR | 28.27 | 29.44 | 154.72 |
| 1945 | MAY | 32.02 | 31.77 | 168.49 |
| 1945 | JUN | 41.31 | 33.45 | 152.82 |
| 1945 | JUL | 19.98 | 30.28 | 128.17 |
| 1945 | AUG | 21.71 | 28.46 | 117.46 |
| 1945 | SEP | 25.13 | 28.07 | 125.96 |
| 1945 | OCT | 35.48 | 29.27 | 127.84 |
| 1945 | NOV | 40.41 | 30.67 | 133.18 |
| 1945 | DEC | 34.15 | 29.47 | 134.26 |
| 1946 | JAN | 51.51 | 34.73 | 146.24 |
| 1946 | FEB | 34.00 | 36.78 | 142.96 |
| 1946 | MAR | 25.66 | 36.86 | 133.34 |
| 1946 | APR | 31.43 | 36.19 | 122.93 |
| 1946 | MAY | 30.41 | 34.52 | 108.66 |
| 1946 | JUN | 21.72 | 32.45 | 97.03 |
| 1946 | JUL | 20.60 | 27.30 | 90.17 |
| 1946 | AUG | 20.81 | 25.10 | 88.20 |
| 1946 | SEP | 43.45 | 28.07 | 100.00 |
| 1946 | OCT | 30.38 | 27.89 | 95.30 |
| 1946 | NOV | 23.82 | 26.79 | 87.37 |
| 1946 | DEC | 29.83 | 28.14 | 95.49 |
| 1947 | JAN | 23.56 | 28.64 | 82.46 |
| 1947 | FEB | 23.76 | 29.13 | 79.21 |
| 1947 | MAR | 19.34 | 25.11 | 68.12 |
| 1947 | APR | 20.62 | 23.48 | 64.89 |
| 1947 | MAY | 20.62 | 22.95 | 66.48 |
| 1947 | JUN | 17.48 | 20.89 | 64.38 |
| 1947 | JUL | 25.47 | 21.21 | 77.70 |
| 1947 | AUG | 14.15 | 19.61 | 78.12 |
| 1947 | SEP | 16.02 | 19.06 | 67.90 |
| 1947 | OCT | 28.64 | 20.39 | 73.12 |
| 1947 | NOV | 16.37 | 19.68 | 73.47 |
| 1947 | DEC | 27.61 | 21.37 | 75.94 |
| 1948 | JAN | 20.22 | 20.50 | 71.58 |
| 1948 | FEB | 16.80 | 20.94 | 71.88 |
| 1948 | MAR | 22.99 | 22.10 | 88.01 |
| 1948 | APR | 34.61 | 23.10 | 98.35 |
| 1948 | MAY | 42.77 | 27.50 | 119.80 |
| 1948 | JUN | 30.92 | 28.05 | 134.25 |
| 1948 | JUL | 24.59 | 28.78 | 135.66 |
| 1948 | AUG | 15.04 | 28.48 | 145.25 |
| 1948 | SEP | 17.56 | 27.58 | 144.72 |
| 1948 | OCT | 20.43 | 25.21 | 123.64 |
| 1948 | NOV | 28.32 | 22.81 | 115.86 |
| 1948 | DEC | 27.96 | 22.31 | 104.40 |

| YEAR | MONTH | NYSTXCH | 6MTH MA | 12RC.6MA |
|---|---|---|---|---|
| 1949 | JAN | 18.83 | 21.35 | 104.17 |
| 1949 | FEB | 17.18 | 21.71 | 103.68 |
| 1949 | MAR | 21.14 | 22.31 | 100.93 |
| 1949 | APR | 19.31 | 22.12 | 95.77 |
| 1949 | MAY | 18.18 | 20.43 | 74.30 |
| 1949 | JUN | 17.77 | 18.73 | 66.78 |
| 1949 | JUL | 18.75 | 18.72 | 65.05 |
| 1949 | AUG | 21.79 | 19.49 | 68.41 |
| 1949 | SEP | 23.84 | 19.94 | 72.29 |
| 1949 | OCT | 28.89 | 21.53 | 85.40 |
| 1949 | NOV | 27.24 | 23.04 | 101.04 |
| 1949 | DEC | 39.29 | 26.63 | 119.35 |
| 1950 | JAN | 42.58 | 30.60 | 143.31 |
| 1950 | FEB | 33.41 | 32.54 | 149.88 |
| 1950 | MAR | 40.41 | 35.30 | 158.25 |
| 1950 | APR | 48.24 | 38.52 | 174.16 |
| 1950 | MAY | 41.60 | 40.92 | 200.29 |
| 1950 | JUN | 45.55 | 41.96 | 224.02 |
| 1950 | JUL | 44.55 | 42.29 | 225.93 |
| 1950 | AUG | 38.47 | 43.13 | 221.35 |
| 1950 | SEP | 38.59 | 42.83 | 214.83 |
| 1950 | OCT | 48.39 | 42.85 | 199.02 |
| 1950 | NOV | 43.09 | 43.10 | 187.05 |
| 1950 | DEC | 59.82 | 45.48 | 170.79 |
| 1951 | JAN | 70.18 | 49.75 | 162.58 |
| 1951 | FEB | 41.23 | 50.21 | 154.32 |
| 1951 | MAR | 35.63 | 49.72 | 140.85 |
| 1951 | APR | 34.29 | 47.37 | 122.96 |
| 1951 | MAY | 38.46 | 46.60 | 113.88 |
| 1951 | JUN | 27.40 | 41.19 | 98.17 |
| 1951 | JUL | 27.99 | 34.16 | 80.78 |
| 1951 | AUG | 33.64 | 32.90 | 76.27 |
| 1951 | SEP | 36.39 | 33.02 | 77.11 |
| 1951 | OCT | 42.53 | 34.40 | 80.27 |
| 1951 | NOV | 25.68 | 32.27 | 74.86 |
| 1951 | DEC | 30.08 | 32.71 | 71.93 |
| 1952 | JAN | 37.14 | 34.24 | 68.82 |
| 1952 | FEB | 27.20 | 33.17 | 66.05 |
| 1952 | MAR | 20.51 | 30.52 | 61.38 |
| 1952 | APR | 28.96 | 28.26 | 59.65 |
| 1952 | MAY | 28.59 | 28.74 | 61.68 |
| 1952 | JUN | 25.52 | 27.98 | 67.93 |
| 1952 | JUL | 24.12 | 25.81 | 75.56 |
| 1952 | AUG | 20.90 | 24.76 | 75.27 |
| 1952 | SEP | 24.14 | 25.37 | 76.82 |
| 1952 | OCT | 25.98 | 24.87 | 72.30 |
| 1952 | NOV | 30.24 | 25.15 | 77.93 |
| 1952 | DEC | 40.52 | 27.65 | 84.51 |

**TABLE A4-3 New York Stock Exchange Volume** (cont.)

| YEAR | MONTH | NYSTXCH | 6MTH MA | 12RC.6MA |
|---|---|---|---|---|
| 1953 | JAN | 34.09 | 29.31 | 85.60 |
| 1953 | FEB | 30.21 | 30.86 | 93.05 |
| 1953 | MAR | 42.47 | 33.91 | 111.12 |
| 1953 | APR | 34.37 | 35.31 | 124.97 |
| 1953 | MAY | 25.77 | 34.57 | 120.27 |
| 1953 | JUN | 26.08 | 32.16 | 114.93 |
| 1953 | JUL | 22.23 | 30.18 | 116.94 |
| 1953 | AUG | 23.89 | 29.13 | 117.64 |
| 1953 | SEP | 27.17 | 26.58 | 104.78 |
| 1953 | OCT | 25.73 | 25.14 | 101.09 |
| 1953 | NOV | 26.68 | 25.29 | 100.58 |
| 1953 | DEC | 36.16 | 26.97 | 97.56 |
| 1954 | JAN | 33.37 | 28.83 | 98.37 |
| 1954 | FEB | 33.29 | 30.40 | 98.50 |
| 1954 | MAR | 44.13 | 33.22 | 97.96 |
| 1954 | APR | 43.87 | 36.25 | 102.64 |
| 1954 | MAY | 41.91 | 38.78 | 112.20 |
| 1954 | JUN | 42.22 | 39.79 | 123.73 |
| 1954 | JUL | 51.85 | 42.87 | 142.04 |
| 1954 | AUG | 56.93 | 46.81 | 160.70 |
| 1954 | SEP | 41.23 | 46.33 | 174.30 |
| 1954 | OCT | 44.17 | 46.38 | 184.43 |
| 1954 | NOV | 63.93 | 50.05 | 197.89 |
| 1954 | DEC | 76.45 | 55.76 | 206.71 |
| 1955 | JAN | 74.65 | 59.56 | 206.55 |
| 1955 | FEB | 60.82 | 60.20 | 198.06 |
| 1955 | MAR | 66.86 | 64.48 | 194.07 |
| 1955 | APR | 53.79 | 66.08 | 182.31 |
| 1955 | MAY | 45.43 | 63.00 | 162.42 |
| 1955 | JUN | 58.15 | 59.95 | 150.64 |
| 1955 | JUL | 48.46 | 55.58 | 129.64 |
| 1955 | AUG | 41.81 | 52.41 | 111.96 |
| 1955 | SEP | 60.10 | 51.29 | 110.69 |
| 1955 | OCT | 42.18 | 49.35 | 106.40 |
| 1955 | NOV | 46.38 | 49.51 | 98.92 |
| 1955 | DEC | 50.99 | 48.32 | 86.66 |
| 1956 | JAN | 47.20 | 48.11 | 80.77 |
| 1956 | FEB | 46.40 | 48.87 | 81.17 |
| 1956 | MAR | 60.36 | 48.91 | 75.86 |
| 1956 | APR | 54.11 | 50.90 | 77.03 |
| 1956 | MAY | 53.23 | 52.04 | 82.61 |
| 1956 | JUN | 37.20 | 49.76 | 82.98 |
| 1956 | JUL | 45.71 | 49.50 | 89.05 |
| 1956 | AUG | 44.53 | 49.19 | 93.84 |
| 1956 | SEP | 37.23 | 45.33 | 88.39 |
| 1956 | OCT | 40.34 | 43.04 | 87.20 |
| 1956 | NOV | 43.55 | 41.42 | 83.67 |
| 1956 | DEC | 46.42 | 42.96 | 88.91 |

| YEAR | MONTH | NYSTXCH | 6MTH MA | 12PC.6MA |
|---|---|---|---|---|
| 1957 | JAN | 48.16 | 43.37 | 90.15 |
| 1957 | FEB | 37.58 | 42.21 | 86.37 |
| 1957 | MAR | 35.65 | 41.95 | 85.75 |
| 1957 | APR | 48.31 | 43.27 | 85.01 |
| 1957 | MAY | 52.56 | 44.78 | 86.03 |
| 1957 | JUN | 44.48 | 44.45 | 89.36 |
| 1957 | JUL | 48.26 | 44.47 | 89.84 |
| 1957 | AUG | 41.41 | 45.11 | 91.71 |
| 1957 | SEP | 36.87 | 45.31 | 99.96 |
| 1957 | OCT | 63.98 | 47.92 | 111.35 |
| 1957 | NOV | 48.22 | 47.20 | 113.95 |
| 1957 | DEC | 54.47 | 48.86 | 113.75 |
| 1958 | JAN | 49.87 | 49.13 | 113.29 |
| 1958 | FEB | 40.20 | 48.93 | 115.92 |
| 1958 | MAR | 46.68 | 50.57 | 120.55 |
| 1958 | APR | 50.31 | 48.29 | 111.58 |
| 1958 | MAY | 54.18 | 49.28 | 110.06 |
| 1958 | JUN | 56.62 | 49.64 | 111.67 |
| 1958 | JUL | 69.50 | 52.91 | 118.93 |
| 1958 | AUG | 62.37 | 56.61 | 125.49 |
| 1958 | SEP | 71.97 | 60.82 | 134.23 |
| 1958 | OCT | 95.09 | 68.28 | 142.49 |
| 1958 | NOV | 74.37 | 71.65 | 151.80 |
| 1958 | DEC | 75.92 | 74.86 | 153.21 |
| 1959 | JAN | 83.25 | 77.16 | 157.04 |
| 1959 | FEB | 65.79 | 77.73 | 158.85 |
| 1959 | MAR | 82.45 | 79.47 | 157.17 |
| 1959 | APR | 75.89 | 76.27 | 157.96 |
| 1959 | MAY | 70.97 | 75.71 | 153.62 |
| 1959 | JUN | 64.35 | 73.78 | 148.63 |
| 1959 | JUL | 69.50 | 71.49 | 135.11 |
| 1959 | AUG | 51.05 | 69.03 | 121.95 |
| 1959 | SEP | 57.52 | 64.87 | 106.67 |
| 1959 | OCT | 61.33 | 62.45 | 91.45 |
| 1959 | NOV | 64.56 | 61.38 | 85.67 |
| 1959 | DEC | 72.24 | 62.69 | 83.74 |
| 1960 | JAN | 63.93 | 61.77 | 80.05 |
| 1960 | FEB | 60.53 | 63.35 | 81.50 |
| 1960 | MAR | 65.72 | 64.71 | 81.43 |
| 1960 | APR | 57.29 | 64.04 | 83.96 |
| 1960 | MAY | 68.83 | 64.75 | 85.53 |
| 1960 | JUN | 76.53 | 65.47 | 88.73 |
| 1960 | JUL | 53.87 | 63.79 | 89.23 |
| 1960 | AUG | 65.35 | 64.59 | 93.57 |
| 1960 | SEP | 60.85 | 63.78 | 98.31 |
| 1960 | OCT | 54.43 | 63.30 | 101.37 |
| 1960 | NOV | 62.00 | 62.17 | 101.28 |
| 1960 | DEC | 77.35 | 62.30 | 99.37 |

**TABLE A4-3 New York Stock Exchange Volume** (cont.)

| YEAR | MONTH | NYSTXCH | 6MTH MA | 12RC.6MA |
|---|---|---|---|---|
| 1961 | JAN | 89.11 | 68.18 | 110.38 |
| 1961 | FEB | 92.80 | 72.75 | 114.85 |
| 1961 | MAR | 118.00 | 82.28 | 127.14 |
| 1961 | APR | 101.80 | 90.17 | 140.80 |
| 1961 | MAY | 96.95 | 96.00 | 148.25 |
| 1961 | JUN | 73.12 | 95.29 | 145.56 |
| 1961 | JUL | 60.90 | 90.59 | 142.01 |
| 1961 | AUG | 81.53 | 88.71 | 137.34 |
| 1961 | SEP | 63.86 | 79.69 | 124.94 |
| 1961 | OCT | 72.99 | 74.89 | 118.29 |
| 1961 | NOV | 87.79 | 73.36 | 118.00 |
| 1961 | DEC | 82.40 | 74.91 | 120.23 |
| 1962 | JAN | 80.88 | 78.24 | 114.76 |
| 1962 | FEB | 66.13 | 75.67 | 104.01 |
| 1962 | MAR | 68.48 | 76.44 | 92.91 |
| 1962 | APR | 65.26 | 75.15 | 83.34 |
| 1962 | MAY | 111.00 | 79.02 | 82.31 |
| 1962 | JUN | 100.20 | 81.99 | 86.04 |
| 1962 | JUL | 74.16 | 80.87 | 89.27 |
| 1962 | AUG | 77.46 | 82.75 | 93.28 |
| 1962 | SEP | 62.89 | 81.82 | 102.68 |
| 1962 | OCT | 78.72 | 84.06 | 112.26 |
| 1962 | NOV | 96.06 | 81.57 | 111.20 |
| 1962 | DEC | 80.96 | 78.37 | 104.62 |
| 1963 | JAN | 100.60 | 82.77 | 105.80 |
| 1963 | FEB | 79.20 | 83.06 | 109.77 |
| 1963 | MAR | 74.80 | 85.05 | 111.27 |
| 1963 | APR | 106.50 | 89.68 | 119.33 |
| 1963 | MAY | 105.10 | 91.19 | 115.40 |
| 1963 | JUN | 90.60 | 92.79 | 113.18 |
| 1963 | JUL | 76.30 | 88.74 | 109.74 |
| 1963 | AUG | 91.60 | 90.81 | 109.74 |
| 1963 | SEP | 106.60 | 96.11 | 117.46 |
| 1963 | OCT | 122.30 | 98.74 | 117.46 |
| 1963 | NOV | 94.00 | 96.89 | 118.78 |
| 1963 | DEC | 98.70 | 98.24 | 125.36 |
| 1964 | JAN | 116.60 | 104.96 | 126.80 |
| 1964 | FEB | 88.10 | 104.38 | 125.66 |
| 1964 | MAR | 114.00 | 105.61 | 124.17 |
| 1964 | APR | 123.50 | 105.81 | 117.99 |
| 1964 | MAY | 99.20 | 106.68 | 116.99 |
| 1964 | JUN | 96.20 | 106.26 | 114.51 |
| 1964 | JUL | 102.60 | 103.93 | 117.11 |
| 1964 | AUG | 82.30 | 102.96 | 113.38 |
| 1964 | SEP | 109.80 | 102.26 | 106.40 |
| 1964 | OCT | 106.60 | 99.44 | 100.71 |
| 1964 | NOV | 93.60 | 98.51 | 101.67 |
| 1964 | DEC | 104.00 | 99.81 | 101.59 |

| YEAR | MONTH | NYSTXCH | 6MTH MA | 12RC.6MA |
|---|---|---|---|---|
| 1965 | JAN | 109.10 | 100.89 | 96.13 |
| 1965 | FEB | 112.30 | 105.89 | 101.45 |
| 1965 | MAR | 124.80 | 108.39 | 102.64 |
| 1965 | APR | 119.10 | 110.48 | 104.41 |
| 1965 | MAY | 110.20 | 113.24 | 106.16 |
| 1965 | JUN | 128.20 | 117.28 | 110.37 |
| 1965 | JUL | 85.20 | 113.29 | 109.01 |
| 1965 | AUG | 109.20 | 112.78 | 109.53 |
| 1965 | SEP | 155.50 | 117.89 | 115.29 |
| 1965 | OCT | 164.20 | 125.41 | 126.11 |
| 1965 | NOV | 147.20 | 131.58 | 133.57 |
| 1965 | DEC | 191.20 | 142.08 | 142.35 |
| 1966 | JAN | 182.90 | 158.36 | 156.96 |
| 1966 | FEB | 166.30 | 167.88 | 158.53 |
| 1966 | MAR | 191.50 | 173.88 | 160.41 |
| 1966 | APR | 186.20 | 177.54 | 160.71 |
| 1966 | MAY | 171.50 | 181.59 | 160.36 |
| 1966 | JUN | 140.60 | 173.16 | 147.65 |
| 1966 | JUL | 119.90 | 162.66 | 143.57 |
| 1966 | AUG | 162.50 | 162.03 | 143.67 |
| 1966 | SEP | 120.20 | 150.14 | 127.36 |
| 1966 | OCT | 146.40 | 143.51 | 114.43 |
| 1966 | NOV | 145.90 | 139.24 | 105.83 |
| 1966 | DEC | 165.50 | 143.39 | 100.93 |
| 1967 | JAN | 207.60 | 158.01 | 99.78 |
| 1967 | FEB | 183.10 | 161.44 | 96.17 |
| 1967 | MAR | 224.80 | 178.87 | 102.88 |
| 1967 | APR | 187.80 | 185.77 | 104.64 |
| 1967 | MAY | 218.50 | 197.87 | 108.97 |
| 1967 | JUN | 212.60 | 205.72 | 118.81 |
| 1967 | JUL | 216.70 | 207.24 | 127.41 |
| 1967 | AUG | 207.90 | 211.37 | 130.46 |
| 1967 | SEP | 205.00 | 208.07 | 138.59 |
| 1967 | OCT | 224.90 | 214.26 | 149.30 |
| 1967 | NOV | 211.60 | 213.11 | 153.05 |
| 1967 | DEC | 229.50 | 215.92 | 150.58 |
| 1968 | JAN | 262.80 | 223.61 | 141.52 |
| 1968 | FEB | 174.50 | 218.04 | 135.06 |
| 1968 | MAR | 192.70 | 215.99 | 120.75 |
| 1968 | APR | 295.60 | 227.77 | 122.61 |
| 1968 | MAY | 292.10 | 241.19 | 121.89 |
| 1968 | JUN | 257.40 | 245.84 | 119.50 |
| 1968 | JUL | 242.50 | 242.46 | 116.99 |
| 1968 | AUG | 194.10 | 245.72 | 116.25 |
| 1968 | SEP | 228.40 | 251.67 | 120.95 |
| 1968 | OCT | 272.00 | 247.74 | 115.63 |
| 1968 | NOV | 252.00 | 241.06 | 113.12 |
| 1968 | DEC | 267.60 | 242.76 | 112.43 |

**TABLE A4-3 New York Stock Exchange Volume** (cont.)

| YEAR | MONTH | NYSTXCH | 6MTH MA | 12RC.6MA |
|---|---|---|---|---|
| 1969 | JAN | 266.70 | 246.79 | 110.37 |
| 1969 | FEB | 210.30 | 249.49 | 114.42 |
| 1969 | MAR | 199.20 | 244.62 | 113.26 |
| 1969 | APR | 237.00 | 238.79 | 104.84 |
| 1969 | MAY | 256.70 | 239.57 | 99.33 |
| 1969 | JUN | 235.30 | 234.19 | 95.26 |
| 1969 | JUL | 228.30 | 227.79 | 93.95 |
| 1969 | AUG | 201.80 | 226.37 | 92.13 |
| 1969 | SEP | 219.20 | 229.71 | 91.27 |
| 1969 | OCT | 310.20 | 241.91 | 97.65 |
| 1969 | NOV | 213.70 | 234.74 | 97.38 |
| 1969 | DEC | 272.50 | 240.94 | 99.25 |
| 1970 | JAN | 221.20 | 239.76 | 97.15 |
| 1970 | FEB | 218.50 | 242.54 | 97.21 |
| 1970 | MAR | 213.00 | 241.51 | 98.73 |
| 1970 | APR | 223.20 | 227.01 | 95.07 |
| 1970 | MAY | 258.30 | 234.44 | 97.86 |
| 1970 | JUN | 226.50 | 226.77 | 96.83 |
| 1970 | JUL | 227.90 | 227.89 | 100.04 |
| 1970 | AUG | 218.80 | 227.94 | 100.69 |
| 1970 | SEP | 302.90 | 242.92 | 105.75 |
| 1970 | OCT | 261.50 | 249.81 | 103.06 |
| 1970 | NOV | 230.40 | 244.66 | 104.22 |
| 1970 | DEC | 335.30 | 262.79 | 109.07 |
| 1971 | JAN | 348.60 | 282.91 | 118.00 |
| 1971 | FEB | 371.30 | 308.32 | 127.12 |
| 1971 | MAR | 390.00 | 322.84 | 133.68 |
| 1971 | APR | 401.70 | 346.21 | 152.51 |
| 1971 | MAY | 303.20 | 358.34 | 152.85 |
| 1971 | JUN | 303.70 | 353.07 | 155.69 |
| 1971 | JUL | 265.30 | 339.19 | 148.84 |
| 1971 | AUG | 320.60 | 330.74 | 145.10 |
| 1971 | SEP | 252.80 | 307.87 | 126.74 |
| 1971 | OCT | 280.10 | 287.61 | 115.36 |
| 1971 | NOV | 276.40 | 283.14 | 115.73 |
| 1971 | DEC | 377.80 | 295.49 | 112.44 |
| 1972 | JAN | 379.50 | 314.52 | 111.18 |
| 1972 | FEB | 376.30 | 323.81 | 105.02 |
| 1972 | MAR | 403.70 | 348.96 | 108.09 |
| 1972 | APR | 368.00 | 363.61 | 105.03 |
| 1972 | MAY | 335.90 | 373.52 | 104.24 |
| 1972 | JUN | 314.60 | 362.99 | 102.81 |
| 1972 | JUL | 289.00 | 347.91 | 102.57 |
| 1972 | AUG | 357.00 | 344.69 | 104.22 |
| 1972 | SEP | 246.30 | 318.46 | 103.44 |
| 1972 | OCT | 317.40 | 310.02 | 107.79 |
| 1972 | NOV | 405.60 | 321.64 | 113.60 |
| 1972 | DEC | 344.80 | 326.67 | 110.55 |

| YEAR | MONTH | NYSTXCH | 6MTH MA | 12RC.6MA |
|---|---|---|---|---|
| 1973 | JAN | 393.80 | 344.14 | 109.42 |
| 1973 | FEB | 318.30 | 337.69 | 104.29 |
| 1973 | MAR | 342.40 | 353.71 | 101.36 |
| 1973 | APR | 278.00 | 347.14 | 95.47 |
| 1973 | MAY | 337.20 | 335.74 | 89.83 |
| 1973 | JUN | 268.70 | 323.06 | 89.00 |
| 1973 | JUL | 307.80 | 308.72 | 88.74 |
| 1973 | AUG | 270.50 | 300.76 | 87.25 |
| 1973 | SEP | 329.00 | 298.52 | 93.74 |
| 1973 | OCT | 422.90 | 322.67 | 104.08 |
| 1973 | NOV | 399.90 | 333.12 | 103.57 |
| 1973 | DEC | 384.50 | 352.42 | 107.88 |
| 1974 | JAN | 363.10 | 361.64 | 105.09 |
| 1974 | FEB | 256.80 | 359.36 | 106.42 |
| 1974 | MAR | 309.60 | 356.12 | 100.68 |
| 1974 | APR | 254.30 | 328.02 | 94.49 |
| 1974 | MAY | 275.30 | 307.26 | 91.52 |
| 1974 | JUN | 245.40 | 284.07 | 87.93 |
| 1974 | JUL | 274.10 | 269.24 | 87.21 |
| 1974 | AUG | 280.10 | 273.12 | 90.81 |
| 1974 | SEP | 280.00 | 268.19 | 89.84 |
| 1974 | OCT | 377.10 | 288.66 | 89.46 |
| 1974 | NOV | 286.80 | 290.57 | 87.23 |
| 1974 | DEC | 315.10 | 302.19 | 85.75 |
| 1975 | JAN | 432.50 | 328.59 | 90.86 |
| 1975 | FEB | 423.90 | 352.56 | 98.11 |
| 1975 | MAR | 453.60 | 381.49 | 107.12 |
| 1975 | APR | 447.30 | 393.19 | 119.87 |
| 1975 | MAY | 457.50 | 421.64 | 137.23 |
| 1975 | JUN | 447.00 | 443.62 | 156.16 |
| 1975 | JUL | 441.70 | 445.16 | 165.34 |
| 1975 | AUG | 281.40 | 421.41 | 154.29 |
| 1975 | SEP | 275.10 | 391.66 | 146.04 |
| 1975 | OCT | 365.50 | 378.02 | 130.96 |
| 1975 | NOV | 318.90 | 354.92 | 122.15 |
| 1975 | DEC | 349.00 | 338.59 | 112.05 |
| 1976 | JAN | 562.90 | 358.79 | 109.19 |
| 1976 | FEB | 640.50 | 418.64 | 118.74 |
| 1976 | MAR | 631.40 | 478.02 | 125.30 |
| 1976 | APR | 631.40 | 522.34 | 132.85 |
| 1976 | MAY | 370.50 | 530.94 | 125.92 |
| 1976 | JUN | 425.70 | 543.72 | 122.56 |
| 1976 | JUL | 451.80 | 525.21 | 117.98 |
| 1976 | AUG | 362.20 | 478.82 | 113.62 |
| 1976 | SEP | 405.00 | 441.09 | 112.62 |
| 1976 | OCT | 407.80 | 403.82 | 106.82 |
| 1976 | NOV | 413.00 | 410.91 | 115.77 |
| 1976 | DEC | 541.10 | 430.14 | 127.04 |

**TABLE A4-3 New York Stock Exchange Volume** (cont.)

| YEAR | MONTH | NYSTXCH | 6MTH MA | 12RC.6MA |
|---|---|---|---|---|
| 1977 | JAN | 509.50 | 439.76 | 122.57 |
| 1977 | FEB | 456.80 | 455.52 | 108.81 |
| 1977 | MAR | 453.50 | 463.61 | 96.98 |
| 1977 | APR | 405.80 | 463.27 | 88.69 |
| 1977 | MAY | 427.80 | 465.74 | 87.72 |
| 1977 | JUN | 484.20 | 456.26 | 83.91 |
| 1977 | JUL | 449.50 | 446.26 | 84.97 |
| 1977 | AUG | 433.10 | 442.31 | 92.37 |
| 1977 | SEP | 371.70 | 428.67 | 97.18 |

**TABLE A4-4 Market Breadth**

| DATE | A/D VAL* | A/D LINE† | 10WK RAT‡ |
|---|---|---|---|
| 28/ 3/1931 | -22.60 | -22.60 | 0.0 |
| 4/ 4/1931 | -16.70 | -39.30 | -16.70 |
| 11/ 4/1931 | -14.10 | -53.40 | -30.80 |
| 18/ 4/1931 | -21.20 | -74.60 | -52.00 |
| 25/ 4/1931 | -24.00 | -98.60 | -76.00 |
| 2/ 5/1931 | -19.10 | -117.70 | -95.10 |
| 9/ 5/1931 | 11.20 | -106.50 | -83.90 |
| 16/ 5/1931 | -19.60 | -126.10 | -103.50 |
| 23/ 5/1931 | -19.90 | -146.00 | -123.40 |
| 30/ 5/1931 | -23.00 | -169.00 | -146.40 |
| 6/ 6/1931 | 1.00 | -168.00 | -122.80 |
| 13/ 6/1931 | 17.00 | -151.00 | -89.10 |
| 20/ 6/1931 | 9.10 | -141.90 | -65.90 |
| 27/ 6/1931 | 35.50 | -106.40 | -9.20 |
| 4/ 7/1931 | 10.90 | -95.50 | 25.70 |
| 11/ 7/1931 | -18.10 | -113.60 | 26.70 |
| 18/ 7/1931 | -14.50 | -128.10 | 1.00 |
| 25/ 7/1931 | -10.10 | -138.20 | 10.50 |
| 1/ 8/1931 | -12.50 | -150.70 | 17.90 |
| 8/ 8/1931 | -13.90 | -164.60 | 27.00 |
| 15/ 8/1931 | 15.10 | -149.50 | 41.10 |
| 22/ 8/1931 | -16.50 | -166.00 | 7.60 |
| 29/ 8/1931 | -4.00 | -170.00 | -5.50 |
| 5/ 9/1931 | -20.60 | -190.60 | -61.60 |
| 12/ 9/1931 | -24.90 | -215.50 | -97.40 |
| 19/ 9/1931 | -32.60 | -248.10 | -111.90 |
| 26/ 9/1931 | -19.10 | -267.20 | -116.50 |
| 3/10/1931 | -31.50 | -298.70 | -137.90 |
| 10/10/1931 | 20.30 | -278.40 | -105.10 |
| 17/10/1931 | -14.00 | -292.40 | -105.20 |
| 24/10/1931 | 15.60 | -276.80 | -104.70 |
| 31/10/1931 | -13.90 | -290.70 | -102.10 |
| 7/11/1931 | 24.00 | -266.70 | -74.10 |
| 14/11/1931 | -12.60 | -279.30 | -66.10 |
| 21/11/1931 | -25.60 | -304.90 | -66.80 |
| 28/11/1931 | -22.30 | -327.20 | -56.50 |
| 5/12/1931 | -14.90 | -342.10 | -52.30 |
| 12/12/1931 | -28.30 | -370.40 | -49.10 |
| 19/12/1931 | -12.60 | -383.00 | -82.00 |
| 26/12/1931 | -16.30 | -399.30 | -84.30 |
| 40 | 40 | | |

* Weekly advance/decline ratio.

† Weekly advance/decline line.

‡ 10-week total of weekly ratios.

**TABLE A4-4 Market Breadth** (cont.)

| DATE | A/D VAL | A/D LINE | 10WK RAT |
|---|---|---|---|
| 2/ 1/1932 | -5.50 | -404.80 | -105.40 |
| 9/ 1/1932 | 23.10 | -381.70 | -68.40 |
| 16/ 1/1932 | 23.50 | -358.20 | -68.90 |
| 23/ 1/1932 | -17.30 | -375.50 | -73.60 |
| 30/ 1/1932 | -15.60 | -391.10 | -63.60 |
| 6/ 2/1932 | -11.00 | -402.10 | -52.30 |
| 13/ 2/1932 | 16.60 | -385.50 | -20.80 |
| 20/ 2/1932 | -1.40 | -386.90 | 6.10 |
| 27/ 2/1932 | -12.40 | -399.30 | 6.30 |
| 5/ 3/1932 | 16.00 | -383.30 | 38.60 |
| 12/ 3/1932 | -12.50 | -395.80 | 31.60 |
| 19/ 3/1932 | -21.50 | -417.30 | -13.00 |
| 26/ 3/1932 | -16.70 | -434.00 | -53.20 |
| 2/ 4/1932 | -23.50 | -457.50 | -59.40 |
| 9/ 4/1932 | -26.80 | -484.30 | -70.60 |
| 16/ 4/1932 | -7.70 | -492.00 | -67.30 |
| 23/ 4/1932 | -12.50 | -504.50 | -96.40 |
| 30/ 4/1932 | -12.40 | -516.90 | -107.40 |
| 7/ 5/1932 | -7.70 | -524.60 | -102.70 |
| 14/ 5/1932 | -16.60 | -541.20 | -135.30 |
| 21/ 5/1932 | -12.00 | -553.20 | -134.80 |
| 28/ 5/1932 | -24.30 | -577.50 | -137.60 |
| 4/ 6/1932 | 7.30 | -570.20 | -113.60 |
| 11/ 6/1932 | -10.10 | -580.30 | -100.20 |
| 18/ 6/1932 | 8.10 | -572.20 | -65.30 |
| 25/ 6/1932 | -14.30 | -586.50 | -71.90 |
| 2/ 7/1932 | -11.10 | -597.60 | -70.50 |
| 9/ 7/1932 | -3.60 | -601.20 | -61.70 |
| 16/ 7/1932 | 18.40 | -582.80 | -35.60 |
| 23/ 7/1932 | 20.20 | -562.60 | 1.20 |
| 30/ 7/1932 | 33.60 | -529.00 | 46.80 |
| 6/ 8/1932 | 28.70 | -500.30 | 99.80 |
| 13/ 8/1932 | 21.70 | -478.60 | 114.20 |
| 20/ 8/1932 | 21.20 | -457.40 | 145.50 |
| 27/ 8/1932 | 36.80 | -420.60 | 174.20 |
| 3/ 9/1932 | 21.80 | -398.80 | 210.30 |
| 10/ 9/1932 | 2.70 | -396.09 | 224.10 |
| 17/ 9/1932 | -38.30 | -434.39 | 189.40 |
| 24/ 9/1932 | 25.30 | -409.09 | 196.30 |
| 1/10/1932 | -21.30 | -430.39 | 154.80 |
| 8/10/1932 | -33.00 | -463.39 | 88.20 |
| 15/10/1932 | 9.50 | -453.89 | 69.00 |
| 22/10/1932 | -16.10 | -469.99 | 31.20 |
| 29/10/1932 | -3.60 | -473.59 | 6.40 |
| 5/11/1932 | -13.80 | -487.39 | -44.20 |
| 12/11/1932 | 27.50 | -459.89 | -38.50 |
| 19/11/1932 | -17.30 | -477.19 | -58.50 |
| 26/11/1932 | -19.30 | -496.49 | -39.50 |
| 3/12/1932 | 11.70 | -484.79 | -53.10 |
| 10/12/1932 | 11.40 | -473.39 | -20.40 |
| 17/12/1932 | -10.70 | -484.09 | 1.90 |
| 24/12/1932 | -20.80 | -504.89 | -28.40 |
| 31/12/1932 | 15.40 | -489.49 | 3.10 |
| 53 | 93 | | |

| DATE | A/D VAL | A/D LINE | 10WK RAT |
|---|---|---|---|
| 7/ 1/1933 | 22.90 | -466.59 | 29.60 |
| 14/ 1/1933 | 9.50 | -457.09 | 52.90 |
| 21/ 1/1933 | -12.80 | -469.89 | 12.60 |
| 28/ 1/1933 | -7.00 | -476.89 | 22.90 |
| 4/ 2/1933 | -16.90 | -493.79 | 25.30 |
| 11/ 2/1933 | 12.10 | -481.69 | 25.70 |
| 18/ 2/1933 | -19.60 | -501.29 | -5.30 |
| 25/ 2/1933 | -20.50 | -521.79 | -15.10 |
| 4/ 3/1933 | -6.90 | -528.69 | -1.20 |
| 11/ 3/1933 | 34.00 | -494.69 | 17.40 |
| 18/ 3/1933 | -21.60 | -516.29 | -27.10 |
| 25/ 3/1933 | -18.40 | -534.69 | -55.00 |
| 1/ 4/1933 | 15.20 | -519.49 | -27.00 |
| 8/ 4/1933 | 21.20 | -498.29 | 1.20 |
| 15/ 4/1933 | 38.50 | -459.79 | 56.60 |
| 22/ 4/1933 | 23.40 | -436.39 | 67.90 |
| 29/ 4/1933 | 26.20 | -410.19 | 113.70 |
| 6/ 5/1933 | 23.80 | -386.39 | 158.00 |
| 13/ 5/1933 | 18.70 | -367.69 | 183.60 |
| 20/ 5/1933 | 33.90 | -333.79 | 183.50 |
| 27/ 5/1933 | 29.40 | -304.39 | 234.50 |
| 3/ 6/1933 | 24.70 | -279.69 | 277.60 |
| 10/ 6/1933 | -30.00 | -309.69 | 232.40 |
| 17/ 6/1933 | 27.00 | -282.69 | 238.20 |
| 24/ 6/1933 | 20.90 | -261.79 | 220.60 |
| 1/ 7/1933 | 30.00 | -231.79 | 227.20 |
| 8/ 7/1933 | 7.80 | -223.99 | 208.80 |
| 15/ 7/1933 | -47.10 | -271.09 | 137.90 |
| 22/ 7/1933 | 26.20 | -244.89 | 145.40 |
| 29/ 7/1933 | -20.30 | -265.19 | 91.20 |
| 5/ 8/1933 | 19.20 | -245.99 | 81.00 |
| 12/ 8/1933 | -9.20 | -255.19 | 47.10 |
| 19/ 8/1933 | 21.90 | -233.29 | 99.00 |
| 26/ 8/1933 | -13.80 | -247.09 | 58.20 |
| 2/ 9/1933 | -23.90 | -270.99 | 13.40 |
| 9/ 9/1933 | 15.60 | -255.39 | -1.00 |
| 16/ 9/1933 | -34.70 | -290.09 | -43.50 |
| 23/ 9/1933 | -25.10 | -315.19 | -21.50 |
| 30/ 9/1933 | 15.50 | -299.69 | -32.20 |
| 7/10/1933 | -16.10 | -315.79 | -28.00 |
| 14/10/1933 | -52.00 | -367.79 | -99.20 |
| 21/10/1933 | 30.30 | -337.49 | -59.70 |
| 28/10/1933 | -6.50 | -343.99 | -88.10 |
| 4/11/1933 | 17.30 | -326.69 | -57.00 |
| 11/11/1933 | -2.40 | -329.09 | -35.50 |
| 18/11/1933 | -3.70 | -332.79 | -54.80 |
| 25/11/1933 | -16.90 | -349.69 | -37.00 |
| 2/12/1933 | 18.60 | -331.09 | 6.70 |
| 9/12/1933 | -14.80 | -345.89 | -23.60 |
| 16/12/1933 | -18.20 | -364.09 | -25.70 |
| 23/12/1933 | 13.80 | -350.29 | 40.10 |
| 30/12/1933 | -10.50 | -360.79 | -0.70 |
| 52 | 145 | | |

**TABLE A4-4 Market Breadth** (cont.)

| DATE | A/D VAL | A/D LINE | 10WK RAT |
|---|---|---|---|
| 6/ 1/1934 | 23.10 | -337.69 | 28.90 |
| 13/ 1/1934 | 50.10 | -287.59 | 61.70 |
| 20/ 1/1934 | 15.30 | -272.29 | 79.40 |
| 27/ 1/1934 | 31.10 | -241.19 | 114.20 |
| 3/ 2/1934 | -21.40 | -262.59 | 109.70 |
| 10/ 2/1934 | 31.30 | -231.29 | 122.40 |
| 17/ 2/1934 | -23.60 | -254.89 | 113.60 |
| 24/ 2/1934 | 10.00 | -244.89 | 141.80 |
| 3/ 3/1934 | -16.60 | -261.49 | 111.40 |
| 10/ 3/1934 | 9.50 | -251.99 | 131.40 |
| 17/ 3/1934 | -14.60 | -266.59 | 93.70 |
| 24/ 3/1934 | -2.20 | -268.79 | 41.40 |
| 31/ 3/1934 | 20.80 | -247.99 | 46.90 |
| 7/ 4/1934 | 12.80 | -235.19 | 28.60 |
| 14/ 4/1934 | 21.10 | -214.09 | 71.10 |
| 21/ 4/1934 | -21.40 | -235.49 | 18.40 |
| 28/ 4/1934 | -33.90 | -269.39 | 8.10 |
| 5/ 5/1934 | -39.30 | -308.69 | -41.20 |
| 12/ 5/1934 | 22.40 | -286.29 | -2.20 |
| 19/ 5/1934 | -13.80 | -300.09 | -25.50 |
| 26/ 5/1934 | -17.30 | -317.39 | -28.20 |
| 2/ 6/1934 | 29.30 | -288.09 | 3.30 |
| 9/ 6/1934 | 11.40 | -276.69 | -6.10 |
| 16/ 6/1934 | -22.50 | -299.19 | -41.40 |
| 23/ 6/1934 | -7.70 | -306.89 | -70.20 |
| 30/ 6/1934 | -4.50 | -311.39 | -53.30 |
| 7/ 7/1934 | 8.40 | -302.99 | -11.00 |
| 14/ 7/1934 | -28.30 | -331.29 | -0.00 |
| 21/ 7/1934 | -39.00 | -370.29 | -61.40 |
| 28/ 7/1934 | 15.80 | -354.49 | -31.80 |
| 4/ 8/1934 | 13.00 | -341.49 | -1.50 |
| 11/ 8/1934 | 13.00 | -328.49 | -17.80 |
| 18/ 8/1934 | 23.20 | -305.29 | -6.00 |
| 25/ 8/1934 | -20.50 | -325.79 | -4.00 |
| 1/ 9/1934 | -15.80 | -341.59 | -12.10 |
| 8/ 9/1934 | -27.80 | -369.39 | -35.40 |
| 15/ 9/1934 | 18.30 | -351.09 | -25.50 |
| 22/ 9/1934 | 15.00 | -336.09 | 17.80 |
| 29/ 9/1934 | -8.90 | -344.99 | 47.90 |
| 6/10/1934 | 16.70 | -328.29 | 48.80 |
| 13/10/1934 | -10.00 | -338.29 | 25.80 |
| 20/10/1934 | -15.60 | -353.89 | -2.80 |
| 27/10/1934 | 9.50 | -344.39 | -16.50 |
| 3/11/1934 | 27.10 | -317.29 | 31.10 |
| 10/11/1934 | 0.0 | -317.29 | 46.90 |
| 17/11/1934 | 14.50 | -302.79 | 89.20 |
| 24/11/1934 | 12.70 | -290.08 | 83.60 |
| 1/12/1934 | 8.90 | -281.18 | 77.50 |
| 8/12/1934 | -17.00 | -298.18 | 69.40 |
| 15/12/1934 | -17.90 | -316.08 | 34.80 |
| 22/12/1934 | 19.70 | -296.38 | 64.50 |
| 29/12/1934 | 17.00 | -279.38 | 97.10 |
| 52 | 197 | | |

| DATE | A/D VAL | A/D LINE | 10WK RAT |
|---|---|---|---|
| 5/ 1/1935 | -19.70 | -299.08 | 67.90 |
| 12/ 1/1935 | 10.50 | -288.58 | 51.30 |
| 19/ 1/1935 | -6.30 | -294.88 | 45.00 |
| 26/ 1/1935 | -18.20 | -313.08 | 12.30 |
| 2/ 2/1935 | -6.30 | -319.38 | -6.70 |
| 9/ 2/1935 | 8.90 | -310.48 | -6.70 |
| 16/ 2/1935 | -12.70 | -323.18 | -2.40 |
| 23/ 2/1935 | -15.80 | -338.98 | -0.30 |
| 2/ 3/1935 | -21.90 | -360.88 | -41.90 |
| 9/ 3/1935 | -24.30 | -385.18 | -83.20 |
| 16/ 3/1935 | 20.00 | -365.18 | -43.50 |
| 23/ 3/1935 | -11.80 | -376.98 | -65.80 |
| 30/ 3/1935 | 18.90 | -358.08 | -40.60 |
| 6/ 4/1935 | 17.30 | -340.78 | -5.10 |
| 13/ 4/1935 | 14.80 | -325.98 | 16.00 |
| 20/ 4/1935 | 10.00 | -315.98 | 17.10 |
| 27/ 4/1935 | 3.90 | -312.08 | 33.70 |
| 4/ 5/1935 | 18.00 | -294.08 | 67.50 |
| 11/ 5/1935 | 13.00 | -281.08 | 102.40 |
| 18/ 5/1935 | 8.30 | -272.78 | 135.00 |
| 25/ 5/1935 | -22.40 | -295.18 | 92.60 |
| 1/ 6/1935 | 17.30 | -277.88 | 121.70 |
| 8/ 6/1935 | 17.60 | -260.28 | 120.40 |
| 15/ 6/1935 | 10.00 | -250.28 | 113.10 |
| 22/ 6/1935 | -14.80 | -265.08 | 83.50 |
| 29/ 6/1935 | 13.00 | -252.08 | 86.50 |
| 6/ 7/1935 | 10.50 | -241.58 | 93.10 |
| 13/ 7/1935 | 12.70 | -228.88 | 87.80 |
| 20/ 7/1935 | 16.70 | -212.18 | 91.50 |
| 27/ 7/1935 | 12.70 | -199.48 | 95.90 |
| 3/ 8/1935 | 19.30 | -180.18 | 137.60 |
| 10/ 8/1935 | 14.80 | -165.38 | 135.10 |
| 17/ 8/1935 | -13.40 | -178.78 | 104.10 |
| 24/ 8/1935 | -9.50 | -188.28 | 84.60 |
| 31/ 8/1935 | 18.40 | -169.88 | 117.80 |
| 7/ 9/1935 | 4.50 | -165.38 | 109.30 |
| 14/ 9/1935 | -24.20 | -189.58 | 74.60 |
| 21/ 9/1935 | 15.50 | -174.08 | 77.40 |
| 28/ 9/1935 | -16.40 | -190.48 | 44.30 |
| 5/10/1935 | 18.40 | -172.08 | 50.00 |
| 12/10/1935 | 15.20 | -156.88 | 45.90 |
| 19/10/1935 | 19.50 | -137.38 | 50.60 |
| 26/10/1935 | 9.50 | -127.88 | 73.50 |
| 2/11/1935 | 13.80 | -114.08 | 96.80 |
| 9/11/1935 | 19.70 | -94.38 | 98.10 |
| 16/11/1935 | 4.50 | -89.88 | 98.10 |
| 23/11/1935 | -14.80 | -104.68 | 107.50 |
| 30/11/1935 | 19.70 | -84.98 | 111.70 |
| 7/12/1935 | -19.70 | -104.68 | 108.40 |
| 14/12/1935 | -9.50 | -114.18 | 80.50 |
| 21/12/1935 | -4.50 | -118.68 | 60.80 |
| 28/12/1935 | 29.50 | -89.18 | 70.80 |

52 249

**TABLE A4-4 Market Breadth** (cont.)

| DATE | A/D VAL | A/D LINE | 10WK RAT |
|---|---|---|---|
| 4/ 1/1936 | 20.30 | -68.88 | 81.60 |
| 11/ 1/1936 | 6.50 | -62.38 | 74.30 |
| 18/ 1/1936 | 16.10 | -46.28 | 70.70 |
| 25/ 1/1936 | 14.10 | -32.18 | 80.30 |
| 1/ 2/1936 | 16.10 | -16.08 | 111.20 |
| 8/ 2/1936 | 13.00 | -3.08 | 104.50 |
| 15/ 2/1936 | 3.20 | 0.12 | 127.40 |
| 22/ 2/1936 | -13.80 | -13.68 | 123.10 |
| 29/ 2/1936 | 14.80 | 1.12 | 142.40 |
| 7/ 3/1936 | -28.20 | -27.08 | 84.70 |
| 14/ 3/1936 | 14.80 | -12.28 | 79.20 |
| 21/ 3/1936 | -12.70 | -24.98 | 60.00 |
| 28/ 3/1936 | 17.30 | -7.68 | 61.20 |
| 4/ 4/1936 | -14.80 | -22.48 | 32.30 |
| 11/ 4/1936 | -23.60 | -46.08 | -7.40 |
| 18/ 4/1936 | -27.40 | -73.48 | -47.80 |
| 25/ 4/1936 | -25.10 | -98.58 | -76.10 |
| 2/ 5/1936 | 11.80 | -86.78 | -50.50 |
| 9/ 5/1936 | 20.50 | -66.28 | -44.80 |
| 16/ 5/1936 | -11.80 | -78.08 | -28.40 |
| 23/ 5/1936 | 13.40 | -64.68 | -29.80 |
| 30/ 5/1936 | -16.40 | -81.08 | -33.50 |
| 6/ 6/1936 | 18.70 | -62.38 | -32.10 |
| 13/ 6/1936 | 8.90 | -53.48 | -8.40 |
| 20/ 6/1936 | 0.0 | -53.48 | 15.20 |
| 27/ 6/1936 | -8.40 | -61.88 | 34.20 |
| 4/ 7/1936 | 18.40 | -43.48 | 77.70 |
| 11/ 7/1936 | 18.70 | -24.78 | 84.60 |
| 18/ 7/1936 | 14.10 | -10.68 | 78.20 |
| 25/ 7/1936 | -4.50 | -15.18 | 85.50 |
| 1/ 8/1936 | 12.70 | -2.48 | 84.80 |
| 8/ 8/1936 | -14.50 | -16.98 | 86.70 |
| 15/ 8/1936 | -18.10 | -35.08 | 49.90 |
| 22/ 8/1936 | 20.00 | -15.08 | 61.00 |
| 29/ 8/1936 | 16.70 | 1.62 | 77.70 |
| 5/ 9/1936 | 7.10 | 8.72 | 93.20 |
| 12/ 9/1936 | 8.40 | 17.12 | 83.20 |
| 19/ 9/1936 | -8.90 | 8.22 | 55.60 |
| 26/ 9/1936 | 22.80 | 31.02 | 64.30 |
| 3/10/1936 | 15.80 | 46.82 | 84.60 |
| 10/10/1936 | 14.80 | 61.62 | 86.70 |
| 17/10/1936 | -13.00 | 48.62 | 88.20 |
| 24/10/1936 | -7.10 | 41.52 | 99.20 |
| 31/10/1936 | 21.00 | 62.52 | 100.20 |
| 7/11/1936 | -10.50 | 52.02 | 73.00 |
| 14/11/1936 | 15.20 | 67.22 | 81.10 |
| 21/11/1936 | 14.50 | 81.72 | 87.20 |
| 28/11/1936 | -8.40 | 73.32 | 87.70 |
| 5/12/1936 | 11.10 | 84.42 | 76.00 |
| 12/12/1936 | -20.40 | 64.02 | 39.80 |
| 19/12/1936 | 10.00 | 74.02 | 35.00 |
| 26/12/1936 | 11.40 | 85.42 | 59.40 |

52 301

| DATE | A/D VAL | A/D LINE | 10WK RAT |
|---|---|---|---|
| 2/ 1/1937 | 24.90 | 110.32 | 91.40 |
| 9/ 1/1937 | 18.40 | 128.72 | 88.80 |
| 16/ 1/1937 | 8.60 | 137.32 | 107.90 |
| 23/ 1/1937 | -15.50 | 121.82 | 77.20 |
| 30/ 1/1937 | 8.60 | 130.42 | 71.30 |
| 6/ 2/1937 | 11.80 | 142.22 | 91.50 |
| 13/ 2/1937 | -11.80 | 130.42 | 68.60 |
| 20/ 2/1937 | -19.00 | 111.42 | 70.00 |
| 27/ 2/1937 | 15.50 | 126.92 | 75.50 |
| 6/ 3/1937 | -21.20 | 105.72 | 42.90 |
| 13/ 3/1937 | -19.20 | 86.52 | -1.20 |
| 20/ 3/1937 | -13.00 | 73.52 | -32.60 |
| 27/ 3/1937 | -14.50 | 59.02 | -55.70 |
| 3/ 4/1937 | -26.20 | 32.82 | -66.40 |
| 10/ 4/1937 | 10.00 | 42.82 | -65.00 |
| 17/ 4/1937 | -19.70 | 23.12 | -96.50 |
| 24/ 4/1937 | -25.50 | -2.38 | -110.20 |
| 1/ 5/1937 | 9.50 | 7.12 | -81.70 |
| 8/ 5/1937 | -34.20 | -27.08 | -131.40 |
| 15/ 5/1937 | 18.40 | -8.68 | -91.80 |
| 22/ 5/1937 | -13.00 | -21.68 | -85.60 |
| 29/ 5/1937 | -11.80 | -33.48 | -84.40 |
| 5/ 6/1937 | -21.60 | -55.08 | -91.50 |
| 12/ 6/1937 | -23.60 | -78.68 | -88.90 |
| 19/ 6/1937 | -8.40 | -87.08 | -107.30 |
| 26/ 6/1937 | 12.20 | -74.88 | -75.40 |
| 3/ 7/1937 | 21.80 | -53.08 | -28.10 |
| 10/ 7/1937 | 8.90 | -44.18 | -28.70 |
| 17/ 7/1937 | 16.10 | -28.08 | 21.60 |
| 24/ 7/1937 | -11.80 | -39.88 | -8.60 |
| 31/ 7/1937 | -5.50 | -45.38 | -1.10 |
| 7/ 8/1937 | 13.00 | -32.38 | 23.70 |
| 14/ 8/1937 | -21.80 | -54.18 | 23.50 |
| 21/ 8/1937 | -27.00 | -81.18 | 20.10 |
| 28/ 8/1937 | -26.40 | -107.58 | 2.10 |
| 4/ 9/1937 | -45.40 | -152.98 | -55.50 |
| 11/ 9/1937 | 6.30 | -146.68 | -71.00 |
| 18/ 9/1937 | -39.30 | -185.98 | -119.20 |
| 25/ 9/1937 | 26.40 | -159.58 | -108.90 |
| 2/10/1937 | -31.30 | -190.88 | -128.40 |
| 9/10/1937 | -45.80 | -236.68 | -168.70 |
| 16/10/1937 | -14.10 | -250.78 | -195.80 |
| 23/10/1937 | 31.80 | -218.98 | -142.20 |
| 30/10/1937 | -32.50 | -251.48 | -147.70 |
| 6/11/1937 | 24.10 | -227.38 | -97.20 |
| 13/11/1937 | -41.30 | -268.68 | -93.10 |
| 20/11/1937 | 15.20 | -253.48 | -84.20 |
| 27/11/1937 | 12.70 | -240.78 | -32.20 |
| 4/12/1937 | -14.50 | -255.28 | -73.10 |
| 11/12/1937 | -19.70 | -274.98 | -61.50 |
| 18/12/1937 | -14.10 | -289.08 | -29.80 |
| 25/12/1937 | -25.10 | -314.18 | -40.80 |

52 353

**TABLE A4-4 Market Breadth** (cont.)

| DATE | A/D VAL | A/D LINE | 10WK RAT |
|---|---|---|---|
| 1/ 1/1938 | 36.70 | -277.48 | -35.90 |
| 8/ 1/1938 | 29.30 | -248.18 | 25.90 |
| 15/ 1/1938 | -22.80 | -270.98 | -21.00 |
| 22/ 1/1938 | -33.10 | -304.08 | -12.80 |
| 29/ 1/1938 | 0.0 | -304.08 | -28.00 |
| 5/ 2/1938 | 16.10 | -287.98 | -24.60 |
| 12/ 2/1938 | 16.70 | -271.28 | 6.60 |
| 19/ 2/1938 | 20.00 | -251.28 | 46.30 |
| 26/ 2/1938 | -26.20 | -277.48 | 34.20 |
| 5/ 3/1938 | -27.40 | -304.88 | 31.90 |
| 12/ 3/1938 | -24.20 | -329.08 | -29.00 |
| 19/ 3/1938 | -38.10 | -367.18 | -96.40 |
| 26/ 3/1938 | 0.0 | -367.18 | -73.60 |
| 2/ 4/1938 | 31.00 | -336.18 | -9.50 |
| 9/ 4/1938 | 19.50 | -316.68 | 10.00 |
| 16/ 4/1938 | -5.50 | -322.18 | -11.60 |
| 23/ 4/1938 | -25.50 | -347.68 | -53.80 |
| 30/ 4/1938 | 19.00 | -328.68 | -54.80 |
| 7/ 5/1938 | 8.90 | -319.78 | -19.70 |
| 14/ 5/1938 | -21.20 | -340.98 | -13.50 |
| 21/ 5/1938 | -24.80 | -365.78 | -14.10 |
| 28/ 5/1938 | 12.20 | -353.58 | 36.20 |
| 4/ 6/1938 | 15.10 | -338.48 | 51.30 |
| 11/ 6/1938 | -8.40 | -346.88 | 11.90 |
| 18/ 6/1938 | 44.70 | -302.18 | 37.10 |
| 25/ 6/1938 | 35.10 | -267.08 | 77.70 |
| 2/ 7/1938 | 3.20 | -263.88 | 106.40 |
| 9/ 7/1938 | 16.10 | -247.78 | 103.50 |
| 16/ 7/1938 | 25.60 | -222.18 | 120.20 |
| 23/ 7/1938 | -19.50 | -241.68 | 121.90 |
| 30/ 7/1938 | 16.70 | -224.98 | 163.40 |
| 6/ 8/1938 | -31.00 | -255.98 | 120.20 |
| 13/ 8/1938 | 17.30 | -238.68 | 122.40 |
| 20/ 8/1938 | 15.50 | -223.18 | 146.30 |
| 27/ 8/1938 | -17.90 | -241.08 | 83.70 |
| 3/ 9/1938 | -20.20 | -261.28 | 28.40 |
| 10/ 9/1938 | -36.60 | -297.88 | -11.40 |
| 17/ 9/1938 | 17.00 | -280.88 | -10.50 |
| 24/ 9/1938 | 32.40 | -248.48 | -3.70 |
| 1/10/1938 | 35.50 | -212.98 | 51.30 |
| 8/10/1938 | 15.50 | -197.48 | 50.10 |
| 15/10/1938 | 15.20 | -182.28 | 96.30 |
| 22/10/1938 | -14.50 | -196.78 | 64.50 |
| 29/10/1938 | 7.10 | -189.68 | 56.10 |
| 5/11/1938 | 25.40 | -164.28 | 99.40 |
| 12/11/1938 | -28.30 | -192.58 | 91.30 |
| 19/11/1938 | -15.80 | -208.38 | 112.10 |
| 26/11/1938 | -15.80 | -224.18 | 79.30 |
| 3/12/1938 | -14.50 | -238.68 | 32.40 |
| 10/12/1938 | 13.80 | -224.88 | 10.70 |
| 17/12/1938 | -12.70 | -237.58 | -17.50 |
| 24/12/1938 | 24.30 | -213.28 | -8.40 |
| 31/12/1938 | -6.30 | -219.58 | -0.20 |
| 53 | 406 | | |

| DATE | A/D VAL | A/D LINE | 10WK RAT |
|---|---|---|---|
| 7/ 1/1939 | -21.40 | -240.98 | -28.70 |
| 14/ 1/1939 | -7.10 | -248.08 | -61.20 |
| 21/ 1/1939 | -36.70 | -284.78 | -69.60 |
| 28/ 1/1939 | 24.70 | -260.08 | -29.10 |
| 4/ 2/1939 | -7.10 | -267.18 | -20.40 |
| 11/ 2/1939 | 12.20 | -254.98 | 6.30 |
| 18/ 2/1939 | 14.10 | -240.88 | 6.60 |
| 25/ 2/1939 | 17.60 | -223.28 | 36.90 |
| 4/ 3/1939 | 14.10 | -209.18 | 26.70 |
| 11/ 3/1939 | -31.80 | -240.98 | 1.20 |
| 18/ 3/1939 | -10.90 | -251.88 | 11.70 |
| 25/ 3/1939 | -39.10 | -290.98 | -20.30 |
| 1/ 4/1939 | -29.20 | -320.18 | -12.80 |
| 8/ 4/1939 | 23.20 | -296.98 | -14.30 |
| 15/ 4/1939 | -5.50 | -302.48 | -12.70 |
| 22/ 4/1939 | 5.50 | -296.98 | -19.40 |
| 29/ 4/1939 | 20.30 | -276.68 | -13.20 |
| 6/ 5/1939 | 11.80 | -264.88 | -19.00 |
| 13/ 5/1939 | -15.20 | -280.08 | -48.30 |
| 20/ 5/1939 | 24.70 | -255.38 | 8.20 |
| 27/ 5/1939 | 0.0 | -255.38 | 19.10 |
| 3/ 6/1939 | 15.50 | -239.88 | 73.70 |
| 10/ 6/1939 | -21.90 | -261.78 | 81.00 |
| 17/ 6/1939 | 13.00 | -248.78 | 70.80 |
| 24/ 6/1939 | -25.40 | -274.18 | 50.90 |
| 1/ 7/1939 | 14.80 | -259.38 | 60.20 |
| 8/ 7/1939 | 21.40 | -237.98 | 61.30 |
| 15/ 7/1939 | 28.30 | -209.68 | 77.80 |
| 22/ 7/1939 | -5.50 | -215.18 | 87.50 |
| 29/ 7/1939 | -14.10 | -229.28 | 48.70 |
| 5/ 8/1939 | -20.00 | -249.28 | 28.70 |
| 12/ 8/1939 | -18.40 | -267.68 | -5.20 |
| 19/ 8/1939 | -17.00 | -284.68 | -0.30 |
| 26/ 8/1939 | -5.50 | -290.18 | -18.80 |
| 2/ 9/1939 | 38.10 | -252.08 | 44.70 |
| 9/ 9/1939 | 17.30 | -234.78 | 47.20 |
| 16/ 9/1939 | 12.70 | -222.08 | 38.50 |
| 23/ 9/1939 | 10.90 | -211.18 | 21.10 |
| 30/ 9/1939 | -20.00 | -231.18 | 6.60 |
| 7/10/1939 | 5.50 | -225.68 | 26.20 |
| 14/10/1939 | 20.20 | -205.48 | 66.40 |
| 21/10/1939 | 8.40 | -197.08 | 93.20 |
| 28/10/1939 | -10.90 | -207.98 | 99.30 |
| 4/11/1939 | -22.80 | -230.78 | 82.00 |
| 11/11/1939 | 15.50 | -215.28 | 59.40 |
| 18/11/1939 | -18.40 | -233.68 | 23.70 |
| 25/11/1939 | -19.20 | -252.88 | -8.20 |
| 2/12/1939 | 13.00 | -239.88 | -6.10 |
| 9/12/1939 | -5.50 | -245.38 | 8.40 |
| 16/12/1939 | -8.90 | -254.28 | -6.00 |
| 23/12/1939 | 8.90 | -245.38 | -17.30 |
| 30/12/1939 | 22.80 | -222.58 | -2.90 |

52 458

**TABLE A4-4 Market Breadth** (cont.)

| DATE | A/D VAL | A/D LINE | 10WK RAT |
|---|---|---|---|
| 6/ 1/1940 | -23.40 | -245.98 | -15.40 |
| 13/ 1/1940 | -8.90 | -254.88 | -1.50 |
| 20/ 1/1940 | 11.00 | -243.88 | -6.00 |
| 27/ 1/1940 | -9.50 | -253.38 | 2.90 |
| 3/ 2/1940 | 18.40 | -234.98 | 40.50 |
| 10/ 2/1940 | 3.10 | -231.88 | 30.60 |
| 17/ 2/1940 | -9.50 | -241.38 | 26.60 |
| 24/ 2/1940 | -9.50 | -250.88 | 26.00 |
| 2/ 3/1940 | 14.50 | -236.38 | 31.60 |
| 9/ 3/1940 | -17.30 | -253.68 | -8.50 |
| 16/ 3/1940 | 12.60 | -241.08 | 27.50 |
| 23/ 3/1940 | 15.80 | -225.28 | 52.20 |
| 30/ 3/1940 | 25.10 | -200.18 | 65.30 |
| 6/ 4/1940 | -16.40 | -216.58 | 59.40 |
| 13/ 4/1940 | -15.80 | -232.38 | 25.20 |
| 20/ 4/1940 | -3.10 | -235.48 | 19.00 |
| 27/ 4/1940 | -10.90 | -246.38 | 17.60 |
| 4/ 5/1940 | -18.40 | -264.78 | 8.70 |
| 11/ 5/1940 | -76.80 | -341.58 | -82.60 |
| 18/ 5/1940 | -29.20 | -370.78 | -94.50 |
| 25/ 5/1940 | 10.90 | -359.88 | -96.20 |
| 1/ 6/1940 | -7.10 | -366.98 | -119.10 |
| 8/ 6/1940 | 25.70 | -341.28 | -113.50 |
| 15/ 6/1940 | 7.10 | -334.18 | -95.00 |
| 22/ 6/1940 | 4.50 | -329.68 | -74.70 |
| 29/ 6/1940 | 5.50 | -324.18 | -66.10 |
| 6/ 7/1940 | 4.50 | -319.68 | -50.70 |
| 13/ 7/1940 | 10.00 | -309.68 | -22.30 |
| 20/ 7/1940 | -5.50 | -315.18 | 49.00 |
| 27/ 7/1940 | 18.40 | -296.78 | 95.50 |
| 3/ 8/1940 | -7.80 | -304.58 | 77.90 |
| 10/ 8/1940 | -19.70 | -324.28 | 65.30 |
| 17/ 8/1940 | 17.00 | -307.28 | 56.60 |
| 24/ 8/1940 | 21.40 | -285.38 | 70.90 |
| 31/ 8/1940 | 25.50 | -260.38 | 91.90 |
| 7/ 9/1940 | -22.30 | -282.68 | 64.10 |
| 14/ 9/1940 | 17.30 | -265.38 | 76.90 |
| 21/ 9/1940 | 6.30 | -259.08 | 73.20 |
| 28/ 9/1940 | 17.30 | -241.78 | 96.00 |
| 5/10/1940 | -17.00 | -258.78 | 60.60 |
| 12/10/1940 | 13.00 | -245.78 | 81.40 |
| 19/10/1940 | 10.50 | -235.28 | 111.60 |
| 26/10/1940 | 17.00 | -218.28 | 111.60 |
| 2/11/1940 | 23.20 | -195.08 | 113.40 |
| 9/11/1940 | -12.20 | -207.28 | 75.70 |
| 16/11/1940 | -14.80 | -222.08 | 83.20 |
| 23/11/1940 | 13.00 | -209.08 | 78.90 |
| 30/11/1940 | 0.0 | -209.08 | 72.60 |
| 7/12/1940 | 12.30 | -196.78 | 67.60 |
| 14/12/1940 | -19.40 | -216.18 | 65.20 |
| 21/12/1940 | 4.50 | -211.68 | 56.70 |
| 28/12/1940 | 17.90 | -193.78 | 64.10 |

52 510

| DATE | A/D VAL | A/D LINE | 10WK RAT |
|---|---|---|---|
| 4/ 1/1941 | 17.00 | -176.78 | 64.10 |
| 11/ 1/1941 | -18.90 | -195.68 | 22.00 |
| 18/ 1/1941 | -3.10 | -198.78 | 31.10 |
| 25/ 1/1941 | -26.40 | -225.18 | 19.50 |
| 1/ 2/1941 | 8.90 | -216.28 | 15.40 |
| 8/ 2/1941 | -27.20 | -243.48 | -11.80 |
| 15/ 2/1941 | 8.40 | -235.08 | -15.70 |
| 22/ 2/1941 | 15.50 | -219.58 | 19.20 |
| 1/ 3/1941 | -6.30 | -225.88 | 8.40 |
| 8/ 3/1941 | 15.20 | -210.68 | 5.70 |
| 15/ 3/1941 | -9.50 | -220.18 | -20.80 |
| 22/ 3/1941 | 3.10 | -217.08 | 1.20 |
| 29/ 3/1941 | 15.80 | -201.28 | 20.10 |
| 5/ 4/1941 | -24.10 | -225.38 | 22.40 |
| 12/ 4/1941 | -17.30 | -242.68 | -3.80 |
| 19/ 4/1941 | 3.10 | -239.58 | 26.50 |
| 26/ 4/1941 | 5.50 | -234.08 | 23.60 |
| 3/ 5/1941 | 14.50 | -219.58 | 22.60 |
| 10/ 5/1941 | -13.00 | -232.58 | 15.90 |
| 17/ 5/1941 | 5.50 | -227.08 | 6.20 |
| 24/ 5/1941 | -7.10 | -234.18 | 8.60 |
| 31/ 5/1941 | 14.50 | -219.68 | 20.00 |
| 7/ 6/1941 | 24.10 | -195.58 | 28.30 |
| 14/ 6/1941 | 3.10 | -192.48 | 55.50 |
| 21/ 6/1941 | 12.60 | -179.88 | 85.40 |
| 28/ 6/1941 | 8.40 | -171.48 | 90.70 |
| 5/ 7/1941 | 27.90 | -143.58 | 113.10 |
| 12/ 7/1941 | -4.50 | -148.08 | 94.10 |
| 19/ 7/1941 | 17.60 | -130.48 | 124.70 |
| 26/ 7/1941 | 10.90 | -119.58 | 130.10 |
| 2/ 8/1941 | -14.10 | -133.68 | 123.10 |
| 9/ 8/1941 | -15.20 | -148.88 | 93.40 |
| 16/ 8/1941 | 12.20 | -136.68 | 81.50 |
| 23/ 8/1941 | 13.80 | -122.88 | 92.20 |
| 30/ 8/1941 | 3.10 | -119.78 | 82.70 |
| 6/ 9/1941 | 5.50 | -114.28 | 79.80 |
| 13/ 9/1941 | 4.50 | -109.78 | 56.40 |
| 20/ 9/1941 | -19.50 | -129.28 | 41.40 |
| 27/ 9/1941 | 10.90 | -118.38 | 34.70 |
| 4/10/1941 | -21.60 | -139.98 | 2.20 |
| 11/10/1941 | -17.60 | -157.58 | -1.30 |
| 18/10/1941 | 13.00 | -144.58 | 26.90 |
| 25/10/1941 | -14.50 | -159.08 | 0.20 |
| 1/11/1941 | 3.10 | -155.98 | -10.50 |
| 8/11/1941 | -15.20 | -171.18 | -28.80 |
| 15/11/1941 | 7.10 | -164.08 | -27.20 |
| 22/11/1941 | -14.50 | -178.58 | -46.20 |
| 29/11/1941 | 12.60 | -165.98 | -14.10 |
| 6/12/1941 | -41.10 | -207.08 | -66.10 |
| 13/12/1941 | -12.60 | -219.68 | -57.10 |
| 20/12/1941 | -7.80 | -227.48 | -47.30 |
| 27/12/1941 | 28.20 | -199.28 | -32.10 |

52 562

**TABLE A4-4 Market Breadth** (cont.)

| DATE | A/D VAL | A/D LINE | 10WK RAT |
|---|---|---|---|
| 3/ 1/1942 | 12.20 | -187.08 | -5.40 |
| 10/ 1/1942 | 12.60 | -174.48 | 4.10 |
| 17/ 1/1942 | -7.10 | -181.58 | 12.20 |
| 24/ 1/1942 | 3.10 | -178.48 | 8.20 |
| 31/ 1/1942 | 7.70 | -170.78 | 30.40 |
| 7/ 2/1942 | -20.20 | -190.98 | -2.40 |
| 14/ 2/1942 | -11.40 | -202.38 | 27.30 |
| 21/ 2/1942 | 9.50 | -192.88 | 49.40 |
| 28/ 2/1942 | -20.00 | -212.88 | 37.20 |
| 7/ 3/1942 | -12.60 | -225.48 | -3.60 |
| 14/ 3/1942 | 9.50 | -215.98 | -6.30 |
| 21/ 3/1942 | -11.40 | -227.38 | -30.30 |
| 28/ 3/1942 | 3.10 | -224.28 | -20.10 |
| 4/ 4/1942 | -11.40 | -235.68 | -34.60 |
| 11/ 4/1942 | -21.20 | -256.88 | -63.50 |
| 18/ 4/1942 | -12.60 | -269.48 | -55.90 |
| 25/ 4/1942 | 9.50 | -259.98 | -35.00 |
| 2/ 5/1942 | 11.80 | -248.18 | -32.70 |
| 9/ 5/1942 | -7.70 | -255.88 | -20.40 |
| 16/ 5/1942 | 6.30 | -249.58 | -1.50 |
| 23/ 5/1942 | 10.00 | -239.58 | -1.00 |
| 30/ 5/1942 | 14.80 | -224.78 | 25.20 |
| 6/ 6/1942 | 4.50 | -220.28 | 26.60 |
| 13/ 6/1942 | 10.00 | -210.28 | 48.00 |
| 20/ 6/1942 | -13.00 | -223.28 | 56.20 |
| 27/ 6/1942 | 12.20 | -211.08 | 81.00 |
| 4/ 7/1942 | 26.40 | -184.68 | 97.90 |
| 11/ 7/1942 | -6.30 | -190.98 | 79.80 |
| 18/ 7/1942 | 7.70 | -183.28 | 95.20 |
| 25/ 7/1942 | -6.30 | -189.58 | 82.60 |
| 1/ 8/1942 | -7.70 | -197.28 | 64.90 |
| 8/ 8/1942 | 13.00 | -184.28 | 63.10 |
| 15/ 8/1942 | 15.20 | -169.08 | 73.80 |
| 22/ 8/1942 | -10.50 | -179.58 | 53.30 |
| 29/ 8/1942 | 7.10 | -172.48 | 73.40 |
| 5/ 9/1942 | -4.50 | -176.98 | 56.70 |
| 12/ 9/1942 | 17.60 | -159.38 | 47.90 |
| 19/ 9/1942 | 18.20 | -141.18 | 72.40 |
| 26/ 9/1942 | 15.80 | -125.38 | 80.50 |
| 3/10/1942 | 23.20 | -102.18 | 110.00 |
| 10/10/1942 | -6.30 | -108.48 | 111.40 |
| 17/10/1942 | 15.20 | -93.28 | 113.60 |
| 24/10/1942 | -7.10 | -100.38 | 91.30 |
| 31/10/1942 | 18.20 | -82.18 | 120.00 |
| 7/11/1942 | -10.00 | -92.18 | 102.90 |
| 14/11/1942 | -11.80 | -103.98 | 95.60 |
| 21/11/1942 | -13.40 | -117.38 | 64.60 |
| 28/11/1942 | 4.50 | -112.88 | 50.90 |
| 5/12/1942 | -6.30 | -119.18 | 28.80 |
| 12/12/1942 | 21.60 | -97.58 | 27.20 |
| 19/12/1942 | -3.10 | -100.68 | 30.40 |
| 26/12/1942 | 10.00 | -90.68 | 25.20 |
| 52 | 614 | | |

| DATE | A/D VAL | A/D LINE | 10WK RAT |
|---|---|---|---|
| 2/ 1/1943 | 19.20 | -71.48 | 51.50 |
| 9/ 1/1943 | 23.40 | -48.08 | 56.70 |
| 16/ 1/1943 | 11.80 | -36.28 | 78.50 |
| 23/ 1/1943 | 26.10 | -10.18 | 116.40 |
| 30/ 1/1943 | 11.80 | 1.62 | 141.60 |
| 6/ 2/1943 | 21.20 | 22.82 | 158.30 |
| 13/ 2/1943 | 12.60 | 35.42 | 177.20 |
| 20/ 2/1943 | 22.30 | 57.72 | 177.90 |
| 27/ 2/1943 | 16.10 | 73.82 | 197.10 |
| 6/ 3/1943 | 17.00 | 90.82 | 204.10 |
| 13/ 3/1943 | -13.70 | 77.12 | 171.20 |
| 20/ 3/1943 | 31.50 | 108.62 | 179.30 |
| 27/ 3/1943 | 17.30 | 125.92 | 184.80 |
| 3/ 4/1943 | -21.00 | 104.92 | 137.70 |
| 10/ 4/1943 | 21.90 | 126.82 | 147.80 |
| 17/ 4/1943 | 15.10 | 141.92 | 141.70 |
| 24/ 4/1943 | 15.10 | 157.02 | 144.20 |
| 1/ 5/1943 | 19.20 | 176.22 | 141.10 |
| 8/ 5/1943 | -16.70 | 159.52 | 108.30 |
| 15/ 5/1943 | 18.70 | 178.22 | 110.00 |
| 22/ 5/1943 | 17.60 | 195.82 | 141.30 |
| 29/ 5/1943 | 10.90 | 206.72 | 120.70 |
| 5/ 6/1943 | -14.80 | 191.92 | 83.60 |
| 12/ 6/1943 | -15.80 | 176.12 | 93.80 |
| 19/ 6/1943 | 19.50 | 195.62 | 91.40 |
| 26/ 6/1943 | 18.20 | 213.82 | 94.50 |
| 3/ 7/1943 | 6.30 | 220.12 | 85.70 |
| 10/ 7/1943 | 13.80 | 233.92 | 80.30 |
| 17/ 7/1943 | -8.40 | 225.52 | 88.60 |
| 24/ 7/1943 | -36.30 | 189.22 | 33.60 |
| 31/ 7/1943 | -15.20 | 174.02 | 0.80 |
| 7/ 8/1943 | 19.50 | 193.52 | 9.40 |
| 14/ 8/1943 | -13.00 | 180.52 | 11.20 |
| 21/ 8/1943 | -9.50 | 171.02 | 17.50 |
| 28/ 8/1943 | 16.10 | 187.12 | 14.10 |
| 4/ 9/1943 | 11.40 | 198.52 | 7.30 |
| 11/ 9/1943 | 21.60 | 220.12 | 22.60 |
| 18/ 9/1943 | -6.30 | 213.82 | 2.50 |
| 25/ 9/1943 | -7.10 | 206.72 | 3.80 |
| 2/10/1943 | -22.30 | 184.42 | 17.80 |
| 9/10/1943 | 9.50 | 193.92 | 42.50 |
| 16/10/1943 | 10.50 | 204.42 | 33.50 |
| 23/10/1943 | 7.80 | 212.22 | 54.30 |
| 30/10/1943 | -27.00 | 185.22 | 36.80 |
| 6/11/1943 | -25.50 | 159.72 | -4.80 |
| 13/11/1943 | 15.50 | 175.22 | -0.70 |
| 20/11/1943 | -17.30 | 157.92 | -39.60 |
| 27/11/1943 | 10.50 | 168.42 | -22.80 |
| 4/12/1943 | 25.30 | 193.72 | 9.60 |
| 11/12/1943 | 10.50 | 204.22 | 42.40 |
| 18/12/1943 | 7.10 | 211.32 | 40.00 |
| 25/12/1943 | 13.00 | 224.32 | 42.50 |

52 666

**TABLE A4-4 Market Breadth** (cont.)

| DATE | A/D VAL | A/D LINE | 10WK RAT |
|---|---|---|---|
| 1/ 1/1944 | 22.90 | 247.22 | 57.60 |
| 8/ 1/1944 | 14.80 | 262.02 | 99.40 |
| 15/ 1/1944 | 3.10 | 265.12 | 128.00 |
| 22/ 1/1944 | -10.00 | 255.12 | 102.50 |
| 29/ 1/1944 | 8.90 | 264.02 | 128.70 |
| 5/ 2/1944 | 10.00 | 274.02 | 128.20 |
| 12/ 2/1944 | 12.70 | 286.72 | 115.60 |
| 19/ 2/1944 | 14.20 | 300.92 | 119.30 |
| 26/ 2/1944 | 4.50 | 305.42 | 116.70 |
| 4/ 3/1944 | 21.40 | 326.82 | 125.10 |
| 11/ 3/1944 | 17.90 | 344.72 | 120.10 |
| 18/ 3/1944 | -12.20 | 332.52 | 93.10 |
| 25/ 3/1944 | -17.90 | 314.62 | 72.10 |
| 1/ 4/1944 | 0.0 | 314.62 | 82.10 |
| 8/ 4/1944 | -13.80 | 300.82 | 59.40 |
| 15/ 4/1944 | -21.20 | 279.62 | 28.20 |
| 22/ 4/1944 | 14.10 | 293.72 | 29.60 |
| 29/ 4/1944 | 16.40 | 310.12 | 31.80 |
| 6/ 5/1944 | -8.90 | 301.22 | 18.40 |
| 13/ 5/1944 | 19.70 | 320.92 | 16.70 |
| 20/ 5/1944 | 15.80 | 336.72 | 14.60 |
| 27/ 5/1944 | 14.10 | 350.82 | 40.90 |
| 3/ 6/1944 | -7.70 | 343.12 | 51.10 |
| 10/ 6/1944 | 35.30 | 378.42 | 86.40 |
| 17/ 6/1944 | 13.40 | 391.82 | 113.60 |
| 24/ 6/1944 | 19.50 | 411.32 | 154.30 |
| 1/ 7/1944 | 21.40 | 432.72 | 161.60 |
| 8/ 7/1944 | 4.50 | 437.22 | 149.70 |
| 15/ 7/1944 | -30.00 | 407.22 | 128.60 |
| 22/ 7/1944 | 14.50 | 421.72 | 123.40 |
| 29/ 7/1944 | -11.40 | 410.32 | 96.20 |
| 5/ 8/1944 | 18.70 | 429.02 | 100.80 |
| 12/ 8/1944 | 20.30 | 449.32 | 128.80 |
| 19/ 8/1944 | -16.70 | 432.62 | 76.80 |
| 26/ 8/1944 | 13.00 | 445.62 | 76.40 |
| 2/ 9/1944 | -31.20 | 414.42 | 25.70 |
| 9/ 9/1944 | -3.10 | 411.32 | 1.20 |
| 16/ 9/1944 | 17.90 | 429.22 | 14.60 |
| 23/ 9/1944 | 18.20 | 447.42 | 62.80 |
| 30/ 9/1944 | 17.90 | 465.32 | 66.20 |
| 7/10/1944 | -4.50 | 460.82 | 73.10 |
| 14/10/1944 | 9.50 | 470.32 | 63.90 |
| 21/10/1944 | -21.20 | 449.12 | 22.40 |
| 28/10/1944 | 14.10 | 463.22 | 53.20 |
| 4/11/1944 | 15.20 | 478.42 | 55.40 |
| 11/11/1944 | -20.70 | 457.72 | 65.90 |
| 18/11/1944 | 16.10 | 473.82 | 85.10 |
| 25/11/1944 | 17.30 | 491.12 | 84.50 |
| 2/12/1944 | 26.10 | 517.22 | 92.40 |
| 9/12/1944 | 16.40 | 533.62 | 90.90 |
| 16/12/1944 | -14.80 | 518.82 | 80.60 |
| 23/12/1944 | 14.50 | 533.32 | 85.60 |
| 30/12/1944 | 16.70 | 550.02 | 123.50 |

53 719

| DATE | A/D VAL | A/D LINE | 10WK RAT |
|---|---|---|---|
| 6/ 1/1945 | 20.00 | 570.02 | 129.40 |
| 13/ 1/1945 | -16.10 | 553.92 | 98.10 |
| 20/ 1/1945 | 16.40 | 570.32 | 135.20 |
| 27/ 1/1945 | 19.50 | 589.82 | 138.60 |
| 3/ 2/1945 | 12.60 | 602.42 | 133.90 |
| 10/ 2/1945 | 23.70 | 626.12 | 131.50 |
| 17/ 2/1945 | 6.30 | 632.42 | 121.40 |
| 24/ 2/1945 | 16.70 | 649.12 | 152.90 |
| 3/ 3/1945 | -26.80 | 622.32 | 111.60 |
| 10/ 3/1945 | 19.20 | 641.52 | 114.10 |
| 17/ 3/1945 | -28.60 | 612.92 | 65.50 |
| 24/ 3/1945 | -9.50 | 603.42 | 72.10 |
| 31/ 3/1945 | 10.00 | 613.42 | 65.70 |
| 7/ 4/1945 | 24.20 | 637.62 | 70.40 |
| 14/ 4/1945 | 23.80 | 661.42 | 81.60 |
| 21/ 4/1945 | 16.10 | 677.52 | 74.00 |
| 28/ 4/1945 | 11.80 | 689.32 | 79.50 |
| 5/ 5/1945 | -15.20 | 674.12 | 47.60 |
| 12/ 5/1945 | 17.30 | 691.42 | 91.70 |
| 19/ 5/1945 | -7.10 | 684.32 | 65.40 |
| 26/ 5/1945 | 16.40 | 700.72 | 110.40 |
| 2/ 6/1945 | 7.80 | 708.52 | 127.70 |
| 9/ 6/1945 | 19.20 | 727.72 | 136.90 |
| 16/ 6/1945 | 14.10 | 741.82 | 126.80 |
| 23/ 6/1945 | -21.90 | 719.92 | 81.10 |
| 30/ 6/1945 | -14.10 | 705.82 | 50.90 |
| 7/ 7/1945 | 10.90 | 716.72 | 50.00 |
| 14/ 7/1945 | -24.30 | 692.42 | 40.90 |
| 21/ 7/1945 | -21.00 | 671.42 | 2.60 |
| 28/ 7/1945 | 16.10 | 687.52 | 25.80 |
| 4/ 8/1945 | 10.90 | 698.42 | 20.30 |
| 11/ 8/1945 | -11.40 | 687.02 | 1.10 |
| 18/ 8/1945 | 18.70 | 705.72 | 0.60 |
| 25/ 8/1945 | 23.20 | 728.92 | 9.70 |
| 1/ 9/1945 | 15.20 | 744.12 | 46.80 |
| 8/ 9/1945 | -7.10 | 737.02 | 53.80 |
| 15/ 9/1945 | 25.50 | 762.52 | 68.40 |
| 22/ 9/1945 | 16.40 | 778.92 | 109.10 |
| 29/ 9/1945 | 17.30 | 796.22 | 147.40 |
| 6/10/1945 | 21.40 | 817.62 | 152.70 |
| 13/10/1945 | 7.10 | 824.72 | 148.90 |
| 20/10/1945 | -7.10 | 817.62 | 153.20 |
| 27/10/1945 | 24.30 | 841.92 | 158.80 |
| 3/11/1945 | 19.20 | 861.12 | 154.80 |
| 10/11/1945 | 14.10 | 875.22 | 153.70 |
| 17/11/1945 | -19.20 | 856.01 | 141.60 |
| 24/11/1945 | 27.90 | 883.91 | 144.00 |
| 1/12/1945 | 26.20 | 910.11 | 153.80 |
| 8/12/1945 | -17.90 | 892.21 | 118.60 |
| 15/12/1945 | -21.40 | 870.81 | 75.80 |
| 22/12/1945 | 12.20 | 883.01 | 80.90 |
| 29/12/1945 | -14.50 | 868.51 | 73.50 |
| 52 | 771 | | |

**TABLE A4-4 Market Breadth** (cont.)

| DATE | A/D VAL | A/D LINE | 10WK RAT |
|---|---|---|---|
| 5/ 1/1946 | 36.10 | 904.61 | 85.30 |
| 12/ 1/1946 | 14.10 | 918.71 | 80.20 |
| 19/ 1/1946 | 12.60 | 931.31 | 78.70 |
| 26/ 1/1946 | 21.20 | 952.51 | 119.10 |
| 2/ 2/1946 | -20.20 | 932.31 | 71.00 |
| 9/ 2/1946 | 21.90 | 954.21 | 56.70 |
| 16/ 2/1946 | -40.80 | 913.41 | 43.80 |
| 23/ 2/1946 | -30.60 | 882.81 | 34.60 |
| 2/ 3/1946 | 21.60 | 904.41 | 44.00 |
| 9/ 3/1946 | -12.20 | 892.21 | 46.30 |
| 16/ 3/1946 | 23.60 | 915.81 | 33.80 |
| 23/ 3/1946 | 18.40 | 934.21 | 38.10 |
| 30/ 3/1946 | 23.40 | 957.61 | 48.90 |
| 6/ 4/1946 | -6.30 | 951.31 | 21.40 |
| 13/ 4/1946 | 18.10 | 969.41 | 59.70 |
| 20/ 4/1946 | -23.40 | 946.01 | 14.40 |
| 27/ 4/1946 | -20.00 | 926.01 | 35.20 |
| 4/ 5/1946 | 25.40 | 951.41 | 91.20 |
| 11/ 5/1946 | -13.00 | 938.41 | 56.60 |
| 18/ 5/1946 | 17.00 | 955.41 | 85.80 |
| 25/ 5/1946 | 20.00 | 975.41 | 82.20 |
| 1/ 6/1946 | -23.00 | 952.41 | 40.80 |
| 8/ 6/1946 | -7.10 | 945.31 | 10.30 |
| 15/ 6/1946 | -40.00 | 905.31 | -23.40 |
| 22/ 6/1946 | 14.10 | 919.41 | -27.40 |
| 29/ 6/1946 | 6.30 | 925.71 | 2.30 |
| 6/ 7/1946 | -16.10 | 909.61 | 6.20 |
| 13/ 7/1946 | -24.40 | 885.21 | -43.60 |
| 20/ 7/1946 | -26.20 | 859.01 | -56.80 |
| 27/ 7/1946 | 22.80 | 881.81 | -51.00 |
| 3/ 8/1946 | 7.10 | 888.91 | -63.90 |
| 10/ 8/1946 | -16.40 | 872.51 | -57.30 |
| 17/ 8/1946 | -20.50 | 852.01 | -70.70 |
| 24/ 8/1946 | -36.80 | 815.21 | -67.50 |
| 31/ 8/1946 | -46.40 | 768.81 | -128.00 |
| 7/ 9/1946 | -35.20 | 733.61 | -169.50 |
| 14/ 9/1946 | -37.00 | 696.61 | -190.40 |
| 21/ 9/1946 | 23.20 | 719.81 | -142.80 |
| 28/ 9/1946 | -24.90 | 694.91 | -141.50 |
| 5/10/1946 | -17.60 | 677.31 | -181.90 |
| 12/10/1946 | 24.90 | 702.21 | -164.10 |
| 19/10/1946 | -17.60 | 684.61 | -165.30 |
| 26/10/1946 | 22.60 | 707.21 | -122.20 |
| 2/11/1946 | -14.80 | 692.41 | -100.20 |
| 9/11/1946 | -19.00 | 673.41 | -72.80 |
| 16/11/1946 | -27.60 | 645.81 | -65.20 |
| 23/11/1946 | 25.10 | 670.91 | -3.10 |
| 30/11/1946 | 12.20 | 683.11 | -14.10 |
| 7/12/1946 | -19.00 | 664.11 | -8.20 |
| 14/12/1946 | 19.70 | 683.81 | 29.10 |
| 21/12/1946 | -20.70 | 663.11 | -16.50 |
| 28/12/1946 | 9.50 | 672.61 | 10.60 |

52 823

| DATE | A/D VAL | A/D LINE | 10WK RAT |
|---|---|---|---|
| 4/ 1/1947 | -17.30 | 655.31 | -29.30 |
| 11/ 1/1947 | 9.50 | 664.81 | -5.00 |
| 18/ 1/1947 | 7.70 | 672.51 | 21.70 |
| 25/ 1/1947 | 28.80 | 701.31 | 78.10 |
| 1/ 2/1947 | 23.60 | 724.91 | 76.60 |
| 8/ 2/1947 | -17.30 | 707.61 | 47.10 |
| 15/ 2/1947 | -8.90 | 698.71 | 57.20 |
| 22/ 2/1947 | -22.80 | 675.91 | 14.70 |
| 1/ 3/1947 | -24.30 | 651.61 | 11.10 |
| 8/ 3/1947 | -23.00 | 628.61 | -21.40 |
| 15/ 3/1947 | 18.90 | 647.51 | 14.80 |
| 22/ 3/1947 | 14.50 | 662.01 | 19.30 |
| 29/ 3/1947 | -17.80 | 644.21 | -5.70 |
| 5/ 4/1947 | -32.40 | 611.81 | -66.90 |
| 12/ 4/1947 | -27.40 | 584.41 | -117.90 |
| 19/ 4/1947 | 5.50 | 589.91 | -95.10 |
| 26/ 4/1947 | 17.90 | 607.81 | -68.30 |
| 3/ 5/1947 | -22.60 | 585.21 | -68.10 |
| 10/ 5/1947 | -39.90 | 545.31 | -83.70 |
| 17/ 5/1947 | 21.90 | 567.21 | -38.80 |
| 24/ 5/1947 | 15.80 | 583.01 | -41.90 |
| 31/ 5/1947 | -8.40 | 574.61 | -64.80 |
| 7/ 6/1947 | 28.30 | 602.91 | -18.70 |
| 14/ 6/1947 | 17.60 | 620.51 | 31.30 |
| 21/ 6/1947 | -5.50 | 615.01 | 53.20 |
| 28/ 6/1947 | 28.60 | 643.61 | 76.30 |
| 5/ 7/1947 | 25.10 | 668.71 | 83.50 |
| 12/ 7/1947 | 7.10 | 675.81 | 113.20 |
| 19/ 7/1947 | 19.00 | 694.81 | 172.10 |
| 26/ 7/1947 | -23.40 | 671.41 | 126.80 |
| 2/ 8/1947 | -20.00 | 651.41 | 91.00 |
| 9/ 8/1947 | 14.80 | 666.21 | 114.20 |
| 16/ 8/1947 | -10.90 | 655.31 | 75.00 |
| 23/ 8/1947 | -12.70 | 642.61 | 44.70 |
| 30/ 8/1947 | -15.80 | 626.81 | 34.40 |
| 6/ 9/1947 | -7.10 | 619.71 | -1.30 |
| 13/ 9/1947 | 17.80 | 637.51 | -8.60 |
| 20/ 9/1947 | -19.70 | 617.81 | -35.40 |
| 27/ 9/1947 | 21.20 | 639.01 | -33.20 |
| 4/10/1947 | 9.50 | 648.51 | -0.30 |
| 11/10/1947 | 20.00 | 668.51 | 39.70 |
| 18/10/1947 | -15.50 | 653.01 | 9.40 |
| 25/10/1947 | -23.20 | 629.81 | -2.90 |
| 1/11/1947 | -20.20 | 609.60 | -10.40 |
| 8/11/1947 | -11.40 | 598.20 | -6.00 |
| 15/11/1947 | 12.60 | 610.80 | 13.70 |
| 22/11/1947 | -25.30 | 585.50 | -29.40 |
| 29/11/1947 | -23.40 | 562.10 | -33.10 |
| 6/12/1947 | 20.70 | 582.80 | -33.60 |
| 13/12/1947 | 15.20 | 598.00 | -27.90 |
| 20/12/1947 | -18.90 | 579.10 | -66.80 |
| 27/12/1947 | 15.20 | 594.30 | -36.10 |

52 875

**TABLE A4-4 Market Breadth** (cont.)

| DATE | A/D VAL | A/D LINE | 10WK RAT |
|---|---|---|---|
| 3/ 1/1948 | 6.30 | 600.60 | -6.60 |
| 10/ 1/1948 | -19.50 | 581.10 | -5.90 |
| 17/ 1/1948 | -24.70 | 556.40 | -19.20 |
| 24/ 1/1948 | 16.10 | 572.50 | -15.70 |
| 31/ 1/1948 | -30.50 | 542.00 | -20.90 |
| 7/ 2/1948 | -29.30 | 512.70 | -26.80 |
| 14/ 2/1948 | 10.90 | 523.60 | -36.60 |
| 21/ 2/1948 | 0.0 | 523.60 | -51.80 |
| 28/ 2/1948 | 18.10 | 541.70 | -14.80 |
| 6/ 3/1948 | -7.70 | 534.00 | -37.70 |
| 13/ 3/1948 | 27.00 | 561.00 | -17.00 |
| 20/ 3/1948 | 17.90 | 578.90 | 20.40 |
| 27/ 3/1948 | 21.70 | 600.60 | 66.80 |
| 3/ 4/1948 | 13.80 | 614.40 | 64.50 |
| 10/ 4/1948 | 14.80 | 629.20 | 109.80 |
| 17/ 4/1948 | 18.70 | 647.90 | 157.80 |
| 24/ 4/1948 | -15.50 | 632.40 | 131.40 |
| 1/ 5/1948 | 14.80 | 647.20 | 146.20 |
| 8/ 5/1948 | 32.20 | 679.40 | 160.30 |
| 15/ 5/1948 | 17.00 | 696.40 | 185.00 |
| 22/ 5/1948 | -6.50 | 689.90 | 151.50 |
| 29/ 5/1948 | -16.10 | 673.80 | 117.50 |
| 5/ 6/1948 | 19.00 | 692.80 | 114.80 |
| 12/ 6/1948 | -13.40 | 679.40 | 87.60 |
| 19/ 6/1948 | -10.00 | 669.40 | 62.80 |
| 26/ 6/1948 | -10.00 | 659.40 | 34.10 |
| 3/ 7/1948 | 10.90 | 670.30 | 60.50 |
| 10/ 7/1948 | -24.90 | 645.40 | 20.80 |
| 17/ 7/1948 | -17.00 | 628.40 | -28.40 |
| 24/ 7/1948 | -22.30 | 606.10 | -67.70 |
| 31/ 7/1948 | 11.80 | 617.90 | -49.40 |
| 7/ 8/1948 | -23.60 | 594.30 | -56.90 |
| 14/ 8/1948 | 15.90 | 610.20 | -60.00 |
| 21/ 8/1948 | -8.90 | 601.30 | -55.50 |
| 28/ 8/1948 | 17.60 | 618.90 | -27.90 |
| 4/ 9/1948 | -26.30 | 592.60 | -44.20 |
| 11/ 9/1948 | -9.50 | 583.10 | -64.60 |
| 18/ 9/1948 | -16.40 | 566.70 | -56.10 |
| 25/ 9/1948 | -7.10 | 559.60 | -46.20 |
| 2/10/1948 | 14.10 | 573.70 | -9.80 |
| 9/10/1948 | 12.60 | 586.30 | -9.00 |
| 16/10/1948 | 25.50 | 611.80 | 40.10 |
| 23/10/1948 | -19.20 | 592.60 | 5.00 |
| 30/10/1948 | -35.90 | 556.70 | -22.00 |
| 6/11/1948 | -24.10 | 532.60 | -63.70 |
| 13/11/1948 | 17.00 | 549.60 | -20.40 |
| 20/11/1948 | -21.90 | 527.70 | -32.80 |
| 27/11/1948 | 16.10 | 543.80 | -0.30 |
| 4/12/1948 | 3.10 | 546.90 | 9.90 |
| 11/12/1948 | -14.80 | 532.10 | -19.00 |
| 18/12/1948 | 10.00 | 542.10 | -21.60 |
| 25/12/1948 | -12.60 | 529.50 | -59.70 |
| 52 | 927 | | |

| DATE | A/D VAL | A/D LINE | 10WK RAT |
|---|---|---|---|
| 1/ 1/1949 | 30.30 | 559.80 | -10.20 |
| 8/ 1/1949 | -15.40 | 544.40 | 10.30 |
| 15/ 1/1949 | 18.10 | 562.50 | 52.50 |
| 22/ 1/1949 | -16.20 | 546.30 | 19.30 |
| 29/ 1/1949 | -19.70 | 526.60 | 21.50 |
| 5/ 2/1949 | -18.80 | 507.80 | -13.40 |
| 12/ 2/1949 | 15.70 | 523.50 | -0.80 |
| 19/ 2/1949 | -19.10 | 504.40 | -5.10 |
| 26/ 2/1949 | 17.80 | 522.20 | 2.70 |
| 5/ 3/1949 | 21.10 | 543.30 | 36.40 |
| 12/ 3/1949 | -10.80 | 532.50 | -4.70 |
| 19/ 3/1949 | -4.60 | 527.90 | 6.10 |
| 26/ 3/1949 | 18.80 | 546.70 | 6.80 |
| 2/ 4/1949 | -6.10 | 540.60 | 16.90 |
| 9/ 4/1949 | -9.70 | 530.90 | 26.90 |
| 16/ 4/1949 | -17.40 | 513.50 | 28.30 |
| 23/ 4/1949 | -9.40 | 504.10 | 3.20 |
| 30/ 4/1949 | 12.10 | 516.20 | 34.40 |
| 7/ 5/1949 | 8.10 | 524.30 | 24.70 |
| 14/ 5/1949 | -13.20 | 511.10 | -9.60 |
| 21/ 5/1949 | -15.40 | 495.70 | -14.20 |
| 28/ 5/1949 | -25.50 | 470.20 | -35.10 |
| 4/ 6/1949 | -17.30 | 452.90 | -71.20 |
| 11/ 6/1949 | -10.70 | 442.20 | -75.80 |
| 18/ 6/1949 | 17.70 | 459.90 | -48.40 |
| 25/ 6/1949 | 10.90 | 470.80 | -20.10 |
| 2/ 7/1949 | 16.50 | 487.30 | 5.80 |
| 9/ 7/1949 | 18.20 | 505.50 | 11.90 |
| 16/ 7/1949 | 16.00 | 521.50 | 19.80 |
| 23/ 7/1949 | 11.60 | 533.10 | 44.60 |
| 30/ 7/1949 | 16.70 | 549.80 | 76.70 |
| 6/ 8/1949 | 18.40 | 568.20 | 120.60 |
| 13/ 8/1949 | 13.90 | 582.10 | 151.80 |
| 20/ 8/1949 | -13.60 | 568.50 | 148.90 |
| 27/ 8/1949 | 7.70 | 576.20 | 138.90 |
| 3/ 9/1949 | 9.70 | 585.90 | 137.70 |
| 10/ 9/1949 | 24.90 | 610.80 | 146.10 |
| 17/ 9/1949 | -7.50 | 603.30 | 120.40 |
| 24/ 9/1949 | 4.00 | 607.30 | 108.40 |
| 1/10/1949 | 22.50 | 629.80 | 119.30 |
| 8/10/1949 | 11.20 | 641.00 | 113.80 |
| 15/10/1949 | -8.40 | 632.60 | 87.00 |
| 22/10/1949 | 13.10 | 645.70 | 86.20 |
| 29/10/1949 | 11.30 | 656.99 | 111.10 |
| 5/11/1949 | -11.50 | 645.49 | 91.90 |
| 12/11/1949 | 17.20 | 662.69 | 99.40 |
| 19/11/1949 | -7.10 | 655.59 | 67.40 |
| 26/11/1949 | 22.10 | 677.69 | 97.00 |
| 3/12/1949 | 9.50 | 687.19 | 102.50 |
| 10/12/1949 | 17.90 | 705.09 | 97.90 |
| 17/12/1949 | -4.70 | 700.39 | 82.00 |
| 24/12/1949 | 13.70 | 714.09 | 104.10 |
| 31/12/1949 | 23.40 | 737.49 | 114.40 |
| 53 | 980 | | |

**TABLE A4-4 Market Breadth** (cont.)

| DATE | A/D VAL | A/D LINE | 10WK RAT |
|---|---|---|---|
| 7/ 1/1950 | -15.40 | 722.09 | 87.70 |
| 14/ 1/1950 | 15.00 | 737.09 | 114.20 |
| 21/ 1/1950 | -9.10 | 727.99 | 87.90 |
| 28/ 1/1950 | 18.10 | 746.09 | 113.10 |
| 4/ 2/1950 | -9.80 | 736.29 | 81.20 |
| 11/ 2/1950 | 10.10 | 746.39 | 81.80 |
| 18/ 2/1950 | 5.80 | 752.19 | 69.70 |
| 25/ 2/1950 | 12.70 | 764.89 | 87.10 |
| 4/ 3/1950 | -15.50 | 749.39 | 57.90 |
| 11/ 3/1950 | 17.60 | 766.99 | 52.10 |
| 18/ 3/1950 | -6.00 | 760.99 | 61.50 |
| 25/ 3/1950 | -18.30 | 742.69 | 28.20 |
| 1/ 4/1950 | 19.20 | 761.89 | 56.50 |
| 8/ 4/1950 | -10.00 | 751.89 | 28.40 |
| 15/ 4/1950 | -1.70 | 750.19 | 36.50 |
| 22/ 4/1950 | 8.40 | 758.59 | 34.80 |
| 29/ 4/1950 | 11.80 | 770.39 | 40.80 |
| 6/ 5/1950 | 8.00 | 778.39 | 36.10 |
| 13/ 5/1950 | 16.90 | 795.29 | 68.50 |
| 20/ 5/1950 | -10.30 | 784.99 | 40.60 |
| 27/ 5/1950 | 5.80 | 790.79 | 52.40 |
| 3/ 6/1950 | 6.90 | 797.69 | 77.60 |
| 10/ 6/1950 | -15.40 | 782.29 | 43.00 |
| 17/ 6/1950 | 10.60 | 792.89 | 63.60 |
| 24/ 6/1950 | -49.30 | 743.59 | 16.00 |
| 1/ 7/1950 | 6.50 | 750.09 | 14.10 |
| 8/ 7/1950 | -22.10 | 727.99 | -19.80 |
| 15/ 7/1950 | 26.40 | 754.39 | -1.40 |
| 22/ 7/1950 | 13.80 | 768.19 | -4.50 |
| 29/ 7/1950 | 17.10 | 785.29 | 22.90 |
| 5/ 8/1950 | 15.40 | 800.69 | 32.50 |
| 12/ 8/1950 | 21.50 | 822.19 | 47.10 |
| 19/ 8/1950 | -9.30 | 812.89 | 53.20 |
| 26/ 8/1950 | 5.90 | 818.79 | 48.50 |
| 2/ 9/1950 | 13.70 | 832.49 | 111.50 |
| 9/ 9/1950 | 23.70 | 856.19 | 128.70 |
| 16/ 9/1950 | 12.00 | 868.19 | 162.80 |
| 23/ 9/1950 | 5.00 | 873.19 | 141.40 |
| 30/ 9/1950 | 21.00 | 894.19 | 148.60 |
| 7/10/1950 | -12.70 | 881.49 | 118.80 |
| 14/10/1950 | 16.50 | 897.99 | 119.90 |
| 21/10/1950 | -18.10 | 879.89 | 80.30 |
| 28/10/1950 | -13.00 | 866.89 | 76.60 |
| 4/11/1950 | 11.20 | 878.09 | 81.90 |
| 11/11/1950 | 18.00 | 896.09 | 86.20 |
| 18/11/1950 | 18.10 | 914.19 | 80.60 |
| 25/11/1950 | -26.70 | 887.49 | 41.90 |
| 2/12/1950 | -9.20 | 878.29 | 27.70 |
| 9/12/1950 | 11.60 | 889.89 | 18.30 |
| 16/12/1950 | 25.40 | 915.29 | 56.40 |
| 23/12/1950 | 18.70 | 933.99 | 58.60 |
| 30/12/1950 | 22.20 | 956.19 | 98.90 |

52 1032

| DATE | A/D VAL | A/D LINE | 10WK RAT |
|---|---|---|---|
| 6/ 1/1951 | 15.90 | 972.09 | 127.80 |
| 13/ 1/1951 | 17.30 | 989.39 | 133.90 |
| 20/ 1/1951 | 6.50 | 995.89 | 122.40 |
| 27/ 1/1951 | 18.60 | 1014.49 | 122.90 |
| 3/ 2/1951 | 13.10 | 1027.59 | 162.70 |
| 10/ 2/1951 | 3.40 | 1030.99 | 175.30 |
| 17/ 2/1951 | -11.80 | 1019.19 | 151.90 |
| 24/ 2/1951 | -4.90 | 1014.29 | 121.60 |
| 3/ 3/1951 | -13.50 | 1000.79 | 89.40 |
| 10/ 3/1951 | -25.60 | 975.19 | 41.60 |
| 17/ 3/1951 | -16.10 | 959.09 | 9.60 |
| 24/ 3/1951 | -15.40 | 943.69 | -23.10 |
| 31/ 3/1951 | 14.50 | 958.19 | -15.10 |
| 7/ 4/1951 | 17.60 | 975.79 | -16.10 |
| 14/ 4/1951 | -13.70 | 962.09 | -42.90 |
| 21/ 4/1951 | 10.00 | 972.09 | -36.30 |
| 28/ 4/1951 | 10.10 | 982.19 | -14.40 |
| 5/ 5/1951 | -14.80 | 967.39 | -24.30 |
| 12/ 5/1951 | -23.90 | 943.49 | -34.70 |
| 19/ 5/1951 | -18.20 | 925.29 | -27.30 |
| 26/ 5/1951 | 15.70 | 940.99 | 4.50 |
| 2/ 6/1951 | 6.90 | 947.89 | 26.80 |
| 9/ 6/1951 | 6.30 | 954.19 | 18.60 |
| 16/ 6/1951 | -23.20 | 930.99 | -22.20 |
| 23/ 6/1951 | -27.00 | 903.99 | -35.50 |
| 30/ 6/1951 | 22.50 | 926.49 | -23.00 |
| 7/ 7/1951 | 14.40 | 940.89 | -18.70 |
| 14/ 7/1951 | 1.70 | 942.59 | -2.20 |
| 21/ 7/1951 | 18.90 | 961.49 | 40.60 |
| 28/ 7/1951 | 17.70 | 979.19 | 76.50 |
| 4/ 8/1951 | 2.80 | 981.99 | 63.60 |
| 11/ 8/1951 | 15.60 | 997.59 | 72.30 |
| 18/ 8/1951 | -9.00 | 988.59 | 57.00 |
| 25/ 8/1951 | -16.90 | 971.69 | 63.30 |
| 1/ 9/1951 | 16.90 | 988.59 | 107.20 |
| 8/ 9/1951 | 13.40 | 1001.99 | 98.10 |
| 15/ 9/1951 | -9.10 | 992.89 | 74.60 |
| 22/ 9/1951 | -10.90 | 981.99 | 62.00 |
| 29/ 9/1951 | 20.20 | 1002.19 | 63.30 |
| 6/10/1951 | -5.30 | 996.89 | 40.30 |
| 13/10/1951 | -22.10 | 974.79 | 15.40 |
| 20/10/1951 | -26.40 | 948.39 | -26.60 |
| 27/10/1951 | -2.60 | 945.79 | -20.20 |
| 3/11/1951 | 10.80 | 956.59 | 7.50 |
| 10/11/1951 | -3.60 | 952.99 | -13.00 |
| 17/11/1951 | -18.70 | 934.28 | -45.10 |
| 24/11/1951 | 20.20 | 954.48 | -15.80 |
| 1/12/1951 | 15.40 | 969.88 | 10.50 |
| 8/12/1951 | -10.00 | 959.88 | -19.70 |
| 15/12/1951 | -9.00 | 950.88 | -23.40 |
| 22/12/1951 | 7.40 | 958.28 | 6.10 |
| 29/12/1951 | 17.50 | 975.78 | 50.00 |

52 1084

**TABLE A4-4 Market Breadth** (cont.)

| DATE | A/D VAL | A/D LINE | 10WK RAT |
|---|---|---|---|
| 5/ 1/1952 | 5.10 | 980.88 | 57.70 |
| 12/ 1/1952 | 11.50 | 992.38 | 58.40 |
| 19/ 1/1952 | 10.10 | 1002.48 | 72.10 |
| 26/ 1/1952 | -5.70 | 996.78 | 85.10 |
| 2/ 2/1952 | -7.30 | 989.48 | 57.60 |
| 9/ 2/1952 | -14.30 | 975.18 | 27.90 |
| 16/ 2/1952 | -19.70 | 955.48 | 18.20 |
| 23/ 2/1952 | 5.30 | 960.78 | 32.50 |
| 1/ 3/1952 | 19.30 | 980.08 | 44.40 |
| 8/ 3/1952 | 5.10 | 985.18 | 32.00 |
| 15/ 3/1952 | 7.70 | 992.88 | 34.60 |
| 22/ 3/1952 | 11.20 | 1004.08 | 34.30 |
| 29/ 3/1952 | -13.70 | 990.38 | 10.50 |
| 5/ 4/1952 | -6.10 | 984.28 | 10.10 |
| 12/ 4/1952 | -19.20 | 965.08 | -1.80 |
| 19/ 4/1952 | -4.70 | 960.38 | 7.80 |
| 26/ 4/1952 | -10.80 | 949.58 | 16.70 |
| 3/ 5/1952 | 15.10 | 964.68 | 26.50 |
| 10/ 5/1952 | -9.50 | 955.18 | -2.30 |
| 17/ 5/1952 | 15.70 | 970.88 | 8.30 |
| 24/ 5/1952 | -4.20 | 966.68 | -3.60 |
| 31/ 5/1952 | 15.40 | 982.08 | 0.60 |
| 7/ 6/1952 | 7.90 | 989.98 | 22.20 |
| 14/ 6/1952 | 8.00 | 997.98 | 36.30 |
| 21/ 6/1952 | 8.60 | 1006.58 | 64.10 |
| 28/ 6/1952 | 8.00 | 1014.58 | 76.80 |
| 5/ 7/1952 | -7.60 | 1006.98 | 80.00 |
| 12/ 7/1952 | -7.00 | 999.98 | 57.90 |
| 19/ 7/1952 | 10.80 | 1010.78 | 78.20 |
| 26/ 7/1952 | 9.80 | 1020.58 | 72.30 |
| 2/ 8/1952 | 7.90 | 1028.48 | 84.40 |
| 9/ 8/1952 | -11.30 | 1017.18 | 57.70 |
| 16/ 8/1952 | -9.90 | 1007.28 | 39.90 |
| 23/ 8/1952 | 11.00 | 1018.28 | 42.90 |
| 30/ 8/1952 | 9.50 | 1027.78 | 43.80 |
| 6/ 9/1952 | -18.20 | 1009.58 | 17.60 |
| 13/ 9/1952 | -3.70 | 1005.88 | 21.50 |
| 20/ 9/1952 | 10.00 | 1015.88 | 38.50 |
| 27/ 9/1952 | -13.30 | 1002.58 | 14.40 |
| 4/10/1952 | 3.90 | 1006.48 | 8.50 |
| 11/10/1952 | -18.70 | 987.78 | -18.10 |
| 18/10/1952 | -11.10 | 976.68 | -17.90 |
| 25/10/1952 | 14.00 | 990.68 | 6.00 |
| 1/11/1952 | 15.40 | 1006.08 | 10.40 |
| 8/11/1952 | 8.90 | 1014.98 | 9.80 |
| 15/11/1952 | 21.60 | 1036.58 | 49.60 |
| 22/11/1952 | 18.90 | 1055.48 | 72.20 |
| 29/11/1952 | 5.70 | 1061.18 | 67.90 |
| 6/12/1952 | 12.30 | 1073.48 | 93.50 |
| 13/12/1952 | -4.20 | 1069.28 | 85.40 |
| 20/12/1952 | 5.70 | 1074.98 | 109.80 |
| 27/12/1952 | 14.60 | 1089.58 | 135.50 |
| 52 | 1136 | | |

| DATE | A/D VAL | A/D LINE | 10WK RAT |
|---|---|---|---|
| 3/ 1/1953 | -6.50 | 1083.08 | 115.00 |
| 10/ 1/1953 | -3.30 | 1079.78 | 96.30 |
| 17/ 1/1953 | 9.80 | 1089.58 | 97.20 |
| 24/ 1/1953 | 13.50 | 1103.08 | 89.10 |
| 31/ 1/1953 | -18.20 | 1084.88 | 52.00 |
| 7/ 2/1953 | 5.10 | 1089.98 | 51.40 |
| 14/ 2/1953 | -4.90 | 1085.08 | 34.20 |
| 21/ 2/1953 | 20.40 | 1105.48 | 58.80 |
| 28/ 2/1953 | 5.50 | 1110.98 | 58.60 |
| 7/ 3/1953 | 15.10 | 1126.08 | 59.10 |
| 14/ 3/1953 | 10.20 | 1136.28 | 75.80 |
| 21/ 3/1953 | -14.00 | 1122.28 | 65.10 |
| 28/ 3/1953 | -27.30 | 1094.98 | 28.00 |
| 4/ 4/1953 | -20.10 | 1074.88 | -5.60 |
| 11/ 4/1953 | -12.50 | 1062.38 | 0.10 |
| 18/ 4/1953 | -20.90 | 1041.48 | -25.90 |
| 25/ 4/1953 | 14.30 | 1055.78 | -6.70 |
| 2/ 5/1953 | 10.00 | 1065.78 | -17.10 |
| 9/ 5/1953 | 2.80 | 1068.58 | -19.80 |
| 16/ 5/1953 | 12.20 | 1080.78 | -22.70 |
| 23/ 5/1953 | -19.90 | 1060.88 | -52.80 |
| 30/ 5/1953 | -23.00 | 1037.88 | -61.80 |
| 6/ 6/1953 | -22.30 | 1015.58 | -56.80 |
| 13/ 6/1953 | -7.60 | 1007.98 | -44.30 |
| 20/ 6/1953 | 15.70 | 1023.68 | -16.10 |
| 27/ 6/1953 | 12.60 | 1036.28 | 17.40 |
| 4/ 7/1953 | 5.80 | 1042.08 | 8.90 |
| 11/ 7/1953 | -4.80 | 1037.28 | -5.90 |
| 18/ 7/1953 | -5.80 | 1031.48 | -14.50 |
| 25/ 7/1953 | 14.80 | 1046.28 | -11.90 |
| 1/ 8/1953 | 14.60 | 1060.88 | 22.60 |
| 8/ 8/1953 | -4.60 | 1056.28 | 41.00 |
| 15/ 8/1953 | -17.10 | 1039.18 | 46.20 |
| 22/ 8/1953 | -23.00 | 1016.18 | 30.80 |
| 29/ 8/1953 | -15.90 | 1000.28 | -0.80 |
| 5/ 9/1953 | -22.40 | 977.88 | -35.80 |
| 12/ 9/1953 | -17.70 | 960.18 | -59.30 |
| 19/ 9/1953 | 18.30 | 978.48 | -36.20 |
| 26/ 9/1953 | 14.30 | 992.78 | -16.10 |
| 3/10/1953 | 6.30 | 999.07 | -24.60 |
| 10/10/1953 | 21.00 | 1020.07 | -18.20 |
| 17/10/1953 | 12.10 | 1032.17 | -1.50 |
| 24/10/1953 | 10.40 | 1042.57 | 26.00 |
| 31/10/1953 | 14.80 | 1057.37 | 63.80 |
| 7/11/1953 | -4.70 | 1052.67 | 75.00 |
| 14/11/1953 | -7.50 | 1045.17 | 89.90 |
| 21/11/1953 | 11.10 | 1056.27 | 118.70 |
| 28/11/1953 | 15.40 | 1071.67 | 115.80 |
| 5/12/1953 | -12.50 | 1059.17 | 89.00 |
| 12/12/1953 | 4.50 | 1063.67 | 87.20 |
| 19/12/1953 | -14.10 | 1049.57 | 52.10 |
| 26/12/1953 | -11.70 | 1037.87 | 28.30 |

52 1188

**TABLE A4-4 Market Breadth** (cont.)

| DATE | A/D VAL | A/D LINE | 10WK RAT |
|---|---|---|---|
| 2/ 1/1954 | 24.10 | 1061.97 | 42.00 |
| 9/ 1/1954 | 21.20 | 1083.17 | 48.40 |
| 16/ 1/1954 | 19.20 | 1102.37 | 72.30 |
| 23/ 1/1954 | 14.40 | 1116.77 | 94.20 |
| 30/ 1/1954 | 15.90 | 1132.67 | 99.00 |
| 6/ 2/1954 | 10.80 | 1143.47 | 94.40 |
| 13/ 2/1954 | -8.40 | 1135.07 | 98.50 |
| 20/ 2/1954 | 10.20 | 1145.27 | 104.20 |
| 27/ 2/1954 | 14.90 | 1160.17 | 133.20 |
| 6/ 3/1954 | 13.40 | 1173.57 | 158.30 |
| 13/ 3/1954 | 11.20 | 1184.77 | 145.40 |
| 20/ 3/1954 | -13.70 | 1171.07 | 110.50 |
| 27/ 3/1954 | 17.70 | 1188.77 | 109.00 |
| 3/ 4/1954 | 10.20 | 1198.97 | 104.20 |
| 10/ 4/1954 | 8.50 | 1207.47 | 97.40 |
| 17/ 4/1954 | -9.70 | 1197.77 | 76.90 |
| 24/ 4/1954 | -1.70 | 1196.07 | 83.60 |
| 1/ 5/1954 | 7.40 | 1203.47 | 80.80 |
| 8/ 5/1954 | 13.00 | 1216.47 | 78.90 |
| 15/ 5/1954 | 11.20 | 1227.67 | 76.70 |
| 22/ 5/1954 | 8.70 | 1236.37 | 74.20 |
| 29/ 5/1954 | 5.00 | 1241.37 | 92.90 |
| 5/ 6/1954 | -17.10 | 1224.27 | 58.10 |
| 12/ 6/1954 | 16.30 | 1240.57 | 64.20 |
| 19/ 6/1954 | 9.30 | 1249.87 | 65.00 |
| 26/ 6/1954 | 9.10 | 1258.97 | 83.80 |
| 3/ 7/1954 | 14.30 | 1273.27 | 99.80 |
| 10/ 7/1954 | 16.00 | 1289.27 | 108.40 |
| 17/ 7/1954 | 15.00 | 1304.27 | 110.40 |
| 24/ 7/1954 | 17.70 | 1321.97 | 116.90 |
| 31/ 7/1954 | -8.20 | 1313.77 | 100.00 |
| 7/ 8/1954 | 21.50 | 1335.27 | 116.50 |
| 14/ 8/1954 | 10.40 | 1345.67 | 144.00 |
| 21/ 8/1954 | -16.70 | 1328.97 | 111.00 |
| 28/ 8/1954 | -9.70 | 1319.27 | 92.00 |
| 4/ 9/1954 | 15.70 | 1334.97 | 98.60 |
| 11/ 9/1954 | 13.90 | 1348.87 | 98.20 |
| 18/ 9/1954 | 11.00 | 1359.87 | 93.20 |
| 25/ 9/1954 | 10.30 | 1370.17 | 88.50 |
| 2/10/1954 | 10.70 | 1380.87 | 81.50 |
| 9/10/1954 | -19.50 | 1361.37 | 70.20 |
| 16/10/1954 | 14.30 | 1375.67 | 63.00 |
| 23/10/1954 | -17.00 | 1358.67 | 35.60 |
| 30/10/1954 | 25.20 | 1383.87 | 77.50 |
| 6/11/1954 | 20.40 | 1404.27 | 107.60 |
| 13/11/1954 | -12.50 | 1391.77 | 79.40 |
| 20/11/1954 | 19.70 | 1411.47 | 85.20 |
| 27/11/1954 | 13.80 | 1425.27 | 88.00 |
| 4/12/1954 | 14.50 | 1439.77 | 92.20 |
| 11/12/1954 | 14.50 | 1454.27 | 96.00 |
| 18/12/1954 | 13.90 | 1468.17 | 129.40 |
| 25/12/1954 | 16.30 | 1484.47 | 131.40 |

52 1240

| DATE | A/D VAL | A/D LINE | 10WK RAT |
|---|---|---|---|
| 1/ 1/1955 | -20.50 | 1463.97 | 127.90 |
| 8/ 1/1955 | 11.40 | 1475.37 | 114.10 |
| 15/ 1/1955 | -5.20 | 1470.17 | 88.50 |
| 22/ 1/1955 | 10.90 | 1481.07 | 111.90 |
| 29/ 1/1955 | 14.10 | 1495.17 | 106.30 |
| 5/ 2/1955 | 19.20 | 1514.37 | 111.70 |
| 12/ 2/1955 | 12.60 | 1526.97 | 109.80 |
| 19/ 2/1955 | -7.70 | 1519.27 | 87.60 |
| 26/ 2/1955 | 18.80 | 1538.07 | 92.50 |
| 5/ 3/1955 | -35.10 | 1502.97 | 41.10 |
| 12/ 3/1955 | 9.80 | 1512.77 | 71.40 |
| 19/ 3/1955 | 19.10 | 1531.87 | 79.10 |
| 26/ 3/1955 | -3.90 | 1527.97 | 80.40 |
| 2/ 4/1955 | 14.30 | 1542.27 | 83.80 |
| 9/ 4/1955 | 15.10 | 1557.37 | 84.80 |
| 16/ 4/1955 | -7.30 | 1550.07 | 58.30 |
| 23/ 4/1955 | -5.30 | 1544.77 | 40.40 |
| 30/ 4/1955 | -7.40 | 1537.37 | 40.70 |
| 7/ 5/1955 | -16.70 | 1520.67 | 5.20 |
| 14/ 5/1955 | 13.30 | 1533.97 | 53.60 |
| 21/ 5/1955 | 11.10 | 1545.07 | 54.90 |
| 28/ 5/1955 | 9.60 | 1554.67 | 45.40 |
| 4/ 6/1955 | 10.80 | 1565.47 | 60.10 |
| 11/ 6/1955 | 12.90 | 1578.37 | 58.70 |
| 18/ 6/1955 | 8.60 | 1586.97 | 52.20 |
| 25/ 6/1955 | 7.50 | 1594.47 | 67.00 |
| 2/ 7/1955 | -16.40 | 1578.07 | 55.90 |
| 9/ 7/1955 | 9.40 | 1587.47 | 72.70 |
| 16/ 7/1955 | 16.50 | 1603.97 | 105.90 |
| 23/ 7/1955 | -12.60 | 1591.37 | 80.00 |
| 30/ 7/1955 | -17.20 | 1574.17 | 51.70 |
| 6/ 8/1955 | -12.90 | 1561.26 | 29.20 |
| 13/ 8/1955 | -9.20 | 1552.06 | 9.20 |
| 20/ 8/1955 | 15.10 | 1567.16 | 11.40 |
| 27/ 8/1955 | 15.30 | 1582.46 | 18.10 |
| 3/ 9/1955 | 13.20 | 1595.66 | 23.80 |
| 10/ 9/1955 | 11.20 | 1606.86 | 51.40 |
| 17/ 9/1955 | 6.90 | 1613.76 | 48.90 |
| 24/ 9/1955 | -28.50 | 1585.26 | 3.90 |
| 1/10/1955 | -19.80 | 1565.46 | -3.30 |
| 8/10/1955 | -21.60 | 1543.86 | -7.70 |
| 15/10/1955 | 19.90 | 1563.76 | 25.10 |
| 22/10/1955 | -6.00 | 1557.76 | 28.30 |
| 29/10/1955 | 17.90 | 1575.66 | 31.10 |
| 5/11/1955 | 14.50 | 1590.16 | 30.30 |
| 12/11/1955 | 5.40 | 1595.56 | 22.50 |
| 19/11/1955 | 12.70 | 1608.26 | 24.00 |
| 26/11/1955 | 11.20 | 1619.46 | 28.30 |
| 3/12/1955 | 11.30 | 1630.76 | 68.10 |
| 10/12/1955 | -16.50 | 1614.26 | 71.40 |
| 17/12/1955 | 11.30 | 1625.56 | 104.30 |
| 24/12/1955 | 4.50 | 1630.06 | 88.90 |
| 31/12/1955 | -6.30 | 1623.76 | 88.60 |

53 1293

**TABLE A4-4 Market Breadth** (cont.)

| DATE | A/D VAL | A/D LINE | 10WK RAT |
|---|---|---|---|
| 7/ 1/1956 | -11.20 | 1612.56 | 59.50 |
| 14/ 1/1956 | -28.60 | 1583.96 | 16.40 |
| 21/ 1/1956 | 5.50 | 1589.46 | 16.50 |
| 28/ 1/1956 | 15.10 | 1604.56 | 18.90 |
| 4/ 2/1956 | -17.00 | 1587.56 | -9.30 |
| 11/ 2/1956 | 18.40 | 1605.96 | -2.20 |
| 18/ 2/1956 | 18.80 | 1624.76 | 33.10 |
| 25/ 2/1956 | 14.20 | 1638.96 | 36.00 |
| 3/ 3/1956 | 19.00 | 1657.96 | 50.50 |
| 10/ 3/1956 | 14.40 | 1672.36 | 71.20 |
| 17/ 3/1956 | 2.80 | 1675.16 | 85.20 |
| 24/ 3/1956 | -9.10 | 1666.06 | 104.70 |
| 31/ 3/1956 | -10.30 | 1655.76 | 88.90 |
| 7/ 4/1956 | -17.20 | 1638.56 | 56.60 |
| 14/ 4/1956 | -11.20 | 1627.36 | 62.40 |
| 21/ 4/1956 | -2.20 | 1625.16 | 41.80 |
| 28/ 4/1956 | 9.90 | 1635.06 | 32.90 |
| 5/ 5/1956 | -15.90 | 1619.16 | 2.80 |
| 12/ 5/1956 | -15.40 | 1603.76 | -31.60 |
| 19/ 5/1956 | -31.10 | 1572.66 | -77.10 |
| 26/ 5/1956 | 12.90 | 1585.56 | -67.00 |
| 2/ 6/1956 | -15.70 | 1569.86 | -73.60 |
| 9/ 6/1956 | 23.10 | 1592.96 | -40.20 |
| 16/ 6/1956 | -3.60 | 1589.36 | -26.60 |
| 23/ 6/1956 | 8.40 | 1597.76 | -7.00 |
| 30/ 6/1956 | 13.30 | 1611.06 | 8.50 |
| 7/ 7/1956 | 16.30 | 1627.36 | 14.90 |
| 14/ 7/1956 | 10.20 | 1637.56 | 41.00 |
| 21/ 7/1956 | -8.60 | 1628.96 | 47.80 |
| 28/ 7/1956 | 10.70 | 1639.66 | 89.60 |
| 4/ 8/1956 | -10.00 | 1629.66 | 66.70 |
| 11/ 8/1956 | -12.10 | 1617.56 | 70.30 |
| 18/ 8/1956 | -19.40 | 1598.16 | 27.80 |
| 25/ 8/1956 | -17.50 | 1580.66 | 13.90 |
| 1/ 9/1956 | 10.60 | 1591.26 | 16.10 |
| 8/ 9/1956 | -16.90 | 1574.36 | -14.10 |
| 15/ 9/1956 | -16.20 | 1558.16 | -46.60 |
| 22/ 9/1956 | -28.40 | 1529.76 | -85.20 |
| 29/ 9/1956 | 14.40 | 1544.16 | -62.20 |
| 6/10/1956 | 13.40 | 1557.56 | -59.50 |
| 13/10/1956 | -8.90 | 1548.66 | -58.40 |
| 20/10/1956 | -10.50 | 1538.16 | -56.80 |
| 27/10/1956 | 8.70 | 1546.86 | -28.70 |
| 3/11/1956 | -9.20 | 1537.66 | -20.40 |
| 10/11/1956 | -11.00 | 1526.66 | -42.00 |
| 17/11/1956 | -18.40 | 1508.26 | -43.50 |
| 24/11/1956 | -13.20 | 1495.06 | -40.50 |
| 1/12/1956 | 17.70 | 1512.76 | 5.60 |
| 8/12/1956 | -7.70 | 1505.06 | -16.50 |
| 15/12/1956 | -11.60 | 1493.46 | -41.50 |
| 22/12/1956 | 2.60 | 1496.06 | -30.00 |
| 29/12/1956 | 19.30 | 1515.36 | -0.20 |

52 1345

| DATE | A/D VAL | A/D LINE | 10WK RAT |
|---|---|---|---|
| 5/ 1/1957 | 14.80 | 1530.16 | 5.90 |
| 12/ 1/1957 | -21.30 | 1508.86 | -6.20 |
| 19/ 1/1957 | 9.00 | 1517.86 | 13.80 |
| 26/ 1/1957 | -11.90 | 1505.96 | 20.30 |
| 2/ 2/1957 | -20.50 | 1485.46 | 13.00 |
| 9/ 2/1957 | -7.50 | 1477.96 | -12.20 |
| 16/ 2/1957 | -4.30 | 1473.66 | -8.80 |
| 23/ 2/1957 | 11.00 | 1484.66 | 13.80 |
| 2/ 3/1957 | 12.80 | 1497.46 | 24.00 |
| 9/ 3/1957 | 7.80 | 1505.26 | 12.50 |
| 16/ 3/1957 | 3.20 | 1508.46 | 0.90 |
| 23/ 3/1957 | 2.20 | 1510.66 | 24.40 |
| 30/ 3/1957 | 10.80 | 1521.46 | 26.20 |
| 6/ 4/1957 | 13.40 | 1534.86 | 51.50 |
| 13/ 4/1957 | 4.30 | 1539.16 | 76.30 |
| 20/ 4/1957 | -6.90 | 1532.26 | 76.90 |
| 27/ 4/1957 | 10.50 | 1542.76 | 91.70 |
| 4/ 5/1957 | 7.50 | 1550.26 | 88.20 |
| 11/ 5/1957 | 8.00 | 1558.26 | 83.40 |
| 18/ 5/1957 | -10.80 | 1547.46 | 64.80 |
| 25/ 5/1957 | -8.90 | 1538.56 | 52.70 |
| 1/ 6/1957 | -9.90 | 1528.66 | 40.60 |
| 8/ 6/1957 | 2.40 | 1531.06 | 32.20 |
| 15/ 6/1957 | -22.90 | 1508.16 | -4.10 |
| 22/ 6/1957 | -5.40 | 1502.75 | -13.80 |
| 29/ 6/1957 | 19.10 | 1521.85 | 12.20 |
| 6/ 7/1957 | 5.00 | 1526.85 | 6.70 |
| 13/ 7/1957 | -14.10 | 1512.75 | -14.90 |
| 20/ 7/1957 | -12.20 | 1500.55 | -35.10 |
| 27/ 7/1957 | -14.80 | 1485.75 | -39.10 |
| 3/ 8/1957 | -18.20 | 1467.55 | -48.40 |
| 10/ 8/1957 | -19.80 | 1447.75 | -58.30 |
| 17/ 8/1957 | -24.40 | 1423.35 | -85.10 |
| 24/ 8/1957 | 10.50 | 1433.85 | -51.70 |
| 31/ 8/1957 | -10.40 | 1423.45 | -56.70 |
| 7/ 9/1957 | 2.20 | 1425.65 | -73.60 |
| 14/ 9/1957 | -20.10 | 1405.55 | -98.70 |
| 21/ 9/1957 | -25.70 | 1379.85 | -110.30 |
| 28/ 9/1957 | 7.90 | 1387.75 | -90.20 |
| 5/10/1957 | -37.10 | 1350.65 | -112.50 |
| 12/10/1957 | -18.00 | 1332.65 | -112.30 |
| 19/10/1957 | -13.80 | 1318.85 | -106.30 |
| 26/10/1957 | -5.10 | 1313.75 | -87.00 |
| 2/11/1957 | -6.90 | 1306.85 | -104.40 |
| 9/11/1957 | 10.00 | 1316.85 | -84.00 |
| 16/11/1957 | 13.70 | 1330.55 | -72.50 |
| 23/11/1957 | 15.80 | 1346.35 | -36.60 |
| 30/11/1957 | -6.00 | 1340.35 | -16.90 |
| 7/12/1957 | -12.00 | 1328.35 | -36.80 |
| 14/12/1957 | -24.90 | 1303.45 | -24.60 |
| 21/12/1957 | 7.50 | 1310.95 | 0.90 |
| 28/12/1957 | 26.30 | 1337.25 | 41.00 |

52 1397

**TABLE A4-4 Market Breadth** (cont.)

| DATE | A/D VAL | A/D LINE | 10WK RAT |
|---|---|---|---|
| 4/ 1/1958 | 12.00 | 1349.25 | 58.10 |
| 11/ 1/1958 | 25.70 | 1374.95 | 90.70 |
| 18/ 1/1958 | 24.60 | 1399.55 | 105.30 |
| 25/ 1/1958 | 9.40 | 1408.95 | 101.00 |
| 1/ 2/1958 | 9.10 | 1418.05 | 94.30 |
| 8/ 2/1958 | -11.30 | 1406.75 | 89.00 |
| 15/ 2/1958 | -10.90 | 1395.85 | 90.10 |
| 22/ 2/1958 | -8.50 | 1387.35 | 106.50 |
| 1/ 3/1958 | 19.90 | 1407.25 | 118.90 |
| 8/ 3/1958 | 13.80 | 1421.05 | 106.40 |
| 15/ 3/1958 | 7.10 | 1428.15 | 101.50 |
| 22/ 3/1958 | -7.30 | 1420.85 | 68.50 |
| 29/ 3/1958 | -18.20 | 1402.65 | 25.70 |
| 5/ 4/1958 | 14.50 | 1417.15 | 30.80 |
| 12/ 4/1958 | 20.60 | 1437.75 | 42.30 |
| 19/ 4/1958 | 16.70 | 1454.45 | 70.30 |
| 26/ 4/1958 | 12.50 | 1466.95 | 93.70 |
| 3/ 5/1958 | 14.60 | 1481.55 | 116.80 |
| 10/ 5/1958 | -10.20 | 1471.35 | 86.70 |
| 17/ 5/1958 | 17.40 | 1488.75 | 90.30 |
| 24/ 5/1958 | 11.80 | 1500.55 | 95.00 |
| 31/ 5/1958 | 16.30 | 1516.85 | 118.60 |
| 7/ 6/1958 | 15.50 | 1532.35 | 152.30 |
| 14/ 6/1958 | -10.30 | 1522.05 | 127.50 |
| 21/ 6/1958 | 9.20 | 1531.25 | 116.10 |
| 28/ 6/1958 | 12.80 | 1544.05 | 112.20 |
| 5/ 7/1958 | 8.40 | 1552.45 | 108.10 |
| 12/ 7/1958 | -9.30 | 1543.15 | 84.20 |
| 19/ 7/1958 | 20.60 | 1563.75 | 115.00 |
| 26/ 7/1958 | 17.10 | 1580.85 | 114.70 |
| 2/ 8/1958 | 11.80 | 1592.65 | 114.70 |
| 9/ 8/1958 | -12.20 | 1580.45 | 86.20 |
| 16/ 8/1958 | 8.30 | 1588.75 | 79.00 |
| 23/ 8/1958 | 6.40 | 1595.15 | 95.70 |
| 30/ 8/1958 | 9.60 | 1604.75 | 96.10 |
| 6/ 9/1958 | 12.20 | 1616.95 | 95.50 |
| 13/ 9/1958 | 14.30 | 1631.25 | 101.40 |
| 20/ 9/1958 | 4.50 | 1635.75 | 115.20 |
| 27/ 9/1958 | 16.00 | 1651.75 | 110.60 |
| 4/10/1958 | 16.50 | 1668.25 | 110.00 |
| 11/10/1958 | -9.30 | 1658.95 | 88.90 |
| 18/10/1958 | -9.90 | 1649.05 | 91.20 |
| 25/10/1958 | 10.30 | 1659.35 | 93.20 |
| 1/11/1958 | 15.30 | 1674.65 | 102.10 |
| 8/11/1958 | 22.00 | 1696.65 | 114.50 |
| 15/11/1958 | 5.70 | 1702.35 | 108.00 |
| 22/11/1958 | -4.60 | 1697.75 | 89.10 |
| 29/11/1958 | 10.40 | 1708.15 | 95.00 |
| 6/12/1958 | 11.50 | 1719.65 | 90.50 |
| 13/12/1958 | 6.50 | 1726.15 | 80.50 |
| 20/12/1958 | -3.00 | 1723.15 | 85.80 |
| 27/12/1958 | 19.90 | 1743.05 | 116.60 |

52 1449

| DATE | A/D VAL | A/D LINE | 10WK RAT |
|---|---|---|---|
| 3/ 1/1959 | 15.30 | 1758.35 | 121.60 |
| 10/ 1/1959 | 20.40 | 1778.75 | 126.70 |
| 17/ 1/1959 | 11.50 | 1790.25 | 116.20 |
| 24/ 1/1959 | -13.00 | 1777.25 | 97.50 |
| 31/ 1/1959 | -15.70 | 1761.55 | 86.40 |
| 7/ 2/1959 | 12.70 | 1774.25 | 88.70 |
| 14/ 2/1959 | 24.80 | 1799.05 | 102.00 |
| 21/ 2/1959 | 8.30 | 1807.35 | 103.80 |
| 28/ 2/1959 | 11.80 | 1819.15 | 118.60 |
| 7/ 3/1959 | 15.80 | 1834.95 | 114.50 |
| 14/ 3/1959 | -9.90 | 1825.05 | 89.30 |
| 21/ 3/1959 | -16.00 | 1809.05 | 52.90 |
| 28/ 3/1959 | 9.10 | 1818.15 | 50.50 |
| 4/ 4/1959 | -6.60 | 1811.55 | 56.90 |
| 11/ 4/1959 | 12.60 | 1824.15 | 85.20 |
| 18/ 4/1959 | -10.90 | 1813.25 | 61.60 |
| 25/ 4/1959 | -12.30 | 1800.95 | 24.50 |
| 2/ 5/1959 | -14.00 | 1786.95 | 2.20 |
| 9/ 5/1959 | 12.30 | 1799.25 | 2.70 |
| 16/ 5/1959 | -7.80 | 1791.45 | -20.90 |
| 23/ 5/1959 | -13.70 | 1777.74 | -24.70 |
| 30/ 5/1959 | -21.00 | 1756.74 | -29.70 |
| 6/ 6/1959 | -11.10 | 1745.64 | -49.90 |
| 13/ 6/1959 | -6.50 | 1739.14 | -49.80 |
| 20/ 6/1959 | 9.90 | 1749.04 | -52.50 |
| 27/ 6/1959 | 20.40 | 1769.44 | -21.20 |
| 4/ 7/1959 | 11.10 | 1780.54 | 2.20 |
| 11/ 7/1959 | -14.80 | 1765.74 | 1.40 |
| 18/ 7/1959 | 13.00 | 1778.74 | 2.10 |
| 25/ 7/1959 | 5.80 | 1784.54 | 15.70 |
| 1/ 8/1959 | -14.10 | 1770.44 | 15.30 |
| 8/ 8/1959 | -15.40 | 1755.04 | 20.90 |
| 15/ 8/1959 | -11.20 | 1743.84 | 20.80 |
| 22/ 8/1959 | 10.30 | 1754.14 | 37.60 |
| 29/ 8/1959 | -24.40 | 1729.74 | 3.30 |
| 5/ 9/1959 | -24.20 | 1705.54 | -41.30 |
| 12/ 9/1959 | -30.00 | 1675.54 | -82.40 |
| 19/ 9/1959 | 11.50 | 1687.04 | -56.10 |
| 26/ 9/1959 | 14.40 | 1701.44 | -54.70 |
| 3/10/1959 | 4.00 | 1705.44 | -56.50 |
| 10/10/1959 | 10.90 | 1716.34 | -31.50 |
| 17/10/1959 | -14.60 | 1701.74 | -30.70 |
| 24/10/1959 | 10.60 | 1712.34 | -8.90 |
| 31/10/1959 | -6.00 | 1706.34 | -25.20 |
| 7/11/1959 | -7.90 | 1698.44 | -8.70 |
| 14/11/1959 | -9.00 | 1689.44 | 6.50 |
| 21/11/1959 | 8.20 | 1697.64 | 44.70 |
| 28/11/1959 | 14.30 | 1711.94 | 47.50 |
| 5/12/1959 | -6.20 | 1705.74 | 26.90 |
| 12/12/1959 | -9.70 | 1696.04 | 13.20 |
| 19/12/1959 | -10.50 | 1685.54 | -8.20 |
| 26/12/1959 | 10.30 | 1695.84 | 16.70 |

52 1501

**TABLE A4-4 Market Breadth** (cont.)

| DATE | A/D VAL | A/D LINE | 10WK RAT |
|---|---|---|---|
| 2/ 1/1960 | 7.00 | 1702.84 | 13.10 |
| 9/ 1/1960 | -11.90 | 1690.94 | 7.20 |
| 16/ 1/1960 | -17.30 | 1673.64 | -2.20 |
| 23/ 1/1960 | -24.50 | 1649.14 | -17.70 |
| 30/ 1/1960 | -3.50 | 1645.64 | -29.40 |
| 6/ 2/1960 | -9.30 | 1636.34 | -53.00 |
| 13/ 2/1960 | 12.40 | 1648.74 | -34.40 |
| 20/ 2/1960 | 4.50 | 1653.24 | -20.20 |
| 27/ 2/1960 | -25.30 | 1627.94 | -35.00 |
| 5/ 3/1960 | -7.30 | 1620.64 | -52.60 |
| 12/ 3/1960 | 16.20 | 1636.84 | -43.40 |
| 19/ 3/1960 | 15.80 | 1652.64 | -15.70 |
| 26/ 3/1960 | -15.30 | 1637.34 | -13.70 |
| 2/ 4/1960 | 15.20 | 1652.54 | 26.00 |
| 9/ 4/1960 | -9.40 | 1643.14 | 20.10 |
| 16/ 4/1960 | -15.00 | 1628.14 | 14.40 |
| 23/ 4/1960 | -20.00 | 1608.14 | -18.00 |
| 30/ 4/1960 | 7.10 | 1615.24 | -15.40 |
| 7/ 5/1960 | 11.80 | 1627.04 | 21.70 |
| 14/ 5/1960 | 4.90 | 1631.94 | 33.90 |
| 21/ 5/1960 | -9.10 | 1622.84 | 8.60 |
| 28/ 5/1960 | 7.80 | 1630.64 | 0.60 |
| 4/ 6/1960 | 21.50 | 1652.14 | 37.40 |
| 11/ 6/1960 | -11.30 | 1640.84 | 10.90 |
| 18/ 6/1960 | 7.10 | 1647.94 | 27.40 |
| 25/ 6/1960 | -10.10 | 1637.84 | 32.30 |
| 2/ 7/1960 | 8.60 | 1646.44 | 60.90 |
| 9/ 7/1960 | -17.90 | 1628.54 | 35.90 |
| 16/ 7/1960 | -17.30 | 1611.24 | 6.80 |
| 23/ 7/1960 | 11.80 | 1623.04 | 13.70 |
| 30/ 7/1960 | 6.00 | 1629.04 | 28.80 |
| 6/ 8/1960 | 24.90 | 1653.94 | 45.90 |
| 13/ 8/1960 | 16.10 | 1670.04 | 40.50 |
| 20/ 8/1960 | 13.30 | 1683.34 | 65.10 |
| 27/ 8/1960 | -13.60 | 1669.74 | 44.40 |
| 3/ 9/1960 | -20.20 | 1649.54 | 34.30 |
| 10/ 9/1960 | -22.60 | 1626.94 | 3.10 |
| 17/ 9/1960 | -21.30 | 1605.64 | -0.30 |
| 24/ 9/1960 | -16.30 | 1589.34 | 0.70 |
| 1/10/1960 | 5.60 | 1594.94 | -5.50 |
| 8/10/1960 | 10.30 | 1605.24 | -1.20 |
| 15/10/1960 | -19.60 | 1585.64 | -45.70 |
| 22/10/1960 | -13.70 | 1571.94 | -75.50 |
| 29/10/1960 | 19.60 | 1591.54 | -69.20 |
| 5/11/1960 | 17.00 | 1608.54 | -38.60 |
| 12/11/1960 | -6.90 | 1601.64 | -25.30 |
| 19/11/1960 | 10.00 | 1611.64 | 7.30 |
| 26/11/1960 | -14.90 | 1596.74 | 13.70 |
| 3/12/1960 | 16.90 | 1613.64 | 46.90 |
| 10/12/1960 | 12.40 | 1626.04 | 53.70 |
| 17/12/1960 | -4.50 | 1621.54 | 38.90 |
| 24/12/1960 | 10.60 | 1632.14 | 69.10 |
| 31/12/1960 | 22.20 | 1654.34 | 105.00 |

53 1554

| DATE | A/D VAL | A/D LINE | 10WK RAT |
|---|---|---|---|
| 7/ 1/1961 | 25.40 | 1679.74 | 110.80 |
| 14/ 1/1961 | 18.30 | 1698.04 | 112.10 |
| 21/ 1/1961 | 13.10 | 1711.14 | 132.10 |
| 28/ 1/1961 | 18.70 | 1729.84 | 140.80 |
| 4/ 2/1961 | 11.20 | 1741.04 | 166.90 |
| 11/ 2/1961 | 19.90 | 1760.94 | 169.90 |
| 18/ 2/1961 | 17.40 | 1778.34 | 174.90 |
| 25/ 2/1961 | 14.00 | 1792.34 | 193.40 |
| 4/ 3/1961 | 3.70 | 1796.04 | 186.50 |
| 11/ 3/1961 | 17.40 | 1813.44 | 181.70 |
| 18/ 3/1961 | -6.40 | 1807.04 | 149.90 |
| 25/ 3/1961 | 11.60 | 1818.64 | 143.20 |
| 1/ 4/1961 | 9.40 | 1828.04 | 139.50 |
| 8/ 4/1961 | 1.40 | 1829.44 | 122.20 |
| 15/ 4/1961 | -7.70 | 1821.74 | 103.30 |
| 22/ 4/1961 | -15.30 | 1806.44 | 68.10 |
| 29/ 4/1961 | 19.20 | 1825.64 | 69.90 |
| 6/ 5/1961 | 9.30 | 1834.94 | 65.20 |
| 13/ 5/1961 | 14.50 | 1849.44 | 76.00 |
| 20/ 5/1961 | -10.00 | 1839.44 | 48.60 |
| 27/ 5/1961 | -8.10 | 1831.33 | 46.90 |
| 3/ 6/1961 | -13.80 | 1817.53 | 21.50 |
| 10/ 6/1961 | -22.60 | 1794.93 | -10.50 |
| 17/ 6/1961 | -12.80 | 1782.13 | -24.70 |
| 24/ 6/1961 | -13.90 | 1768.23 | -30.90 |
| 1/ 7/1961 | 18.30 | 1786.53 | 2.70 |
| 8/ 7/1961 | -15.80 | 1770.73 | -32.30 |
| 15/ 7/1961 | -16.90 | 1753.83 | -58.50 |
| 22/ 7/1961 | 18.50 | 1772.33 | -54.50 |
| 29/ 7/1961 | 15.70 | 1788.03 | -28.80 |
| 5/ 8/1961 | 9.40 | 1797.43 | -11.30 |
| 12/ 8/1961 | -2.80 | 1794.63 | -0.30 |
| 19/ 8/1961 | -13.90 | 1780.73 | 8.40 |
| 26/ 8/1961 | 12.50 | 1793.23 | 33.70 |
| 2/ 9/1961 | -15.30 | 1777.93 | 32.30 |
| 9/ 9/1961 | -11.00 | 1766.93 | 3.00 |
| 16/ 9/1961 | -16.90 | 1750.03 | 1.90 |
| 23/ 9/1961 | -10.10 | 1739.93 | 8.70 |
| 30/ 9/1961 | 17.50 | 1757.43 | 7.70 |
| 7/10/1961 | 4.50 | 1761.93 | -3.50 |
| 14/10/1961 | -4.00 | 1757.93 | -16.90 |
| 21/10/1961 | -10.80 | 1747.13 | -24.90 |
| 28/10/1961 | 17.10 | 1764.23 | 6.10 |
| 4/11/1961 | 20.00 | 1784.23 | 13.60 |
| 11/11/1961 | 10.90 | 1795.13 | 39.80 |
| 18/11/1961 | 7.20 | 1802.33 | 58.00 |
| 25/11/1961 | 7.70 | 1810.03 | 82.60 |
| 2/12/1961 | -8.00 | 1802.03 | 84.70 |
| 9/12/1961 | -12.00 | 1790.03 | 55.20 |
| 16/12/1961 | -16.70 | 1773.33 | 34.00 |
| 23/12/1961 | 8.70 | 1782.03 | 46.70 |
| 30/12/1961 | -13.00 | 1769.03 | 44.50 |

52 1606

**TABLE A4-4 Market Breadth** (cont.)

| DATE | A/D VAL | A/D LINE | 10WK RAT |
|---|---|---|---|
| 6/ 1/1962 | 8.40 | 1777.43 | 35.80 |
| 13/ 1/1962 | -11.40 | 1766.03 | 4.40 |
| 20/ 1/1962 | -10.60 | 1755.43 | -17.10 |
| 27/ 1/1962 | 17.50 | 1772.93 | -6.80 |
| 3/ 2/1962 | 16.70 | 1789.63 | 2.20 |
| 10/ 2/1962 | 10.40 | 1800.03 | 20.60 |
| 17/ 2/1962 | -13.10 | 1786.93 | 19.50 |
| 24/ 2/1962 | 6.90 | 1793.83 | 43.10 |
| 3/ 3/1962 | 5.70 | 1799.53 | 40.10 |
| 10/ 3/1962 | 10.30 | 1809.83 | 63.40 |
| 17/ 3/1962 | -12.20 | 1797.63 | 42.80 |
| 24/ 3/1962 | -16.80 | 1780.83 | 37.40 |
| 31/ 3/1962 | -16.40 | 1764.43 | 31.60 |
| 7/ 4/1962 | -21.00 | 1743.43 | -6.90 |
| 14/ 4/1962 | 14.80 | 1758.23 | -8.80 |
| 21/ 4/1962 | -24.10 | 1734.13 | -43.30 |
| 28/ 4/1962 | -9.40 | 1724.73 | -39.60 |
| 5/ 5/1962 | -29.60 | 1695.13 | -76.10 |
| 12/ 5/1962 | 15.90 | 1711.03 | -65.90 |
| 19/ 5/1962 | -36.20 | 1674.83 | -112.40 |
| 26/ 5/1962 | -12.20 | 1662.63 | -112.40 |
| 2/ 6/1962 | -18.90 | 1643.73 | -114.50 |
| 9/ 6/1962 | -32.90 | 1610.83 | -131.00 |
| 16/ 6/1962 | -29.40 | 1581.43 | -139.40 |
| 23/ 6/1962 | 20.20 | 1601.63 | -134.00 |
| 30/ 6/1962 | 19.60 | 1621.23 | -90.30 |
| 7/ 7/1962 | 29.40 | 1650.63 | -51.50 |
| 14/ 7/1962 | -19.70 | 1630.93 | -41.60 |
| 21/ 7/1962 | 4.10 | 1635.03 | -53.40 |
| 28/ 7/1962 | 12.30 | 1647.33 | -4.90 |
| 4/ 8/1962 | -11.30 | 1636.03 | -4.00 |
| 11/ 8/1962 | 21.60 | 1657.63 | 36.50 |
| 18/ 8/1962 | 15.80 | 1673.43 | 85.20 |
| 25/ 8/1962 | -9.40 | 1664.03 | 105.20 |
| 1/ 9/1962 | -16.40 | 1647.63 | 68.60 |
| 8/ 9/1962 | 6.80 | 1654.43 | 55.80 |
| 15/ 9/1962 | -22.40 | 1632.03 | 4.00 |
| 22/ 9/1962 | -24.90 | 1607.13 | -1.20 |
| 29/ 9/1962 | 9.90 | 1617.03 | 4.60 |
| 6/10/1962 | 4.80 | 1621.83 | -2.90 |
| 13/10/1962 | -24.50 | 1597.33 | -16.10 |
| 20/10/1962 | -25.90 | 1571.43 | -63.60 |
| 27/10/1962 | 30.70 | 1602.13 | -48.70 |
| 3/11/1962 | 23.20 | 1625.33 | -16.10 |
| 10/11/1962 | 28.10 | 1653.43 | 28.40 |
| 17/11/1962 | 20.70 | 1674.13 | 42.30 |
| 24/11/1962 | 22.40 | 1696.53 | 87.10 |
| 1/12/1962 | 11.70 | 1708.23 | 123.70 |
| 8/12/1962 | -18.70 | 1689.53 | 95.10 |
| 15/12/1962 | -12.60 | 1676.93 | 77.70 |
| 22/12/1962 | 8.70 | 1685.63 | 110.90 |
| 29/12/1962 | 31.50 | 1717.13 | 168.30 |

52 1658

| DATE | A/D VAL | A/D LINE | 10WK RAT |
|---|---|---|---|
| 5/ 1/1963 | 20.40 | 1737.53 | 158.00 |
| 12/ 1/1963 | 11.30 | 1748.83 | 146.10 |
| 19/ 1/1963 | 18.20 | 1767.03 | 136.20 |
| 26/ 1/1963 | 10.20 | 1777.23 | 125.70 |
| 2/ 2/1963 | -9.60 | 1767.63 | 93.70 |
| 9/ 2/1963 | 12.30 | 1779.93 | 94.30 |
| 16/ 2/1963 | -12.10 | 1767.83 | 100.90 |
| 23/ 2/1963 | -23.80 | 1744.03 | 89.70 |
| 2/ 3/1963 | 17.20 | 1761.23 | 98.20 |
| 9/ 3/1963 | 8.80 | 1770.03 | 75.50 |
| 16/ 3/1963 | 7.80 | 1777.83 | 62.90 |
| 23/ 3/1963 | 11.80 | 1789.63 | 63.40 |
| 30/ 3/1963 | 19.50 | 1809.13 | 64.70 |
| 6/ 4/1963 | 8.50 | 1817.63 | 63.00 |
| 13/ 4/1963 | 8.80 | 1826.43 | 81.40 |
| 20/ 4/1963 | 11.70 | 1838.13 | 80.80 |
| 27/ 4/1963 | 9.20 | 1847.33 | 102.10 |
| 4/ 5/1963 | 15.20 | 1862.53 | 141.10 |
| 11/ 5/1963 | 13.00 | 1875.53 | 136.90 |
| 18/ 5/1963 | 4.00 | 1879.53 | 132.10 |
| 25/ 5/1963 | 11.10 | 1890.62 | 135.40 |
| 1/ 6/1963 | -9.50 | 1881.12 | 114.10 |
| 8/ 6/1963 | -7.10 | 1874.02 | 87.50 |
| 15/ 6/1963 | -3.20 | 1870.82 | 75.80 |
| 22/ 6/1963 | -19.30 | 1851.52 | 47.70 |
| 29/ 6/1963 | 12.20 | 1863.72 | 48.20 |
| 6/ 7/1963 | -12.50 | 1851.22 | 26.50 |
| 13/ 7/1963 | -19.90 | 1831.32 | -8.60 |
| 20/ 7/1963 | -8.40 | 1822.92 | -30.00 |
| 27/ 7/1963 | 9.50 | 1832.42 | -24.50 |
| 3/ 8/1963 | 16.30 | 1848.72 | -19.30 |
| 10/ 8/1963 | 17.00 | 1865.72 | 7.20 |
| 17/ 8/1963 | 11.80 | 1877.52 | 26.10 |
| 24/ 8/1963 | 16.30 | 1893.82 | 45.60 |
| 31/ 8/1963 | 3.60 | 1897.42 | 68.50 |
| 7/ 9/1963 | -8.00 | 1889.42 | 48.30 |
| 14/ 9/1963 | -9.10 | 1880.32 | 51.70 |
| 21/ 9/1963 | -21.40 | 1858.92 | 50.20 |
| 28/ 9/1963 | 12.30 | 1871.22 | 70.90 |
| 5/10/1963 | -8.20 | 1863.02 | 53.20 |
| 12/10/1963 | 12.70 | 1875.72 | 49.60 |
| 19/10/1963 | -8.00 | 1867.72 | 24.60 |
| 26/10/1963 | -11.50 | 1856.22 | 1.30 |
| 2/11/1963 | -8.90 | 1847.32 | -23.90 |
| 9/11/1963 | -10.80 | 1836.52 | -38.30 |
| 16/11/1963 | -29.60 | 1806.92 | -59.90 |
| 23/11/1963 | 24.10 | 1831.02 | -26.70 |
| 30/11/1963 | 11.30 | 1842.32 | 6.00 |
| 7/12/1963 | -8.40 | 1833.92 | -14.70 |
| 14/12/1963 | -13.50 | 1820.42 | -20.00 |
| 21/12/1963 | 8.00 | 1828.42 | -24.70 |
| 28/12/1963 | 19.10 | 1847.52 | 2.40 |

52 1710

**TABLE A4-4 Market Breadth** (cont.)

| DATE | A/D VAL | A/D LINE | 10WK RAT |
|---|---|---|---|
| 4/ 1/1964 | 13.30 | 1860.82 | 27.20 |
| 11/ 1/1964 | -3.20 | 1857.52 | 32.90 |
| 18/ 1/1964 | 6.20 | 1863.82 | 49.90 |
| 25/ 1/1964 | -14.50 | 1849.32 | 65.00 |
| 1/ 2/1964 | 10.00 | 1859.32 | 50.90 |
| 8/ 2/1964 | 9.30 | 1868.62 | 48.90 |
| 15/ 2/1964 | 9.90 | 1878.52 | 67.20 |
| 22/ 2/1964 | 11.30 | 1889.82 | 92.00 |
| 29/ 2/1964 | 12.80 | 1902.62 | 96.80 |
| 7/ 3/1964 | 11.30 | 1913.92 | 88.99 |
| 14/ 3/1964 | -2.40 | 1911.52 | 73.29 |
| 21/ 3/1964 | 5.70 | 1917.22 | 82.19 |
| 28/ 3/1964 | 12.60 | 1929.82 | 88.59 |
| 4/ 4/1964 | 7.80 | 1937.62 | 110.89 |
| 11/ 4/1964 | 4.40 | 1942.02 | 105.29 |
| 18/ 4/1964 | -16.50 | 1925.52 | 79.49 |
| 25/ 4/1964 | -11.50 | 1914.02 | 58.09 |
| 2/ 5/1964 | 14.10 | 1928.12 | 60.39 |
| 9/ 5/1964 | -8.90 | 1919.22 | 39.19 |
| 16/ 5/1964 | -8.00 | 1911.22 | 19.89 |
| 23/ 5/1964 | -8.10 | 1903.12 | 14.19 |
| 30/ 5/1964 | -19.70 | 1883.42 | -11.21 |
| 6/ 6/1964 | 10.20 | 1893.62 | -13.51 |
| 13/ 6/1964 | 17.10 | 1910.72 | -4.31 |
| 20/ 6/1964 | 14.00 | 1924.72 | 5.29 |
| 27/ 6/1964 | 14.50 | 1939.22 | 36.29 |
| 4/ 7/1964 | 15.20 | 1954.42 | 62.99 |
| 11/ 7/1964 | 12.40 | 1966.82 | 61.29 |
| 18/ 7/1964 | -6.50 | 1960.32 | 63.69 |
| 25/ 7/1964 | -7.70 | 1952.62 | 63.99 |
| 1/ 8/1964 | -19.60 | 1933.02 | 52.49 |
| 8/ 8/1964 | 16.50 | 1949.52 | 88.69 |
| 15/ 8/1964 | -8.70 | 1940.82 | 69.79 |
| 22/ 8/1964 | -6.90 | 1933.92 | 45.79 |
| 29/ 8/1964 | 15.80 | 1949.72 | 47.59 |
| 5/ 9/1964 | 12.60 | 1962.32 | 45.69 |
| 12/ 9/1964 | -3.90 | 1958.42 | 26.59 |
| 19/ 9/1964 | 12.90 | 1971.32 | 27.09 |
| 26/ 9/1964 | -2.40 | 1968.92 | 31.19 |
| 3/10/1964 | 14.60 | 1983.52 | 53.49 |
| 10/10/1964 | -8.50 | 1975.02 | 64.59 |
| 17/10/1964 | 10.10 | 1985.12 | 58.19 |
| 24/10/1964 | -8.10 | 1977.02 | 58.79 |
| 31/10/1964 | -2.80 | 1974.22 | 62.89 |
| 7/11/1964 | 10.00 | 1984.22 | 57.09 |
| 14/11/1964 | 12.80 | 1997.02 | 57.29 |
| 21/11/1964 | -15.70 | 1981.32 | 45.49 |
| 28/11/1964 | -16.90 | 1964.42 | 15.69 |
| 5/12/1964 | -15.90 | 1948.52 | 2.19 |
| 12/12/1964 | 10.70 | 1959.22 | -1.71 |
| 19/12/1964 | -7.60 | 1951.62 | -0.81 |
| 26/12/1964 | 7.10 | 1958.72 | -3.81 |

52 1762

| DATE | A/D VAL | A/D LINE | 10WK RAT |
|---|---|---|---|
| 2/ 1/1965 | 19.70 | 1978.42 | 23.99 |
| 9/ 1/1965 | 18.80 | 1997.22 | 45.59 |
| 16/ 1/1965 | 11.10 | 2008.32 | 46.69 |
| 23/ 1/1965 | 11.20 | 2019.52 | 45.09 |
| 30/ 1/1965 | 9.30 | 2028.82 | 70.09 |
| 6/ 2/1965 | -14.70 | 2014.12 | 72.29 |
| 13/ 2/1965 | 14.30 | 2028.42 | 102.49 |
| 20/ 2/1965 | 15.70 | 2044.12 | 107.49 |
| 27/ 2/1965 | -10.00 | 2034.12 | 105.09 |
| 6/ 3/1965 | 13.00 | 2047.12 | 110.99 |
| 13/ 3/1965 | -9.30 | 2037.82 | 81.99 |
| 20/ 3/1965 | -12.20 | 2025.61 | 50.99 |
| 27/ 3/1965 | 6.30 | 2031.91 | 46.19 |
| 3/ 4/1965 | 15.80 | 2047.71 | 50.79 |
| 10/ 4/1965 | 12.70 | 2060.41 | 54.19 |
| 17/ 4/1965 | 9.20 | 2069.61 | 78.09 |
| 24/ 4/1965 | -6.60 | 2063.01 | 57.19 |
| 1/ 5/1965 | 6.60 | 2069.61 | 48.09 |
| 8/ 5/1965 | 5.70 | 2075.31 | 63.79 |
| 15/ 5/1965 | -15.20 | 2060.11 | 35.59 |
| 22/ 5/1965 | -17.60 | 2042.51 | 27.29 |
| 29/ 5/1965 | -23.50 | 2019.01 | 15.99 |
| 5/ 6/1965 | -28.10 | 1990.91 | -18.41 |
| 12/ 6/1965 | -8.60 | 1982.31 | -42.81 |
| 19/ 6/1965 | -31.50 | 1950.81 | -87.01 |
| 26/ 6/1965 | 19.30 | 1970.11 | -76.91 |
| 3/ 7/1965 | 14.30 | 1984.41 | -56.01 |
| 10/ 7/1965 | 7.10 | 1991.51 | -55.51 |
| 17/ 7/1965 | -20.20 | 1971.31 | -81.41 |
| 24/ 7/1965 | 17.40 | 1988.71 | -48.81 |
| 31/ 7/1965 | 15.80 | 2004.51 | -15.41 |
| 7/ 8/1965 | 16.00 | 2020.51 | 24.09 |
| 14/ 8/1965 | 4.10 | 2024.61 | 56.29 |
| 21/ 8/1965 | 13.20 | 2037.81 | 78.09 |
| 28/ 8/1965 | 18.30 | 2056.11 | 127.39 |
| 4/ 9/1965 | 13.70 | 2069.81 | 122.29 |
| 11/ 9/1965 | 7.40 | 2077.21 | 115.39 |
| 18/ 9/1965 | -6.60 | 2070.61 | 101.69 |
| 25/ 9/1965 | -10.20 | 2060.41 | 111.69 |
| 2/10/1965 | 16.50 | 2076.91 | 110.79 |
| 9/10/1965 | 13.60 | 2090.51 | 108.59 |
| 16/10/1965 | 3.50 | 2094.01 | 96.09 |
| 23/10/1965 | 2.60 | 2096.61 | 94.59 |
| 30/10/1965 | 7.30 | 2103.91 | 88.69 |
| 6/11/1965 | -7.70 | 2096.21 | 62.69 |
| 13/11/1965 | -2.60 | 2093.61 | 46.39 |
| 20/11/1965 | 8.50 | 2102.11 | 47.49 |
| 27/11/1965 | -3.90 | 2098.21 | 50.19 |
| 4/12/1965 | 7.70 | 2105.91 | 68.09 |
| 11/12/1965 | 12.00 | 2117.91 | 63.59 |
| 18/12/1965 | -11.90 | 2106.01 | 38.09 |
| 25/12/1965 | 8.80 | 2114.81 | 43.39 |

52 1814

**TABLE A4-4 Market Breadth** (cont.)

| DATE | A/D VAL | A/D LINE | 10WK RAT |
|---|---|---|---|
| 1/ 1/1966 | 15.50 | 2130.31 | 56.29 |
| 8/ 1/1966 | 15.40 | 2145.71 | 64.39 |
| 15/ 1/1966 | 2.00 | 2147.71 | 74.09 |
| 22/ 1/1966 | 6.20 | 2153.91 | 82.89 |
| 29/ 1/1966 | -15.40 | 2138.51 | 58.99 |
| 5/ 2/1966 | 11.60 | 2150.11 | 74.49 |
| 12/ 2/1966 | -20.20 | 2129.91 | 46.59 |
| 19/ 2/1966 | -19.60 | 2110.31 | 14.99 |
| 26/ 2/1966 | -26.20 | 2084.11 | 0.69 |
| 5/ 3/1966 | -17.00 | 2067.11 | -25.11 |
| 12/ 3/1966 | -20.80 | 2046.31 | -61.41 |
| 19/ 3/1966 | 14.10 | 2060.41 | -62.71 |
| 26/ 3/1966 | -7.70 | 2052.71 | -72.41 |
| 2/ 4/1966 | 19.50 | 2072.21 | -59.11 |
| 9/ 4/1966 | 4.60 | 2076.81 | -39.11 |
| 16/ 4/1966 | 7.70 | 2084.51 | -43.01 |
| 23/ 4/1966 | -16.80 | 2067.71 | -39.61 |
| 30/ 4/1966 | -31.00 | 2036.71 | -51.01 |
| 7/ 5/1966 | -26.20 | 2010.51 | -51.01 |
| 14/ 5/1966 | -18.50 | 1992.01 | -52.51 |
| 21/ 5/1966 | 22.80 | 2014.81 | -8.91 |
| 28/ 5/1966 | -16.50 | 1998.31 | -39.51 |
| 4/ 6/1966 | 7.70 | 2006.01 | -24.11 |
| 11/ 6/1966 | 8.50 | 2014.51 | -35.11 |
| 18/ 6/1966 | -6.60 | 2007.91 | -46.31 |
| 25/ 6/1966 | -23.00 | 1984.91 | -77.01 |
| 2/ 7/1966 | 19.70 | 2004.61 | -40.51 |
| 9/ 7/1966 | -10.40 | 1994.21 | -19.91 |
| 16/ 7/1966 | -21.00 | 1973.21 | -14.71 |
| 23/ 7/1966 | -27.70 | 1945.51 | -23.91 |
| 30/ 7/1966 | -7.30 | 1938.21 | -54.01 |
| 6/ 8/1966 | -15.30 | 1922.91 | -52.81 |
| 13/ 8/1966 | -39.60 | 1883.31 | -100.11 |
| 20/ 8/1966 | -42.00 | 1841.31 | -150.61 |
| 27/ 8/1966 | -12.90 | 1828.41 | -156.91 |
| 3/ 9/1966 | -20.30 | 1808.11 | -154.21 |
| 10/ 9/1966 | 28.80 | 1836.91 | -145.11 |
| 17/ 9/1966 | -26.10 | 1810.81 | -160.81 |
| 24/ 9/1966 | -19.80 | 1791.01 | -159.61 |
| 1/10/1966 | -34.80 | 1756.21 | -166.71 |
| 8/10/1966 | 26.30 | 1782.51 | -133.11 |
| 15/10/1966 | 14.80 | 1797.31 | -103.01 |
| 22/10/1966 | 23.00 | 1820.31 | -40.41 |
| 29/10/1966 | 13.10 | 1833.41 | 14.69 |
| 5/11/1966 | 18.10 | 1851.51 | 45.69 |
| 12/11/1966 | 3.00 | 1854.51 | 68.99 |
| 19/11/1966 | -13.30 | 1841.21 | 26.89 |
| 26/11/1966 | -10.30 | 1830.91 | 42.69 |
| 3/12/1966 | 17.40 | 1848.31 | 79.89 |
| 10/12/1966 | 8.50 | 1856.81 | 123.19 |
| 17/12/1966 | 8.70 | 1865.51 | 105.59 |
| 24/12/1966 | -20.20 | 1845.31 | 70.59 |
| 31/12/1966 | 30.60 | 1875.91 | 78.19 |

53 1867

| DATE | A/D VAL | A/D LINE | 10WK RAT |
|---|---|---|---|
| 7/ 1/1967 | 37.90 | 1913.81 | 102.99 |
| 14/ 1/1967 | 28.10 | 1941.91 | 112.99 |
| 21/ 1/1967 | 12.40 | 1954.31 | 122.39 |
| 28/ 1/1967 | 19.80 | 1974.11 | 155.49 |
| 4/ 2/1967 | 11.40 | 1985.51 | 177.19 |
| 11/ 2/1967 | 8.90 | 1994.41 | 168.69 |
| 18/ 2/1967 | -8.00 | 1986.41 | 152.19 |
| 25/ 2/1967 | 6.20 | 1992.61 | 149.69 |
| 4/ 3/1967 | 15.80 | 2008.41 | 185.69 |
| 11/ 3/1967 | 11.40 | 2019.80 | 166.49 |
| 18/ 3/1967 | 12.20 | 2032.00 | 140.79 |
| 25/ 3/1967 | 5.70 | 2037.70 | 118.39 |
| 1/ 4/1967 | -13.50 | 2024.20 | 92.49 |
| 8/ 4/1967 | 9.20 | 2033.40 | 81.89 |
| 15/ 4/1967 | 17.40 | 2050.80 | 87.89 |
| 22/ 4/1967 | 13.30 | 2064.10 | 92.29 |
| 29/ 4/1967 | 14.70 | 2078.80 | 114.99 |
| 6/ 5/1967 | -12.00 | 2066.80 | 96.79 |
| 13/ 5/1967 | -14.90 | 2051.90 | 66.09 |
| 20/ 5/1967 | -21.10 | 2030.80 | 33.59 |
| 27/ 5/1967 | -23.80 | 2007.00 | -2.41 |
| 3/ 6/1967 | 19.20 | 2026.20 | 11.09 |
| 10/ 6/1967 | 16.90 | 2043.10 | 41.49 |
| 17/ 6/1967 | -7.70 | 2035.40 | 24.59 |
| 24/ 6/1967 | -13.40 | 2022.00 | -6.21 |
| 1/ 7/1967 | 17.90 | 2039.90 | -1.61 |
| 8/ 7/1967 | 19.40 | 2059.30 | 3.09 |
| 15/ 7/1967 | 10.90 | 2070.20 | 25.99 |
| 22/ 7/1967 | 15.10 | 2085.30 | 55.99 |
| 29/ 7/1967 | 17.30 | 2102.60 | 94.39 |
| 5/ 8/1967 | -12.30 | 2090.30 | 105.89 |
| 12/ 8/1967 | -13.30 | 2077.00 | 73.39 |
| 19/ 8/1967 | -23.20 | 2053.80 | 33.29 |
| 26/ 8/1967 | 21.00 | 2074.80 | 61.99 |
| 2/ 9/1967 | 11.00 | 2085.80 | 86.39 |
| 9/ 9/1967 | 15.10 | 2100.90 | 83.59 |
| 16/ 9/1967 | 6.60 | 2107.50 | 70.79 |
| 23/ 9/1967 | -14.10 | 2093.40 | 45.79 |
| 30/ 9/1967 | 5.90 | 2099.30 | 36.59 |
| 7/10/1967 | -22.70 | 2076.60 | -3.41 |
| 14/10/1967 | -22.20 | 2054.40 | -13.31 |
| 21/10/1967 | -19.50 | 2034.90 | -19.51 |
| 28/10/1967 | -29.90 | 2005.00 | -26.21 |
| 4/11/1967 | -9.50 | 1995.50 | -56.71 |
| 11/11/1967 | -8.10 | 1987.40 | -75.81 |
| 18/11/1967 | 9.70 | 1997.10 | -81.21 |
| 25/11/1967 | 15.10 | 2012.20 | -72.71 |
| 2/12/1967 | 14.40 | 2026.60 | -44.21 |
| 9/12/1967 | -8.30 | 2018.30 | -53.41 |
| 16/12/1967 | -2.60 | 2015.70 | -38.31 |
| 23/12/1967 | 20.30 | 2036.00 | 4.19 |
| 30/12/1967 | 14.20 | 2050.20 | 37.89 |

52 1919

**TABLE A4-4 Market Breadth** (cont.)

| DATE | A/D VAL | A/D LINE | 10WK RAT |
|---|---|---|---|
| 6/ 1/1968 | 28.10 | 2078.30 | 95.89 |
| 13/ 1/1968 | -9.90 | 2068.40 | 95.49 |
| 20/ 1/1968 | -26.50 | 2041.90 | 77.09 |
| 27/ 1/1968 | -18.10 | 2023.80 | 49.29 |
| 3/ 2/1968 | -29.10 | 1994.70 | 5.09 |
| 10/ 2/1968 | -9.80 | 1984.90 | -19.11 |
| 17/ 2/1968 | 15.20 | 2000.10 | 4.39 |
| 24/ 2/1968 | -26.50 | 1973.60 | -19.51 |
| 2/ 3/1968 | -16.90 | 1956.70 | -56.71 |
| 9/ 3/1968 | -10.70 | 1946.00 | -81.61 |
| 16/ 3/1968 | -16.20 | 1929.80 | -125.91 |
| 23/ 3/1968 | 16.40 | 1946.20 | -99.61 |
| 30/ 3/1968 | 30.10 | 1976.30 | -43.01 |
| 6/ 4/1968 | 28.10 | 2004.40 | 3.19 |
| 13/ 4/1968 | 16.60 | 2021.00 | 48.89 |
| 20/ 4/1968 | 18.20 | 2039.20 | 76.89 |
| 27/ 4/1968 | 19.10 | 2058.30 | 80.79 |
| 4/ 5/1968 | 14.10 | 2072.40 | 121.39 |
| 11/ 5/1968 | -11.80 | 2060.60 | 126.49 |
| 18/ 5/1968 | 11.00 | 2071.60 | 148.19 |
| 25/ 5/1968 | 19.00 | 2090.60 | 183.39 |
| 1/ 6/1968 | 26.60 | 2117.20 | 193.59 |
| 8/ 6/1968 | -7.00 | 2110.20 | 156.49 |
| 15/ 6/1968 | -13.40 | 2096.80 | 114.99 |
| 22/ 6/1968 | -17.50 | 2079.30 | 80.89 |
| 29/ 6/1968 | 18.60 | 2097.90 | 81.29 |
| 6/ 7/1968 | 21.10 | 2119.00 | 83.29 |
| 13/ 7/1968 | -19.60 | 2099.40 | 49.59 |
| 20/ 7/1968 | -29.90 | 2069.50 | 31.49 |
| 27/ 7/1968 | -21.60 | 2047.90 | -1.11 |
| 3/ 8/1968 | 16.80 | 2064.70 | -3.31 |
| 10/ 8/1968 | 20.10 | 2084.80 | -9.81 |
| 17/ 8/1968 | 11.20 | 2096.00 | 8.39 |
| 24/ 8/1968 | 6.90 | 2102.90 | 28.69 |
| 31/ 8/1968 | 20.70 | 2123.60 | 66.89 |
| 7/ 9/1968 | 7.20 | 2130.80 | 55.49 |
| 14/ 9/1968 | 17.60 | 2148.40 | 51.99 |
| 21/ 9/1968 | 16.00 | 2164.40 | 87.59 |
| 28/ 9/1968 | 17.20 | 2181.60 | 134.69 |
| 5/10/1968 | -12.20 | 2169.40 | 144.09 |
| 12/10/1968 | 18.70 | 2188.10 | 145.99 |
| 19/10/1968 | -5.70 | 2182.40 | 120.19 |
| 26/10/1968 | -17.70 | 2164.70 | 91.29 |
| 2/11/1968 | 13.80 | 2178.50 | 98.19 |
| 9/11/1968 | 28.20 | 2206.70 | 105.69 |
| 16/11/1968 | 14.10 | 2220.80 | 112.59 |
| 23/11/1968 | 22.20 | 2243.00 | 117.19 |
| 30/11/1968 | -10.50 | 2232.50 | 90.69 |
| 7/12/1968 | 6.90 | 2239.40 | 80.39 |
| 14/12/1968 | -19.90 | 2219.50 | 72.69 |
| 21/12/1968 | -20.50 | 2199.00 | 33.49 |
| 28/12/1968 | -14.10 | 2184.90 | 25.09 |

52 1971

| DATE | A/D VAL | A/D LINE | 10WK RAT |
|---|---|---|---|
| 4/ 1/1969 | -35.40 | 2149.50 | 7.39 |
| 11/ 1/1969 | 14.50 | 2164.00 | 8.09 |
| 18/ 1/1969 | 14.80 | 2178.80 | -5.31 |
| 25/ 1/1969 | 2.60 | 2181.40 | -16.81 |
| 1/ 2/1969 | 4.40 | 2185.80 | -34.61 |
| 8/ 2/1969 | -9.90 | 2175.90 | -34.01 |
| 15/ 2/1969 | -43.70 | 2132.19 | -84.61 |
| 22/ 2/1969 | -33.70 | 2098.49 | -98.41 |
| 1/ 3/1969 | -14.20 | 2084.29 | -92.11 |
| 8/ 3/1969 | -16.90 | 2067.39 | -94.91 |
| 15/ 3/1969 | 15.50 | 2082.89 | -44.01 |
| 22/ 3/1969 | 16.90 | 2099.79 | -41.61 |
| 29/ 3/1969 | -16.20 | 2083.59 | -72.61 |
| 5/ 4/1969 | 3.30 | 2086.89 | -71.91 |
| 12/ 4/1969 | -13.50 | 2073.39 | -89.81 |
| 19/ 4/1969 | -6.50 | 2066.89 | -86.41 |
| 26/ 4/1969 | 23.20 | 2090.09 | -19.51 |
| 3/ 5/1969 | 12.90 | 2102.99 | 27.09 |
| 10/ 5/1969 | 1.40 | 2104.39 | 42.69 |
| 17/ 5/1969 | -18.30 | 2086.09 | 41.29 |
| 24/ 5/1969 | -19.70 | 2066.39 | 6.09 |
| 31/ 5/1969 | -22.70 | 2043.69 | -33.51 |
| 7/ 6/1969 | -42.90 | 2000.79 | -60.21 |
| 14/ 6/1969 | -31.10 | 1969.69 | -94.61 |
| 21/ 6/1969 | -16.10 | 1953.59 | -97.21 |
| 28/ 6/1969 | 24.00 | 1977.59 | -66.71 |
| 5/ 7/1969 | -32.00 | 1945.59 | -121.91 |
| 12/ 7/1969 | -19.80 | 1925.79 | -154.61 |
| 19/ 7/1969 | -30.80 | 1894.99 | -186.81 |
| 26/ 7/1969 | 9.70 | 1904.69 | -158.81 |
| 2/ 8/1969 | 10.90 | 1915.59 | -128.21 |
| 9/ 8/1969 | -9.80 | 1905.79 | -115.31 |
| 16/ 8/1969 | 19.70 | 1925.49 | -52.71 |
| 23/ 8/1969 | -8.50 | 1916.99 | -30.11 |
| 30/ 8/1969 | -23.90 | 1893.09 | -37.91 |
| 6/ 9/1969 | -5.20 | 1887.89 | -67.11 |
| 13/ 9/1969 | 15.90 | 1903.79 | -19.21 |
| 20/ 9/1969 | -15.00 | 1888.79 | -14.41 |
| 27/ 9/1969 | -16.50 | 1872.29 | -0.11 |
| 4/10/1969 | -6.00 | 1866.29 | -15.81 |
| 11/10/1969 | 36.40 | 1902.69 | 9.69 |
| 18/10/1969 | 21.50 | 1924.19 | 40.99 |
| 25/10/1969 | -11.20 | 1912.99 | 10.09 |
| 1/11/1969 | 12.90 | 1925.89 | 31.49 |
| 8/11/1969 | -22.10 | 1903.79 | 33.29 |
| 15/11/1969 | -33.30 | 1870.49 | 5.19 |
| 22/11/1969 | -16.40 | 1854.09 | -27.11 |
| 29/11/1969 | -31.00 | 1823.09 | -43.11 |
| 6/12/1969 | -24.60 | 1798.49 | -51.21 |
| 13/12/1969 | -8.70 | 1789.79 | -53.91 |
| 20/12/1969 | 8.80 | 1798.59 | -81.51 |
| 27/12/1969 | 24.10 | 1822.69 | -78.91 |

52 2023

**TABLE A4-4 Market Breadth** (cont.)

| DATE | A/D VAL | A/D LINE | 10WK RAT |
|---|---|---|---|
| 3/ 1/1970 | 13.80 | 1836.49 | -53.91 |
| 10/ 1/1970 | -26.70 | 1809.79 | -93.51 |
| 17/ 1/1970 | -17.10 | 1792.69 | -88.51 |
| 24/ 1/1970 | -31.10 | 1761.59 | -86.31 |
| 31/ 1/1970 | 13.70 | 1775.29 | -56.21 |
| 7/ 2/1970 | 10.10 | 1785.39 | -15.11 |
| 14/ 2/1970 | 16.00 | 1801.39 | 25.49 |
| 21/ 2/1970 | 22.40 | 1823.79 | 56.59 |
| 28/ 2/1970 | 15.60 | 1839.39 | 63.39 |
| 7/ 3/1970 | -22.80 | 1816.59 | 16.49 |
| 14/ 3/1970 | -20.10 | 1796.49 | -17.41 |
| 21/ 3/1970 | 26.70 | 1823.19 | 35.99 |
| 28/ 3/1970 | -8.80 | 1814.39 | 44.29 |
| 4/ 4/1970 | -22.50 | 1791.89 | 52.89 |
| 11/ 4/1970 | -30.40 | 1761.49 | 8.79 |
| 18/ 4/1970 | -32.20 | 1729.29 | -33.51 |
| 25/ 4/1970 | -23.90 | 1705.39 | -73.41 |
| 2/ 5/1970 | -24.30 | 1681.09 | -120.11 |
| 9/ 5/1970 | -34.20 | 1646.89 | -169.91 |
| 16/ 5/1970 | -41.90 | 1604.99 | -189.01 |
| 23/ 5/1970 | 28.80 | 1633.79 | -140.11 |
| 30/ 5/1970 | 12.80 | 1646.59 | -154.01 |
| 6/ 6/1970 | -26.60 | 1619.99 | -171.81 |
| 13/ 6/1970 | 19.40 | 1639.39 | -129.91 |
| 20/ 6/1970 | -35.40 | 1603.99 | -134.91 |
| 27/ 6/1970 | -18.70 | 1585.29 | -121.41 |
| 4/ 7/1970 | 17.20 | 1602.49 | -80.31 |
| 11/ 7/1970 | 24.50 | 1626.99 | -31.51 |
| 18/ 7/1970 | 12.90 | 1639.89 | 15.59 |
| 25/ 7/1970 | 11.80 | 1651.69 | 69.29 |
| 1/ 8/1970 | -14.50 | 1637.19 | 25.99 |
| 8/ 8/1970 | -21.10 | 1616.09 | -7.91 |
| 15/ 8/1970 | 23.70 | 1639.79 | 42.39 |
| 22/ 8/1970 | 36.20 | 1675.99 | 59.19 |
| 29/ 8/1970 | 21.70 | 1697.69 | 116.29 |
| 5/ 9/1970 | 10.20 | 1707.89 | 145.19 |
| 12/ 9/1970 | 14.60 | 1722.49 | 142.59 |
| 19/ 9/1970 | 21.90 | 1744.39 | 139.99 |
| 26/ 9/1970 | 18.40 | 1762.79 | 145.49 |
| 3/10/1970 | -4.50 | 1758.29 | 129.19 |
| 10/10/1970 | -19.90 | 1738.39 | 123.79 |
| 17/10/1970 | -19.40 | 1718.99 | 125.49 |
| 24/10/1970 | -17.50 | 1701.49 | 84.29 |
| 31/10/1970 | 15.50 | 1716.99 | 63.59 |
| 7/11/1970 | -12.10 | 1704.89 | 29.79 |
| 14/11/1970 | -12.80 | 1692.09 | 6.79 |
| 21/11/1970 | 22.70 | 1714.79 | 14.89 |
| 28/11/1970 | 38.10 | 1752.89 | 31.09 |
| 5/12/1970 | 18.20 | 1771.09 | 30.89 |
| 12/12/1970 | -3.00 | 1768.09 | 32.39 |
| 19/12/1970 | 18.70 | 1786.79 | 70.99 |
| 26/12/1970 | 23.00 | 1809.79 | 113.39 |

52 2075

| DATE | A/D VAL | A/D LINE | 10WK RAT |
|---|---|---|---|
| 2/ 1/1971 | 24.50 | 1834.29 | 155.39 |
| 9/ 1/1971 | 29.20 | 1863.49 | 169.09 |
| 16/ 1/1971 | 26.50 | 1889.99 | 207.69 |
| 23/ 1/1971 | 19.00 | 1908.99 | 239.49 |
| 30/ 1/1971 | 23.10 | 1932.09 | 239.89 |
| 6/ 2/1971 | 20.70 | 1952.79 | 222.49 |
| 13/ 2/1971 | -20.80 | 1931.99 | 183.49 |
| 20/ 2/1971 | -11.00 | 1920.99 | 175.49 |
| 27/ 2/1971 | 21.20 | 1942.19 | 177.99 |
| 6/ 3/1971 | 14.00 | 1956.19 | 168.99 |
| 13/ 3/1971 | 18.30 | 1974.49 | 162.79 |
| 20/ 3/1971 | -17.40 | 1957.09 | 116.19 |
| 27/ 3/1971 | 9.30 | 1966.39 | 98.99 |
| 3/ 4/1971 | 16.30 | 1982.68 | 96.29 |
| 10/ 4/1971 | 14.50 | 1997.18 | 87.69 |
| 17/ 4/1971 | -7.10 | 1990.08 | 59.89 |
| 24/ 4/1971 | -7.60 | 1982.48 | 73.09 |
| 1/ 5/1971 | -18.70 | 1963.78 | 65.39 |
| 8/ 5/1971 | -11.10 | 1952.68 | 33.09 |
| 15/ 5/1971 | -22.60 | 1930.08 | -3.51 |
| 22/ 5/1971 | -19.90 | 1910.18 | -41.71 |
| 29/ 5/1971 | 20.00 | 1930.18 | -4.31 |
| 5/ 6/1971 | -16.40 | 1913.78 | -30.01 |
| 12/ 6/1971 | -26.10 | 1887.68 | -72.41 |
| 19/ 6/1971 | -16.90 | 1870.78 | -103.81 |
| 26/ 6/1971 | 21.20 | 1891.98 | -75.51 |
| 3/ 7/1971 | 16.20 | 1908.18 | -51.71 |
| 10/ 7/1971 | -16.40 | 1891.78 | -49.41 |
| 17/ 7/1971 | -11.40 | 1880.38 | -49.71 |
| 24/ 7/1971 | -39.30 | 1841.08 | -66.41 |
| 31/ 7/1971 | -15.60 | 1825.48 | -62.11 |
| 7/ 8/1971 | 14.00 | 1839.48 | -68.11 |
| 14/ 8/1971 | 27.40 | 1866.88 | -24.31 |
| 21/ 8/1971 | 23.60 | 1890.48 | 25.39 |
| 28/ 8/1971 | 11.70 | 1902.18 | 53.99 |
| 4/ 9/1971 | 6.40 | 1908.58 | 39.19 |
| 11/ 9/1971 | -14.80 | 1893.78 | 8.19 |
| 18/ 9/1971 | -21.70 | 1872.08 | 2.89 |
| 25/ 9/1971 | 5.40 | 1877.48 | 19.69 |
| 2/10/1971 | 13.20 | 1890.68 | 72.19 |
| 9/10/1971 | -18.00 | 1872.68 | 69.79 |
| 16/10/1971 | -23.90 | 1848.78 | 31.89 |
| 23/10/1971 | -22.60 | 1826.18 | -13.11 |
| 30/10/1971 | -4.80 | 1821.38 | -46.51 |
| 6/11/1971 | -24.80 | 1796.58 | -83.01 |
| 13/11/1971 | -16.70 | 1779.88 | -106.11 |
| 20/11/1971 | -13.80 | 1766.08 | -105.11 |
| 27/11/1971 | 40.60 | 1806.68 | -42.81 |
| 4/12/1971 | 16.10 | 1822.78 | -32.11 |
| 11/12/1971 | 21.00 | 1843.78 | -24.31 |
| 18/12/1971 | 11.90 | 1855.68 | 5.59 |
| 25/12/1971 | 20.20 | 1875.88 | 49.69 |

52 2127

**TABLE A4-4 Market Breadth** (cont.)

| DATE | A/D VAL | A/D LINE | 10WK RAT |
|---|---|---|---|
| 1/ 1/1972 | 27.20 | 1903.08 | 99.49 |
| 8/ 1/1972 | 12.30 | 1915.38 | 116.59 |
| 15/ 1/1972 | 8.40 | 1923.78 | 149.79 |
| 22/ 1/1972 | 14.20 | 1937.98 | 180.69 |
| 29/ 1/1972 | 13.80 | 1951.78 | 208.29 |
| 5/ 2/1972 | -3.20 | 1948.58 | 164.49 |
| 12/ 2/1972 | -1.00 | 1947.58 | 147.39 |
| 19/ 2/1972 | 11.70 | 1959.28 | 138.09 |
| 26/ 2/1972 | 15.50 | 1974.78 | 141.69 |
| 4/ 3/1972 | 12.10 | 1986.88 | 133.59 |
| 11/ 3/1972 | -15.90 | 1970.98 | 90.49 |
| 18/ 3/1972 | -17.20 | 1953.78 | 60.99 |
| 25/ 3/1972 | -13.20 | 1940.58 | 39.39 |
| 1/ 4/1972 | 18.10 | 1958.68 | 43.29 |
| 8/ 4/1972 | 5.30 | 1963.98 | 34.79 |
| 15/ 4/1972 | -15.20 | 1948.78 | 22.79 |
| 22/ 4/1972 | -19.40 | 1929.38 | 4.39 |
| 29/ 4/1972 | -21.50 | 1907.88 | -28.81 |
| 6/ 5/1972 | -11.10 | 1896.78 | -55.41 |
| 13/ 5/1972 | 16.90 | 1913.68 | -50.61 |
| 20/ 5/1972 | 11.00 | 1924.68 | -23.71 |
| 27/ 5/1972 | -10.70 | 1913.98 | -17.21 |
| 3/ 6/1972 | -24.70 | 1889.28 | -28.71 |
| 10/ 6/1972 | 4.00 | 1893.28 | -42.81 |
| 17/ 6/1972 | -13.20 | 1880.08 | -61.31 |
| 24/ 6/1972 | -15.30 | 1864.78 | -61.41 |
| 1/ 7/1972 | 14.20 | 1878.98 | -27.81 |
| 8/ 7/1972 | -20.60 | 1858.38 | -26.91 |
| 15/ 7/1972 | -13.30 | 1845.08 | -29.11 |
| 22/ 7/1972 | -4.20 | 1840.88 | -50.21 |
| 29/ 7/1972 | 15.70 | 1856.58 | -45.51 |
| 5/ 8/1972 | 13.50 | 1870.08 | -21.31 |
| 12/ 8/1972 | 10.10 | 1880.18 | 13.49 |
| 19/ 8/1972 | -6.70 | 1873.48 | 2.79 |
| 26/ 8/1972 | 5.30 | 1878.78 | 21.29 |
| 2/ 9/1972 | -15.90 | 1862.88 | 20.69 |
| 9/ 9/1972 | -18.90 | 1843.98 | -12.41 |
| 16/ 9/1972 | -13.60 | 1830.38 | -5.41 |
| 23/ 9/1972 | 4.50 | 1834.88 | 12.39 |
| 30/ 9/1972 | -15.20 | 1819.68 | 1.39 |
| 7/10/1972 | -14.10 | 1805.58 | -28.41 |
| 14/10/1972 | 8.40 | 1813.98 | -33.51 |
| 21/10/1972 | 17.30 | 1831.28 | -26.31 |
| 28/10/1972 | 23.80 | 1855.08 | 4.19 |
| 4/11/1972 | 10.70 | 1865.78 | 9.59 |
| 11/11/1972 | 18.60 | 1884.38 | 44.09 |
| 18/11/1972 | 17.30 | 1901.68 | 80.29 |
| 25/11/1972 | 14.40 | 1916.08 | 108.29 |
| 2/12/1972 | 10.30 | 1926.38 | 114.09 |
| 9/12/1972 | -18.30 | 1908.08 | 110.99 |
| 16/12/1972 | -23.70 | 1884.38 | 101.39 |
| 23/12/1972 | 13.60 | 1897.98 | 106.59 |
| 30/12/1972 | 19.60 | 1917.57 | 108.89 |

53 2180

| DATE | A/D VAL | A/D LINE | 10WK RAT |
|---|---|---|---|
| 6/ 1/1973 | -14.50 | 1903.07 | 70.59 |
| 13/ 1/1973 | -19.10 | 1883.97 | 40.79 |
| 20/ 1/1973 | -26.30 | 1857.67 | -4.11 |
| 27/ 1/1973 | -21.40 | 1836.27 | -42.81 |
| 3/ 2/1973 | -13.90 | 1822.37 | -71.11 |
| 10/ 2/1973 | -6.50 | 1815.87 | -87.91 |
| 17/ 2/1973 | -19.60 | 1796.27 | -89.21 |
| 24/ 2/1973 | -20.70 | 1775.57 | -86.21 |
| 3/ 3/1973 | 15.20 | 1790.77 | -84.61 |
| 10/ 3/1973 | -16.60 | 1774.17 | -120.81 |
| 17/ 3/1973 | -32.40 | 1741.77 | -138.71 |
| 24/ 3/1973 | 18.00 | 1759.77 | -101.61 |
| 31/ 3/1973 | -20.30 | 1739.47 | -95.61 |
| 7/ 4/1973 | 17.00 | 1756.47 | -57.21 |
| 14/ 4/1973 | -9.80 | 1746.67 | -53.11 |
| 21/ 4/1973 | -26.40 | 1720.27 | -73.01 |
| 28/ 4/1973 | 17.80 | 1738.07 | -35.61 |
| 5/ 5/1973 | -17.00 | 1721.07 | -31.91 |
| 12/ 5/1973 | -32.20 | 1688.87 | -79.31 |
| 19/ 5/1973 | 10.10 | 1698.97 | -52.61 |
| 26/ 5/1973 | -24.50 | 1674.47 | -44.71 |
| 2/ 6/1973 | 10.10 | 1684.57 | -52.61 |
| 9/ 6/1973 | -10.60 | 1673.97 | -42.91 |
| 16/ 6/1973 | -21.10 | 1652.87 | -81.01 |
| 23/ 6/1973 | -6.60 | 1646.27 | -77.81 |
| 30/ 6/1973 | -15.00 | 1631.27 | -66.41 |
| 7/ 7/1973 | 25.20 | 1656.47 | -59.01 |
| 14/ 7/1973 | 22.30 | 1678.77 | -19.71 |
| 21/ 7/1973 | 13.60 | 1692.37 | 26.09 |
| 28/ 7/1973 | -23.60 | 1668.77 | -7.61 |
| 4/ 8/1973 | -19.60 | 1649.17 | -2.71 |
| 11/ 8/1973 | -19.00 | 1630.17 | -31.81 |
| 18/ 8/1973 | -14.00 | 1616.17 | -35.21 |
| 25/ 8/1973 | 22.50 | 1638.67 | 8.39 |
| 1/ 9/1973 | 19.60 | 1658.27 | 34.59 |
| 8/ 9/1973 | -10.60 | 1647.67 | 38.99 |
| 15/ 9/1973 | 28.80 | 1676.47 | 42.59 |
| 22/ 9/1973 | 23.40 | 1699.87 | 43.69 |
| 29/ 9/1973 | 25.10 | 1724.97 | 55.19 |
| 6/10/1973 | 12.60 | 1737.57 | 91.39 |
| 13/10/1973 | -20.00 | 1717.57 | 90.99 |
| 20/10/1973 | -12.00 | 1705.57 | 97.99 |
| 27/10/1973 | -27.90 | 1677.67 | 84.09 |
| 3/11/1973 | -21.50 | 1656.17 | 40.09 |
| 10/11/1973 | -27.70 | 1628.47 | -7.21 |
| 17/11/1973 | -30.40 | 1598.07 | -27.01 |
| 24/11/1973 | -29.10 | 1568.97 | -84.91 |
| 1/12/1973 | -15.30 | 1553.67 | -123.61 |
| 8/12/1973 | -22.00 | 1531.67 | -170.71 |
| 15/12/1973 | -8.30 | 1523.37 | -191.61 |
| 22/12/1973 | 19.50 | 1542.87 | -152.11 |
| 29/12/1973 | 37.40 | 1580.27 | -102.71 |

52 2232

**TABLE A4-4 Market Breadth** (cont.)

| DATE | A/D VAL | A/D LINE | 10WK RAT |
|---|---|---|---|
| 5/ 1/1974 | -18.90 | 1561.37 | -93.71 |
| 12/ 1/1974 | 17.60 | 1578.97 | -54.61 |
| 19/ 1/1974 | 11.30 | 1590.27 | -15.61 |
| 26/ 1/1974 | -7.80 | 1582.47 | 6.99 |
| 2/ 2/1974 | -18.50 | 1563.97 | 17.59 |
| 9/ 2/1974 | -5.70 | 1558.27 | 27.19 |
| 16/ 2/1974 | 20.80 | 1579.07 | 69.99 |
| 23/ 2/1974 | 14.10 | 1593.17 | 92.39 |
| 2/ 3/1974 | 19.20 | 1612.37 | 92.09 |
| 9/ 3/1974 | 13.90 | 1626.27 | 68.59 |
| 16/ 3/1974 | -21.90 | 1604.37 | 65.59 |
| 23/ 3/1974 | -26.50 | 1577.87 | 21.49 |
| 30/ 3/1974 | -16.00 | 1561.87 | -5.81 |
| 6/ 4/1974 | -14.50 | 1547.37 | -12.51 |
| 13/ 4/1974 | 10.80 | 1558.17 | 16.79 |
| 20/ 4/1974 | -32.30 | 1525.87 | -9.81 |
| 27/ 4/1974 | 10.40 | 1536.27 | -20.21 |
| 4/ 5/1974 | -13.90 | 1522.37 | -48.21 |
| 11/ 5/1974 | -26.30 | 1496.07 | -93.71 |
| 18/ 5/1974 | -15.60 | 1480.47 | -123.21 |
| 25/ 5/1974 | -15.00 | 1465.47 | -116.31 |
| 1/ 6/1974 | 30.20 | 1495.67 | -59.61 |
| 8/ 6/1974 | -17.80 | 1477.87 | -61.41 |
| 15/ 6/1974 | -27.80 | 1450.07 | -74.71 |
| 22/ 6/1974 | -20.30 | 1429.77 | -105.81 |
| 29/ 6/1974 | -19.60 | 1410.17 | -93.11 |
| 6/ 7/1974 | -15.30 | 1394.87 | -118.81 |
| 13/ 7/1974 | 13.10 | 1407.97 | -91.81 |
| 20/ 7/1974 | 6.90 | 1414.87 | -58.61 |
| 27/ 7/1974 | -25.40 | 1389.47 | -68.41 |
| 3/ 8/1974 | 21.40 | 1410.87 | -32.01 |
| 10/ 8/1974 | -26.90 | 1383.97 | -89.11 |
| 17/ 8/1974 | -26.90 | 1357.07 | -98.21 |
| 24/ 8/1974 | -19.00 | 1338.07 | -89.41 |
| 31/ 8/1974 | -13.70 | 1324.37 | -82.81 |
| 7/ 9/1974 | -36.30 | 1288.07 | -99.51 |
| 14/ 9/1974 | 23.20 | 1311.27 | -61.01 |
| 21/ 9/1974 | -16.20 | 1295.07 | -90.31 |
| 28/ 9/1974 | -17.90 | 1277.17 | -115.11 |
| 5/10/1974 | 40.20 | 1317.37 | -49.51 |
| 12/10/1974 | 11.00 | 1328.37 | -59.91 |
| 19/10/1974 | -16.80 | 1311.57 | -49.81 |
| 26/10/1974 | 14.40 | 1325.97 | -8.51 |
| 2/11/1974 | 14.90 | 1340.87 | 25.39 |
| 9/11/1974 | -14.80 | 1326.07 | 24.29 |
| 16/11/1974 | -25.70 | 1300.37 | 34.89 |
| 23/11/1974 | 4.60 | 1304.97 | 16.29 |
| 30/11/1974 | -34.00 | 1270.97 | -1.51 |
| 7/12/1974 | -6.60 | 1264.37 | 9.79 |
| 14/12/1974 | -12.90 | 1251.47 | -43.31 |
| 21/12/1974 | -2.80 | 1248.67 | -57.11 |
| 28/12/1974 | 36.00 | 1284.67 | -4.31 |

52 2284

| DATE | A/D VAL | A/D LINE | 10WK RAT |
|---|---|---|---|
| 4/ 1/1975 | 40.00 | 1324.67 | 21.29 |
| 11/ 1/1975 | 12.30 | 1336.97 | 18.69 |
| 18/ 1/1975 | 18.10 | 1355.07 | 51.59 |
| 25/ 1/1975 | 32.60 | 1387.67 | 109.89 |
| 1/ 2/1975 | 18.40 | 1406.07 | 123.69 |
| 8/ 2/1975 | 16.20 | 1422.27 | 173.89 |
| 15/ 2/1975 | 9.20 | 1431.47 | 189.69 |
| 22/ 2/1975 | -15.40 | 1416.06 | 187.19 |
| 1/ 3/1975 | 24.00 | 1440.06 | 213.99 |
| 8/ 3/1975 | 15.10 | 1455.16 | 193.09 |
| 15/ 3/1975 | -12.10 | 1443.06 | 140.99 |
| 22/ 3/1975 | -9.40 | 1433.66 | 119.29 |
| 29/ 3/1975 | -20.50 | 1413.16 | 80.69 |
| 5/ 4/1975 | 12.60 | 1425.76 | 60.69 |
| 12/ 4/1975 | 16.30 | 1442.06 | 58.59 |
| 19/ 4/1975 | 4.50 | 1446.56 | 46.89 |
| 26/ 4/1975 | 9.00 | 1455.56 | 46.69 |
| 3/ 5/1975 | 17.00 | 1472.56 | 79.09 |
| 10/ 5/1975 | 7.60 | 1480.16 | 62.69 |
| 17/ 5/1975 | -5.50 | 1474.66 | 42.09 |
| 24/ 5/1975 | 10.70 | 1485.36 | 64.89 |
| 31/ 5/1975 | 17.70 | 1503.06 | 91.99 |
| 7/ 6/1975 | -14.80 | 1488.26 | 97.69 |
| 14/ 6/1975 | 16.90 | 1505.16 | 101.99 |
| 21/ 6/1975 | 20.20 | 1525.36 | 105.89 |
| 28/ 6/1975 | 0.0 | 1525.36 | 101.39 |
| 5/ 7/1975 | 13.90 | 1539.26 | 106.29 |
| 12/ 7/1975 | -3.50 | 1535.76 | 85.79 |
| 19/ 7/1975 | -29.70 | 1506.06 | 48.49 |
| 26/ 7/1975 | -17.80 | 1488.26 | 36.19 |
| 2/ 8/1975 | -23.40 | 1464.86 | 2.09 |
| 9/ 8/1975 | -10.60 | 1454.26 | -26.21 |
| 16/ 8/1975 | -20.40 | 1433.86 | -31.81 |
| 23/ 8/1975 | 22.00 | 1455.86 | -26.71 |
| 30/ 8/1975 | -14.20 | 1441.66 | -61.11 |
| 6/ 9/1975 | -18.90 | 1422.76 | -80.01 |
| 13/ 9/1975 | 11.50 | 1434.26 | -82.41 |
| 20/ 9/1975 | 4.60 | 1438.86 | -74.31 |
| 27/ 9/1975 | -13.30 | 1425.56 | -57.91 |
| 4/10/1975 | 14.60 | 1440.16 | -25.51 |
| 11/10/1975 | 11.80 | 1451.96 | 9.69 |
| 18/10/1975 | 14.40 | 1466.36 | 34.69 |
| 25/10/1975 | -10.30 | 1456.06 | 44.79 |
| 1/11/1975 | 10.30 | 1466.36 | 33.09 |
| 8/11/1975 | 16.40 | 1482.76 | 63.69 |
| 15/11/1975 | -12.40 | 1470.36 | 70.19 |
| 22/11/1975 | 14.10 | 1484.46 | 72.79 |
| 29/11/1975 | -25.80 | 1458.66 | 42.39 |
| 6/12/1975 | -6.40 | 1452.26 | 49.29 |
| 13/12/1975 | 12.80 | 1465.06 | 47.49 |
| 20/12/1975 | 16.20 | 1481.26 | 51.89 |
| 27/12/1975 | 19.30 | 1500.56 | 56.79 |

52     2336

**TABLE A4-4 Market Breadth** (cont.)

| DATE | A/D VAL | A/D LINE | 10WK RAT |
|---|---|---|---|
| 3/ 1/1976 | 40.50 | 1541.06 | 107.59 |
| 10/ 1/1976 | 24.70 | 1565.76 | 121.99 |
| 17/ 1/1976 | 23.30 | 1589.06 | 128.89 |
| 24/ 1/1976 | 20.40 | 1609.46 | 161.69 |
| 31/ 1/1976 | 12.20 | 1621.66 | 159.79 |
| 7/ 2/1976 | 19.30 | 1640.96 | 204.89 |
| 14/ 2/1976 | 22.20 | 1663.16 | 233.49 |
| 21/ 2/1976 | -18.90 | 1644.26 | 201.79 |
| 28/ 2/1976 | -8.00 | 1636.26 | 177.59 |
| 6/ 3/1976 | 10.50 | 1646.76 | 168.79 |
| 13/ 3/1976 | -13.20 | 1633.56 | 115.09 |
| 20/ 3/1976 | 13.40 | 1646.96 | 103.79 |
| 27/ 3/1976 | -11.60 | 1635.36 | 68.89 |
| 3/ 4/1976 | -18.20 | 1617.16 | 30.29 |
| 10/ 4/1976 | -4.00 | 1613.16 | 14.09 |
| 17/ 4/1976 | 17.80 | 1630.96 | 12.59 |
| 24/ 4/1976 | -8.80 | 1622.16 | -18.41 |
| 1/ 5/1976 | -5.70 | 1616.46 | -5.21 |
| 8/ 5/1976 | 5.70 | 1622.16 | 8.49 |
| 15/ 5/1976 | -7.70 | 1614.46 | -9.71 |
| 22/ 5/1976 | -19.10 | 1595.36 | -15.61 |
| 29/ 5/1976 | -10.40 | 1584.96 | -39.41 |
| 5/ 6/1976 | 11.60 | 1596.56 | -16.21 |
| 12/ 6/1976 | 21.70 | 1618.26 | 23.69 |
| 19/ 6/1976 | 7.40 | 1625.66 | 35.09 |
| 26/ 6/1976 | 12.60 | 1638.26 | 29.89 |
| 3/ 7/1976 | 14.10 | 1652.36 | 52.79 |
| 10/ 7/1976 | 8.80 | 1661.16 | 67.29 |
| 17/ 7/1976 | -12.00 | 1649.16 | 49.59 |
| 24/ 7/1976 | -13.50 | 1635.66 | 43.79 |
| 31/ 7/1976 | 11.30 | 1646.96 | 74.19 |
| 7/ 8/1976 | 7.10 | 1654.06 | 91.69 |
| 14/ 8/1976 | -16.10 | 1637.96 | 63.99 |
| 21/ 8/1976 | -11.40 | 1626.56 | 30.89 |
| 28/ 8/1976 | 20.80 | 1647.36 | 44.29 |
| 4/ 9/1976 | 10.50 | 1657.86 | 42.19 |
| 11/ 9/1976 | 14.00 | 1671.86 | 42.09 |
| 18/ 9/1976 | 12.80 | 1684.66 | 46.09 |
| 25/ 9/1976 | -17.50 | 1667.16 | 40.59 |
| 2/10/1976 | -15.40 | 1651.76 | 38.69 |
| 9/10/1976 | -16.60 | 1635.16 | 10.79 |
| 16/10/1976 | -7.10 | 1628.06 | -3.41 |
| 23/10/1976 | 17.10 | 1645.16 | 29.79 |
| 30/10/1976 | -3.30 | 1641.86 | 37.89 |
| 6/11/1976 | -16.20 | 1625.66 | 0.89 |
| 13/11/1976 | 21.20 | 1646.86 | 11.59 |
| 20/11/1976 | 18.40 | 1665.26 | 15.99 |
| 27/11/1976 | 12.80 | 1678.06 | 15.99 |
| 4/12/1976 | 26.70 | 1704.76 | 60.19 |
| 11/12/1976 | 11.20 | 1715.96 | 86.79 |
| 18/12/1976 | 1.70 | 1717.66 | 105.09 |
| 25/12/1976 | 26.60 | 1744.26 | 138.79 |

52 2388

| DATE | A/D VAL | A/D LINE | 10WK RAT |
|---|---|---|---|
| 1/ 1/1977 | 5.70 | 1749.96 | 127.39 |
| 8/ 1/1977 | -10.90 | 1739.06 | 119.79 |
| 15/ 1/1977 | 5.10 | 1744.16 | 141.09 |
| 22/ 1/1977 | -9.30 | 1734.85 | 110.59 |
| 29/ 1/1977 | 1.40 | 1736.25 | 93.59 |
| 5/ 2/1977 | -13.50 | 1722.75 | 67.29 |
| 12/ 2/1977 | 4.50 | 1727.25 | 45.09 |
| 19/ 2/1977 | -15.60 | 1711.65 | 18.29 |
| 26/ 2/1977 | 16.30 | 1727.95 | 32.89 |
| 5/ 3/1977 | 1.40 | 1729.35 | 7.69 |
| 12/ 3/1977 | 13.60 | 1742.95 | 15.59 |
| 19/ 3/1977 | -17.10 | 1725.85 | 9.39 |
| 26/ 3/1977 | -8.60 | 1717.25 | -4.31 |
| 2/ 4/1977 | -13.20 | 1704.05 | -8.21 |
| 9/ 4/1977 | 20.40 | 1724.45 | 10.79 |
| 16/ 4/1977 | -13.00 | 1711.45 | 11.29 |
| 23/ 4/1977 | 0.0 | 1711.45 | 6.79 |
| 30/ 4/1977 | 16.70 | 1728.15 | 39.09 |
| 7/ 5/1977 | 4.40 | 1732.55 | 27.19 |
| 14/ 5/1977 | 11.50 | 1744.05 | 37.29 |
| 21/ 5/1977 | -23.10 | 1720.95 | 0.59 |
| 28/ 5/1977 | 11.90 | 1732.85 | 29.59 |
| 4/ 6/1977 | 14.30 | 1747.15 | 52.49 |
| 11/ 6/1977 | 17.30 | 1764.45 | 82.99 |
| 18/ 6/1977 | 18.20 | 1782.65 | 80.79 |
| 25/ 6/1977 | -2.00 | 1780.65 | 91.79 |
| 2/ 7/1977 | 9.20 | 1789.85 | 100.99 |
| 9/ 7/1977 | 6.30 | 1796.15 | 90.59 |
| 16/ 7/1977 | 13.90 | 1810.05 | 100.09 |
| 23/ 7/1977 | -23.20 | 1786.85 | 65.39 |
| 30/ 7/1977 | -3.80 | 1783.05 | 84.69 |
| 6/ 8/1977 | -11.30 | 1771.75 | 61.49 |
| 13/ 8/1977 | -10.00 | 1761.75 | 37.19 |

**TABLE A4-5 Raw Data Listing of Yields**

| YEAR | MONTH | COMMP* | CORP AAA† | S & P‡ |
|---|---|---|---|---|
| 1919 | JAN | 5.25 | 5.35 | 0.0 |
| 1919 | FEB | 5.13 | 5.35 | 0.0 |
| 1919 | MAR | 5.50 | 5.39 | 0.0 |
| 1919 | APR | 5.38 | 5.44 | 0.0 |
| 1919 | MAY | 5.25 | 5.39 | 0.0 |
| 1919 | JUN | 5.38 | 5.40 | 0.0 |
| 1919 | JUL | 5.38 | 5.44 | 0.0 |
| 1919 | AUG | 5.38 | 5.56 | 0.0 |
| 1919 | SEP | 5.38 | 5.60 | 0.0 |
| 1919 | OCT | 5.25 | 5.54 | 0.0 |
| 1919 | NOV | 5.25 | 5.66 | 0.0 |
| 1919 | DEC | 5.88 | 5.73 | 0.0 |
| 1920 | JAN | 6.00 | 5.75 | 0.0 |
| 1920 | FEB | 6.38 | 5.86 | 0.0 |
| 1920 | MAR | 6.88 | 5.92 | 0.0 |
| 1920 | APR | 6.88 | 6.04 | 0.0 |
| 1920 | MAY | 7.38 | 6.25 | 0.0 |
| 1920 | JUN | 7.88 | 6.38 | 0.0 |
| 1920 | JUL | 8.13 | 6.34 | 0.0 |
| 1920 | AUG | 8.13 | 6.30 | 0.0 |
| 1920 | SEP | 8.13 | 6.22 | 0.0 |
| 1920 | OCT | 8.13 | 6.05 | 0.0 |
| 1920 | NOV | 8.13 | 6.08 | 0.0 |
| 1920 | DEC | 8.00 | 6.26 | 0.0 |
| 1921 | JAN | 7.88 | 6.14 | 0.0 |
| 1921 | FEB | 7.63 | 6.08 | 0.0 |
| 1921 | MAR | 7.63 | 6.08 | 0.0 |
| 1921 | APR | 7.63 | 6.06 | 0.0 |
| 1921 | MAY | 6.88 | 6.11 | 0.0 |
| 1921 | JUN | 6.75 | 6.18 | 0.0 |
| 1921 | JUL | 6.38 | 6.12 | 0.0 |
| 1921 | AUG | 6.13 | 5.99 | 0.0 |
| 1921 | SEP | 6.00 | 5.93 | 0.0 |
| 1921 | OCT | 5.88 | 5.84 | 0.0 |
| 1921 | NOV | 5.50 | 5.60 | 0.0 |
| 1921 | DEC | 5.13 | 5.50 | 0.0 |
| 1922 | JAN | 5.00 | 5.34 | 0.0 |
| 1922 | FEB | 4.88 | 5.29 | 0.0 |
| 1922 | MAR | 4.75 | 5.23 | 0.0 |
| 1922 | APR | 4.50 | 5.15 | 0.0 |
| 1922 | MAY | 4.38 | 5.13 | 0.0 |
| 1922 | JUN | 4.38 | 5.08 | 0.0 |
| 1922 | JUL | 4.13 | 5.00 | 0.0 |
| 1922 | AUG | 4.13 | 4.96 | 0.0 |
| 1922 | SEP | 4.13 | 4.93 | 0.0 |
| 1922 | OCT | 4.38 | 4.97 | 0.0 |
| 1922 | NOV | 4.75 | 5.09 | 0.0 |
| 1922 | DEC | 4.88 | 5.08 | 0.0 |

* 4- to 6-Month Commercial Paper Yield.

† Moody's Corporate AAA Bond Yields.

‡ Standard and Poor's 500 Stock Yield.

| YEAR | MONTH | COMMP | CORP AAA | S & P |
|---|---|---|---|---|
| 1923 | JAN | 4.50 | 5.04 | 0.0 |
| 1923 | FEB | 4.63 | 5.07 | 0.0 |
| 1923 | MAR | 5.13 | 5.18 | 0.0 |
| 1923 | APR | 5.38 | 5.22 | 0.0 |
| 1923 | MAY | 5.00 | 5.16 | 0.0 |
| 1923 | JUN | 5.00 | 5.15 | 0.0 |
| 1923 | JUL | 5.13 | 5.14 | 0.0 |
| 1923 | AUG | 5.13 | 5.08 | 0.0 |
| 1923 | SEP | 5.38 | 5.12 | 0.0 |
| 1923 | OCT | 5.38 | 5.11 | 0.0 |
| 1923 | NOV | 5.13 | 5.09 | 0.0 |
| 1923 | DEC | 5.00 | 5.09 | 0.0 |
| 1924 | JAN | 4.88 | 5.09 | 0.0 |
| 1924 | FEB | 4.88 | 5.09 | 0.0 |
| 1924 | MAR | 4.88 | 5.10 | 0.0 |
| 1924 | APR | 4.63 | 5.08 | 0.0 |
| 1924 | MAY | 4.50 | 5.04 | 0.0 |
| 1924 | JUN | 4.13 | 4.99 | 0.0 |
| 1924 | JUL | 3.50 | 4.95 | 0.0 |
| 1924 | AUG | 3.25 | 4.95 | 0.0 |
| 1924 | SEP | 3.13 | 4.95 | 0.0 |
| 1924 | OCT | 3.13 | 4.92 | 0.0 |
| 1924 | NOV | 3.25 | 4.94 | 0.0 |
| 1924 | DEC | 3.63 | 4.95 | 0.0 |
| 1925 | JAN | 3.63 | 4.95 | 0.0 |
| 1925 | FEB | 3.63 | 4.95 | 0.0 |
| 1925 | MAR | 4.00 | 4.91 | 0.0 |
| 1925 | APR | 4.00 | 4.87 | 0.0 |
| 1925 | MAY | 3.88 | 4.83 | 0.0 |
| 1925 | JUN | 3.88 | 4.83 | 0.0 |
| 1925 | JUL | 3.88 | 4.87 | 0.0 |
| 1925 | AUG | 4.00 | 4.90 | 0.0 |
| 1925 | SEP | 4.25 | 4.87 | 0.0 |
| 1925 | OCT | 4.38 | 4.85 | 0.0 |
| 1925 | NOV | 4.38 | 4.84 | 0.0 |
| 1925 | DEC | 4.38 | 4.85 | 0.0 |
| 1926 | JAN | 4.38 | 4.82 | 4.46 |
| 1926 | FEB | 4.25 | 4.77 | 4.51 |
| 1926 | MAR | 4.38 | 4.79 | 4.92 |
| 1926 | APR | 4.38 | 4.74 | 5.17 |
| 1926 | MAY | 4.00 | 4.71 | 5.26 |
| 1926 | JUN | 4.00 | 4.72 | 5.24 |
| 1926 | JUL | 4.00 | 4.71 | 5.09 |
| 1926 | AUG | 4.38 | 4.72 | 4.92 |
| 1926 | SEP | 4.63 | 4.72 | 4.86 |
| 1926 | OCT | 4.63 | 4.71 | 4.97 |
| 1926 | NOV | 4.50 | 4.68 | 4.93 |
| 1926 | DEC | 4.50 | 4.68 | 4.94 |

**TABLE A4-5 Raw Data Listing of Yields** (cont.)

| YEAR | MONTH | COMMP | CORP AAA | S & P |
|---|---|---|---|---|
| 1927 | JAN | 4.25 | 4.66 | 5.02 |
| 1927 | FEB | 4.13 | 4.67 | 5.01 |
| 1927 | MAR | 4.13 | 4.62 | 5.07 |
| 1927 | APR | 4.13 | 4.58 | 4.97 |
| 1927 | MAY | 4.13 | 4.57 | 4.85 |
| 1927 | JUN | 4.25 | 4.58 | 4.97 |
| 1927 | JUL | 4.25 | 4.60 | 4.91 |
| 1927 | AUG | 4.00 | 4.56 | 4.68 |
| 1927 | SEP | 4.00 | 4.54 | 4.42 |
| 1927 | OCT | 4.00 | 4.51 | 4.47 |
| 1927 | NOV | 4.00 | 4.49 | 4.37 |
| 1927 | DEC | 4.00 | 4.46 | 4.32 |
| 1928 | JAN | 4.00 | 4.46 | 4.31 |
| 1928 | FEB | 4.00 | 4.46 | 4.39 |
| 1928 | MAR | 4.13 | 4.46 | 4.20 |
| 1928 | APR | 4.38 | 4.46 | 4.05 |
| 1928 | MAY | 4.50 | 4.49 | 3.93 |
| 1928 | JUN | 4.75 | 4.57 | 4.13 |
| 1928 | JUL | 5.13 | 4.61 | 4.14 |
| 1928 | AUG | 5.38 | 4.64 | 4.01 |
| 1928 | SEP | 5.63 | 4.61 | 3.78 |
| 1928 | OCT | 5.50 | 4.61 | 3.72 |
| 1928 | NOV | 5.38 | 4.58 | 3.50 |
| 1928 | DEC | 5.38 | 4.61 | 3.60 |
| 1929 | JAN | 5.38 | 4.62 | 3.36 |
| 1929 | FEB | 5.50 | 4.66 | 3.38 |
| 1929 | MAR | 5.88 | 4.70 | 3.31 |
| 1929 | APR | 6.00 | 4.69 | 3.35 |
| 1929 | MAY | 6.00 | 4.70 | 3.37 |
| 1929 | JUN | 6.00 | 4.77 | 3.38 |
| 1929 | JUL | 6.00 | 4.77 | 3.16 |
| 1929 | AUG | 6.13 | 4.79 | 3.01 |
| 1929 | SEP | 6.25 | 4.80 | 2.92 |
| 1929 | OCT | 6.25 | 4.77 | 3.33 |
| 1929 | NOV | 5.75 | 4.76 | 4.54 |
| 1929 | DEC | 5.00 | 4.67 | 4.48 |
| 1930 | JAN | 4.88 | 4.66 | 4.38 |
| 1930 | FEB | 4.75 | 4.69 | 4.14 |
| 1930 | MAR | 4.25 | 4.62 | 3.99 |
| 1930 | APR | 3.88 | 4.60 | 3.78 |
| 1930 | MAY | 3.75 | 4.60 | 4.02 |
| 1930 | JUN | 3.50 | 4.57 | 4.50 |
| 1930 | JUL | 3.25 | 4.52 | 4.44 |
| 1930 | AUG | 3.00 | 4.47 | 4.42 |
| 1930 | SEP | 3.00 | 4.42 | 4.33 |
| 1930 | OCT | 3.00 | 4.42 | 5.05 |
| 1930 | NOV | 2.88 | 4.47 | 5.43 |
| 1930 | DEC | 2.88 | 4.52 | 5.62 |

| YEAR | MONTH | COMMP | CORP AAA | S & P |
|---|---|---|---|---|
| 1931 | JAN | 2.88 | 4.42 | 5.50 |
| 1931 | FEB | 2.63 | 4.43 | 5.12 |
| 1931 | MAR | 2.50 | 4.39 | 4.95 |
| 1931 | APR | 2.38 | 4.40 | 5.43 |
| 1931 | MAY | 2.25 | 4.37 | 5.95 |
| 1931 | JUN | 2.00 | 4.36 | 5.96 |
| 1931 | JUL | 2.00 | 4.36 | 5.66 |
| 1931 | AUG | 2.00 | 4.40 | 5.74 |
| 1931 | SEP | 2.00 | 4.55 | 6.51 |
| 1931 | OCT | 3.13 | 4.99 | 7.28 |
| 1931 | NOV | 4.00 | 4.94 | 7.06 |
| 1931 | DEC | 3.88 | 5.32 | 8.68 |
| 1932 | JAN | 3.88 | 5.20 | 8.22 |
| 1932 | FEB | 3.88 | 5.23 | 8.04 |
| 1932 | MAR | 3.63 | 4.98 | 7.16 |
| 1932 | APR | 3.50 | 5.17 | 9.13 |
| 1932 | MAY | 3.13 | 5.36 | 9.57 |
| 1932 | JUN | 2.75 | 5.41 | 10.30 |
| 1932 | JUL | 2.50 | 5.26 | 8.85 |
| 1932 | AUG | 2.25 | 4.91 | 5.65 |
| 1932 | SEP | 2.13 | 4.70 | 4.94 |
| 1932 | OCT | 2.00 | 4.64 | 5.73 |
| 1932 | NOV | 1.63 | 4.63 | 5.84 |
| 1932 | DEC | 1.50 | 4.59 | 5.68 |
| 1933 | JAN | 1.38 | 4.44 | 5.42 |
| 1933 | FEB | 1.38 | 4.48 | 6.04 |
| 1933 | MAR | 3.00 | 4.68 | 6.18 |
| 1933 | APR | 2.63 | 4.78 | 5.47 |
| 1933 | MAY | 2.13 | 4.63 | 3.90 |
| 1933 | JUN | 1.88 | 4.46 | 3.20 |
| 1933 | JUL | 1.63 | 4.36 | 2.95 |
| 1933 | AUG | 1.50 | 4.30 | 3.19 |
| 1933 | SEP | 1.38 | 4.36 | 3.30 |
| 1933 | OCT | 1.25 | 4.34 | 3.59 |
| 1933 | NOV | 1.25 | 4.54 | 3.65 |
| 1933 | DEC | 1.38 | 4.50 | 3.59 |
| 1934 | JAN | 1.50 | 4.35 | 3.36 |
| 1934 | FEB | 1.38 | 4.20 | 3.10 |
| 1934 | MAR | 1.25 | 4.13 | 3.33 |
| 1934 | APR | 1.00 | 4.07 | 3.25 |
| 1934 | MAY | 1.00 | 4.01 | 3.58 |
| 1934 | JUN | 0.88 | 3.93 | 3.55 |
| 1934 | JUL | 0.88 | 3.89 | 3.67 |
| 1934 | AUG | 0.88 | 3.93 | 4.00 |
| 1934 | SEP | 0.88 | 3.96 | 4.21 |
| 1934 | OCT | 0.88 | 3.90 | 4.22 |
| 1934 | NOV | 0.88 | 3.86 | 4.14 |
| 1934 | DEC | 0.88 | 3.81 | 4.25 |

**TABLE A4-5 Raw Data Listing of Yields** (cont.)

| YEAR | MONTH | COMMP | CORP AAA | S & P |
|---|---|---|---|---|
| 1935 | JAN | 0.88 | 3.77 | 4.24 |
| 1935 | FEB | 0.75 | 3.69 | 4.24 |
| 1935 | MAR | 0.75 | 3.67 | 4.51 |
| 1935 | APR | 0.75 | 3.66 | 4.35 |
| 1935 | MAY | 0.75 | 3.65 | 4.00 |
| 1935 | JUN | 0.75 | 3.61 | 3.82 |
| 1935 | JUL | 0.75 | 3.56 | 3.58 |
| 1935 | AUG | 0.75 | 3.60 | 3.48 |
| 1935 | SEP | 0.75 | 3.59 | 3.51 |
| 1935 | OCT | 0.75 | 3.52 | 3.50 |
| 1935 | NOV | 0.75 | 3.47 | 3.24 |
| 1935 | DEC | 0.75 | 3.44 | 3.38 |
| 1936 | JAN | 0.75 | 3.37 | 3.27 |
| 1936 | FEB | 0.75 | 3.32 | 3.14 |
| 1936 | MAR | 0.75 | 3.29 | 3.11 |
| 1936 | APR | 0.75 | 3.29 | 3.19 |
| 1936 | MAY | 0.75 | 3.27 | 3.46 |
| 1936 | JUN | 0.75 | 3.24 | 3.51 |
| 1936 | JUL | 0.75 | 3.23 | 3.37 |
| 1936 | AUG | 0.75 | 3.21 | 3.40 |
| 1936 | SEP | 0.75 | 3.18 | 3.56 |
| 1936 | OCT | 0.75 | 3.18 | 3.41 |
| 1936 | NOV | 0.75 | 3.15 | 3.52 |
| 1936 | DEC | 0.75 | 3.10 | 4.36 |
| 1937 | JAN | 0.75 | 3.10 | 4.34 |
| 1937 | FEB | 0.75 | 3.22 | 3.83 |
| 1937 | MAR | 0.75 | 3.32 | 3.47 |
| 1937 | APR | 1.00 | 3.42 | 3.74 |
| 1937 | MAY | 1.00 | 3.33 | 4.04 |
| 1937 | JUN | 1.00 | 3.28 | 4.41 |
| 1937 | JUL | 1.00 | 3.25 | 4.18 |
| 1937 | AUG | 1.00 | 3.24 | 4.23 |
| 1937 | SEP | 1.00 | 3.28 | 5.06 |
| 1937 | OCT | 1.00 | 3.27 | 6.01 |
| 1937 | NOV | 1.00 | 3.24 | 7.01 |
| 1937 | DEC | 1.00 | 3.21 | 7.94 |
| 1938 | JAN | 1.00 | 3.17 | 7.84 |
| 1938 | FEB | 1.00 | 3.20 | 7.52 |
| 1938 | MAR | 0.88 | 3.22 | 6.02 |
| 1938 | APR | 0.88 | 3.30 | 5.84 |
| 1938 | MAY | 0.88 | 3.22 | 5.68 |
| 1938 | JUN | 0.88 | 3.26 | 5.40 |
| 1938 | JUL | 0.75 | 3.22 | 4.35 |
| 1938 | AUG | 0.75 | 3.18 | 4.26 |
| 1938 | SEP | 0.69 | 3.21 | 4.30 |
| 1938 | OCT | 0.69 | 3.15 | 3.77 |
| 1938 | NOV | 0.69 | 3.10 | 3.59 |
| 1938 | DEC | 0.63 | 3.08 | 3.58 |

| YEAR | MONTH | COMMP | CORP AAA | S & P |
|---|---|---|---|---|
| 1939 | JAN | 0.56 | 3.01 | 3.49 |
| 1939 | FEB | 0.56 | 3.00 | 3.61 |
| 1939 | MAR | 0.56 | 2.99 | 3.69 |
| 1939 | APR | 0.56 | 3.02 | 4.23 |
| 1939 | MAY | 0.56 | 2.97 | 4.25 |
| 1939 | JUN | 0.56 | 2.92 | 4.33 |
| 1939 | JUL | 0.56 | 2.89 | 4.26 |
| 1939 | AUG | 0.56 | 2.93 | 4.33 |
| 1939 | SEP | 0.69 | 3.25 | 3.91 |
| 1939 | OCT | 0.69 | 3.15 | 3.87 |
| 1939 | NOV | 0.63 | 3.00 | 4.12 |
| 1939 | DEC | 0.56 | 2.94 | 4.54 |
| 1940 | JAN | 0.56 | 2.88 | 4.58 |
| 1940 | FEB | 0.56 | 2.86 | 4.65 |
| 1940 | MAR | 0.56 | 2.84 | 4.70 |
| 1940 | APR | 0.56 | 2.82 | 4.73 |
| 1940 | MAY | 0.56 | 2.93 | 5.69 |
| 1940 | JUN | 0.56 | 2.96 | 6.31 |
| 1940 | JUL | 0.56 | 2.88 | 6.08 |
| 1940 | AUG | 0.56 | 2.85 | 6.11 |
| 1940 | SEP | 0.56 | 2.82 | 5.95 |
| 1940 | OCT | 0.56 | 2.79 | 5.92 |
| 1940 | NOV | 0.56 | 2.75 | 5.98 |
| 1940 | DEC | 0.56 | 2.71 | 6.37 |
| 1941 | JAN | 0.56 | 2.75 | 6.38 |
| 1941 | FEB | 0.56 | 2.78 | 6.61 |
| 1941 | MAR | 0.56 | 2.80 | 6.52 |
| 1941 | APR | 0.56 | 2.82 | 6.77 |
| 1941 | MAY | 0.56 | 2.81 | 6.91 |
| 1941 | JUN | 0.54 | 2.77 | 6.79 |
| 1941 | JUL | 0.50 | 2.74 | 6.43 |
| 1941 | AUG | 0.50 | 2.74 | 6.51 |
| 1941 | SEP | 0.50 | 2.75 | 6.56 |
| 1941 | OCT | 0.50 | 2.73 | 6.90 |
| 1941 | NOV | 0.50 | 2.72 | 7.37 |
| 1941 | DEC | 0.55 | 2.80 | 8.13 |
| 1942 | JAN | 0.59 | 2.83 | 7.88 |
| 1942 | FEB | 0.63 | 2.85 | 8.05 |
| 1942 | MAR | 0.63 | 2.86 | 8.15 |
| 1942 | APR | 0.63 | 2.83 | 8.47 |
| 1942 | MAY | 0.63 | 2.85 | 8.21 |
| 1942 | JUN | 0.67 | 2.85 | 7.37 |
| 1942 | JUL | 0.69 | 2.83 | 7.07 |
| 1942 | AUG | 0.69 | 2.81 | 6.98 |
| 1942 | SEP | 0.69 | 2.80 | 6.67 |
| 1942 | OCT | 0.69 | 2.80 | 6.18 |
| 1942 | NOV | 0.69 | 2.79 | 6.00 |
| 1942 | DEC | 0.69 | 2.81 | 5.84 |

**TABLE A4-5 Raw Data Listing of Yields** (cont.)

| YEAR | MONTH | COMMP | CORP AAA | S & P |
|---|---|---|---|---|
| 1943 | JAN | 0.69 | 2.79 | 5.54 |
| 1943 | FEB | 0.69 | 2.77 | 5.21 |
| 1943 | MAR | 0.69 | 2.76 | 5.04 |
| 1943 | APR | 0.69 | 2.76 | 4.86 |
| 1943 | MAY | 0.69 | 2.74 | 4.69 |
| 1943 | JUN | 0.69 | 2.72 | 4.64 |
| 1943 | JUL | 0.69 | 2.69 | 4.57 |
| 1943 | AUG | 0.69 | 2.69 | 4.78 |
| 1943 | SEP | 0.69 | 2.69 | 4.79 |
| 1943 | OCT | 0.69 | 2.70 | 4.84 |
| 1943 | NOV | 0.69 | 2.71 | 5.07 |
| 1943 | DEC | 0.69 | 2.74 | 5.07 |
| 1944 | JAN | 0.69 | 2.72 | 4.95 |
| 1944 | FEB | 0.69 | 2.74 | 4.99 |
| 1944 | MAR | 0.69 | 2.74 | 4.88 |
| 1944 | APR | 0.72 | 2.74 | 4.97 |
| 1944 | MAY | 0.75 | 2.73 | 4.99 |
| 1944 | JUN | 0.75 | 2.73 | 4.81 |
| 1944 | JUL | 0.75 | 2.72 | 4.68 |
| 1944 | AUG | 0.75 | 2.71 | 4.77 |
| 1944 | SEP | 0.75 | 2.72 | 4.91 |
| 1944 | OCT | 0.75 | 2.72 | 4.75 |
| 1944 | NOV | 0.75 | 2.72 | 4.82 |
| 1944 | DEC | 0.75 | 2.70 | 4.77 |
| 1945 | JAN | 0.75 | 2.69 | 4.60 |
| 1945 | FEB | 0.75 | 2.65 | 4.43 |
| 1945 | MAR | 0.75 | 2.62 | 4.48 |
| 1945 | APR | 0.75 | 2.61 | 4.36 |
| 1945 | MAY | 0.75 | 2.62 | 4.21 |
| 1945 | JUN | 0.75 | 2.61 | 4.12 |
| 1945 | JUL | 0.75 | 2.60 | 4.21 |
| 1945 | AUG | 0.75 | 2.61 | 4.27 |
| 1945 | SEP | 0.75 | 2.62 | 4.01 |
| 1945 | OCT | 0.75 | 2.62 | 3.85 |
| 1945 | NOV | 0.75 | 2.62 | 3.73 |
| 1945 | DEC | 0.75 | 2.61 | 3.71 |
| 1946 | JAN | 0.75 | 2.54 | 3.56 |
| 1946 | FEB | 0.75 | 2.48 | 3.62 |
| 1946 | MAR | 0.75 | 2.47 | 3.72 |
| 1946 | APR | 0.75 | 2.46 | 3.51 |
| 1946 | MAY | 0.75 | 2.51 | 3.48 |
| 1946 | JUN | 0.75 | 2.49 | 3.53 |
| 1946 | JUL | 0.77 | 2.48 | 3.62 |
| 1946 | AUG | 0.81 | 2.51 | 3.65 |
| 1946 | SEP | 0.81 | 2.58 | 4.21 |
| 1946 | OCT | 0.89 | 2.60 | 4.33 |
| 1946 | NOV | 0.94 | 2.59 | 4.50 |
| 1946 | DEC | 1.00 | 2.61 | 4.47 |

| YEAR | MONTH | COMMP | CORP AAA | S & P |
|---|---|---|---|---|
| 1947 | JAN | 1.00 | 2.57 | 4.49 |
| 1947 | FEB | 1.00 | 2.55 | 4.38 |
| 1947 | MAR | 1.00 | 2.55 | 4.61 |
| 1947 | APR | 1.00 | 2.53 | 4.75 |
| 1947 | MAY | 1.00 | 2.53 | 5.05 |
| 1947 | JUN | 1.00 | 2.55 | 5.03 |
| 1947 | JUL | 1.00 | 2.55 | 4.81 |
| 1947 | AUG | 1.00 | 2.56 | 4.97 |
| 1947 | SEP | 1.02 | 2.61 | 5.16 |
| 1947 | OCT | 1.06 | 2.70 | 5.10 |
| 1947 | NOV | 1.10 | 2.77 | 5.24 |
| 1947 | DEC | 1.22 | 2.86 | 5.55 |
| 1948 | JAN | 1.30 | 2.86 | 5.71 |
| 1948 | FEB | 1.38 | 2.85 | 6.01 |
| 1948 | MAR | 1.38 | 2.83 | 5.90 |
| 1948 | APR | 1.38 | 2.78 | 5.50 |
| 1948 | MAY | 1.38 | 2.76 | 5.32 |
| 1948 | JUN | 1.38 | 2.76 | 5.02 |
| 1948 | JUL | 1.38 | 2.81 | 4.96 |
| 1948 | AUG | 1.47 | 2.84 | 5.17 |
| 1948 | SEP | 1.54 | 2.84 | 5.29 |
| 1948 | OCT | 1.56 | 2.84 | 5.29 |
| 1948 | NOV | 1.56 | 2.84 | 6.09 |
| 1948 | DEC | 1.56 | 2.79 | 6.24 |
| 1949 | JAN | 1.56 | 2.71 | 6.20 |
| 1949 | FEB | 1.56 | 2.71 | 6.54 |
| 1949 | MAR | 1.56 | 2.70 | 6.57 |
| 1949 | APR | 1.56 | 2.70 | 6.65 |
| 1949 | MAY | 1.56 | 2.71 | 6.72 |
| 1949 | JUN | 1.56 | 2.71 | 7.09 |
| 1949 | JUL | 1.56 | 2.67 | 6.73 |
| 1949 | AUG | 1.43 | 2.62 | 6.54 |
| 1949 | SEP | 1.38 | 2.60 | 6.36 |
| 1949 | OCT | 1.38 | 2.61 | 6.26 |
| 1949 | NOV | 1.38 | 2.60 | 6.61 |
| 1949 | DEC | 1.33 | 2.58 | 6.75 |
| 1950 | JAN | 1.31 | 2.57 | 6.65 |
| 1950 | FEB | 1.31 | 2.58 | 6.62 |
| 1950 | MAR | 1.31 | 2.58 | 6.53 |
| 1950 | APR | 1.31 | 2.60 | 6.36 |
| 1950 | MAY | 1.31 | 2.61 | 6.16 |
| 1950 | JUN | 1.31 | 2.62 | 6.12 |
| 1950 | JUL | 1.31 | 2.65 | 6.63 |
| 1950 | AUG | 1.42 | 2.61 | 6.54 |
| 1950 | SEP | 1.65 | 2.64 | 6.65 |
| 1950 | OCT | 1.72 | 2.67 | 6.39 |
| 1950 | NOV | 1.69 | 2.67 | 6.97 |
| 1950 | DEC | 1.72 | 2.67 | 7.24 |

**TABLE A4-5 Raw Data Listing of Yields** (cont.)

| YEAR | MONTH | COMMP | CORP AAA | S & P |
|---|---|---|---|---|
| 1951 | JAN | 1.86 | 2.66 | 6.34 |
| 1951 | FEB | 1.96 | 2.66 | 6.20 |
| 1951 | MAR | 2.04 | 2.78 | 6.34 |
| 1951 | APR | 2.11 | 2.87 | 6.27 |
| 1951 | MAY | 2.16 | 2.89 | 6.23 |
| 1951 | JUN | 2.31 | 2.94 | 6.30 |
| 1951 | JUL | 2.31 | 2.94 | 6.25 |
| 1951 | AUG | 2.26 | 2.88 | 5.96 |
| 1951 | SEP | 2.19 | 2.84 | 5.75 |
| 1951 | OCT | 2.22 | 2.89 | 5.77 |
| 1951 | NOV | 2.25 | 2.96 | 6.18 |
| 1951 | DEC | 2.30 | 3.01 | 5.99 |
| 1952 | JAN | 2.38 | 2.98 | 5.82 |
| 1952 | FEB | 2.38 | 2.93 | 5.99 |
| 1952 | MAR | 2.38 | 2.96 | 5.94 |
| 1952 | APR | 2.35 | 2.93 | 5.96 |
| 1952 | MAY | 2.31 | 2.93 | 5.99 |
| 1952 | JUN | 2.31 | 2.94 | 5.83 |
| 1952 | JUL | 2.31 | 2.95 | 5.65 |
| 1952 | AUG | 2.31 | 2.94 | 5.66 |
| 1952 | SEP | 2.31 | 2.95 | 5.74 |
| 1952 | OCT | 2.31 | 3.01 | 5.89 |
| 1952 | NOV | 2.31 | 2.98 | 5.69 |
| 1952 | DEC | 2.31 | 2.97 | 5.47 |
| 1953 | JAN | 2.31 | 3.02 | 5.45 |
| 1953 | FEB | 2.31 | 3.07 | 5.52 |
| 1953 | MAR | 2.36 | 3.12 | 5.48 |
| 1953 | APR | 2.44 | 3.23 | 5.75 |
| 1953 | MAY | 2.67 | 3.34 | 5.77 |
| 1953 | JUN | 2.75 | 3.40 | 5.99 |
| 1953 | JUL | 2.75 | 3.28 | 5.90 |
| 1953 | AUG | 2.75 | 3.24 | 5.87 |
| 1953 | SEP | 2.74 | 3.29 | 6.14 |
| 1953 | OCT | 2.55 | 3.16 | 5.99 |
| 1953 | NOV | 2.31 | 3.11 | 5.92 |
| 1953 | DEC | 2.25 | 3.13 | 5.83 |
| 1954 | JAN | 2.11 | 3.06 | 5.69 |
| 1954 | FEB | 2.00 | 2.95 | 5.62 |
| 1954 | MAR | 2.00 | 2.86 | 5.46 |
| 1954 | APR | 1.76 | 2.85 | 5.24 |
| 1954 | MAY | 1.58 | 2.88 | 5.02 |
| 1954 | JUN | 1.56 | 2.90 | 4.99 |
| 1954 | JUL | 1.45 | 2.89 | 4.77 |
| 1954 | AUG | 1.33 | 2.87 | 4.63 |
| 1954 | SEP | 1.31 | 2.89 | 4.57 |
| 1954 | OCT | 1.31 | 2.87 | 4.43 |
| 1954 | NOV | 1.31 | 2.89 | 4.52 |
| 1954 | DEC | 1.31 | 2.90 | 4.45 |

| YEAR | MONTH | COMMP | CORP AAA | S & P |
|---|---|---|---|---|
| 1955 | JAN | 1.47 | 2.93 | 4.38 |
| 1955 | FEB | 1.68 | 2.93 | 4.32 |
| 1955 | MAR | 1.69 | 3.02 | 4.33 |
| 1955 | APR | 1.90 | 3.01 | 4.19 |
| 1955 | MAY | 2.00 | 3.04 | 4.23 |
| 1955 | JUN | 2.00 | 3.05 | 4.01 |
| 1955 | JUL | 2.11 | 3.06 | 3.72 |
| 1955 | AUG | 2.33 | 3.11 | 3.87 |
| 1955 | SEP | 2.54 | 3.13 | 3.72 |
| 1955 | OCT | 2.70 | 3.10 | 3.94 |
| 1955 | NOV | 2.81 | 3.10 | 4.05 |
| 1955 | DEC | 2.99 | 3.15 | 4.15 |
| 1956 | JAN | 3.00 | 3.11 | 4.24 |
| 1956 | FEB | 3.00 | 3.08 | 4.24 |
| 1956 | MAR | 3.00 | 3.10 | 3.97 |
| 1956 | APR | 3.14 | 3.24 | 3.94 |
| 1956 | MAY | 3.27 | 3.28 | 4.09 |
| 1956 | JUN | 3.38 | 3.26 | 4.09 |
| 1956 | JUL | 3.27 | 3.28 | 3.89 |
| 1956 | AUG | 3.28 | 3.43 | 3.92 |
| 1956 | SEP | 3.50 | 3.56 | 4.07 |
| 1956 | OCT | 3.63 | 3.59 | 4.12 |
| 1956 | NOV | 3.63 | 3.69 | 4.27 |
| 1956 | DEC | 3.63 | 3.75 | 4.24 |
| 1957 | JAN | 3.63 | 3.77 | 4.31 |
| 1957 | FEB | 3.63 | 3.67 | 4.54 |
| 1957 | MAR | 3.63 | 3.66 | 4.47 |
| 1957 | APR | 3.63 | 3.67 | 4.36 |
| 1957 | MAY | 3.63 | 3.74 | 4.18 |
| 1957 | JUN | 3.79 | 3.91 | 4.04 |
| 1957 | JUL | 3.88 | 3.99 | 3.95 |
| 1957 | AUG | 3.98 | 4.10 | 4.17 |
| 1957 | SEP | 4.00 | 4.12 | 4.31 |
| 1957 | OCT | 4.10 | 4.10 | 4.54 |
| 1957 | NOV | 4.07 | 4.08 | 4.67 |
| 1957 | DEC | 3.81 | 3.81 | 4.64 |
| 1958 | JAN | 3.49 | 3.60 | 4.48 |
| 1958 | FEB | 2.63 | 3.59 | 4.47 |
| 1958 | MAR | 2.33 | 3.63 | 4.37 |
| 1958 | APR | 1.90 | 3.60 | 4.33 |
| 1958 | MAY | 1.71 | 3.57 | 4.19 |
| 1958 | JUN | 1.54 | 3.57 | 4.08 |
| 1958 | JUL | 1.50 | 3.67 | 3.98 |
| 1958 | AUG | 1.96 | 3.85 | 3.78 |
| 1958 | SEP | 2.93 | 4.09 | 3.69 |
| 1958 | OCT | 3.23 | 4.11 | 3.54 |
| 1958 | NOV | 3.08 | 4.09 | 3.42 |
| 1958 | DEC | 3.33 | 4.08 | 3.33 |

**TABLE A4-5 Raw Data Listing of Yields** (cont.)

| YEAR | MONTH | COMMP | CORP AAA | S & P |
|---|---|---|---|---|
| 1959 | JAN | 3.30 | 4.12 | 3.24 |
| 1959 | FEB | 3.26 | 4.14 | 3.32 |
| 1959 | MAR | 3.35 | 4.13 | 3.25 |
| 1959 | APR | 3.42 | 4.23 | 3.26 |
| 1959 | MAY | 3.56 | 4.37 | 3.21 |
| 1959 | JUN | 3.83 | 4.46 | 3.23 |
| 1959 | JUL | 3.98 | 4.47 | 3.11 |
| 1959 | AUG | 3.97 | 4.43 | 3.14 |
| 1959 | SEP | 4.63 | 4.52 | 3.26 |
| 1959 | OCT | 4.73 | 4.57 | 3.26 |
| 1959 | NOV | 4.67 | 4.56 | 3.24 |
| 1959 | DEC | 4.88 | 4.58 | 3.18 |
| 1960 | JAN | 4.91 | 4.61 | 3.27 |
| 1960 | FEB | 4.66 | 4.56 | 3.44 |
| 1960 | MAR | 4.49 | 4.49 | 3.51 |
| 1960 | APR | 4.16 | 4.45 | 3.47 |
| 1960 | MAY | 4.25 | 4.46 | 3.51 |
| 1960 | JUN | 3.81 | 4.45 | 3.40 |
| 1960 | JUL | 3.39 | 4.41 | 3.49 |
| 1960 | AUG | 3.34 | 4.28 | 3.43 |
| 1960 | SEP | 3.39 | 4.25 | 3.55 |
| 1960 | OCT | 3.30 | 4.30 | 3.60 |
| 1960 | NOV | 3.28 | 4.31 | 3.51 |
| 1960 | DEC | 3.23 | 4.35 | 3.41 |
| 1961 | JAN | 2.98 | 4.32 | 3.28 |
| 1961 | FEB | 3.03 | 4.27 | 3.13 |
| 1961 | MAR | 3.03 | 4.22 | 3.03 |
| 1961 | APR | 2.91 | 4.25 | 2.95 |
| 1961 | MAY | 2.76 | 4.27 | 2.92 |
| 1961 | JUN | 2.91 | 4.33 | 2.99 |
| 1961 | JUL | 2.72 | 4.41 | 3.00 |
| 1961 | AUG | 2.92 | 4.45 | 2.91 |
| 1961 | SEP | 3.05 | 4.45 | 2.93 |
| 1961 | OCT | 3.00 | 4.42 | 2.91 |
| 1961 | NOV | 2.98 | 4.39 | 2.84 |
| 1961 | DEC | 3.19 | 4.42 | 2.85 |
| 1962 | JAN | 3.26 | 4.42 | 2.97 |
| 1962 | FEB | 3.22 | 4.42 | 2.95 |
| 1962 | MAR | 3.25 | 4.39 | 2.95 |
| 1962 | APR | 3.20 | 4.33 | 3.05 |
| 1962 | MAY | 3.16 | 4.28 | 3.32 |
| 1962 | JUN | 3.25 | 4.28 | 3.78 |
| 1962 | JUL | 3.36 | 4.34 | 3.68 |
| 1962 | AUG | 3.30 | 4.35 | 3.57 |
| 1962 | SEP | 3.34 | 4.32 | 3.60 |
| 1962 | OCT | 3.27 | 4.28 | 3.71 |
| 1962 | NOV | 3.23 | 4.25 | 3.50 |
| 1962 | DEC | 3.29 | 4.24 | 3.40 |

| YEAR | MONTH | COMMP | CORP AAA | S & P |
|---|---|---|---|---|
| 1963 | JAN | 3.34 | 4.21 | 3.31 |
| 1963 | FEB | 3.25 | 4.19 | 3.27 |
| 1963 | MAR | 3.34 | 4.19 | 3.28 |
| 1963 | APR | 3.32 | 4.21 | 3.15 |
| 1963 | MAY | 3.25 | 4.22 | 3.13 |
| 1963 | JUN | 3.38 | 4.23 | 3.16 |
| 1963 | JUL | 3.49 | 4.26 | 3.20 |
| 1963 | AUG | 3.72 | 4.29 | 3.13 |
| 1963 | SEP | 3.88 | 4.31 | 3.06 |
| 1963 | OCT | 3.88 | 4.32 | 3.05 |
| 1963 | NOV | 3.88 | 4.33 | 3.14 |
| 1963 | DEC | 3.96 | 4.35 | 3.13 |
| 1964 | JAN | 3.97 | 4.37 | 3.05 |
| 1964 | FEB | 3.88 | 4.36 | 3.05 |
| 1964 | MAR | 4.00 | 4.38 | 3.03 |
| 1964 | APR | 3.91 | 4.40 | 3.00 |
| 1964 | MAY | 3.89 | 4.41 | 3.01 |
| 1964 | JUN | 4.00 | 4.41 | 3.05 |
| 1964 | JUL | 3.96 | 4.40 | 2.96 |
| 1964 | AUG | 3.88 | 4.41 | 3.03 |
| 1964 | SEP | 3.89 | 4.42 | 3.00 |
| 1964 | OCT | 4.00 | 4.42 | 2.95 |
| 1964 | NOV | 4.02 | 4.43 | 2.96 |
| 1964 | DEC | 4.17 | 4.44 | 3.05 |
| 1965 | JAN | 4.25 | 4.43 | 2.99 |
| 1965 | FEB | 4.27 | 4.41 | 2.99 |
| 1965 | MAR | 4.38 | 4.42 | 2.99 |
| 1965 | APR | 4.38 | 4.43 | 2.95 |
| 1965 | MAY | 4.38 | 4.44 | 2.92 |
| 1965 | JUN | 4.38 | 4.46 | 3.07 |
| 1965 | JUL | 4.38 | 4.48 | 3.09 |
| 1965 | AUG | 4.38 | 4.49 | 3.06 |
| 1965 | SEP | 4.38 | 4.52 | 2.98 |
| 1965 | OCT | 4.38 | 4.56 | 2.91 |
| 1965 | NOV | 4.38 | 4.60 | 2.96 |
| 1965 | DEC | 4.65 | 4.68 | 3.05 |
| 1966 | JAN | 4.82 | 4.74 | 3.02 |
| 1966 | FEB | 4.88 | 4.78 | 3.06 |
| 1966 | MAR | 5.21 | 4.92 | 3.23 |
| 1966 | APR | 5.38 | 4.96 | 3.15 |
| 1966 | MAY | 5.39 | 4.98 | 3.30 |
| 1966 | JUN | 5.51 | 5.07 | 3.36 |
| 1966 | JUL | 5.63 | 5.16 | 3.37 |
| 1966 | AUG | 5.85 | 5.31 | 3.60 |
| 1966 | SEP | 5.89 | 5.49 | 3.75 |
| 1966 | OCT | 6.00 | 5.41 | 3.76 |
| 1966 | NOV | 6.00 | 5.35 | 3.66 |
| 1966 | DEC | 6.00 | 5.39 | 3.59 |

**TABLE A4-5 Raw Data Listing of Yields** (cont.)

| YEAR | MONTH | COMMP | CORP AAA | S & P |
|---|---|---|---|---|
| 1967 | JAN | 5.73 | 5.20 | 3.51 |
| 1967 | FEB | 5.38 | 5.03 | 3.36 |
| 1967 | MAR | 5.24 | 5.13 | 3.29 |
| 1967 | APR | 4.83 | 5.11 | 3.24 |
| 1967 | MAY | 4.67 | 5.24 | 3.19 |
| 1967 | JUN | 4.65 | 5.44 | 3.19 |
| 1967 | JUL | 4.92 | 5.58 | 3.15 |
| 1967 | AUG | 5.00 | 5.62 | 3.11 |
| 1967 | SEP | 5.00 | 5.65 | 3.07 |
| 1967 | OCT | 5.07 | 5.82 | 3.07 |
| 1967 | NOV | 5.28 | 6.07 | 3.18 |
| 1967 | DEC | 5.56 | 6.19 | 3.09 |
| 1968 | JAN | 5.60 | 6.17 | 3.13 |
| 1968 | FEB | 5.50 | 6.10 | 3.28 |
| 1968 | MAR | 5.64 | 6.11 | 3.34 |
| 1968 | APR | 5.81 | 6.21 | 3.12 |
| 1968 | MAY | 6.18 | 6.27 | 3.07 |
| 1968 | JUN | 6.25 | 6.28 | 3.00 |
| 1968 | JUL | 6.19 | 6.24 | 3.00 |
| 1968 | AUG | 5.88 | 6.02 | 3.09 |
| 1968 | SEP | 5.82 | 5.97 | 3.01 |
| 1968 | OCT | 5.80 | 6.09 | 2.94 |
| 1968 | NOV | 5.92 | 6.19 | 2.92 |
| 1968 | DEC | 6.17 | 6.45 | 2.93 |
| 1969 | JAN | 6.53 | 6.59 | 3.06 |
| 1969 | FEB | 6.62 | 6.66 | 3.10 |
| 1969 | MAR | 6.82 | 6.85 | 3.17 |
| 1969 | APR | 7.04 | 6.89 | 3.11 |
| 1969 | MAY | 7.35 | 6.79 | 3.02 |
| 1969 | JUN | 8.23 | 6.98 | 3.18 |
| 1969 | JUL | 8.65 | 7.08 | 3.34 |
| 1969 | AUG | 8.33 | 6.97 | 3.37 |
| 1969 | SEP | 8.48 | 7.14 | 3.33 |
| 1969 | OCT | 8.57 | 7.33 | 3.33 |
| 1969 | NOV | 8.46 | 7.35 | 3.31 |
| 1969 | DEC | 8.84 | 7.72 | 3.52 |
| 1970 | JAN | 8.78 | 7.91 | 3.56 |
| 1970 | FEB | 8.55 | 7.93 | 3.68 |
| 1970 | MAR | 8.33 | 7.84 | 3.60 |
| 1970 | APR | 8.06 | 7.83 | 3.70 |
| 1970 | MAY | 8.23 | 8.11 | 4.20 |
| 1970 | JUN | 8.21 | 8.48 | 4.17 |
| 1970 | JUL | 8.29 | 8.44 | 4.20 |
| 1970 | AUG | 7.90 | 8.13 | 4.07 |
| 1970 | SEP | 7.32 | 8.09 | 3.82 |
| 1970 | OCT | 6.85 | 8.03 | 3.74 |
| 1970 | NOV | 6.30 | 8.05 | 3.72 |
| 1970 | DEC | 5.73 | 7.64 | 3.46 |

| YEAR | MONTH | COMMP | CORP AAA | S & P |
|---|---|---|---|---|
| 1971 | JAN | 5.11 | 7.36 | 3.32 |
| 1971 | FEB | 4.47 | 7.08 | 3.18 |
| 1971 | MAR | 4.19 | 7.21 | 3.10 |
| 1971 | APR | 4.57 | 7.25 | 2.99 |
| 1971 | MAY | 5.10 | 7.53 | 3.04 |
| 1971 | JUN | 5.45 | 7.64 | 3.10 |
| 1971 | JUL | 5.75 | 7.64 | 3.13 |
| 1971 | AUG | 5.73 | 7.59 | 3.18 |
| 1971 | SEP | 5.75 | 7.44 | 3.09 |
| 1971 | OCT | 5.54 | 7.39 | 3.16 |
| 1971 | NOV | 4.92 | 7.26 | 3.31 |
| 1971 | DEC | 4.74 | 7.25 | 3.10 |
| 1972 | JAN | 4.08 | 7.19 | 2.96 |
| 1972 | FEB | 3.93 | 7.29 | 2.92 |
| 1972 | MAR | 4.17 | 7.24 | 2.86 |
| 1972 | APR | 4.58 | 7.30 | 2.83 |
| 1972 | MAY | 4.51 | 7.30 | 2.88 |
| 1972 | JUN | 4.64 | 7.23 | 2.87 |
| 1972 | JUL | 4.85 | 7.21 | 2.90 |
| 1972 | AUG | 4.82 | 7.19 | 2.80 |
| 1972 | SEP | 5.14 | 7.22 | 2.83 |
| 1972 | OCT | 5.30 | 7.21 | 2.82 |
| 1972 | NOV | 5.25 | 7.12 | 2.73 |
| 1972 | DEC | 5.45 | 7.08 | 2.70 |
| 1973 | JAN | 5.78 | 7.15 | 2.69 |
| 1973 | FEB | 6.22 | 7.22 | 2.80 |
| 1973 | MAR | 6.85 | 7.29 | 2.83 |
| 1973 | APR | 7.14 | 7.26 | 2.90 |
| 1973 | MAY | 7.27 | 7.29 | 3.01 |
| 1973 | JUN | 7.99 | 7.37 | 3.05 |
| 1973 | JUL | 9.18 | 7.45 | 3.04 |
| 1973 | AUG | 10.21 | 7.68 | 3.16 |
| 1973 | SEP | 10.23 | 7.63 | 3.13 |
| 1973 | OCT | 8.92 | 7.60 | 3.05 |
| 1973 | NOV | 8.94 | 7.67 | 3.36 |
| 1973 | DEC | 9.08 | 7.68 | 3.70 |
| 1974 | JAN | 8.66 | 7.83 | 3.64 |
| 1974 | FEB | 7.82 | 7.85 | 3.81 |
| 1974 | MAR | 8.42 | 8.01 | 3.65 |
| 1974 | APR | 9.79 | 8.25 | 3.86 |
| 1974 | MAY | 10.62 | 8.37 | 4.00 |
| 1974 | JUN | 10.96 | 8.47 | 4.02 |
| 1974 | JUL | 11.72 | 8.72 | 4.42 |
| 1974 | AUG | 11.65 | 9.00 | 4.90 |
| 1974 | SEP | 11.23 | 9.24 | 5.45 |
| 1974 | OCT | 9.36 | 9.27 | 5.38 |
| 1974 | NOV | 8.81 | 8.89 | 5.13 |
| 1974 | DEC | 8.98 | 8.89 | 5.43 |

**TABLE A4-5 Raw Data Listing of Yields** (cont.)

| YEAR | MONTH | COMMP | CORP AAA | S & P |
|---|---|---|---|---|
| 1975 | JAN | 7.30 | 8.83 | 5.07 |
| 1975 | FEB | 6.33 | 8.62 | 4.61 |
| 1975 | MAR | 6.06 | 8.67 | 4.42 |
| 1975 | APR | 6.15 | 8.95 | 4.34 |
| 1975 | MAY | 5.82 | 8.90 | 4.08 |
| 1975 | JUN | 5.79 | 8.77 | 4.02 |
| 1975 | JUL | 6.44 | 8.84 | 4.02 |
| 1975 | AUG | 6.70 | 8.95 | 4.36 |
| 1975 | SEP | 6.86 | 8.95 | 4.39 |
| 1975 | OCT | 6.48 | 8.86 | 4.22 |
| 1975 | NOV | 5.91 | 8.78 | 4.07 |
| 1975 | DEC | 5.97 | 8.79 | 4.14 |
| 1976 | JAN | 5.27 | 8.60 | 3.80 |
| 1976 | FEB | 5.23 | 8.55 | 3.67 |
| 1976 | MAR | 5.37 | 8.52 | 3.65 |
| 1976 | APR | 5.23 | 8.40 | 3.66 |
| 1976 | MAY | 5.54 | 8.58 | 3.76 |
| 1976 | JUN | 5.94 | 8.62 | 3.75 |
| 1976 | JUL | 5.67 | 8.56 | 3.64 |
| 1976 | AUG | 5.47 | 8.45 | 3.74 |
| 1976 | SEP | 5.45 | 8.38 | 3.71 |
| 1976 | OCT | 5.22 | 8.32 | 3.85 |
| 1976 | NOV | 5.05 | 8.25 | 4.04 |
| 1976 | DEC | 4.70 | 7.98 | 3.98 |
| 1977 | JAN | 4.74 | 7.96 | 3.99 |
| 1977 | FEB | 4.82 | 8.04 | 4.21 |
| 1977 | MAR | 4.87 | 8.10 | 4.37 |
| 1977 | APR | 4.87 | 8.04 | 4.47 |
| 1977 | MAY | 5.34 | 8.05 | 4.39 |
| 1977 | JUN | 5.51 | 7.96 | 4.60 |
| 1977 | JUL | 5.41 | 7.93 | 4.59 |
| 1977 | AUG | 5.84 | 7.99 | 4.72 |
| 1977 | SEP | 6.17 | 7.92 | 4.82 |
| 1977 | OCT | 6.55 | 8.04 | 4.97 |
| 1977 | NOV | 6.60 | 8.08 | 0.0 |
| 1977 | DEC | 6.63 | 8.18 | 0.0 |
| 1978 | JAN | 6.79 | 8.40 | 0.0 |
| 1978 | FEB | 6.79 | 8.47 | 0.0 |

# GLOSSARY

**Advisory Services** Privately circulated publications which comment upon the future course of financial markets, and for which a subscription is usually required.

**Bear Trap** A signal which suggests that the rising trend of an index or stock has reversed but which proves to be false.

**(Market) Breadth** Breadth relates to the number of issues participating in a move. A rally is considered suspect if the number of advancing issues is diminishing as the rally develops. Conversely, a decline which is associated with fewer and fewer stocks falling is considered to be a bullish sign.

**Bull Trap** A signal which suggests that the declining trend of an index or stock has reversed but which proves to be false.

**Customer Free Balances** The total amount of unused money on deposit in brokerage accounts. These are "free" funds representing cash which may be employed in the purchase of securities.

**Cyclical Investing** The process of buying and selling stocks based on a longer-term or primary market move. The cycle approximates to the 4-year business cycle, to which such primary movements in stock prices are normally related.

**Insider** Any person who directly or indirectly owns more than 10 percent of any class of stock listed on a national exchange, or who is an officer or director of the company in question.

**Margin** Occurs when an investor pays part of the purchase price of a security and borrows the balance, usually from a broker; the "margin" is the difference between the market value of the stock and the loan which is made against it.

**Margin Call** The demand upon a customer to put up money or securities with a broker. The call is made if a customer's equity in a margin account declines below a minimum standard set by the exchange or brokerage firm. This happens when there is a drop in price of the securities being held as collateral.

**Members** Members of a stock exchange who are empowered to buy and sell securities on the floor of the exchange either for a client or their own account.

**Odd Lots** Units of stock of less than 100 shares; these do not customarily appear on the tape.

**Odd-Lot Shorts** Odd lots that are sold short. Since odd lots are usually the vehicle of uninformed traders, a high level of odd-lot shorts in relation to total odd-lot sales often characterizes a major market bottom. A low level of odd-lot shorts compared with total odd-lot sales is a sign of a market top.

**Option** The right to buy or sell specific securities at a specified price within a specified time. A "put" gives the holder the right to sell the stock, a "call" the right to buy the stock. In recent years options on specific stocks have been listed on several exchanges, so that it is now possible to trade these instruments in the same way that the underlying stocks can be bought and sold.

**Overbought** An opinion as to the level of prices. It may refer to a specific indicator or to the market as a whole after a period of vigorous buying, following which it may be argued that prices are overextended for the time being and are in need of a period of downward or horizontal adjustment.

**Oversold** The opposite of overbought, i.e., a price move that has overextended itself on the downside.

**Over-the-Counter Market (OTC)** An informal collection of brokers and dealers. Securities traded include almost all federal, state, municipal and corporate bonds and all widely owned equity issues not listed on the stock exchanges.

**Price/Earnings Ratio** The ratio of the price of a stock to the earnings per share, i.e., the total annual profit of a company divided by the number of shares outstanding.

**Rally** A brisk rise following a decline or consolidation of the general price level of the market.

**Reaction** A temporary price weakness following an upswing.

**Secondary Distribution or Offering** The redistribution of a block of stock some time after it has been sold by the issuing company. The sale is handled off the exchanges by a securities firm or group of firms, and the shares are usually offered at a fixed price which is related to the current market price of the stock.

**Short Covering** The process of buying back stock that has already been sold short.

**Short-Interest Ratio** The ratio of the short position to the average daily trading volume of the month in question. A high short-interest ratio is considered bullish (above 1.8) and a low one below 1.15 is considered bearish.

**Short Position (Interest)** The total amount of short sales outstanding on a specific exchange at a particular time. The short position is published monthly.

**Short Selling** Short selling is normally a speculative operation undertaken in the belief that the price of the shares will fall. It is accomplished by selling shares one does not own by borrowing stock from a broker. Most stock exchanges prohibit the short sale of a security below the price at which the last board lot was traded.

**Specialist** A member of a stock exchange who acts as a specialist in a listed issue and who is registered with the exchange for that purpose. He agrees to efficiently execute all orders left with him and, insofar as is reasonably practical, to maintain a fair and orderly market in the issue or issues for which he is a specialist.

**Yield Curve** The structure of the level of interest rates through various maturities. Usually the shorter the maturity, the lower the interest rate. Thus, 3-month Treasury bills usually yield less than 20-year government bonds. The slope of the yield curve relates to the speed with which rates rise as the maturity increases. In periods of tight money, short-term rates usually yield more than longer-term rates, and the curve is then called an inverse yield curve.

# BIBLIOGRAPHY

Appel, Gerald: *Winning Stock Market Systems*. Signalert Corp., Great Neck, N.Y., 1974.

Ayres, Leonard P.: *Turning Points in Business Cycles*. Augustus M. Kelly, New York, 1967.

Bretz, W. G.: *Juncture Recognition in the Stock Market*. Vantage Press, New York, 1972.

Coppock, E. S. C.: *Practical Relative Strength Charting*. Trendex Corp., San Antonio, Tex., 1960.

Dewey, E. R.: *Cycles: The Mysterious Forces That Trigger Events*. Hawthorne Books, New York, 1971.

——— and E. F. Dakin: *Cycles: The Science of Prediction*. Henry Holt, New York, 1947.

Drew, Garfield: *New Methods for Profit in the Stock Market*. Metcalfe Press, Boston, 1968.

Edwards, Robert D., and John Magee: *Technical Analysis of Stock Trends*. John Magee, Springfield, Mass., 1957.

Eiteman, W. J., C. A. Dice, and D. K. Eiteman: *The Stock Market*. McGraw-Hill, New York, 1966.

Fosback, Norman G.: *Stock Market Logic: A Sophisticated Approach to Profits on Wall Street*. The Institute for Econometric Research, Fort Lauderdale, Fla., 1976.

Frost, A. J., and Robert R. Prechter, Jr.: *Elliot Wave Principle—Key to Stock Market Profits*. New Classics Library, Chappaqua., N.Y., 1978.

Gann, W. D.: *Truth of the Stock Tape*. Financial Guardian, New York, 1932.

**Gordon, William:** *The Stock Market Indicators*. Investors Press, Palisades Park, N.J., 1968.

**Granville, Joseph:** *Strategy of Daily Stock Market Timing*. Prentice-Hall, Englewood Cliffs, N.J., 1960.

**Greiner, Perry, and Hale C. Whitcomb:** *Dow Theory*. Investors Intelligence, New York, 1969.

**Hamilton, W. D.:** *The Stock Market Barometer*. Harper Brothers, New York, 1922.

**Hurst, J. M.:** *The Profit Magic of Stock Transaction Timing*. Prentice-Hall, Englewood Cliffs, N.J., 1970.

**Jiler, William:** *How Charts Can Help You in the Stock Market*. Commodity Research Publication, New York, 1962.

**Krow, Harvey:** *Stock Market Behavior*. Random House, New York, 1969.

**Merrill, Arthur A.:** *Filtered Waves: Basic Theory*. Analysis Press, Chappaqua, N.Y., 1977.

**Nelson, Samuel:** *ABC of Stock Market Speculation*. Taylor, New York, 1934.

**Rhea, Robert:** *Dow Theory*. Barrons, New York, 1932.

**Shuman, James B., and David Roseneau:** *The Kondratieff Wave*. World Publishing, New York, 1972.

**Smith, Edgar Lawrence:** *Common Stocks and Business Cycles*. William Frederick Press, New York, 1959.

# INDEX